D1090854

Introduction to plant physiology

Introduction to plant physiology

JACOB LEVITT
Professor of Botany
University of Missouri
Columbia, Missouri

SECOND EDITION

with 181 illustrations

The C. V. Mosby Company
Saint Louis 1974

SECOND EDITION

Printed in the United States of America

Distributed in Great Britain by Henry Kimpton, London

Library of Congress Cataloging in Publication Data

Levitt, Jacob, 1911-
Introduction to plant physiology.

1. Plant physiology. I. Title.
QK711.2.L48 1974 581.1 73-13940
ISBN 0-8016-2993-4

CB/CB/B 9 8 7 6 5 4 3 2 1

Preface
to second edition

Any author must be prepared to defend the release of a second edition, just five short years after the first. Simply to correct the typographical errors in the first edition or to make the few changes necessary to satisfy the criticisms of the reviewers is insufficient reason. The chief culprit is the information explosion. Each year, *Biological Abstracts* lists some 250,000 new publications in biology, appearing mainly in some 7,500 scientific journals. The main reason for the second edition is therefore the avalanche of new information in many areas of plant physiology—translocation, stomatal opening, photorespiration, C_4 photosynthesis, growth and development, etc. Some new material must also be added simply because it was overlooked in the first edition. But the textbook writer must also take his cue from the plant, which knows when to eliminate enzymes that are no longer useful or may even be injurious. When hypotheses prove to be incorrect or inadequate, they are no longer useful and may be harmful to research. When scientists make mistakes or, in rare cases, are dishonest (as in the case of auxins *a* and *b*), their conclusions are injurious to the science. Wherever possible, such misinformation must be removed from the first edition.

Some physiological problems have been attacked by veritable armies of investigators, without yielding categorical answers. In such cases the author must attempt to evaluate and summarize the results. This requires generous documentation so that the student will know where he can obtain further information on the subject. The documentation also informs him of how recent the information is, the degree of uncertainty that still exists, and how much research activity is going on at the time of this writing. For a first smooth reading of such chapters, the student must learn not to see the references in parentheses, just as an intelligent television viewer learns to turn off his ears for the commercials. When he is interested in consulting the original literature for more information, he may return to the text for the references. Fortunately, except for Chapter 2, such generous documentation does not appear until the later chapters in the text. By this time it is hoped that the student will be far enough advanced to tolerate it. These later chapters are written somewhat in the nature of reviews and will therefore introduce the student to the kind of reading he will have to do after completing the course, so that he may remain up to date.

I wish to thank all reviewers, colleagues, and students who have helped to improve on the first edition by their constructive criticisms.

Jacob Levitt

Preface
to first edition

This text is intended for use in a first course in plant physiology. But what excuse is there for adding another to the already long list of plant physiology texts? In recent years I have tried several of these texts without satisfying either myself or my students. A basic reason is the unprecedented information explosion, which has left all of them woefully out of date. True, there is a rapidly expanding list of monographs, symposium volumes, and short or long treatises, each of which covers selected topics in depth. They are invaluable to the instructor. But what of the student? Can he be expected to go to the twenty-odd volume *Handbook of Plant Physiology* and to the *Annual Review of Plant Physiology?* Or should we be satisfied to have him study in depth selected topics and ignore others? I reject both these alternatives. Symposia, encyclopedias, annual reviews, and treatises never develop a consistent viewpoint and always fail to achieve a balance in covering a whole field. Furthermore, knowledge is expanding at an explosive rate in all fields. It is, therefore, unrealistic to expect the student to read large numbers of volumes for each course. On the other hand, most students of plant physiology are exposed to only a single course. Therefore, they will be forever ignorant of those topics not covered in this introduction to the science.

There is no question, then, of the student's need. To fill this need, I have attempted to build on a course taught for twenty-five years. Though originally founded on standard works (Maximov, Meyer and Anderson, Thomas, Scarth and Lloyd, Gortner), the course has evolved steadily by incorporating the results of modern research. An earlier text that I authored (*Plant Physiology,* Prentice-Hall, 1954), now out of print and out of date, has served as the core; but it has been completely rewritten, new chapters have been added, and new terminology and concepts introduced in place of the outdated ones. The attempt has been made, on the one hand, to give the known facts and, on the other, to leave the student with an appreciation of the many uncertainties that still remain. Since so many texts must be read by the student, the objective has been conciseness without loss of thoroughness. To achieve conciseness, lengthy discussions of irrelevant questions have been omitted. To achieve thoroughness without loss of conciseness, the attempt has been made to combine selectively and critically both the old and the new; for the true is not always new and the new is not always true.

Nevertheless, because of the many new advances, certain sacrifices had to be made: (1) the subject matter has been confined, with few exceptions, to the higher plants, which is logical in view of the tremendous growth of microbiology as a separate science and the paucity of information on the physiology of the "macro" lower plants; (2) in opposition to some plant physiology texts, which include whole chapters on plant anatomy, neither this nor any other branch of botany is included; and (3) a detailed consideration of each topic is left for more advanced courses (for example, the many products and enzymes of intermediate metabolism).

The prime goal of this text is the presentation and elucidation of the principles and laws of plant physiology, without including much of the scientific literature, since it is not possible in a first course to include all the history of a science. In some cases this goal has not been achieved, since the principles have not yet been adequately established, and it is necessary to attack the problem by a "recent advances" approach—that is, a lengthier consideration of the recent literature. This approach has the advantage of emphasizing to the student that each small advance of science is the result of the painstaking work of many dedicated scientists. It is therefore a pleasure to give credit to many of these scientists in the following pages. But unfortunately, all the important contributions cannot be mentioned, and some of the most important have been omitted. Apologies are due these investigators as well as the reader for not acquainting him with their work. Such slights are not intentional, but are largely the result of circumstances. Some less important contributions are mentioned simply to indicate the present state of an unsolved problem. Some more important contributions are omitted for the sake of brevity and because the problem is sufficiently well understood to explain without reference to these contributions. The length of the bibliography is therefore not to be taken as evidence of the state of knowledge of a subject, since, unfortunately, the understanding of a subject is sometimes inversely related to the volume of its literature.

Another apology is due an unknown number of investigators. Sometimes a concept may be read or heard and then lie dormant in one's brain. At a later date, this concept may germinate into what is apparently one's own idea, but really originated with someone else. Consequently, although I have attempted to give credit in all cases where it is due, I can only apologize in advance for any borrowings that may not be acknowledged. Many of these ideas have their source in the teachings of my former professor, the late G. W. Scarth.

No man can complete a work unaided by others. So many at one time or another have influenced and aided me that I can only say this book is inevitably the creation of the many people in my life, and I am only their mouthpiece. For direct aid in the preparation of the manuscript, I should like to thank my partner in all things, my wife; my artist, Mrs. Judy Ban; and my office helper, Miss Cheryl Zdazinsky.

Jacob Levitt

Contents

PART ONE

Introduction

1

Basic principles of plant physiology

PHYSIOLOGY AS A SCIENCE

The first question the student is likely to ask is, "What need does the study of plant physiology fulfill?" The answer, however, will vary with the needs of the student himself.

To some it is of interest primarily as a source of basic information that can be applied practically in such allied fields as horticulture, agronomy, forestry, soil science, etc. They would prefer to study physiology as a physicist studies mathematics—purely as a tool to aid in attacking and understanding the science in which he is really interested. But physiology has nowhere approached the exactness of mathematics and therefore cannot be taught in this way. Plant physiology is not a satisfying study for one who must have pat answers to all questions. Any attempt to write a text that proposes to give categorical answers to most physiological problems will surely, although unintentionally, lead to a falsification of the facts and to a complete denial of the very spirit of the science.

For those students with exploratory and analytical minds, a course in plant physiology has much more than bare facts to offer. It can help them to discover the problems of life that physiology attempts to solve and the methods used to solve them, the facts that have been discovered by these methods, and most important, the hypotheses that have been suggested to explain why the plant behaves in the way it does. By describing experiments currently being performed and suggesting experiments that should be attempted, the course can give these students a feeling of privileged observers watching explorations into unknown fields and learning the techniques and reasoning that are being used to attack the unsolved problems. It should encourage them to examine the known facts critically, to compare them carefully with any conclusions reached, and if reason so dictates, to reject these conclusions, and if sufficiently inspired, to launch attacks of their own in an attempt to add another grain of truth to the small pile.

Where does plant physiology fit into the field of human knowledge? This question can be simply and graphically answered as follows:

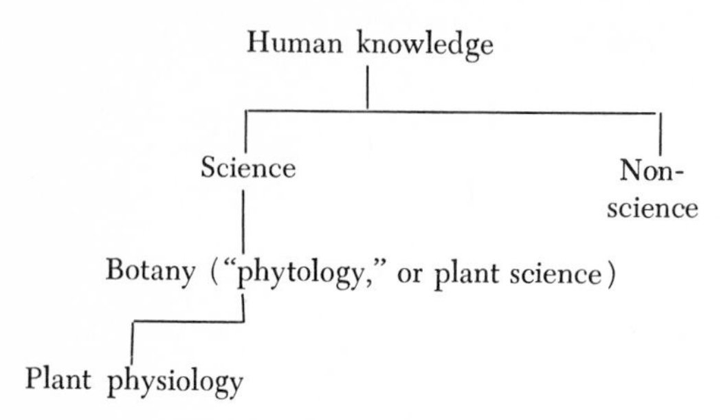

As a science, plant physiology must also face the accusations raised against all science: that it is inhuman, irrelevant, and without value judgments. In the choice of his profession, the plant physiologist proves these accusations to be false. He must be motivated to succeed in his chosen field, and the commonest motive is undoubtedly to better the lot of his fellow man. One way is by increasing the supply of food and other useful plant products. But there is a more fundamental and far-reaching significance of plant physiology. Man himself is a living organism, and no solution to his problems is possible without a clear understanding of the physiological principles that apply to him as certainly and unambiguously as to other organisms. Many of these principles are more readily discovered by experiments with other organisms than with man himself. Plant physiology has repeatedly proved its usefulness in this respect. For example, the existence of circadian rhythms (Chapter 21) was first discovered in plants. As a result of this discovery, it has now been shown that man is subject to the same circadian rhythms, and we now know one reason for the impaired judgment of our diplomats who are quickly transported to countries where day and night are reversed for their bodies.

By definition, plant physiology is the science that attempts to explain all plant processes by means of physical and chemical principles. This means that the basic assumption of the plant physiologist is that all plant processes *can* be explained by physical and chemical principles. So many processes have been successfully explained in this way that the universality of the assumption is nowadays seldom questioned. From time to time, however, scientists as well as nonscientists have rejected the assumption that physics and chemistry can explain all aspects of plant life, although no one has produced an acceptable reason for discarding it (Schaffner, 1967). Nevertheless, the true scientist will always remember that it is an assumption, although he adopts it in practice. On this basis all explanations not founded on physical and chemical principles are beyond the realm of plant physiology.

Since plant physiology is a science, it cannot be understood without a clear concept of the meaning of science. The aim of science is to discover all the laws of nature. In practice, however, a science is most simply defined as a body of knowledge obtained by the scientific method. The classical, although incomplete, concept of the scientific method is as follows (Simpson, 1963):

1. A problem is stated.
2. Observations relative to the problem are collected. (This includes a search of the literature for recorded observations, as well as additional observations by the investigator.)
3. A hypothetical solution of the problem consistent with the observations is formulated. (Frequently, this may also be found in the literature.)
4. Predictions of other observable phenomena are deduced from the hypothesis.
5. Occurrence or nonoccurrence of the predicted phenomena is observed by means of experiment.
6. The hypothesis is accepted, modified, or rejected in accordance with the degree of fulfillment of the predictions. If accepted, it becomes established as theory and finally as law.

This description of the scientific method is, of course, greatly oversimplified. It does not, for instance, explain the methods that must be used to obtain scientifically valid observations or to test the hypothesis: the controls that must always be included in an experiment, the statistical analysis of data to determine whether the differences are significant, the critical analysis of the data, etc. Nor does it explain how, from among several hypotheses that agree with the known facts, to select the one most likely to be correct. One method is to apply the principle of Occam's razor, according to which the hypothesis with the fewest assumptions (preferably only one) is most likely to be correct. If such a choice cannot be made, it may be possible to devise crucial experiments that will, as nearly as possible, exclude one or more of the hypotheses (Platt, 1964). It may not, in fact, be possible to prove any one hypothesis, since a later, alternative explanation may be just as good or better. Furthermore, it must also be remembered that every hypothesis involves some assumptions, and no matter how reasonable these may be, their existence must never be forgotten. It must be remembered, also, that science is the creation of scientists, who are human beings with all the emotional characteristics and deficiencies of other human beings. Both experimentalist and theorist frequently rely on unspecified arts and intuitive judgments, which may aid them or lead them astray. They frequently forget or overlook the deficiencies of the tools they have created. Life scientists, in general, tend to be overawed by statistics (Feller, 1969), and sometimes forget that this technique can be used only to refute hypotheses, and not to prove them (Davis, 1969).

Although this method is not always strictly followed, it does emphasize that at least part of the body of scientific knowledge is not necessarily the truth, since it includes calculated guesses (hypotheses). In other words, we must clearly distinguish between the aims and the actual state of a science. This deficiency of science must be faced squarely; but, on the other hand, it does not detract from the usefulness and even the necessity of such hypotheses, without which science would be at a standstill. Unfortunately, not only students but even sophisticated scientists themselves frequently speak contemptuously of hypotheses and plead, "Let us be practical. Let us talk sense and give only the experimentally determined facts." But these so-called facts by themselves are at best half truths and may even be completely false if the experiments were not performed carefully. Furthermore, the most carefully performed experiments may not reveal the truth, since as Heisenberg's uncertainty principle points out, the very performance of an experiment changes the conditions; and the observed phenomena may be different from what they were before the test.

Thus the deficiencies of both hypothesis and experiment must be recognized. It is only when the two are analyzed critically and fused into a logical explanation, that the hypothesis can begin to approach scientific truth. New facts may appear to contradict the hypothesis, but it must not be discarded hurriedly. The wavelength theory of light has continued to be used successfully long after the discovery of the photon and quantum, which seemed to oppose this theory. The hypothesis can therefore be considered as a bridge that must be crossed in order to arrive at the truth, and experiments can be considered as pathways that lead to either new or old bridges.

Science consists of many guesses that are constantly being tested by experimental research. For this reason we cannot expect cut-and-dried answers to all questions. Nevertheless, a good scientist should be expected to make only educated and critically sound guesses. He should not suggest a hypothesis if evidence is already available that proves the hypothesis untenable. This requires familiarity with, and understanding of, the already existing body of knowledge—that is, the literature on the subject. Unfortunately, scientists are human, and sometimes their hypotheses can be proved untenable on the basis of already available information without performing a single new experiment.

The purpose of research is therefore twofold: (1) to uncover the untruths and (2) to discover new truths. But every new research is in danger of adding not only truths but untruths. Thus of all the pursuits of scientists, the most important (and usually, alas, least practiced) is the critical analysis of the methods, the data, and the conclusions or hypotheses proposed to explain the data. If a scientist devotes 1 year to accumulating the data, he may have to devote 2 years to a critical analysis of them in relation to the accumulated body of knowledge; but he seldom does so. It is thus possible for some scientists to make their contributions to science by critically analyzing the work of others without performing experiments themselves, although, of course, there is a limit to this kind of contribution, and one who has not himself performed experiments is usually unable to understand the experiments of others.

The older the science the more closely it may be expected to approach its aim. Yet even in the case of the one most generally accepted to have progressed the farthest, physics, some confusion still exists (Hubbert, 1963). For, as Hubbert points out, in many cases the truths are available but the scientists themselves are unaware of them. This is because of the great concentration of efforts on the discovery and mastery of the newer truths, leading to a neglect of the older truths.

The point that must be driven home to every student is that the body of knowledge in this text, as well as in any other, must not be accepted as absolute truth. Eternal vigilance is needed by all scientists, and each must receive every hypothesis (including his own) with skepticism. It behooves the student to show this same skepticism. Every statement should be analyzed critically by the student and accepted or rejected on the basis of such careful analysis. Since the author and other scientists are doing this constantly, a new edition of any science text differs from the old not only by including new truths (and perhaps untruths) but also by having weeded out old untruths. It does not necessarily follow, however, that the newer is true and the older false. Truths must be identified on their own merits without regard to whether they are new or old. But in every case a principle, law, or hypothesis must be thoroughly understood before exposing it to the test of experiment or criticism. Only too often rejection of a principle has resulted from its misinterpretation by student and mature scientist alike. It is for this reason that a text must concentrate on presenting principles, laws, and even hypotheses clearly, thoroughly, and quantitatively.

BASIC LAWS OF PLANT PHYSIOLOGY

In the physical sciences, rigorous application of the scientific method has established many natural laws, for example, Newton's laws of motion, the gas laws, and the laws of thermodynamics. In most of the biological sciences, theory is still in its early stages, and fewer natural laws have been established. In this respect, physiology has some advantages over other biological sciences, since its main purpose is to explain all biological phenomena in terms of physics and chemistry. Thus on the basis of our definition of plant physiology, the physiologist in search of the laws of physiology has two main sources of information: (1) physical and chemical laws and (2) direct experimental evidence. In many cases the physical and chemical laws by themselves clearly indicate which of several conclusions are plausible and which are untenable, without the necessity of the scientist having to perform a single experiment. But this does not, by any means, free the physiologist of the necessity of turning to experimental evidence. On the contrary, in order to determine which laws control a physiological process, it is usually necessary to perform many carefully planned experiments. For example, anyone with the merest rudiments of physical knowledge is certain to explain freezing injury in plants as a result of cell rupture because everyone knows that water expands on freezing. Yet experiment has shown that not only does no cell rupture occur on freezing but that the cell actually contracts because of ice formation in the intercellular (air-filled) spaces at the expense of water drawn out of the cells. This does not mean that the plant changes the physical properties of water. It means that other properties of water, besides its expansion on freezing, must be considered in order to explain the behavior of plants on freezing. The properties that must be applied can be discovered only by experiment. In another series of experiments, physiologists succeeded in discovering that the plant actually pumps substances into itself instead of waiting for them to diffuse in. Again, this must not be taken to mean that the plant evades the laws of diffusion. On the contrary, both the laws of diffusion and the laws that control the operation of pumps must be made use of to explain the process of absorption by plants.

Thus by applying physical and chemical laws and by performing experiments, the physiologist is gradually succeeding in his aim to develop physiological laws. As a by-product of his experiments, he has also discovered purely physical and chemical phenomena not previously known to the physicist or chemist, for example, osmosis, brownian movement, and chromatography.

Unfortunately, despite the yeoman service of physiologists, much experimental work remains to be done before we can hope for a complete understanding of most physiological processes. Consequently, most of them can be discussed only somewhat superficially in the light of the incomplete evidence that is available, and it must always be remembered that the explanations are usually only working hypotheses, the main purposes of which are (1) to explain the known facts and (2) to predict the plant's behavior. These predictions point to the direction in which further experimental work must be pushed.

It is not, then, scientifically permissible to teach even an elementary physiology course in the same cut-and-dried fashion as is often used in a course in the physical sciences. This deficiency of physiology, however, has the advantage of exposing the student not only to the precepts of science but to its methods as well. Consequently, a well-taught course in physiology may conceivably give a better insight into scientific methods than those that are usually considered to be well-taught courses in physics and chemistry. For example, the scientific method, as just described, involves both inductive and deductive reasoning—inductive in proceeding from the individual observations to the more general hypothesis and deductive in proceeding from the hypothesis to the individual predictions. One of the major deficiencies of science courses is the lack of this twofold approach. Time does not permit it. Frequently, both the inductive and deductive reasoning are eliminated, and the student memorizes the theory or law as dogma. Furthermore, the more highly developed the field the more attention is devoted by a text to theory and the less to experiment, since a sound theory is able to predict experimental results correctly. But even in a plant physiology course, time does not permit use of the scientific method in all phases of the teaching. The basic physical and chemical principles must be taught as dogma on the assumption that they have been firmly established by a rigorous application of the scientific method. For the student who is not satisfied by this dogmatic approach, the only solution is to take those basic physical and chemical courses in which the derivation of these principles is taught.

Must we, then, be physicists and chemists first before studying plant physiology? Unfortunately, the answer to this question is "yes." But, fortunately, a relatively small number of physical and chemical laws and principles may be sufficient for a relatively large fraction of present-day plant physiology. As pointed out by Hubbert (1963), in the whole field of science the "master generalizations number at most but a few tens." He includes among these the "three Newtonian laws of motion, the law of universal gravitation, the three laws of thermodynamics and the associated thermodynamics of irreversible processes, the two Maxwellian laws of electromagnetism, the law of conservation of matter, and the concept of the atomic and molecular nature of the chemical elements and their compounds." These can serve as a basis for both student and teacher of plant physiology.

To determine which of these physical and chemical principles are most important to the physiologist, physiology itself must be reduced to its essentials. Plant physiology is really the study of the plant at work. One basic aim is to determine the precise chemical nature of all the materials found in the plant and the way in which the plant succeeds in obtaining or making them. This is the *biochemistry* of the plant. The other aim is to understand all the kinds of work performed by the plant (mechanical, chemical, osmotic, electrical) and the nature of the energy involved in performing this work. This is the *biophysics* of the plant. Even the biochemistry of the plant always involves energy changes. Therefore the one common denominator of all aspects of plant physiology is work and energy. This fact has been recognized by many physiologists. According to Calvin and Calvin (1964, cited by Amen, 1966), two fundamental attributes distinguish living systems from inanimate ones: the ability to transfer and transform energy in a directed way and the ability to remember and communicate information. According to Hawkins (1964, cited by Amen, 1966), a living system is one that contains a complete differential description of itself and one in which the free energy of reproduction is comparable to the total free energy of the system.

It is obvious from this fundamental importance of work and energy in plant physiology, that the first laws that must be mastered by the physiologist are those pertaining to work and energy, that is, the laws of thermodynamics. The first and second laws are particularly simple and useful. According to the first law, energy cannot be destroyed or created but is simply converted from one form to another. It is, of course, now known that energy can also be converted into mass and vice versa as follows:

$$E = mc^2$$

where E = energy
m = mass
c = speed of light

Because of the tremendous speed of light, it follows that a small amount of mass would yield a tremendous amount of energy. But the conversion of mass to energy does not occur appreciably in the plant, since it requires the breakdown of atoms, and nearly all the atoms normally found in the plant are stable. On the other hand, it requires the conversion of tremendous amounts of energy to produce an appreciable increase in mass. Therefore the relatively small amounts of energy absorbed or released by the plant have no measurable effects on its mass.

The first law of thermodynamics (or the Principle of the Conservation of Energy) is represented by the equation (Getman and Daniels, 1937):

$$\Delta E = q - w$$

where ΔE = change in energy of a body
q = heat or other energy absorbed by the body
w = work done by the body on its environment

In the case of the living plant, the first law can be represented schematically (Fig. 1-1). An increase in the plant's energy results if there is an excess of energy absorbed over work done. The plant does work on its environment by growing (e.g., by pushing downward through the soil or by lifting its new cells upward against gravity), by evaporation of its water, etc. But this is more than compensated for by the radiant energy absorbed by the plant. Part of this absorbed radiant energy is converted into heat energy, which is used to evaporate water from the plant; a second part is converted into mechanical energy, which lifts the water up the plant; part of the remainder is converted into chemical energy in the process of photosynthesis. The first law also tells us about work done within the plant. Energy-consuming processes such as growth must be fed by energy-releasing processes such as respiration; and the work done within the plant that results in growth can be no greater than the preceding net increase in energy (ΔE) absorbed from the environment. Similarly, the amount of water evaporated is limited by the net amount of energy absorbed, and the yield of the plant is limited by that component of the energy absorbed which is available for photosynthesis (Chapter 14).

The second law of thermodynamics goes beyond the first law. It tells us that not all of this increase in energy is "free" or available for useful work because it consists of two components:

$$\Delta E = \Delta A + \Delta TS$$

where ΔA = isothermally available energy
ΔTS = isothermally unavailable energy
(T = absolute temperature, S = entropy)

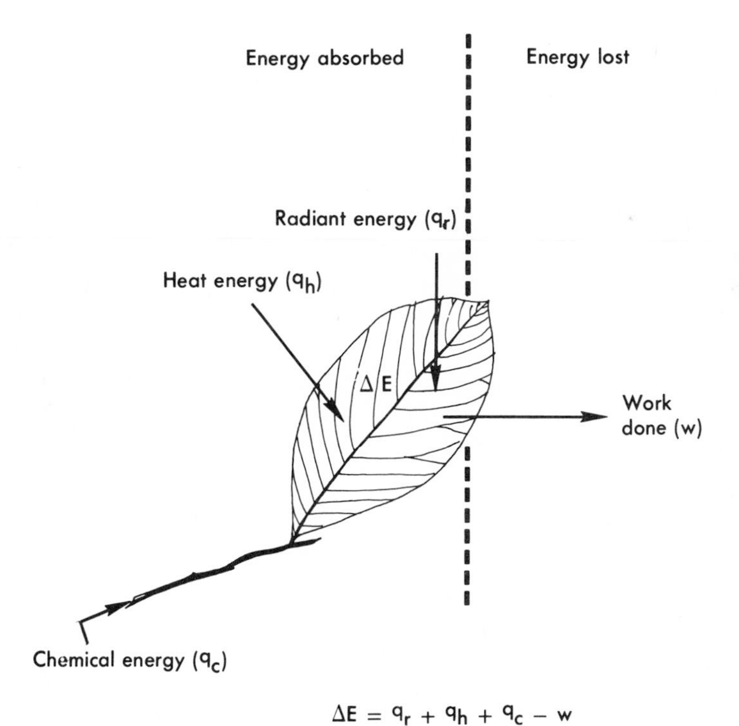

$$\Delta E = q_r + q_h + q_c - w$$

Fig. 1-1. Application of the first law of thermodynamics to the plant. The increase in the plant's energy is equal to the sum of all the forms of energy absorbed minus the work done by the plant on its environment. Each quantity represents the net value; for example, q_h is the heat energy absorbed minus the heat energy given off.

Since only systems at constant pressure and temperature can be considered, this equation becomes:

$$E = \Delta G - P\Delta V + T\Delta S$$

where P = pressure
ΔV = change in volume
ΔG = change in free energy

The quantity G is frequently called the "Gibbs' free energy." Only this portion of the increase in energy is available to do the work within the plant—the movement of substances and bodies within the plant, the maintenance of concentration, electrical, and other gradients, and most important of all, the synthesis of all the thousands of substances produced in the cell's chemical factories.

According to the second law, heat can pass of itself only from a warmer to a colder body, and therefore only this fraction of the heat of one body is "free" or available to do work on the other. This law can be applied to all other kinds of energy. The second law is made use of in many cases in the following chapters. It will be seen that no substances can move into a cell of themselves unless they are present in higher activities (or concentrations) in the medium than in the cell. Thus the laws of diffusion follow naturally from the second law of thermodynamics. It is only when the cell expends free energy that it can absorb a nutrient from a medium of lower activity or concentration (with respect to that nutrient) than its own; that is, the cell must then pump the nutrient into itself. According to a corollary of the second law of thermodynamics, free energy tends toward a minimum (in the process of doing work). This explains the fact that many substances tend to concentrate at surfaces (a phenomenon known as adsorption) because in so doing they reduce the free surface energy. Thus adsorption follows from the second law.

The second law can be expressed in another way. Free energy tends to be converted into unavailable energy by doing work. In other words, free energy tends to decrease to a minimum. Yet the growing plant steadily increases its free energy by synthesizing and accumulating new complex chemical substances. Some biologists (Fox, 1971) have therefore concluded that the second law does not apply to living organisms. But this conclusion is due to a misunderstanding of the second law. It applies only to *closed* systems, which are of two types (Katchalsky and Curran, 1965). *Adiabatic* systems are enclosed by walls that prevent exchange of both matter and energy with their environment. *Diathermal* systems are surrounded by walls that allow an exchange of energy but not of matter with their environment. A plant enclosed in a light-tight box would behave as such a closed system, and would steadily lose free energy, converting its more ordered, complex organic substances into less ordered and simpler inorganic substances. In opposition to these two closed systems, *open* systems allow exchange of both energy and matter with their environment. The plant is, of course, an open system, absorbing both radiant energy and matter (oxygen, carbon dioxide, water, and mineral nutrients) from its environment and releasing both heat energy and matter (carbon dioxide, water, etc.) into its environment. It is therefore able to show a net daily increase in energy, matter, and order, apparently but not actually in opposition to the second law of thermodynamics.

It is evident from the above that although the second law of classical or *reversible* thermodynamics applies fully to those processes in the plant that may be treated as closed systems, it is inadequate for processes involving an exchange of matter with the environment, such as the absorption of nutrients. In order to understand such transfers of mass, it is necessary to apply the laws of *irreversible* (or steady state) *thermodynamics*. The most useful tool supplied to the plant physiologist by this discipline is the *Onsager equation,* which has been used to describe and explain the transfer of substances into and out of the plant (Nobel, 1970).

The dependence of all branches of plant physiology on thermodynamics is shown by the following subtitles for the remaining chapters in this text in terms of the kinds of energy and energy changes involved:

Chapter	*Title*	*Energy subtitles*
2	The living cell	The machinery that performs the plant's work
3,4,5	Acidity Specific surface and adsorption Colloids	The work performed in the plant by the electrostatic energy of molecules and particles
6,7,8,11	Diffusion and osmosis Permeability Absorption* Exchange of gases	The work performed in the plant by molecular kinetic energy (or the energy of thermal agitation)
9	Ascent of sap	The conversion of radiant (solar) energy into mechanical work (bulk flow)
8,10	Absorption Translocation of solutes	The conversion of chemical energy into mechanical transfer of solutes
12	Nutrition	The elements needed to control energy transformations in the plant
13,14, 15,16	Metabolism Respiration Photosynthesis Other metabolic paths	Energy changes caused by electron transport between molecules
15	Photosynthesis	The conversion of radiant energy into chemical energy
17,18	Growth Irritability and movement	The conversion of chemical energy into external work
19,20	Growth regulators Development	The regulation by the plant of the release of chemical energy
21	Plant rhythms	The rhythmic control by the plant of the utilization of its energy
22	Stress resistance	The effect of the environment on the energy-producing machinery of the plant

*Electrostatic and electron transport energy are also involved.

As new information continues to accumulate in plant physiology, instead of enlarging our texts, it should be possible to compress them by describing the main principles concisely, without all the data and controversial material that was so necessary in the developmental stages. This goal, as yet, has not been achieved. It is possible, however, to compress the subject matter into a small number of topics because the major problems covered by plant physiology fall more or less naturally into three categories: (1) the transfer of substances, (2) nutrition and metabolism, and (3) growth and development.

The first category is mainly biophysical in nature, the second is biochemical, and the third depends on the intimate interrelations between these two. But before these three categories of plant physiology can be investigated, it is first necessary to understand (1) certain physiological characteristics of the living cell and (2) some of the main physicochemical principles that must be applied in attempting to explain the physiology of the plant. These two areas are described in Part 1, the previous three in Parts 2 to 4.

SPECIFIC REFERENCES

Amen, R. D. 1966. A biological systems concept. BioScience **16**:396-401.

Davis, H. L. 1969. Objectivity in science—a dangerous illusion. Scientific Research **28**:24-26.

Feller, W. 1969. Are life scientists overawed by statistics? Scientific Research **3**:24-29.

Fox, R. F. 1971. Entropy reduction in open systems. J. Theor. Biol. **31**:43-46.
Getman, F. H., and F. Daniels. 1937. Outlines of theoretical chemistry. John Wiley & Sons, Inc., New York.
Hubbert, M. K. 1963. Are we retrogressing in science? Science **139**:884-890.
Katchalsky, A., and P. F. Curran. 1965. Non-equilibrium thermodynamics in biophysics. Harvard University Press, Cambridge, Mass.
Nobel, P. S. 1970. Plant cell physiology; a physicochemical approach. W. H. Freeman & Co., Publishers, San Francisco.
Platt, J. R. 1964. Strong inference. Science **146**: 347-352.
Schaffner, K. F. 1967. Antireductionism and molecular biology. Science **157**:644-647.
Simpson, G. G. 1963. Biology and nature of science. Science **139**:81-88.

GENERAL REFERENCES

Annual review of plant physiology. 1950-.
Biological abstracts.
Chemical abstracts.
Lehninger, A. L. 1965. Bioenergetics. W. A. Benjamin, Inc., New York.
Linford, J. H. 1966. An introduction to energetics. Butterworth & Co. (Publishers), Ltd., London.
Ruhland, W. (Ed.). 1955-. Handbook of plant physiology. [Many volumes.] Springer-Verlag, Berlin.
Spector, W. S. (Ed.). 1956-. Handbook of biological data. W. B. Saunders Co., Philadelphia.
Steward, F. C. (Ed.). 1959-. Plant physiology. [Several volumes.] Academic Press, Inc., New York.

2

The living cell

THE CELL AS A WHOLE—THE PROTOPLAST

Just as the molecule is the basic unit of a chemical substance, the cell is both the functional and the structural unit of the plant. Thus the molecule is the smallest unit of a substance which still retains the properties of that substance, and the living cell is the smallest unit of the plant which so far has been shown to have all the properties necessary for regenerating the whole plant (Fig. 2-1). That is what is meant when it is said that the living cell is *totipotent.*

Even the nonliving cells usually have definite physiological functions. For instance, the living cells obtain a good share of their raw materials (minerals, water, etc.) by transport through dead cells (vessels and tracheids). But since the life processes are the center of interest for the physiologist, it is primarily these living cells that must be studied to understand the physiology of the plant. How, then, can we distinguish between living and nonliving cells? First we must know the characteristics of living cells and, second, we must develop methods of identifying these. Two characteristics are particularly important to identify.

A living cell does work

It is obvious that a living cell works when a plant is growing, since it is expanding against its environment. But how can we know whether a mature, nongrowing cell is still alive and doing work or is dead and not doing work? There are many sophisticated methods of measuring chemical or even electrical work, but these require special equipment and complicated techniques. There is one kind of work, however, that can be readily detected by simple observation under the microscope. The protoplasm of a living cell performs mechanical work by continuously moving in streams that carry particles with them. This movement is known as protoplasmic, or *cytoplasmic, streaming*, and it is dependent on a continuous supply of free energy by the living cell. As soon as the cell dies, it ceases to produce this free energy and is no longer able to perform work. Therefore cytoplasmic streaming stops. Unfortunately, the absence of streaming is not proof that the cell is doing no work, so that negative evidence in this case is no evidence. In both living and dead cells, particles show another kind of movement called *brownian movement.* This does not depend on a supply of free energy released by the cell. It is simply

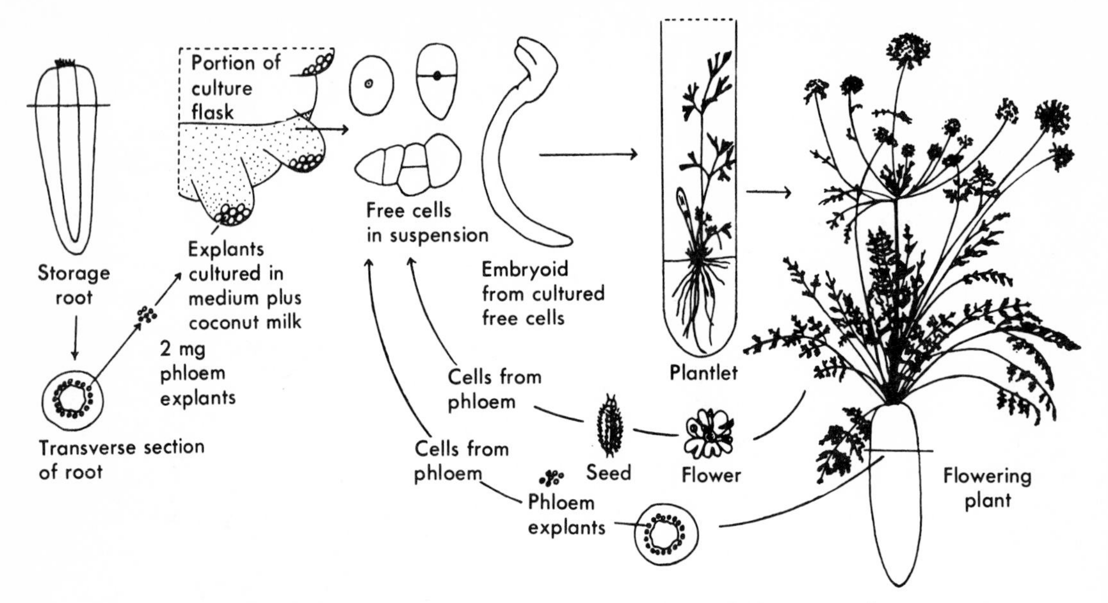

Fig. 2-1. Diagrammatic representation of the cycle of growth of the carrot plant; successive cycles of growth are linked through free, cultured cells derived from phloem or from the embryo. (From Steward, F. C., et al. 1964. Science **143**:20-24. Copyright 1964 by the American Association for the Advancement of Science.)

caused by the kinetic energy (or thermal agitation) of the molecules, and since this energy is on the average identical on all sides of a particle, there will be no net movement of the particle in any one direction, as occurs in cytoplasmic streaming, but only a haphazard movement in all directions about a constant average position. Brownian movement is, of course, no evidence of life.

A living cell has semipermeable (differentially permeable) membranes

Approximately 90% of the living cell consists of water in which many substances are dissolved. Some of these substances have been absorbed from the outside; most have been synthesized by the cell. When the cell is placed in water (or aqueous solutions of other substances), we would expect (from the second law of thermodynamics) that the cell solutes would become equally distributed in the water inside and outside the cell. Since the solutes are essential for supplying the cell with the free energy necessary for the performance of its work, this loss would result in death of the cell. Actually, if the cell is living, there is little or no loss of solutes from it. This is possible only if a barrier surrounds the cell preventing the loss. Since water moves freely into and out of the cell, the barrier must be essentially *semipermeable*—permeable to solvent but not to solutes. When the cell loses its property of semipermeability, all soluble substances quickly leak out, and the cell immediately loses its ability to do work and is therefore dead. There are two simple methods of determining whether the cell is semipermeable.

Vital staining. Basic dyes (salts of a dye base and an inorganic acid, Chapter 4), such as neutral red or methylene blue, penetrate the living cell in the molecular form and pass through the protoplasm into the vacuole, where they are converted to the ionic form (Chapter 8). Since the semipermeable membranes are impermeable to the ionic form, it accumulates in the vacuole, which thus becomes colored or *stained* (Fig. 2-2). Such cells may remain alive in the vitally stained state for hours or days. If the cell

Fig. 2-2. Onion epidermis cell "vitally stained" with neutral red. The dark region occupying most of the cell is the stained vacuole (also called the cell sap). The unstained, granular, terminal caps are the cytoplasm surrounded by a thin cell wall. The lateral cytoplasm is so thin as to be all but invisible. Only the cytoplasm and nucleus (not seen in the diagram) are alive.

is subsequently killed, its membranes become freely permeable and the dye quickly escapes. The vacuole of the dead cell is therefore no longer colored. The cell wall may be stained whether the cell is living or dead. Vital staining of the protoplasm by means of other dyes is also possible, but it must be carried out in the dark because the dye sensitizes the protoplasm to light, leading to injury and death within a few hours. Since light must be used to observe the cells, the least injury results when the staining is produced by a dye that can be observed by means of fluorescent light (e.g., K fluorescein), since this enables the detection of much weaker (and therefore less injurious) staining (Strugger, 1952). A vital stain may therefore be defined as a stain (or dye) that colors living but not dead protoplasts (i.e., protoplasm or its inclusions).

Another method of identifying living cells by staining is of practical importance in estimating the "germination potential" of seed lots without waiting for them to germinate (Lindenbein, 1965). Tetrazolium dyes (e.g., TTC, or triphenyltetrazolium chloride) are used for this purpose, since they are converted from the colorless to the colored (e.g., red) form by freshly cut surfaces of living cells. This is not a true vital staining, since the cells are killed in the process, nor is it evidence of the existence of semipermeable membranes but rather of a chemical reduction of the dye by freshly cut living cells. Cells that are dead before cutting are unable to produce this chemical reaction.

Plasmolysis. If a living cell (Fig. 2-3, *A*) is placed in a weak solution, it may expand because of uptake of water (Fig. 2-3, *B*). If transferred to a solution of, for example, about 10% sugar (Fig. 2-3, *C*), it shrinks and *plasmolysis* (a separation of the protoplasm from the cell wall) may occur. Transfer to a still stronger solution (Fig. 2-3, *D*) fails to cause any further shrinkage of the wall, although the plasmolyzed protoplast contracts still further. If transferred back to a weak solution, the protoplast expands (it deplasmolyzes) to the size it originally possessed in that solution. If the cell shrinks to its minimum size with little or no plasmolysis when placed in a solution, this solution is *isotonic* (of the same strength as the cell sap). Solutions that cause plasmolysis are *hypertonic;* those that fail to produce any trace of plasmolysis (although they may cause a slight shrinkage of the whole cell) are *hypotonic.* Some cells (e.g., many marine algae) do not plasmolyze because their walls swell so readily (Gessner, 1968). Some moss cells fail to plasmolyze because their cell walls adhere to and collapse with the protoplasts. Even in the case of cells of higher plants,

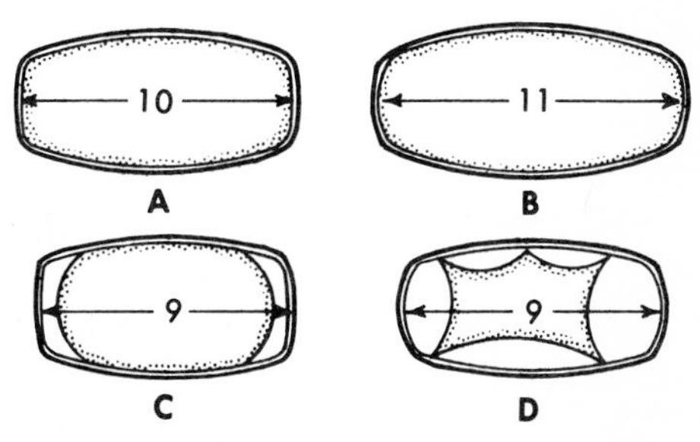

Fig. 2-3. Behavior of a living cell originally 10 units long in solutions of different concentrations. **A,** 9% sugar (slightly hypotonic). **B,** 2% sugar (hypotonic). **C,** 10% sugar (hypertonic). **D,** 20% sugar (hypertonic). Isotonic sugar solution is between 9% and 10%. Plasmolysis is apparent in **C** and **D.**

such adhesion may be favored under certain conditions (e.g., presence of Al^{+++}).

If the cell wall is cut during strong plasmolysis and the cell is then transferred to a weak solution, the *protoplast* (the protoplasm plus its vacuole) may swell enough to come completely out of the wall (Fig. 2-4). Protoplasts can also be isolated by dissolving the cell wall with the enzyme cellulase in the presence of 0.59 M sucrose (Cocking, 1960) or of other solutes (Ruesink and Thimann, 1966). In the case of some tissues (*Avena* coleoptiles, tobacco leaves), mannitol is a suitable solute but sucrose is not (Fodil et al., 1971). It is usually necessary to use a mixture of pectinases and cellulases (Power and Cocking, 1970) or even a "wall modifying enzyme" (Karr and Albersheim, 1970). Most commonly, the enzymes are obtained from snails or fungi. The same process occurs naturally in some fleshy fruit during ripening. Therefore free protoplasts can be found in the juice of ripe tomatoes and grapes.

In spite of this ready separation of protoplasts from their cell walls and thus from each other, the living cells of plant tissues are always connected with each other by protoplasmic strands about 0.2 μm in diameter (Livingston, 1964) known as *plasmodesmata* (Fig. 2-5), which pass through openings in the cell walls known as pits. The electron microscope has revealed that there are many more of these than can be seen with the optical microscope. Plasmolysis stretches the plasmodesmata, producing cytoplasmic ("Hechtian") strands during the pulling apart of the adjacent protoplasts. The preparation of free protoplasts, of course, breaks these plasmodesmal connections. Therefore even though the protoplast may deplasmolyze without visible signs of injury, quantitative measurements reveal that its metabolism may be greatly affected (Falk et al., 1966). Similarly, free protoplasts are frequently injured, and in some cases consist only of the vacuole sur-

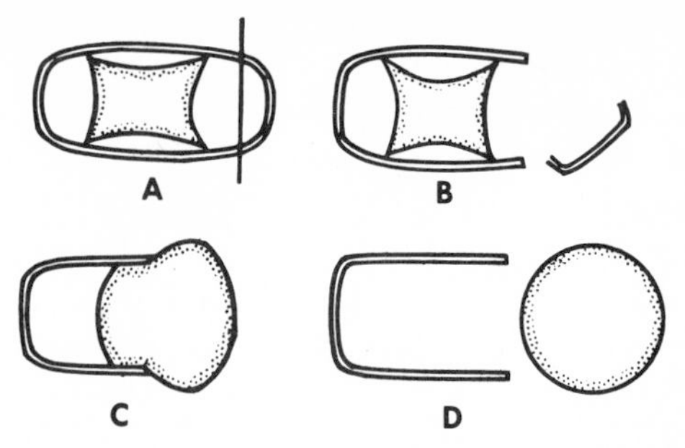

Fig. 2-4. Method of obtaining free protoplast (see Levitt et al., 1936). **A,** Onion cell is plasmolyzed in 20% sugar, then cut with a razor blade. **B,** Open cell. **C,** Protoplast swelling in 10% sugar. **D,** Empty cell wall and spherical free protoplast.

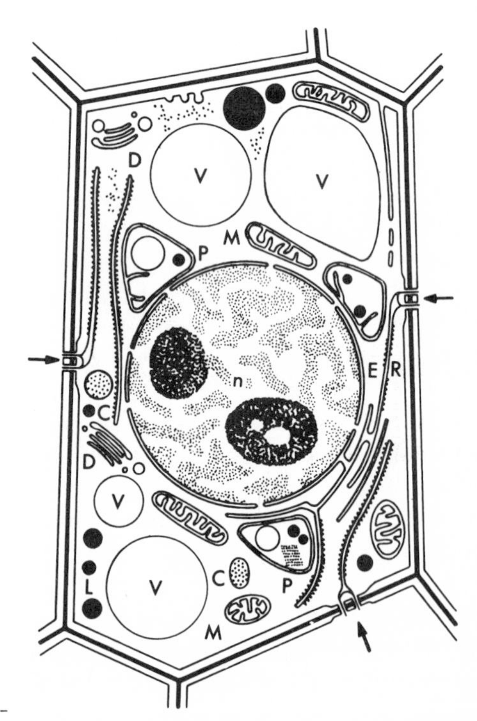

Fig. 2-5. Schematic representation of meristematic (young) plant cell. **n,** Nucleus containing two nucleoli; **V,** vacuole; **M,** mitochondrion; **P,** proplastid (incompletely developed plastid); **C,** lysosome; **ER,** endoplasmic reticulum; **D,** Golgi apparatus containing dictyosomes; **L,** lipid drops. Ribosomes are shown attached to **ER** or free in the groundplasm (at top). Three plasmodesmal connections through the cell walls are shown by the arrows. (From Sitte, P. 1965. Bau und Feinbau der Pflanzenzelle. Gustav Fischer Verlag, Stuttgart.)

rounded by the inner protoplasmic membrane (the tonoplast), the rest of the protoplasm having coagulated. Nevertheless, several investigators have been able to prepare free protoplasts capable of regenerating new cell walls. These results have been finally crowned by a successful reproduction of a whole plant from a free protoplast (Takebe et al., 1971; Nitsch and Ohyama, 1971; Nagata and Takebe, 1971), demonstrating that the protoplast is totipotent.

By use of one or more of the foregoing methods it is possible in all or nearly all cases to determine whether or not a plant cell is alive. Air-dry cells (e.g., of seeds) must be allowed to swell in water before applying any of these methods. Even then, it may be difficult to observe the response of the cells. As mentioned earlier, the most successful method for seeds is staining with tetrazolium dyes.

CELL COMPONENTS

As shown previously, it is possible to release the inside of a living cell from its surrounding wall without loss of life. The empty wall shows no signs of life, but the freed protoplast will expand in weak solutions and contract in strong solutions (Levitt et al., 1936) just as in the plasmolysis and deplasmolysis of the intact cell. Even the living plant cell consists of a nonliving part (the cell wall) and a living part (the protoplast). The protoplast itself consists of a protoplasmic layer surrounding the vacuole, and since some living cells (e.g., the youngest in meristems) exist without any vacuoles, the protoplasm is the only essential component of a living cell. Direct proof of the nonliving nature of the vacuole has been obtained by washing out the vacuole sap and replacing it by an artificial sap without interfering with the normal activities of the cell (Tazawa, 1964). Yet the nonliving cell wall and vacuole have pronounced physiological significance, since they greatly affect the protoplasm, for example, its water content or growth.

Table 2-1. Average analysis for primary cell walls*

Substance	Percent of fresh weight
Water	60
Hemicelluloses	5-15
Pectic substances	2-8
Cellulose	10-15
Protein	1-2
Lipids	0.5-3.0

*From Setterfield, G., and S. T. Bayley. 1961. Ann. Rev. Plant Physiol. **12**:35-62.

Cell wall

The cell wall is a continuous matrix of polysaccharides in which are embedded a system of microfibrils of α-cellulose (Northcote, 1969). Therefore it consists mainly of these carbohydrates or their derivatives such as pectins (Table 2-1). The central portion of the fused walls of adjacent cells is called the *middle lamella.* It is here that most of the pectic substances are concentrated. With few exceptions cell walls also contain a small amount of protein, more in young cells than in old cells. The name *extensin* has been proposed for the protein found in primary cell walls (Lamport, 1963). It is richer in proline and hydroxyproline than other plant proteins (Olson, 1964) and is glycoprotein in nature (Dashek, 1970). Besides this structural protein (or proteins), there is also evidence for the presence of several with enzyme (Chapter 13) activity: β-glycerophosphatase in barley root tips (Hall and Butt, 1968), β-glucosidase in bean hypocotyl (Nevins, 1970), and peroxidase in soybean hypocotyls (Barnett and Curtis, 1970). Several other enzymes have been demonstrated cytochemically in the walls of pollen grains (Knox and Heslop-Harrison, 1970). Even RNA (Chapter 16) has been isolated from cell walls of peas, beans, and barley, although only to the extent of 0.05% to 0.6% of the dry weight (Phethean et al., 1968). Mature cell walls may contain many substances not found in appreciable quantities in young cells, for example, lignin in wood cells, cutin in epidermal cells of leaves, suberin in cork cells, tannin in bark cells, etc. Cellulose-protein complexes may occur during cellulose deposition, after which most of the protein is resorbed (Preston, see Edds, 1961). But the composition varies greatly from one kind of cell to another in the same plant and even in the same cells under different conditions and at different stages of development. In the normal, turgid cell the wall is essentially saturated with water, which may account for 90% or more of the wall (Frey-Wyssling, 1952). This water is nearly, but not entirely pure (Bernstein, 1971), containing 2 to 10 meq ions/liter, mainly Ca^{++} (30%), Mg^{++} (10%), and Na^{+} + K^{+} (60%).

Despite this high water content, the cell wall has the physical properties of a solid. It is both elastically (reversibly) and plastically (irreversibly) extensible. Cell contraction on water removal before plasmolysis (Fig. 2-3) is a result of the elastic extensibility of the cell wall. In the turgid cell the wall is stretched; in the flaccid cell it returns to the unstretched state. Plastic extensibility of the cell wall results in permanent cell enlargement.

Protoplasm

The most important part of the cell to the physiologist is the living protoplasm, which occupies all the space between the cell wall and the vacuole and includes strands that may traverse both the wall and the vacuole. It consists of cytoplasm, nucleoplasm, and many organelles embedded in the cytoplasm. Here are synthesized all the multitude of organic substances found in the plant. Many of these substances may subsequently be broken down again in the protoplasm.

The gross analysis of protoplasm is possible in the case of animals and of some lower plants (e.g., slime molds), which yield masses of protoplasm without a cell wall or vacuole (Table 2-2). This old analysis includes the nucleic acids with the proteins. The ribonucleic acid (RNA) content of root tips is 4.5 to 13.8 μg/mg fresh weight, and the deoxyribonucleic acid (DNA) is 10% to 20% as high (Martin, 1966). On this basis the total nucleic acid content would be 0.5% to 1.5% of the root tips. Since these root tips are probably about one half protoplasm, this would mean a nucleic acid content of 1% to 3% in the protoplasm.

Table 2-2. Generalized protoplasmic analysis*

Substance	Percent of fresh weight	Approximate relative number of molecules
Water	85-90	18,000
Proteins†	7-10	1
Fatty substances	1-2	10
Other organic substances	1-1.5	20
Inorganic ions	1-1.5	100

*Adapted from Sponsler, D. L., and J. D. Bath. 1942. *In* W. Seifriz. The structure of protoplasm. Iowa State University Press, Ames, Iowa.

†This undoubtedly includes the nucleic acids, which account for 1% to 3% of the protoplasm.

Of course, this does not mean that it would be possible to synthesize protoplasm by mixing these substances in the preceding proportions. In other words, the essential physical and chemical properties of protoplasm are not revealed by such gross analyses. For instance, if the cell sap is mixed with the protoplasm by grinding up cells of leaves or fruit, at least some of the proteins will usually coagulate. This is partly because of the high acidity of the cell sap. Yet this same acidic cell sap is surrounded by the nonacidic protoplasm in the living cell. Therefore the protoplasm must be able in some way to maintain its own low acidity although in contact with the acid vacuole. It does this by forming a barrier, the semipermeable membrane, which keeps the vacuole contents from penetrating into it.

Vacuole

The term *vacuole* is used here for the large central portion of mature plant cells (and its smaller progenitors in meristematic cells), usually consisting of structureless, water-clear sap surrounded by a thin membrane. This term has also been used for a variety of structures, such as the peripheral bodies in some cells, derived from invaginations of the plasma membrane, which may project later into the above central vacuole (Mahlberg et al., 1970). It would seem better to call these vesicles. Since the vacuole is surrounded by cytoplasm, all the substances in it must get there by way of the cytoplasm layer. Despite its origin, the vacuole differs markedly from the protoplasm, not only in acidity (see preceding discussion), but also in its chemical composition and physical properties. Its water content is generally higher than that of the protoplasm. Unlike the protoplasm, it usually does not contain proteins and fatty substances. Salts of organic and inorganic acids and carbohydrates such as sugars, mucilages, etc. are probably always present in the vacuole. It also usually contains some so-called secondary metabolites (Chapter 16), such as tannins, flavonols, anthocyanins, alkaloids, etc. But the vacuole contents are much more variable than those of the protoplasm and therefore no general gross analysis can be given. This is because the vacuole is the repository of (1) any substances that accumulate in excess from the outside and (2) any substances produced in excess by the protoplasm. Heavily fertilized plants may accumulate nitrates, phosphates, or potassium salts in their vacuoles; halophytes (plants that grow in high-salt soils) store large quantities of sodium salts in their vacuoles; and, as just seen, vital staining with neutral red involves accumulation in the vacuole. The nature of the solutes varies with the species. Nitrates, for instance, are low in the Leguminosae but high in the Urticaceae (von Schnurbein, 1967). Though characteristically absent from vacuoles, protein bodies occur in the vacuoles of tomato leaf cells (Shumway et al., 1970). Even intravacuolar membrane systems have been reported in some vacuoles (Bobak and Herich, 1970). Many plants deposit excess calcium and organic acid as insoluble crystals in their vacuoles, or excess of the breakdown products of protein in the form of asparagine, or in the case of plants with specialized metabolism, alkaloids. Some of the stored substances may be reutilized later (e.g., nitrates, organic acids, asparagine, sugars); others may remain permanently in the vacuoles (e.g., alkaloids, sodium salts). Even cells in the same tissue may vary greatly in their vacuolar contents of such substances. This can be clearly shown by staining a section vitally with neutral red. Cells that stain heavily are called "full," and those that stain lightly are "empty." The former have high tannin contents in their vacuoles; the latter have low contents or none. Even the color of the staining may differ from vacuole to vacuole in the same tissue.

The vacuoles of young cells have been reported to contain substances not found in the vacuoles of mature cells. Vacuoles isolated from rootlets of corn seedlings, for instance, contain a number of hydrolytic enzymes (Matile, 1966). These, however, are probably lysosomes (see following discussion) rather than true vacuoles. They must be clearly distinguished from the provacuoles or vesicles of meristematic cells, which appear to develop from the rough endoplasmic reticulum (Mesquita, 1969) and that later enlarge to form the central vacuole of the mature cell. Therefore vacuoles are best defined in terms of their function as storehouses for (1) waste products and (2) water and dissolved substances that may later be reutilized (Buvat, 1968). Vacuolar material may be obtained for analysis in different ways.

1. In the case of mature, higher plants, the juice squeezed out of the killed tissue

is mainly vacuolar sap contaminated by relatively small quantities of soluble materials from the protoplasm and cell wall. This may seem surprising, since the cell wall and protoplasm may contain nearly as high a percentage of water as the vacuole (usually 60% to 90% vs. 90% to 95%, respectively). However, the juice must come mainly from that part of the cell which accounts for the largest percent of its volume. This proportion will vary, in the case of the vacuole, from essentially zero in meristematic cells, to over 90% of the cell in mature parenchyma cells (Fig. 2-2). But since meristematic cells account for only a small fraction of the mature plant and parenchyma cells make up the largest part of the mature (herbaceous) plant, most of the juice obtained from herbaceous plants must come from the cell vacuoles. This is also true of resting organs with high water contents (e.g., bulbs, tubers, corms).

2. If free protoplasts are damaged, their protoplasm may slough off, leaving free vacuoles surrounded only by the extremely thin, semipermeable tonoplast (Chapter 7). In the case of yeast cells, it is possible to convert 75% of the free protoplasts into free vacuoles by adding glucose and a chelating agent (Chapter 12) such as EDTA (Indge, 1968). This would seem to be the best method of obtaining essentially uncontaminated vacuole sap from higher plants, but it has not yet been used in practice.

3. The best sources of uncontaminated vacuolar sap are the large-celled algae (*Nitella, Chara,* etc.) A sample of vacuolar sap large enough for analysis may be obtained from a single such cell simply by inserting a microsyringe through the cell wall and protoplasm layer. Analyses of such samples have revealed a higher concentration of total inorganic ions (K^+, Na^+, and Cl^-) than in the cytoplasm (Dainty, 1968).

Organelles (Fig. 2-5)

Physically, the protoplasm is heterogeneous, possessing microscopically visible structure as opposed to the optical homogeneity of the vacuole. Discrete bodies can always be seen in it under the optical microscope. The largest is the nucleus and its included nucleoli, the next largest are the plastids (commonly 2 by 5 μm), and the smallest are the mitochondria (commonly 0.5 by 2 μm). In some cells, other structures such as spherosomes and Golgi bodies (see following discussion) may also be visible under the optical microscope. Because of their small size, it is not surprising that there may be 1000 mitochondria per meristematic cell of 2500 μm^3 volume, about 170 proplastids (incompletely developed plastids), and 225 Golgi bodies (Avers, 1962). Other bodies may be visible in some cells, but for further observation of protoplasmic structure the electron microscope must be used.

When observed with the electron microscope, these protoplasmic bodies may be so highly structured (Fig. 2-6) that they look like little organs and are therefore called *organelles* by the electron microscopist. The biochemist has developed methods for separating these organelles, or pieces of them (e.g., *microsomes,* which are pieces of endoplasmic reticulum), and has called them *protoplasmic particulates.*

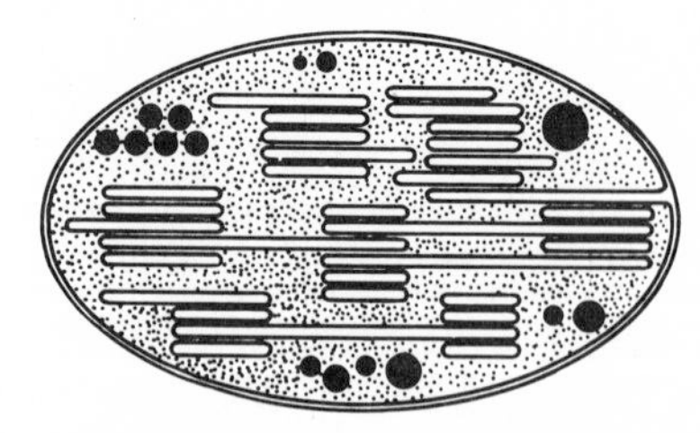

Fig. 2-6. Fine structure of chloroplasts showing schematically the outer membrane, internal lamellae, granular matrix, and several globules. (From Sitte, P. 1965. Bau und Feinbau der Pflanzenzelle. Gustav Fischer Verlag, Stuttgart.)

The methods used are now highly sophisticated (Birnie and Fox, 1969), and sufficiently pure preparations of each organelle have been tested biochemically. As a result, it is now clear that all or nearly all the cell's activities occur in or on its organelles. The importance of both their observable structure and their biochemical nature is now recognized, and according to Bell (1965) an organelle should be defined as part of a cell with certain definite functions, distinctive chemical constituents, and characteristic morphological features. Thus each of the bodies performs a specific function in the living cell, although in some cases it is not yet fully understood (Table 2-3). They may even continue to perform these functions when separated from the cell, although it may then be necessary to supply them with substances normally provided by other parts of the protoplasm or lost from them during extraction.

The larger organelles are the main centers of cell metabolism. Because of its role as the storehouse of genetic information, the nucleus is the main center of nucleic acid synthesis. Some proteins are also synthesized in it, apparently principally in the nucleolus (Lyndon, 1968). Although other organelles may show a considerable degree of independence, and may even multiply in the absence of nuclei (Brachet, 1965), the nucleus is essential for synthesis of the ribosomes in most cells, and even organelles that multiply in its absence may be produced originally by the nucleus (Frey-Wyssling and Mühlethaler, 1965). The chloroplasts, however, are believed by Yuasa (1969) to be self-replicating and independent of the nucleus.

The complete photosynthetic process occurs in the chloroplast, even when it is separated from the rest of the cell (Chapter 15). Aerobic respiration takes place in the mitochondria (Chapter 14), which are therefore found in all aerobic cells. They

Table 2-3. Protoplasmic structures (or organelles) and their physiological functions

Protoplasmic structures	Physiological functions
A. Visible with the optical microscope	
1. Nucleus	DNA synthesis or control of code
2. Nucleolus	RNA synthesis for protein synthesis
3. Chloroplast*	Photosynthesis
4. Mitochondrion	Respiration
5. Spherosome, lysosome	Fat storage, digestion
B. Sometimes visible with optical microscope	
6. Endoplasmic reticulum (ER)	Membranes, plasmodesmal connections, translocation of solutes
7. Golgi bodies or apparatus (GA) or dictyosomes	Cell plate, pectins, cellulose of cell wall, origin of vacuoles and vesicles
C. Resolvable only with the electron microscope	
8. Plasma membranes	Control of semipermeability, absorption
9. Microbodies, peroxisomes, glyoxysomes	Photorespiration, fat synthesis
10. Microtubules	Orientation of cellulose microfibrils
11. Ribosomes†	Protein synthesis
12. Particulate components of organelle structure (oxysome, quantasome, plasmalemmasome)	Not established
13. Groundplasm	Glycolysis

*In nongreen cells other kinds of plastids may occur, for example, leucoplasts that synthesize starch from sugar.

†The earlier microsomes of the biochemist were probably pieces of endoplasmic reticulum containing ribosomes.

may, however, also play a biosynthetic role (Flavell, 1971). Both the chloroplast (Fig. 2-6) and the mitochondrion (Fig. 14-3) have a highly developed, specific internal structure. Both are enclosed in a membrane, and can swell and contract osmotically, in the same way as the protoplast itself (Overman et al., 1970). Both are believed to have been independent organisms (prokaryotes), which early in evolution penetrated and became part of the cells of eukaryotes (Raven, 1970). They may, however, be only semiautonomous. The chloroplast, for instance, appears to depend on the nucleus only for its DNA polymerase (Giles and Taylor, 1971).

A second group of organelles are responsible for the formation of the two nonliving components of living cells, the cell wall and the vacuole. The *endoplasmic reticulum* (ER), despite its name, is not a reticulum (a net) but is a system of interconnecting canals within the protoplasm (Frey-Wyssling and Mühlethaler, 1965). The central strand traversing the plasmodesmata, which connect adjacent cells, is apparently in open continuity with the ER of the adjacent cells (Robards, 1971). This strand has been called the *desmotubule.* The Golgi apparatus (GA) consists of a series of double membranes, sometimes called the dictyosomes and contains "cisternae" (cavities surrounded by membranes), which may enlarge to form the cell vacuole (Marinos, 1963) or perhaps may form the cell plate (Whaley and Mollenhauer, 1963). Yet they are also believed to be involved in the synthesis of pectins for cell wall formation (Dashek and Rosen, 1966), as well as the cellulose fibrils (Fig. 2-7). They even secrete the mucilage by means of which an insectivorous plant captures insects (Schnepf, 1963). They are clearly responsible for wall formation in the case of the free protoplasts of yeast cells because a large number of dictyosomes appear in the cytoplasm simultaneously with new wall formation (Havelkova, 1969). Similarly, in young, actively growing cells of a marine alga, wall fragments are synthesized within the cisternae, which are distended and oriented toward the cell surface (Brown, 1969). In a single cell, however, there may be two functionally different types of dictyosomes, one contributing vesicles, lacking fibrillar content, to the vacuole and the other contributing vesicles, with fibrillar content, to the cell wall (Brown and Weier, 1970). A fraction rich in Golgi bodies has been isolated from roots of pea seedlings (Harris and Northcote, 1971). Incubation with radioactive glucose revealed a synthesis of pectic substances and hemicelluloses by these Golgi bodies and their associated vesicles. The Golgi cisternae are produced at the rate of 1 per 2 min in the above marine alga, and it requires 41 to 82 Golgi generations to synthesize the cell wall. In the case of xylem cells, both endoplasmic reticulum and Golgi bodies were involved in wall incorporation (Picket-Heaps, 1968). This was also true of root tip cells of *Vicia faba* (Bobak, 1970).

The evidence now seems to indicate that the Golgi apparatus is the center of synthesis of the polysaccharides (Ray et al., 1969; Brown and Weier, 1970), and the ER is the center of synthesis of the glycoprotein component of the cell wall (Dashek, 1970). The orientation of the cellulose microfibrils in the cell wall seems to be determined by the *microtubules* (Chafe and Wardrop, 1970), which are organelles that are found adjacent to the growing wall (Fig. 2-7). Colchicine inhibits the activity of the microtubules. By use of this inhibitor, evidence has been produced that the microtubules play a major role in defining the patterns of secondary wall formation and in directing the orientation of the cellulose microfibrils in the wall (Hepler and Fosket, 1971). Nevertheless, they are not involved in the actual synthesis of cellulose (Marx-Figini, 1971).

Spherical, highly refractive bodies have

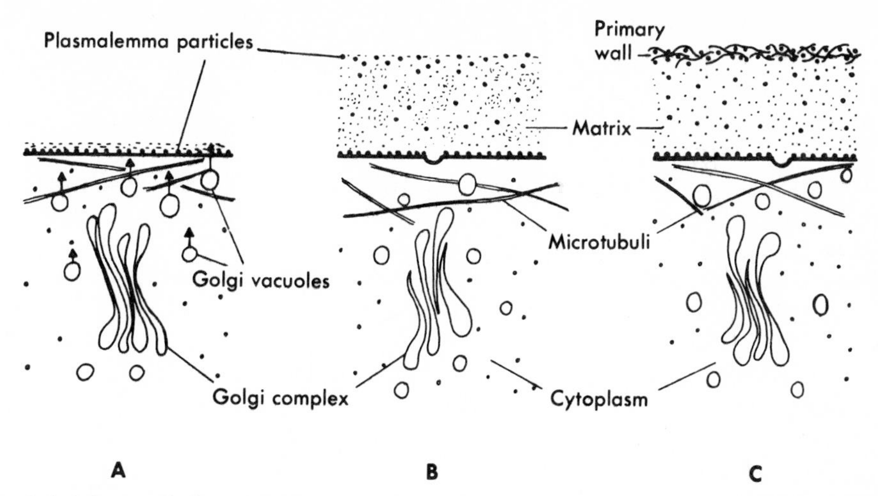

Fig. 2-7. Scheme of the subsequent steps in cell wall formation of the green alga *Chlorella* sp. **A,** Cortical region of the cell with the plasmalemma covered with particles and a subjacent Golgi complex. **B,** Accumulation of matrix material, carried to the cell surface by Golgi vacuoles. Plasmalemma particles become detached and move to the outer periphery of the matrix. **C,** Primary wall formation in the region where the particles are concentrated. (From Mühlethaler, K. 1967. Ann. Rev. Plant Physiol. **18**:1-24.)

long been observed with the optical microscope in plant cells, and have been called *spherosomes.* Some of these appear to be oil droplets, and when seen in chloroplasts, have been called *plastoglobuli* (Lichtenthaler, 1969). Others seem to be equivalent to the *lysosomes* of animal cells because they have been found to contain hydrolytic enzymes (Semadeni, 1967; Matile et al., 1965), and their function is therefore thought to be digestion of the cell contents when these enzymes are released from the lysosomes (Gahan and Maple, 1966). In the case of *Euglena,* during starvation, portions of the cytoplasm become encapsulated in membrane-bound cavities, similar to cytolysomes of mammalian cells. Progressive degradation of the encapsulated material may represent a mechanism for providing the cell with breakdown products for utilization in continued maintenance of basic metabolic processes. Strong evidence indicates participation of the Golgi apparatus in the formation of the cytolysomes. Yet the spherosomes of higher plants appear to originate from the ER. In healthy cells of pea root meristem (Coulomb, 1971) the lysosomes hydrolyze aging organelles without injuring the cell. In crown gall tumors, on the other hand, they may completely lyse certain cells, acting as "instruments of suicide." They have also been found in the latex of laticiferous plants (Matile et al., 1970) and contain alkaloids. Lysosome-like bodies have been identified in root tips, pollen grains, and stigma hairs (Malik et al., 1969), as well as in differentiating, lignified collenchyma (Wardrop, 1968).

Three names have been given to small structures with similar functions. *Microbodies* are organelles 0.5 to 1.5 μm in diameter, limited by a simple membrane and occurring abundantly in chlorophyllous cells of leaves (Frederick and Newcomb, 1969a). They are probably identical with leaf particles called *glyoxysomes* (Cooper and Beevers, 1969) that are apparently involved in the glyoxylate cycle and therefore in fat synthesis (Chapter 16). The presence of the enzyme catalase in them also indicates that they are identical with *peroxisomes* (Tolbert et al., 1968; Frederick and Newcomb, 1969b), which are particles that contain the catalase of green leaves, an enzyme previously believed to be present in the chloroplasts (Price, 1970). These two conclusions are not contradictory, for the glyoxysome belongs to the peroxisome family (de Duve, 1969). Unlike mitochondria, the peroxisome catalyzes an apparently wasteful form of respiration called *photorespiration* (Chapter 14). It is perhaps a protection against oxygen toxicity (de Duve, 1969). In the case of watermelon seedlings, the glyoxysomes and peroxisomes are different organelles (Kagawa and Beevers, 1970). It has been suggested that all these microbodies (used in a general sense) are derived from the endoplasmic reticulum (Gerhardt and Beevers, 1969).

The remaining organelles are one or more orders of magnitude smaller than those just mentioned and may even be components of the larger organelles. The *ribosomes*, are simple particles about 15 nm in diameter, occurring either free in the cytoplasm or adhering to the ER. They frequently occur as aggregates called polysomes. Although the ribosome is the main seat of protein synthesis, the nucleus (James and Richens, 1963), the nucleolus (Birnstiel and Hyde, 1963), the chloroplast (Spencer, 1965), and the mitochondrion (Das and Mukherjee, 1964) also synthesize proteins. In the case of the chloroplast (Lyttleton, 1962) and the mitochondrion (Küntzel and Noll, 1967), although apparently not in the case of the nucleus (Flamm and Birnstiel, 1964), this is a result of the presence of ribosomes within it.

The mitochondrial ribosomes of yeast differ from the cytoplasmic ribosomes in nucleotide composition and other properties (Morimoto and Halvorson, 1971). The membranes are also sometimes considered as separate organelles, although, like the ribosomes, they are components of other organelles (the nucleus, plastid, mitochondrion) that they enclose. The whole protoplasm layer is also bounded by its membranes.

It is not always certain whether the organelles are self-producing or originate de novo. The Golgi apparatus (dictyosomes), for instance, differentiates de novo from perinuclear bodies, which are synthesized under the action of ribonucleoprotein and are therefore nucleus-dependent (at least in *Acetabularia*, Werz, 1964).

The peroxisome family of organelles apparently occur free in the protoplasm, independent of the other organelles; the ribosomes may occur free, attached to the ER, or within larger organelles (chloroplasts or mitochondria). Another group of particles are not free organelles but are component parts of the structure of larger organelles.

High resolution electron microscopy has identified an "elementary particle," or oxysome, in the mitochondrion (Lehninger, 1964). Similarly, the *quantasome* is an oblate sphere with a molecular weight of 2×10^6 (Park and Biggins, 1964) and is the structural and functional unit of the chloroplast. Besides these quantasomes, there are other particles attached to the chloroplast structure (Park and Pfeifhofer, 1969). There are even particles attached to the outer surface of the plasmalemma, which have been called *plasmalemmasomes* (Cole and Lin, 1970). All of these subunits have been called *multimolecular assemblies* (MMAs) (Schmitt, 1963). The relationship to the cell is shown in Fig. 2-8. The large numbers of such particles is illustrated by the ribosomes. There are about $6000/\mu m^3$, and they occupy about 0.5% of the protoplasm (Ts'o, 1962).

The fundamental importance of the organelles is emphasized by Palade's (1963) statement that the basic requirements of a living cell are (1) the codes (in the nucleus), (2) the ribosomes, (3) the cell membranes, and (4) an energy-supplying system (e.g., the mitochondria or the chloroplasts). The dynamic state of these organelles cannot be revealed by the electron microscope, since it observes the dead, or

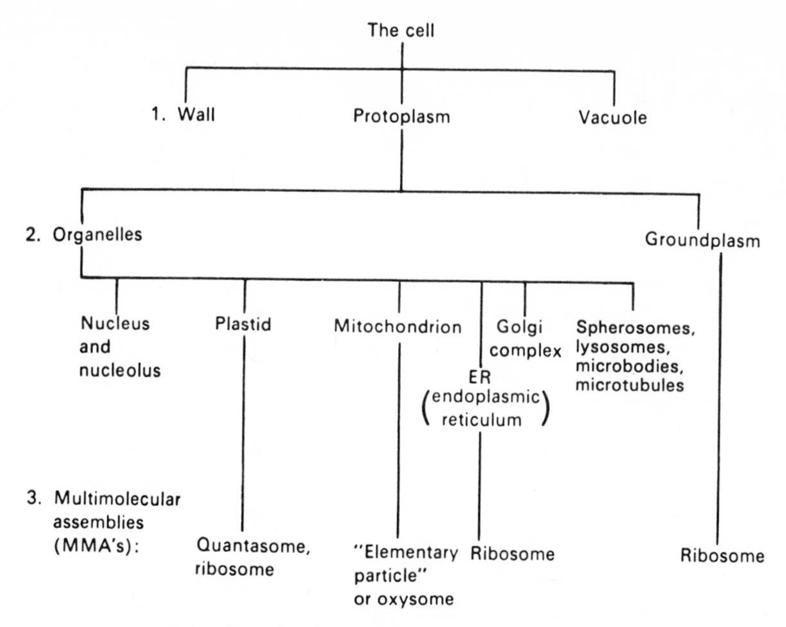

Fig. 2-8. Three levels of organization of cell constituents.

at least nonfunctioning, cell. Not only does streaming occur in the cytoplasmic strands of living cells but even the structural components may undergo visible changes. The mitochondria change their shape, fragment, and migrate to the outer rim of the cytoplasm. Even the plasma membrane may show an undulating motion (Gropp, 1963).

It is often forgotten that all the organelles are embedded in a granular matrix, the "groundplasm," which must therefore also have some function. It has a fine, granular structure, just detectable under the electron microscope, yet it is also fibrillar (Sitte, 1965). It contains soluble enzymes, which are not bound to the organelles and which function at least in the process of glycolysis (Chapter 14). The chemical complexity of the structural components of protoplasm is illustrated by analyses of chloroplasts. Each chloroplast (Fig. 2-6) consists of a lamellar phase (mainly in disc-shaped grana) embedded in a matrix (stroma). The whole chloroplast is surrounded by a membrane. The lamellae may be separated experimentally from the matrix (or stroma) and are then found to contain all the chlorophyll and 50% of the protein nitrogen. The complete analysis of the lamellae is given in Table 2-4.

Chemical nature. Despite this chemical complexity of the protoplasmic structures, they consist primarily of three groups of substances: (1) proteins, (2) nucleic acids

Table 2-4. Substances and their proportions in spinach chloroplast lamellae*

	Moles	Parts per million (app.)
I. Lipids		
Chlorophylls	115	103,200
Carotenoids	24	13,700
Quinones	23	15,900
Phospholipids	58	45,400
Glycerides	245	201,000
Sulfolipids	24	20,500
Others		95,300
II. Proteins		464,000
Manganese in protein	1	55
Iron in protein	6	336
Copper in protein	3	159

*From Lichtenthaler, H. K., and R. B. Park. 1963. Nature (London) **198**:1070-1072.

Table 2-5. Distribution (percent) of three main groups of organic substances in some protoplasmic structures*

	Protein	Nucleic acid	Lipid
Nucleus	74	26	Insignificant
Mitochondrion	67	1-3	30-33
Ribosome	50	50	

*Adapted from Frey-Wyssling, A., and K. Mühlethaler. 1965. Ultrastructural plant cytology. American Elsevier Publishing Co., Inc., New York.

(RNA, DNA), and (3) lipids (Table 2-5). It must be emphasized, however, that even the organelles may be largely aqueous; for instance, 80% of the enclosed volume of wheat embryo ribosomes was found to consist of water (Wolfe and Kay, 1967). Although DNA is primarily found in the nucleus, it also occurs in the chloroplast (Goffeau and Brachet, 1965), to the extent of one five thousandth the protein content (Gibor and Izawa, 1963), or about 9% of the total DNA of the tobacco leaf (Tewari and Wildman, 1966). Its molecular weight is approximately 4×10^7. There is evidence that DNA is synthesized in young (but not old) chloroplasts (Wollgiehn and Mothes, 1964), and although RNA is synthesized in the nucleus, as much as 25% to 35% of the total RNA of a leaf cell may be in its chloroplasts (Heber, 1963). A small amount of DNA is also present in the mitochondria. In *Euglena* it is less than 5% of the total DNA (Holm-Hansen, 1969). Some of the nucleic acids also occur free in the groundplasm. That the nucleoproteins are the most fundamental substances of life is indicated by the fact that the smallest living entity, the virus, consists solely of nucleoproteins. Similarly, the units of heredity in all organisms, the genes, are also nucleoproteins. Ribosomes are actually ribonucleoprotein particles (30% RNA and 70% protein in the ribosomes of spinach chloroplasts [Biswas and Biswas, 1966]). The published proportions of these two substances vary (Table 2-5), perhaps depending on their source.

The phospholipids are of particular importance because they are largely confined to the surface layers (the membranes) of the structural components (the plastids, mitochondria, etc.) and the protoplasm, and they therefore control the semipermeability of the cell. However, the actual proportions of these substances are not the same in the different structures. Furthermore, even the groundplasm in which the organelles are embedded also contains some phospholipids. Because of this reserve, the groundplasm is capable of instantly forming new membranes on wounding of the protoplasm (Sitte, 1965).

The matrix of the cytoplasm (the groundplasm) is essentially made up of a protein solution, or gel. This matrix is usually considerably more viscous than the vacuole. Since viscosity and gel formation are characteristic of the colloidal state, protoplasm is obviously a colloid. The nonliving parts of the cell also show colloidal properties. The vacuole, although optically homogeneous and mobile, usually possesses enough colloidal material (e.g., tannin) to accumulate dyes such as neutral red, staining more or less strongly if the cell is alive. The wall is a colloidal gel of cellulose and other large-moleculed organic substances. Although these substances are insoluble in water, they imbibe it readily in the same way that a gelatin gel does.

THE STATE OF WATER IN THE LIVING CELL

All three components of the living cell—wall, protoplasm, and vacuole—contain ample water that can pass readily from one to the other. The physical chemistry of aqueous systems (e.g., diffusion, osmosis, etc.) is therefore of paramount importance in plant physiology. The lines of separation between these three components of the cell are known as interfaces. Special physical forces known as surface tension, or interfacial tension, operate at such interfaces. Substances tend to become concentrated there, producing the phenomenon of adsorption. However, this approach assumes that the aqueous system in the cell is identical to any aqueous system in nonliving bodies. In opposition to this assumption, the suggestion has frequently been made that although water accounts for up to 90% or more of the cell, it differs in some mysterious way from water in nonliving systems. Physiologists long have thought that this concept was finally put to rest by the classic quantitative investigations which showed that the cell's water obeys the simple Boyle-van't Hoff laws of aqueous solutions (Lucké and McCutcheon, 1932; Nobel, 1969). Nevertheless, the concept of "vital" or "icelike" water, which is different from the ordinary water of nonliving systems, continues to be championed (Tait and Franks, 1971). The recent "discovery" of "polywater" (or polymerized water inside glass microcapillaries) for a while gave new credence to these concepts. However, so many thorough and independent investigations in so many laboratories have failed to corroborate the existence of polywater (Barnes et al., 1971), and so many of the best trained physical chemists have produced evidence that the earlier results were due to contamination by a number of substances (silicates, borates, and even sweat!) that we are forced to reject the earlier positive results on which the concept of polywater was based.

Such negative evidence, of course, does not by itself prove that the concept of "vital" or even "icelike" water in living cells is incorrect. However, the previously mentioned classic investigations of the osmotic behavior of cells have been corroborated by results with free, living protoplasm squeezed out of *Nitella* cells (Yoneda and Kamiya, 1969), which showed that practically all the space in the drop of protoplasm is osmotically active. Similarly, no osmotically inactive compartments could be found in yeast cells (de Bruijne and van Steveninck, 1970). Finally, freezing experiments with living cells have definitely shown that living cells have a freezing point, which would not be expected if the cell's water were already "icelike." When determined carefully (Levitt, 1956), this freezing point is within about 0.1° C. of the freezing point when the cell is dead, which is as close as can be expected from the experimental conditions imposed by the use of living cells (e.g., the absence of stirring to obtain thorough mixing). It has also been shown that the amount of ice formed at any one freezing temperature is exactly the same in living leaves as in dead mash obtained by grinding them (Johansson and Krull, 1970).

In view of all these quantitative results, the plant physiologist must conclude that the water in living cells is no different from water in nonliving systems. Consequently, the laws of aqueous systems can be applied directly to the living cells.

It is now apparent that in order to understand more about the properties of protoplasm and of the cell as a whole, we must consider such purely physicochemical concepts as acidity, colloids, permeability, diffusion, osmosis, and adsorption.

QUESTIONS

1. What is the smallest portion of the plant that is capable of regenerating the whole plant?
2. Does the living cell do work when growing or when not growing?
3. What is the theoretical basis for this conclusion?
4. What experimental evidence is there for this conclusion?
5. Does brownian movement occur in living or dead cells?
6. How can living cells be distinguished from dead cells?
7. If a cell shows cytoplasmic streaming at room temperature but not when its

temperature drops to +1° C does this indicate it is now dead?
8. When a living cell is placed in water, do the solutes leak out? Explain.
9. How can the semipermeability of living cells be demonstrated?
10. What is vital staining?
11. Do vital stains color cell walls of living or dead cells?
12. Are tetrazolium dyes vital stains? Explain.
13. What is meant by plasmolysis?
14. Does the cell wall remain unchanged when a cell plasmolyzes?
15. If a cell plasmolyzes, is the solution hypertonic, hypotonic, or isotonic?
16. What is the protoplast?
17. What is a free protoplast?
18. How can free protoplasts be obtained?
19. What chemical substances occur in the cell wall?
20. Of the cell's wall, protoplasm, and vacuole, which part is alive?
21. How does a cell wall react to a stretching force?
22. What chemical substances does protoplasm consist of?
23. What chemical substances does the vacuole consist of?
24. If excess fertilizer is absorbed, in what part of the cell is it accumulated?
25. If an excess of some substances is produced by the cell, in what part of the cell is it most likely to accumulate?
26. From what part of the cell does plant juice mainly originate?
27. What are the protoplasmic particulates?
28. What are the cell organelles?
29. What are their functions?
30. What are the basic structural requirements for a living cell?
31. What chemical substances do organelles consist of?
32. What is the chemical nature of the protoplasmic surface?
33. Which organelles are components of other organelles?
34. In what protoplasmic material do the organelles occur?
35. What are MMAs?
36. With what cell function are phospholipids associated?
37. What is required for regeneration of wounded membranes?

SPECIFIC REFERENCES

Avers, C. J. 1962. Fine structure studies of *Phleum* root meristem cells. I. Mitochondria. Amer. J. Bot. **49**:996-1003.

Barnes, P., I. Cherry, J. L. Finney, and S. Petersen. 1971. Polywater and polypollutants. Nature **230**:31-33.

Barnett, N. M., and C. R. Curtis. 1970. Release of peroxidase and hydroxyproline protein from cell walls by hydrolytic enzymes. Plant Physiol. **46**(suppl.): 74.

Bell, P. R. 1965. The structure and origin of mitochondria. Sci. Progr. (London) **53**(209):33-44.

Bernstein, L. 1971. Method for determining solutes in the cell walls of leaves. Plant Physiol. **47**:361-365.

Birnie, G. D., and S. M. Fox. 1969. Subcellular components: preparation and fractionation. Plenum Publishing Corp., New York, and Butterworth & Co. (Publishers), Ltd., London.

Birnstiel, M. L., and B. B. Hyde. 1963. Protein synthesis by isolated pea nucleoli. J. Cell Biol. **18**:41-50.

Bishop, C. T., S. T. Bayley, and G. Setterfield. 1958. Chemical constitution of the primary cell walls of *Avena* coleoptiles. Plant Physiol. **33**: 283-289.

Biswas, S., and B. B. Biswas. 1966. Characterization of ribonucleoprotein particles and protein synthesis in chloroplasts. Indian J. Biochem. **3**: 96-100.

Bobak, M. 1970. A study of the initial development phases of plant cell walls. Biologia **25** (1):57-60.

Bobak, M., and R. Herich. 1970. On intravacuolar formations in plant cells. Acta Biol. Acad. Sci. Hung. **21**:193-196.

Brachet, J. L. A. 1965. *Acetabularia*. Endeavour **24**(93):155-161.

Brown, D. L., and T. E. Weier. 1970. Ultrastructure of the freshwater alga *Batrachospermum:* 1. Thin-section and freeze-etch analysis of juvenile and photosynthetic filament vegetative cells. Phycologia **9**:217-235.

Brown, R. M., Jr. 1969. Observations on the relationship of the Golgi apparatus to wall formation in the marine chrysophycean alga *Pleuro-*

chrysis scherfelti Pringsheim. J. Cell Biol. **41**: 109-123.

Buvat, R. 1968. Diversity of vacuoles in the cells of the root of barley. Compt. Rend. Acad. Sci. (Paris) **267**:296-298.

Chafe, S. C., and A. B. Wardrop. 1970. Microfibril orientation in plant cell walls. Planta **92**: 13-24.

Cocking, E. C. 1960. A method for the isolation of plant protoplasts and vacuoles. Nature (London) **187**:927-929.

Cole, K., and S.-C. Lin. 1970. Plasmalemmasomes in sporelings of the brown alga *Petalonia debilis*. Can. J. Bot. **48**:265-268.

Cooper, T. G., and H. Beevers. 1969. Mitochondria and glyoxysomes from castor bean endosperm. J. Biol. Chem. **244**:3507-3513.

Coulomb, P. 1971. Sur la présence de phytolysosomes dans les cellules de tumeurs de la plantule de pois *(Pisum sativum L.)* induites par l'*Agrobacterium tumefasciens*. Compt. Rend. Acad. Sci. (Paris) **272**:1229-1231.

Dainty, J. 1968. The structure and possible function of the vacuole, p. 40-46. *In* J. B. Pridham, editor: Plant cell organelles. Academic Press, Inc., New York.

Das, H. K., and T. Mukherjee. 1964. Protein synthesis in plant mitochondria. III. Characterization of mitochondria and the microsomal fraction of the seedlings of *Vigna sinensis*. Biochim. Biophys. Acta **93**:304-310.

Dashek, W. V. 1970. Synthesis and transport of hydroxyproline-rich components in suspension cultures of sycamore-maple cells. Plant Physiol. **46**:831-838.

Dashek, W. V., and W. G. Rosen. 1966. Electron microscopical localization of chemical components in the growth zones of lily pollen tubes. Protoplasma **61**:192-204.

de Bruijne, A. W., and J. van Steveninck. 1970. Apparent nonsolvent water and osmotic behavior of yeast cells. Biochim. Biophys. Acta **196**:45-52.

de Duve, C. 1969. The peroxisome: a new cytoplasmic organelle. Proc. R. Soc. Lond. [Biol.] **173**:71-83.

Falk, H., U. Lüttge, and J. Weigl. 1966. Research on the physiology of plasmolysed cells. II. Ion uptake, oxygen exchange, transport. Z. Pflanzenphysiol. **54**:446-462.

Flamm, W. G., and M. L. Birnstiel. 1964. The nuclear synthesis of ribosomes in cell cultures. Biochim. Biophys. Acta **87**:101-110.

Flavell, R. B. 1971. Mitochondria also play a biosynthetic role. Nature **230**:504-506.

Fodil, Y., R. Esnault, and G. Trapy. 1971. Remarques sur l'isolement de protoplastes végétaux: étude de l'influence de l'agent de plasmolyse. Compt. Rend. Acad. Sci. (Paris) **272**:948-951.

Frederick, S. E., and E. H. Newcomb. 1969a. Microbody-like organelles in leaf cells. Science **163**:1353-1355.

Frederick S. E., and E. H. Newcomb. 1969b. Cytochemical localization of catalase in leaf microbodies (peroxisomes). J. Cell Biol. **43**:343-353.

Frey-Wyssling, A. 1952. Growth of plant cell walls. Sympos. Soc. Exp. Biol. **6**:320-328.

Gahan, P. B., and A. J. Maple. 1966. The behaviour of lysosome-like particles during cell differentiation. J. Exp. Bot. **17**:151-155.

Gerhardt, B. P., and H. Beevers. 1969. Occurrence of RNA in glyoxysomes from castor bean endosperm. Plant Physiol. **44**:1475-1477.

Gessner, F. 1968. The cell wall of marine phanerogams. Mar. Biol. **1**:191-200.

Gibor, A., and M. Izawa. 1963. The DNA content of the chloroplasts of *Acetabularia*. Proc. Nat. Acad. Sci. U. S. A. **50**:1164-1169.

Giles, K. L., and A. O. Taylor. 1971. The control of chloroplast division in *Funaria hygrometrica*. I. Patterns of nucleic acid, protein and lipid synthesis. Plant Cell Physiol. **12**:437-445.

Goffeau, A., and J. Brachet. 1965. Deoxyribonucleic acid-dependent incorporation of amino acids into the proteins of chloroplasts isolated from anucleate *Acetabularia* fragments. Biochim. Biophys. Acta **95**:302-313.

Gropp, A. 1963. Morphologie und Verhalten lebender Zellen. Med. Welt **1**:20-23.

Harris, P. J., and D. H. Northcote. 1971. Polysaccharide formation in plant Golgi bodies. Biochim. Biophys. Acta **237**:56-64.

Hall, J. L., and V. S. Butt. 1968. Localization and kinetic properties of B-glycerophosphatase in barley roots. J. Exp. Bot. **19**:276-287.

Havelkova, M. 1969. Electron microscopy study of cell structures and their changes during growth and regeneration of *Schizosaccharomyces pombe* protoplasts. Folia Microbiol. **14**: 155-164.

Heber, U. 1963. Ribonucleic acids in the chloroplasts of the leaf cell. [Transl. title.] Planta **59**: 600-616.

Helper, P. K., and D. E. Fosket. 1971. The role of microtubules in vessel member differentiation in *Coleus*. Protoplasma **72**:213-236.

Holm-Hansen, O. 1969. Algae: amounts of DNA and organic carbon in single cells. Science **163**: 87-88.

Indge, K. J. 1968. The isolation and properties of the yeast cell vacuole. J. Gen. Microbiol. **51**:441-446.

James, W. O., and A. M. Richens. 1963. Energy transport from mitochondria to nuclei. Proc. R. Soc. Lond. [Biol.] **157**:149-159.

Johansson, N.-O., and E. Krull. 1970. Ice formation, cell contraction, and frost killing of wheat plants. Nat. Swed. Inst. Plant Protection Contrib. **14**(131):343-362.

Kagawa, T., and H. Beevers, 1970. Glyoxysomes and peroxisomes in watermelon seedlings. Plant Physiol. **46** (suppl.):207.

Karle, I. L., J. W. Daly, and B. Witkop. 1969. 2,3-cis-3,4-trans-3,4-dihydroxy-L-proline: mass spectrometry and x-ray analysis. Science **164**: 1401-1402.

Karr, A. L., Jr., and P. Albersheim. 1970. Polysaccharide-degrading enzymes are unable to attack plant cell walls without prior action by a "wall-modifying enzyme." Plant Physiol. **46**: 69-80.

Knox, R. B., and J. Heslop-Harrison. 1970. Pollen wall proteins: localization and enzymatic activity. J. Cell. Sci. **6**:1-27.

Küntzel, H., and H. Noll. 1967. Mitochondrial and cytoplasmic polysomes from *Neurospora crassa*. Nature (London) **215**:1340-1345.

Lamport, D. T. A. 1963. Oxygen fixation into hydroxyproline of plant cell wall protein. J. Biol. Chem. **238**:1438-1440.

Levitt, J., G. W. Scarth, and R. D. Gibbs. 1936. Water permeability of isolated protoplasts in relation to volume change. Protoplasma **26**:237-248.

Levitt, J. 1956. The hardiness of plants. Academic Press, Inc., New York.

Lichtenthaler, H. K. 1969. Plastoglobuli and lipoquinone content of chloroplasts from *Cereus peruvianus* (L.) Mill. Planta **87**:304-310.

Lichtenthaler, H. K., and R. B. Park. 1963. Chemical composition of chloroplast lamellae from spinach, Nature (London) **198**:1070-1072.

Lindenbein, W. 1965. Tetrazolium testing. Proc. Int. Seed Test. Ass. **30**:89-97.

Livingston, L. G. 1964. The nature of plasmodesmata in normal (living) plant tissue. Amer. J. Bot. **51**:950-957.

Lucké, B., and M. McCutcheon. 1932. The living cell as an osmotic system and its permeability to water. Physiol. Rev. **12**:68-139.

Lyndon, R. F. 1968. The structure, function and development of the nucleus, p. 16-39. *In* J. B. Pridham (Ed.). Plant cell organelles. Academic Press, Inc., New York.

Lyttleton, J. W. 1962. Isolation of ribosomes from spinach chloroplasts. Exp. Cell Res. **26**:312-317.

Mahlberg, P. G., K. Olson, and C. Walkinshaw. 1970. Development of peripheral vacuoles in plant cells. Amer. J. Bot. **57**:962-968.

Malik, C. P., P. P. Sood, and H. B. Tewari. 1969. Occurrence of lysosome-like bodies in plant cells-acid phosphatase reaction. Zeitschr. f. Biol. **116**:264-268.

Marinos, N. G. 1963. Vacuolation in plant cells. J. Ultrastruc. Res. **9**:177-185.

Martin, P. G. 1966. Variation in the amounts of nucleic acids in the cells of different species of higher plants. Exp. Cell Res. **44**:84-94.

Marx-Figini, M. 1971. Investigations on biosynthesis of cellulose. DPw and yield of cellulose of the alga *Valonia* in the presence of colchicine. Biochim. Biophys. Acta **237**:75-77.

Matile, P. 1966. Enzyme der Vakuolen aus Wurzelzellen von Maize keimlingen. Ein Beitrag zur funktionellen Bedeutung der Vakuole bei der intrazellulären Verdauung. Z. Naturforsch. **21b**: 871-878.

Matile, P., P. J. Balz, E. Semadeni, and M. Jost. 1965. Isolation of spherosomes with lysosome characteristics from seedlings. Z. Naturforsch. [B] **20**:693-698.

Matile, P., B. Jans, and R. Rickenbacher. 1970. Vacuoles of *Chelidonium* latex: lysosomal property and accumulation of alkaloids. Biochem. Physiol. Pflanzen. **161**:447-458.

Mesquita, J. F. 1969. Electron microscope study of the origin and development of the vacuoles in root-tip cells of *Lupinus albus* L. J. Ultrastruct. Res. **26**:242-250.

Morimoto, H., and H. O. Halvorson. 1971. Characterization of mitochondrial ribosomes from yeast. Proc. Nat. Acad. Sci. USA **68**:324-328.

Mühlethaler, K. 1967. Ultrastructure and formation of plant cell walls. Ann. Rev. Plant Physiol. **18**:1-24.

Nagata, T., and I. Takebe. 1971. Plating of isolated tobacco mesophyll protoplasts on agar medium. Planta **99**:12-20.

Nevins, Donald J. 1970. Relation of glycosidases to bean hypocotyl growth. Plant Physiol. **46**:458-462.

Newcomb, E. H. 1963. Cytoplasm—cell wall relationships. Ann. Rev. Plant Physiol. **14**:43-64.

Nitsch, J. P., and K. Ohyama. 1971. Obtention de plantes à partir de protoplastes haploides cultivés in vitro. Compt. Rend. Acad. Sci. (Paris) **273**:801-804.

Nobel, P. S. 1969. The Boyle-Van't Hoff relation. J. Theor. Biol. **23**:375-379.

Northcote, D. H. 1969. Fine structure of cytoplasm in relation to synthesis and secretion in plant cells. Proc. R. Soc. Lond. [Biol.] **173**: 21-30.

Olson, A. C. 1964. Proteins and plant cell walls; proline to hydroxyproline in tobacco suspension cultures. Plant Physiol. **39**:543-550.

Overman, A. R., G. H. Lorimer, and R. J. Miller. 1970. Diffusion and osmotic transfer in corn mitochondria. Plant Physiol. **45**:126-132.

Palade, G. E. 1963. Cell structure, p. 57-74. *In* M. Fishbein. Birth defects. J. B. Lippincott Co., Philadelphia.

Park, R. B., and J. Biggins. 1964. Quantasome; size and composition. Science **144**:1009-1011.

Park, R. B., and A. O. Pfeifhofer. 1969. The effect of ethylenediamine tetraacetate washing on the structure of spinach thylakoids. J. Cell Sci. **5**:313-219.

Phethean, P. D., L. Jervis, and M. Hallaway. 1968. The presence of ribonucleic acid in the cell walls of higher plants. Biochem. J. **108**:25-31.

Picket-Heaps, J. D. 1969. Xylem wall deposition: radioautographic investigation using lignin precursors. Protoplasma **65**:181-205.

Power, J. B., and E. C. Cocking. 1970. Isolation of leaf protoplasts: macromolecule uptake and growth substance response. J. Exp. Bot. **21**:64-70.

Price, C. A. 1970. Separation of plant particles. Science **168**:282-283.

Raven, P. H. 1970. A multiple origin for plastids and mitochondria. Science **169**:641-646.

Ray, P. M., T. L. Shininger, and M. M. Ray. 1969. Isolation of β-glucan synthetase particles from plant cells and identification with Golgi membranes. Proc. Nat. Acad. Sci. USA **64**:605-612.

Robards, A. W. 1971. The ultrastructure of plasmodesmata. Protoplasma **72**:315-323.

Ruesink, A. W., and K. V. Thimann. 1966. Protoplasts: prepartion from higher plants. Science **154**:280-281.

Ruppel, H. G. 1964. Nucleic acids in chloroplasts of *Allium porrum* and *Antirrhinum majus*. Biochim. Biophys. Acta **80**:63-72.

Schmitt, F. O. 1963. The macromolecular assembly, a hierarchical entity in cellular organization. Develop. Biol. **7**:546-559.

Schnepf, E. 1963. Zur Cytologie und Physiologie pflanzlicher Drüsen. I. Über den Fangschleim der Insektivoren. Flora **153**:1-22.

Schnurbein, C. von. 1967. Über den Anteil von Nitrat und Chlorid an der Zusammensetzung des Zellsaftes von Blutenpflanzen. Flora **158**:575-593.

Semadeni, E. G. 1967. Enzymatische Charakterisierung der Lysosomenäquivalente (Sphärosomen) von Maiskeimlingen. Planta **72**:91-118.

Setterfield, G., and S. T. Bayley. 1961. Structure and physiology of cell walls. Ann. Rev. Plant Physiol. **12**:35-62.

Shumway, L. K., J. M. Rancour, and C. A. Ryan. 1970. Vacuolar protein bodies in tomato leaf cells and their relationship to storage of chymotrypsin inhibitor I protein. Planta **93**:1-14.

Spencer, D. 1965. Protein synthesis by isolated spinach chloroplasts. Arch. Biochem. **111**:381-390.

Sponsler, O. L., and J. D. Bath. 1942. Molecular structure in protoplasm, p. 41-85. *In* W. Seifriz. The structure of protoplasm. Monogr. Amer. Soc. Plant Physiol. Iowa State University Press, Ames, Iowa.

Steward, F. C., M. O. Mapes, A. E. Kent, and R. D. Holstein. 1964. Growth and development of cultured plant cells. Science **143**:20-27.

Tait, M. J., and F. Franks. 1971. Water in biological systems. Nature **230**:91-94.

Takebe, I., G. Labib, and G. Melchers. 1971. Regeneration of whole plants from isolated mesophyll protoplasts of tobacco. Naturwiss. **58**:319-320.

Tazawa, M. 1964. Studies on *Nitella* having artificial cell sap. I. Replacement of the cell sap with artificial solutions. Plant Cell Physiol. **5**:33-43.

Tewari, K. K., and S. G. Wildman. 1966. Chloroplast DNA from tobacco leaves. Science **153**:1269-1271.

Tolbert, N. E., A. Oeser, T. Kisaki, R. H. Hageman, and R. K. Yamazaki. 1968. Peroxisomes from spinach leaves containing enzymes related to glycolate metabolism. J. Biol. Chem. **243**:5179-5184.

Ts'o, P. O. P. 1962. The ribosomes-ribonucleoprotein particles. Ann. Rev. Plant Physiol **13**:45-80.

Wardrop, A. B. 1968. Occurrence of structure with lysosome-like function in plant cells. Nature **218**:978-980.

Werz, G. 1964. Elektronenmikroskopische Untersuchungen zur Genese des Golgi-Apparates (Dictyosomen) und ihrer Kernabhängigkeit bei *Acetabularia*. Planta **63**:366-381.

Whaley, W. G., and H. H. Mollenhauer. 1963. The Golgi apparatus and cell plate; a postulate. J. Cell Biol. **17**:216-221.

Wolfe, F. H., and C. M. Kay 1967. Physicochemical, chemical, and biological studies on wheat embryo ribosomes. Biochemistry (Wash.) **6**:2853-2860.

Wollgiehn, R., and K. Mothes. 1964. On the incorporation of ^{3}H thymidine into DNA of chloroplasts of *Nicotiana rustica*. Exp. Cell Res. **35**:52-57.

Yoneda, M., and N. Kamiya. 1969. Osmotic volume change of cytoplasmic drops isolated in vitro from internodal cells of Characeae. Plant Cell Physiol. **10**:821-826.

Yuasa, A. 1969. The behavior and distribution

of plastids during the life-cycle of plants. Bot. Mag. Tokyo **82**:76-88.

GENERAL REFERENCES

Edds, M. V., Jr. (Ed.) 1961. Macromolecular complexes. The Ronald Press Co., New York.

Frey-Wyssling, A. 1964. Ultrastructural cell organelles. Proceedings 10th International Botanical Congress, p. 57-68.

Frey-Wyssling, A., and K. Mühlethaler. 1965. Ultrastructural plant cytology. American Elsevier Publishing Co., Inc., New York.

Ledbetter, M. C., and K. R. Porter. 1970. Introduction to the fine structure of plant cells. Springer-Verlag New York, Inc.

Lehninger, A. L. 1964. The mitochondrion; molecular basis of structure and function. W. A. Benjamin, Inc., New York.

Seifriz, W. 1936. Protoplasm. McGraw-Hill Book Co., New York.

Siegel, S. M. 1962. The plant cell wall. Pergamon Press, Inc., New York.

Sitte, P. 1965. Bau und Feinbau der Pflanzenzelle. Gustav Fischer Verlag, Stuttgart.

Stadelmann, E. J., and H. Kinzel. 1972. Vital staining of plant cells. Methods in Cell Physiol. **5**:325-372.

Strugger, S. 1952. Praktikum der Zell- und Gewebephysiologie der Pflanze. Springer-Verlag, Berlin.

PART TWO

Biophysics

3

Acidity

Actual acidity

Titratable acidity and buffers

pH measurements

ACTUAL ACIDITY

Measurements have revealed that, in nearly all plant cells, the vacuole sap is more acid than the protoplasm. But how are such measurements made, and how is the cell's acidity expressed quantitatively?

According to Bronsted, an acid is a proton donor—a substance capable of releasing a proton (H^+); and a base is a proton acceptor—a substance capable of combining with a proton. Water itself is both a proton donor, since it dissociates H^+,

$$H_2O \rightleftharpoons H^+ + OH^-$$

and a proton acceptor, since these H^+'s do not remain free but combine with a water molecule to form the hydronium ion (H_3O^+):

$$H_2O + H^+ \rightleftharpoons H_3O^+$$

For simplicity, however, the H^+ can be written in this way if we realize that it actually exists in the hydrated (hydronium ion) form. This notation is, in fact, necessary for the sake of uniformity, since other ions are also normally hydrated yet they are always referred to in the nonhydrated form (e.g., Na^+ and K^+).

There are, of course, many different proton donors (acids) and proton acceptors (bases). In an aqueous system such as exists in the plant if an acid is added, it will increase the concentration of protons in aqueous solution; if a base is added, it will combine with the protons, leading to an increase in OH^- concentration, because of dissociation of more and more water molecules to replace the H^+ removed. Therefore the following quantitative relations hold in aqueous systems:

$$\text{Acidity} \propto [OH^-]$$
$$\text{Basicity (or alkalinity)} \propto [H^+]$$

where the square brackets = concentration in gram atoms per liter.

In aqueous solutions:

$$[H^+] \times [OH^-] = K$$

Therefore the H^+ alone reveals both the acidity and alkalinity. Since K in the preceding equation is 10^{-14}, a solution that is *neutral* (i.e., acidity is equal to alkalinity) and therefore has equal concentrations of hydrogen and hydroxyl ions must have $[H^+] = 10^{-7}$ (1/10,000,000) gram atom per liter.

The range of acidity for the cell sap of plants is from about 10^{-1} to 10^{-7} gram atom of hydrogen ions per liter, the one extreme being a million times as acid as the other. To cover such a large range conveniently, it is necessary to use an exponential method of expressing the values. Consequently, the negative logarithm of the hydrogen ion concentration is used and is called the pH. Thus pH is $-\log [H^+]$ or $\log (1/[H^+])$; conversely $[H^+] = 1/(\text{antilog pH})$. Expressed in another way, if $[H^+] = 10^{-x}$, then pH = x. Although this explanation of the meaning of pH is historically correct, it is not a correct interpretation of the significance of measured pH values. What is really measured is the difference in H^+ *activity* (not concentration) between an unknown solution and a standard, buffered solution (see following discussion). Although this measured value may be used in the original sense as an approximate measure of H^+ concentration, such conversions must not be taken too literally. In the case of a small cell organelle, for instance, although there may be 30,000 proton donors and acceptors, there may actually be no free protons, and the H^+ concentration may therefore be zero (Butler, 1973). The true significance of pH is therefore in its use as an index of the chemical potential of the proton, which is dependent not only on the proton existing free (or hydrated) but also on the dissociable protons incorporated into the proton-donor molecules. Consequently, whenever pH is interpreted below in terms of $[H^+]$, this must be taken only as a rough, relative value.

In terms of pH, then, the range of values for cell sap is from about 1 to 7 (in rare

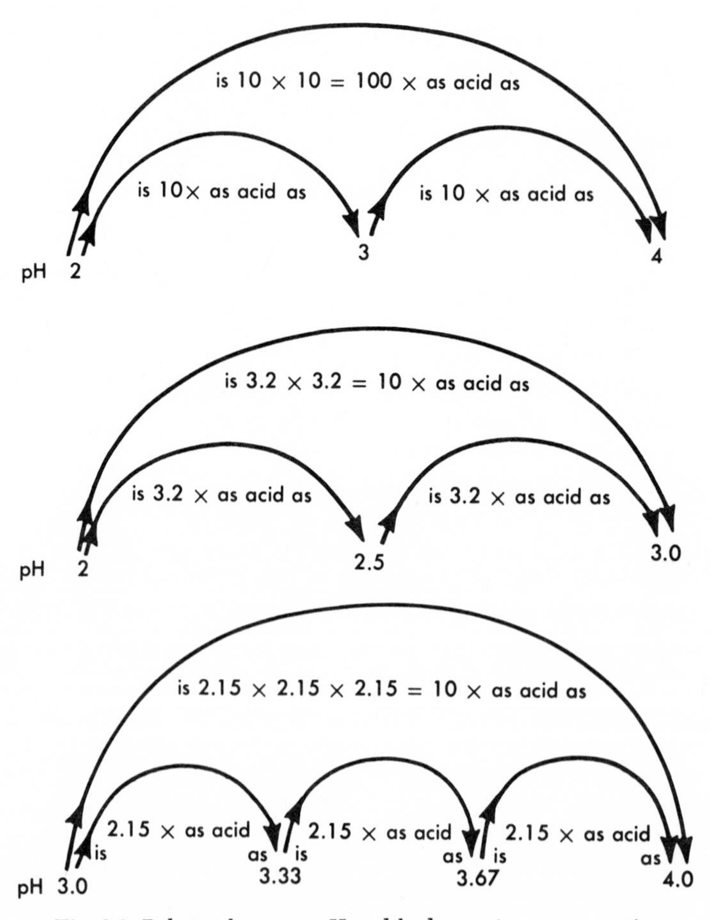

Fig. 3-1. Relation between pH and hydrogen ion concentration.

cases 8), the former being a million times as acid as the latter. Since each whole number represents an acidity 10 times as great as the succeeding number, the unit change in pH represents a smaller and smaller change in acidity as neutrality (pH 7) is approached. Thus a change in acidity on advancing from pH 1 to 2 and from pH 6 to 7 in each case represents a reduction in acidity to a tenth (0.1) of the original value; yet the actual decrease in acidity in the first case is $0.1 - 0.01 = 0.09$ gram atom of hydrogen ions per liter. In the second case the decrease is only $0.000001 - 0.0000001 = 0.0000009$ gram atom of hydrogen ions per liter. Consequently, the former unit increase in pH represents 100,000 times as great a drop in acidity as the latter. Just as a unit decrease in pH represents a 10 times increase in acidity, so a fraction of a unit is a definite multiple of a higher one (Fig. 3-1).

TITRATABLE ACIDITY AND BUFFERS

Acidity, as just defined, is a measure of actual H ion concentration. Besides this true acidity, there is the *titratable acidity*, or the normality of the acid as determined by adding just enough base of known normality to bring the pH to the neutral point. The titratable acidity cannot be predicted from the actual H^+ concentration (Table 3-1). Thus the normality (or titratable acidity) of lemon juice (mainly citric acid) is 10 times that of *Begonia* leaf juice (mainly oxalic acid), but the latter has a greater hydrogen ion concentration and therefore a lower pH. The titratable acidity includes not only the actual but also the potential hydrogen ion concentration, that is, the concentration of hydrogen atoms that may change to hydrogen ions if, for instance, some of the hydrogen ions already present are deionized by neutralization or titration with hydroxyl ions. The true acidity is expressed as gram atoms of hydrogen ions per liter, and the titratable acidity as normality of acid. In the case of a strong (completely or nearly completely dissociated) monoprotic acid such as hydrochloric acid, the two are almost identical; but in the case of a weak acid (with only a small fraction of the molecules dissociated), the true acidity is much less than the titratable acidity. Since pH is related to true acidity, a strong acid has a much lower pH than a weak acid of the same normality. Thus the pH of 1 N hydrochloric acid is 0.10, which is close to the ideal value of 0.00. The pH of 1 N acetic acid is 2.37 because only 0.42% of the acid molecules are dissociated. This means that the 1 N hydrochloric acid is nearly 200 times as acidic as the 1 N acetic acid, although the same amount of alkali would be required to neutralize equal volumes of the two solutions. Therefore the strength of an acid depends on how much of it is ionized, and this is expressed quantitatively by the ionization constant (Table 3-2).

Since the weak acid has nearly all its titratable hydrogen atoms in the form of

Table 3-1. Titratable acidity (normality) and hydrogen ion concentration (expressed as the negative logarithm or pH) of plant juices*

Organ from which sap was expressed	Normality of sap	pH of sap
Lemon fruit	0.95	2.4
Red-black fruits of blackberry	0.23	2.7
Rhubarb petioles	0.22	3.2
Unripe grapes	0.21	3.0
Oxalis leaves	0.16	2.3
Green cooking apple	0.13	3.2
Begonia rex leaves	0.11	2.2
Begonia tuberosa leaves	0.10	2.2
Ripe tomato fruit	0.063	4.4
Ripe Worcester Pearnian apple	0.045	3.9
Celery petioles	0.025	5.2
Root of white beet	0.025	5.8

*Adapted from Thomas, M. 1951. Endeavour **10**(39):160-165.

Table 3-2. Dissociation or ionization constants $(K = \frac{[H^+] \times [A^-]}{[HA]})$ of some acids found in plants and their respective pK values (the pH at the middle of their buffering zones)

		K	pK
Inorganic acids			
Phosphoric	1st H to ionize	7.5×10^{-3}	2.1
	2nd H	6.2×10^{-8}	7.2
	3rd H	1.0×10^{-12}	12.0
Carbonic	1st H	0.45×10^{-6}	6.3
Organic acids			
Oxalic	1st H	5.9×10^{-2}	1.2
	2nd H	6.4×10^{-5}	4.2
Malic	1st H	4.0×10^{-4}	3.4
Tartaric	1st H	9.7×10^{-4}	3.0
Citric	1st H	8.7×10^{-4}	3.1
Acetic		1.8×10^{-5}	4.7

potential hydrogen ions, when mixed with its salt it maintains a nearly constant pH, even though large amounts of hydrogen or hydroxyl ions are added to it. Such a mixture of weak acid plus its salt (or weak base plus its salt) is called a *buffered solution* because of its ability to prevent changes in pH.

The acid dissociates as follows:

$$HA \rightleftharpoons H^+ + A^-$$

Buffering is based on the relation

$$\frac{[H^+] \times [A^-]}{[HA]} = K$$

where K = dissociation constant of the acid (HA).

At equilibrium ($\rightleftharpoons$) there is a constant proportion of undissociated acid molecules and of ions, the proportion depending on the value of K for the particular acid. In the case of a strong acid, for every ten molecules the dissociation may result in:

$$10H^+ + 10A^-$$

In the case of a weak acid, it may be:

$$9HA + 1H^+ + 1A^-$$

Addition of a foreign base disturbs the equilibrium by adding OH^-, which combines with H^+s. The equilibrium ratio is immediately regained by dissociation of more HA, as a result of which the $[H^+]$ is nearly unaltered. The pH of a weak acid is therefore buffered against the addition of base because of its large reserve of HA.

But in order for a solution to be buffered against the addition of both H^+ and OH^-, it must have a reserve of both A^- and HA. This is achieved by mixing a weak acid with its salt. For every ten molecules of each, dissociation may lead to

$$\text{Acid} = 9HA + 1H^+ + 1A^-$$
$$\text{Salt} = 1NaA + 9Na^+ + 9A^-$$

since salts of even weak acids are highly dissociated. (Strong electrolytes such as inorganic salts are actually completely dissociated.) If the two are mixed, a new equilibrium will be produced, but the concentrations of HA and A^- will remain high. Consequently, a mixture of equal quantities (of equal normality) of a weak acid and its salt has a pH that tends to remain constant because it is buffered against the addition of both H^+ (by the high $[A^-]$) and OH^- (by the potential H^+ of the high [HA]). A good example is a mixture of equimolar quantities of acetic acid and sodium acetate. A drop of concentrated hydrochloric acid added to a liter of such a buffered solution has a negligible effect on the pH. In contrast, a drop of concentrated hydrochloric acid added to a liter of unbuffered water lowers its pH about 3.7 units (i.e., increases its acidity about 5000 times).

The buffer range of an acid can be observed from its titration curve because it is that portion of the curve which shows the most gradual rise in pH (Fig. 3-2). A strong acid (0.1 N HCl) also shows a region of gradual rise with only a unit increase in pH, although enough alkali is added to neutralize 90% of it. But this is not as much of a buffering action as it appears to be, since the $[H^+]$ is reduced by about 0.09 gram atom per liter. The buffer zone of the weak acid is much higher on the pH scale; therefore although a similar rise in pH occurs, it represents only a fraction of

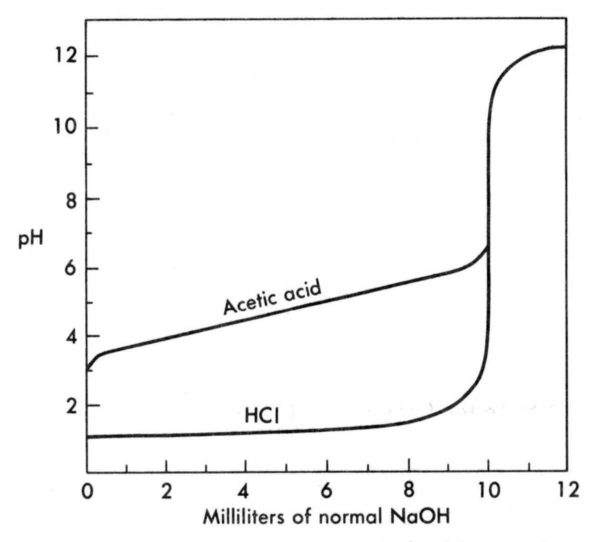

Fig. 3-2. Titration curves of 100 ml portions of 0.1 N hydrochloric acid and 0.1 N acetic acid with N sodium hydroxide. (From Michaelis, L. 1926. Hydrogen ion concentration. The Williams & Wilkins Co., Baltimore.)

Table 3-3. Some buffer solutions and their pH ranges*

	Acidic component	Alkaline component	pH range
1.	Hydrochloric acid	Glycine	1.0- 3.7
2.	Hydrochloric acid	Potassium hydrogen phthalate	2.2- 4.0
3.	Citric acid	Disodium hydrogen phosphate	2.2- 8.0
4.	Acetic acid	Sodium acetate	3.7- 5.6
5.	Potassium hydrogen phthalate	Sodium hydroxide	4.0- 6.2
6.	Potassium dihydrogen phosphate	Sodium hydroxide	5.8- 8.0
7.	Hydrochloric acid	Tris(hydroxymethyl)aminomethane	7.0- 9.0
8.	Diethylbarbituric acid	Sodium diethylbarbiturate	7.0- 9.2
9.	Hydrochloric acid or boric acid	Borax	8.0- 9.2
10.	Glycine	Sodium hydroxide	8.2-10.1
11.	Borax	Sodium hydroxide	9.2-11.0
12.	Sodium bicarbonate	Sodium hydroxide	9.6-11.0
13.	Disodium hydrogen phosphate	Sodium hydroxide	11.0-12.0

*From Altman, P. L., and D. S. Dittmer. 1964. Federation of American Societies for Experimental Biology, Washington, D. C.

1% as much of a decrease in $[H^+]$ for the same addition of OH^- as in the case of the hydrochloric acid. This is because of the large store of potential hydrogen ions that ionize as more alkali is added.

Each buffer mixture is effective only within a certain range of pH (Table 3-3). The maximum buffering of any mixture is obtained when half the acid (or base) molecules are dissociated, that is, when equal quantities of undissociated acid (or base) and hydrogen (or hydroxyl) ions are present. This occurs at an $[H^+]$ that is numerically equal to the dissociation constant K of the acid because if an acid HA dissociates into H^+ and A^-, then:

$$K = \frac{[H^+] \times [A^-]}{[HA]}$$

When the acid is half dissociated:

$$[A^-] = [HA] \text{ and } K = [H^+]$$

Since the dissociation constants of most acids are known, the optimum buffer zone for each can be easily calculated because the midpoint of this zone will occur at the pH equal to –log K (the so-called pK).

In plant juices several organic and inorganic salts and acids (Table 3-2) are responsible for the buffering action (Small, 1954). Around the neutral point, phosphates (best between 5.5 and 7.5) and bicarbonates are the commonest buffers; on the acid side, organic salts and acids such as malates (apples, peas, asparagus), citrates (tomato, citrus fruits), oxalates (tomato, plum, strawberry) tartrates (grapes), etc. buffer best between pH 3.6 and 6. Usually more than one buffer is present, for example, malate and acetate in onion or malate and oxalate in lettuce. Most of the preceding substances are in the vacuole. The protoplasm is buffered by its proteins, which are known to be effective over a pH range of 4 to 10 (Webb, 1963), although other substances may also have some effect, and metabolic control of pH is probably decisive. Some plant saps are not well buffered, and their pH may be greatly shifted by addition of carbon dioxide (Fig. 3-3). In some cases excess carbon dioxide may produce indirect effects leading to a rise in pH.

Buffer solutions are important in all preparative procedures, for example, in separating organelles from plant cells, in preparing proteins for enzymatic work, etc. It is a more crucial problem when dealing with plants than with animals, since the plant cell consists largely of vacuole sap, which is much more acidic than the protoplasm. If plant cells are broken open without the addition of a buffer solution, the extracted protoplasm will immediately take on the pH of the vacuolar sap because of the much larger volume of the latter. The increased acidity is usually sufficient to destroy the normal properties of both the organelles and the proteins. Many buffer solutions have been developed to prevent such injurious changes. They are usually prepared with pHs of 7 to 8. Phosphate, Tris, and borate buffer solutions are among the best (Good et al., 1966).

pH MEASUREMENTS

There are two main methods of measuring pH: one by the use of indicators (dyes that change color at more or less specific pHs), the other by pH meters. Each has its own virtues, although the modern, line-operated pH meters with glass electrodes

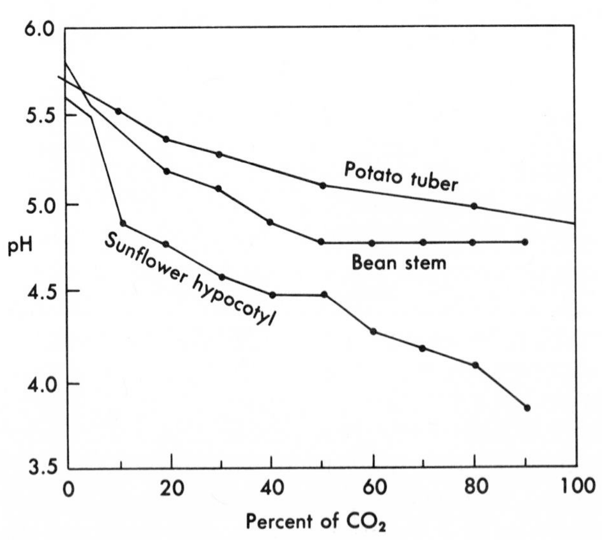

Fig. 3-3. Effect of carbon dioxide on pH of plant juices. (From Small, J. 1954. pH and plants. Baillière, Tindall, Cassell, Ltd., London.)

are unequaled for precision, speed, and simplicity. The indicator method is still sometimes used, for instance in field work. It is accurate enough for most biological purposes. In fact, it is usually just as *accurate* a measure of biological pHs as the far more *precise* pH meter, because although the pH meter enables measurement of pH to 0.01 units and indicators cannot do better than 0.2 units, the procedure of obtaining the plant juice for measurement is certain to change the pH at least 0.1 to 0.2 unit. For determining the pH of individual cells it is the only method available, since electrodes cannot be made small enough to insert into most living cells. In the case of large cells, however, microelectrodes have been used. The sap of a single cell of *Nitella flexilis* was in this way found to have a pH of 5.6 (Hirakawa and Yoshimura, 1964). Naturally occurring indicators inside the cell may sometimes be made use of: flavones turn from colorless to yellow above pH 8; anthocyanins are usually pink or red below pH 5 and blue or purple above it.

By the use of these methods the pHs of many juices (mainly from cell vacuoles) have been determined. Fruit juices usually range from pH 2.5 to 4.5 (Table 3-1), although there are some exceptions (e.g., lime, pH 1.7). Juices expressed from other plant parts also show wide variations, although they commonly fall within the range of pH 5 to 6.5. Phloem exudate has the exceptionally high value of 7 to 8. In plants with acid metabolism (e.g., succulents) the pH may be around 3. Even in the case of a single plant it may vary greatly. In many plants it drops at night and rises during the day. Extreme examples of such diurnal pH changes are found among succulents: in *Opuntia phaeacantha* (a cactus) from pH 5.5 at 4:00 P.M. to 1.4 at 6:45 A.M. Where the change in pH is small, it may at least partly result from increases in carbon dioxide content at night (from respiration), and decreases during the day (because of use in photosynthesis). But such extreme

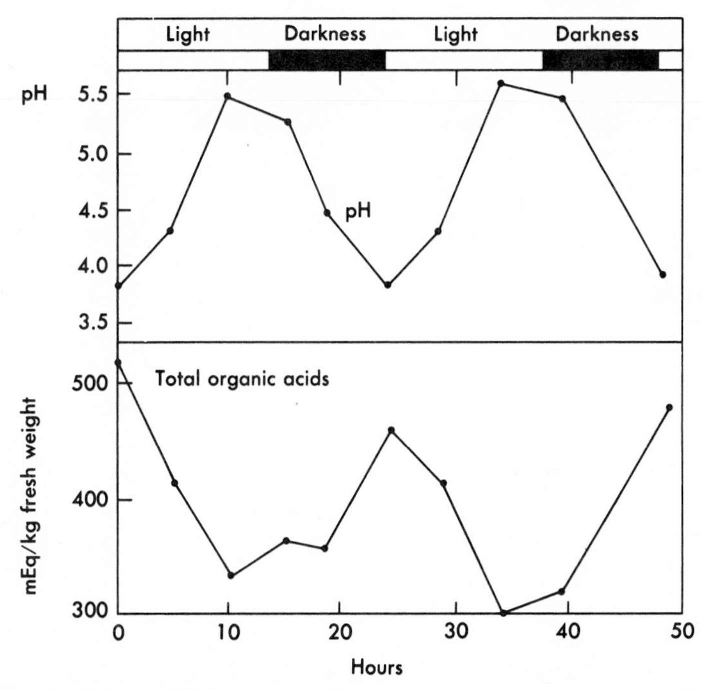

Fig. 3-4. Effect of light and darkness on pH and organic acid content of *Bryophyllum* leaves. (From Vickery, H. B. 1952. Plant Physiol. **27**:9-17.)

changes as in the *Opuntia* and other succulents result from accumulation of organic acids in the dark and their disappearance in the light (Fig. 3-4). In these cases the CO_2, rather than combining with water to yield the weak carbonic acid (H_2CO_3), instead combines with an organic carbon chain to form the stronger organic acids (e.g., malic). The acidity of most fruit can be explained in a similar manner. Because of their small specific surface (Chapter 4) and in many cases their impervious skins, the CO_2 released by respiration accumulates in the cells. It is then converted to organic acids that are excreted into the cell vacuole. Other bulky organs such as the potato tuber are able to respire without accumulating as much CO_2 because of the presence of permanently open pores known as *lenticels*, through which the CO_2 can diffuse out of the tissue.

The pH of plant protoplasm is much more difficult to determine, partly because of the far greater quantity of acid vacuolar sap adjacent to it. It is perhaps because of this and the acid of injury which often arises when protoplasm is damaged that values as low as pH 5.6 have been recorded for plant protoplasm. Because of these sources of error and because protoplasm of other organisms has usually been found to be around the neutral point, it is highly probable that higher plant protoplasm has a pH around neutrality. Indirect evidence of this is the fact that when protoplasmic proteins are isolated from plants, some of them fail to remain in solution unless the pH is maintained above 6 to 6.5. In the case of the large-celled alga *Acetabularia mediterranea*, cytoplasmic pH was found to be 8 to 8.4, using indicators (Dodd and Bidwell, 1971). This measurement was made possible by centrifuging the cells so as to obtain a large mass of protoplasm at one end.

Judging from the range of pHs found in the vacuole (1 to 7), the plant is apparently much more tolerant of acidity than of alkalinity. Even the tolerance of externally applied pH is not so great in the alkaline range. The most acidophilic plants grow at soil pHs down to 4 or even 3.5 in extreme cases, and the basophilic plants grow at soil pHs to 8 or 8.4. The critical upper limit for tomato plants occurs at pH 8.8 for plants established in nutrient solution and at pH 8.2 for young developing roots. New growth occurs readily below pH 8 (Thorup, 1969). Thus the tolerance of alkalinity is greater than it appears from vacuole measurements, but it is still less than the tolerance of acidity. This is at least partly associated with indirect effects of pH, for example, the unavailability of some soil nutrients at high pH, rather than a direct damage of protoplasm by OH ions.

Similarly, plants that do poorly in acid soils (such as hemp, mustard, asparagus, cucumber, spinach, squash, and alfalfa) are, at least in some cases, injured not by the H^+ directly but by the excess iron absorbed at low pH (Wallace and Bhan, 1962). It has long been known, however, that the roots of a plant can change an unfavorable pH to a somewhat more favorable one.

A highly acid-resistant *Chlorella ellipsoida* has been found that is capable of growth within the entire range of pH 2 to 10 (Kessler and Kramer, 1960). According to Olson (1953), a direct damage by hydroxyl ions does not occur until pH 10.5, exactly the same distance from the neutral point as the pH (3.5) causing direct damage because of hydrogen ions.

The role of pH in many physiological processes will be discussed in later sections (e.g., protein hydration, stomatal movement, and enzyme action).

QUESTIONS

1. What is meant by acidity? By alkalinity?
2. What is the acidity at the neutral point?
3. Is it necessary to measure both?

4. Does an increase in pH represent an increase or decrease in acidity?
5. How much of a change in acidity is represented by one pH unit?
6. How does titratable acidity differ from true acidity?
7. What are the units for each?
8. If two plants change their pHs from 6.5 to 6, does this necessarily mean an equal change in true or titratable acidity?
9. Does a change from 6.5 to 6 represent the same, a greater, or a smaller change in true acidity than a change from 5.5 to 5?
10. If the pHs of a strong and a weak acid are the same, which (if either) will have the greater titratable acidity?
11. What is a buffered solution a buffer against?
12. What does it consist of?
13. Do all buffer solutions cover the same range?
14. What substances in plants produce a buffering effect?
15. What is the acidity of the vacuole?
16. What is the acidity of protoplasm?
17. What pH range will living cells tolerate?
18. How do the night and day pHs of some plants compare?
19. How do the pHs of fruit compare with those of other plant parts?
20. How do the pHs of succulents compare with those of other plants?
21. What is the effect of carbon dioxide on pH?

SPECIFIC REFERENCES

Butler, T. C. 1973. pH: another view. Science **179**:854.

Dodd, W. A., and R. G. S. Bidwell. 1971. The effect of pH on the products of photosynthesis in $^{14}CO_2$ by chloroplast preparations from *Acetabularia mediterranea*. Plant Physiol. **47**: 779-783.

Good, N. E., G. D. Winget, W. Winter, T. N. Connolly, S. Izawa, and R. M. M. Singh. 1966. Hydrogen ion buffers for biological research. Biochemistry (Wash.) **5**:467-477.

Hirakawa, S., and H. Yoshimura. 1964. Measurements of the intracellular pH in a single cell of *Nitella flexilis* by means of micro-glass pH electrodes. Jap. J. Physiol. **14**:45-55.

Kessler, E., and H. Kramer. 1960. Physiological research on a highly acid resistant *Chlorella*. [Transl. title.] Arch. Mikrobiol. **37**:245-255.

Olson, C. 1953. The significance of concentration for the rate of ion absorption by higher plants in water culture. IV. The influence of hydrogen ion concentration. Physiol. Plant. **6**:848-858.

Thomas, M. 1951. Vegetable acids in higher plants. Endeavour **10**(39):160-165.

Thorup, J. T. 1969. pH effect on root growth and water uptake by plants. Agron. J. **61**:225-227.

Vickery, H. B. 1952. The behavior of isocitric acid in excised leaves of *Bryophyllum calycinum* during culture in alternating light and darkness. Plant Physiol. **27**:9-17.

Wallace, A., and K. C. Bhan. 1962. Plants that do poorly in acid soils, p. 36-38. *In* A. Wallace. A decade of synthetic chelating agents in inorganic plant nutrition. A. Wallace, Los Angeles.

Webb, J. L. 1963. Enzyme and metabolic inhibitors. Vol. I. General principles of enzyme inhibition. Academic Press, Inc., New York.

GENERAL REFERENCES

Altman, P. L., and D. S. Dittmer. 1964. Biology data book. Federation of American Societies for Experiment Biology, Washington, D. C.

Christensen, H. N. 1964. pH and dissociation. W. B. Saunders Co., Philadelphia.

Clark, W. M. 1922. The determination of hydrogen ions. The Williams & Wilkins Co., Baltimore.

Michaelis, L. 1926. Hydrogen ion concentration. The Williams & Wilkins Co., Baltimore.

Pauling, L. 1947. General chemistry. W. H. Freeman & Co., Publishers, San Francisco.

Small, J. 1954. Modern aspects of pH. Baillière, Tindall, & Cassell, Ltd., London.

Willard, H. H., L. L. Merritt, and J. A. Dean. 1951. Instrumental methods of analysis, p. 179-205. D. Van Nostrand Co., Inc., Princeton, N. J.

4

Specific surface and adsorption

The surface of cells, as well as that of their organelles, possesses special properties and plays special roles in the life of the organism. It has, in fact, been suggested (Ling, 1962) that life processes are under the control of specific "sites," which reside at such surfaces. These "cardinal sites" are supposed to be small in number but of such strategic importance that they can control and modulate the behavior of a large number of "fixed sites." They must owe their importance to the presence of specific molecules or ions (adenosine triphosphate [ATP], hormones, drugs, etc.), which may be held, at least temporarily, at these sites by the forces of adsorption. In order to comprehend the properties of living cells, it is, therefore, necessary to understand the nature of surfaces and their ability to adsorb substances.

SPECIFIC SURFACE AND SURFACE ENERGY

The surface of a sphere is given by the formula:

$$S = 4\pi r^2$$

This means that if spheres of different sizes are compared, the surface increases with the square of the radius. The larger the volume of a sphere, the larger its surface. On the other hand, the specific surface is the amount of surface area per unit mass or volume:

$$S_v = \frac{S}{V}$$

where S_v = specific surface (on a volume basis)
S = actual surface area
V = volume

In the case of a sphere:

$$S_v = \frac{4\pi r^2}{4/3\pi r^3} = \frac{3}{r}$$

If r is measured in centimeters, the unit for specific surface is cm^2/cm^3 or cm^{-1}. It follows that the specific surface of a sphere varies inversely with its radius. In contrast to actual surface, the *larger* the volume of a sphere, the *smaller* is its specific surface. This becomes obvious when a potato tuber is compared with its individual cells (Table 4-1). Although the actual tuber surface is a million times that of a single cell, its specific surface is one thousandth that of the cell. This means that the total surface of *all* of the tuber cells is 1000 times that of the tuber itself. Going beyond the cell, the

Table 4-1. Actual as opposed to specific surface of a potato tuber and its cells

	Diameter	Relative radius	Relative actual surface (S)	Relative actual volume (V)	Relative specific surface (S_v)
Tuber	10 cm	10^3	10^6	10^9	10^{-3}
Cell	0.1 mm (0.01 cm)	1	1	1	1

specific surface of its organelles will be still greater, since this value increases with the decrease in volume of the body.

The importance of surface and specific surface is a result of the existence of surface energy. This follows from the first law of thermodynamics, which states that energy cannot be created or destroyed but can be converted from one form to another. If a body is cut into smaller pieces, new surfaces are created, and part of the energy used to cut it up is converted into free surface energy. This energy is either in the form of electric charge (solids) or surface tension (liquids). Any other method of creating new surfaces will produce the same result. Consequently, if a potato sprouts and produces a leafy shoot, it is producing new surfaces and new surface energy at the expense of the chemical energy released by respiration. If, for instance, the sprout has 100 times the specific surface of the tuber (and other factors are equal), it is able to absorb oxygen and give off carbon dioxide at 100 times as great a rate per unit volume as can the tuber. In other words, the active, growing body of the plant (the shoot and root systems) is characterized by high specific surface and surface energy; the inactive, nongrowing plant body (the tuber, bulb, etc.) is characterized by low specific surface and surface energy. This relation has been found to hold at all levels from the macroscopic to the electron microscopic (Table 4-2).

Mitochondria, for instance, are induced to increase in number and therefore in surface per cell by wounding (activating) potato tubers (Verleur, 1969). Similarly, active mitochondria from developing barley contain increased surface because of their many cristae (membranes) and numerous ribosomes at the stage of maximum increase in grain dry weight and maximum respiration rate (Abdul-Baki and Baker, 1969). When grain growth was complete, both respiratory activity and mitochondrial ultrastructure (surface) decreased rapidly. The question of whether the high specific surface can be thought of as the cause or the effect of the activity can only be considered after the active and inactive states of the plant are discussed in more detail. There is, of course, a limit to the generality of this relation. Deciduous trees, for instance, reduce their outer surface tremendously when they become dormant. Evergreens do not, but this may partly account for their success in climates having low enough temperatures during their dormant period to reduce their metabolism to a minimum without any reduction in surface.

Even when plants are compared in the active state, the degree of activity may be closely related to specific surface. In the case of many crops, such as wheat and sugar beet, the rate of photosynthesis under field conditions, and therefore the growth rate of the crop, is closely related to the leaf area index (LAI), which is a measure of the leaf area per unit area of soil surface (Fig. 4-1). This relationship is easily understood, since the degree of light interception by the stand is proportional to the leaf area index (Puckridge, 1971).

As shown in Table 4-1, even more im-

Table 4-2. Direct relationship between specific surface and plant activity

Surface involved	Active state (large specific surface)	Inactive state (small specific surface)
1. External plant surface		
a. Deciduous trees	Leaves	No leaves
b. Herbaceous stems	Branched and foliate	Spherical (tubers, corms)
2. External cell surface	Meristematic cells (small)	Parenchyma cells (large)
3. Internal protoplasmic surface	Many organelles	Few organelles
4. Internal organelle surface		
a. Mitochondria	Cristae	Few or no cristae
b. Chloroplasts	Lamellae	No lamellae

portant and much larger than the external surface of a plant is its internal surface. It is not surprising, therefore, that the major metabolism of the plant is controlled by the organelles (Table 2-3). Much of this no doubt occurs at their surfaces (e.g., the surfaces of the cristae in mitochondria and of the lamellae in chloroplasts).

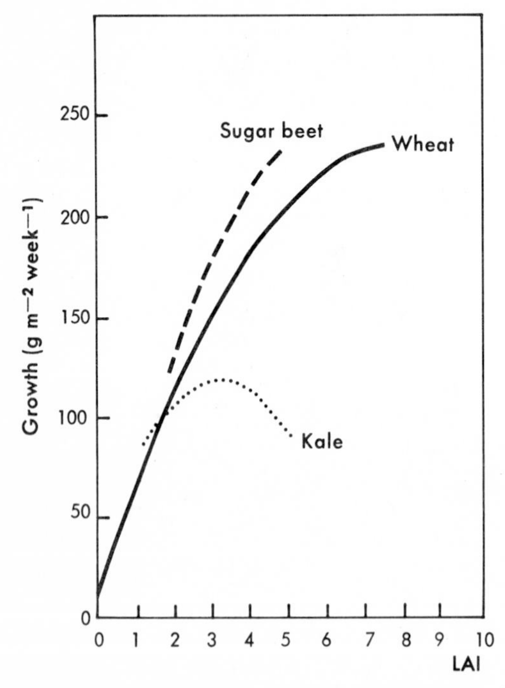

Fig. 4-1. Dependence of crop growth rate on leaf area index (LAI). (From Stoy. *In* J. D. Eastin et al. [Eds.]. Physiological aspects of crop yield. Copyright 1969 by the American Society of Agronomy and reprinted by permission of the copyright owner.)

POLARITY AND ADSORPTION

From the second law of thermodynamics it follows that the free surface energy tends to be converted into unavailable energy. In other words, the surface charge will tend to be neutralized by opposite charges, and the surface tension will tend to be lowered by other substances. In both cases the reduction in surface energy can be brought about only if molecules or particles that reduce surface energy accumulate at the surface in higher concentration than in the surrounding medium. This concentration at a surface is known as *adsorption*. It is more pronounced at some surfaces than at others and from some media than from others. In some cases there may even be negative adsorption if this results in a lower free energy of the surface (i.e., if the only available particles have the same sign of charge as the surface or if they raise surface tension).

Table 4-3. Electronegativity values for some biologically important elements; relative positions as in the periodic table*

H 2.1				
	C 2.5	**N** 3.0	**O** 3.5	**F** 4.0
		P 2.1	**S** 2.5	**Cl** 3.0

*From Pauling, L. 1970. General chemistry. W. H. Freeman & Co., Publishers, San Francisco.

Even though most substances are not ionized in the plant, there may be an asymmetrical distribution of electrons within the molecule, leading to a polarization of the molecule with an electrically positive and an electrically negative end. Oxygen is more electronegative (i.e., has a greater attraction for electrons) than hydrogen (Table 4-3). Therefore when the two combine in the water molecule, the electrons of the hydrogen atoms are displaced toward the oxygen atom:

$$:\ddot{\underset{\cdot}{O}}\cdot + .H = \overset{-}{:}\ddot{\underset{\cdot\cdot}{O}}:\!\!-\!H \quad \text{(with .H below the reactants; H bonded below O, marked +)}$$

The oxygen end of the molecule becomes negative to the hydrogen end and the molecule is thus a dipole. Carbon and hydrogen are nearly equally electronegative (Table 4-3); therefore the methane molecule (CH_4) is unpolarized. Water is thus a *polar* substance; methane is *apolar*. But there are all degrees of polarization of molecules from essentially zero, as in methane, to a complete splitting into positive and negative ions, as in salts (Fig. 4-2). The greater the number of apolar groups ($-CH_3$, $-C_6H_5$, etc.) the more apolar is the substance. The greater the number of polar groups (-OH, -COOH, -CO, $-NH_2$, metals, etc.) the more polar is the substance (Table 4-4).

The division of substances into (relatively) polar and (relatively) apolar substances is practically useful, since polar substances attract each other and are mutually soluble. The small degree of polarity in apolar substances results in their ready solubility in each other. Polar and apolar substances, on the other hand, are not soluble in each other because the mutual attraction of the molecules of a polar substance results in a repulsion of apolar substances. Even individual chemical groups on large molecules show the same relation. The long protein chains in the aqueous medium of the cell are therefore folded in specific patterns, and the polar groups are held in place by other polar groups and by the external polar water molecules that attract them electrically. The apolar groups are pushed together mostly inside the molecule because of the polar nature of the aqueous medium. It is this attraction of polar groups for each other, and the much smaller attraction of apolar groups for each other, that controls the phenomenon of adsorption.

The adsorption of particles at a surface reduces the surface energy in the following two ways:

1. By lowering the surface tension. The surface tension of a liquid depends on the polarity of its molecules. Water possesses a high surface tension; alcohols possess much lower surface tensions. But the surface tension of the pure liquid may be much different from that of the impure liquid, that is, solutions of substances in the liquid. Because of the already high surface tension of water, the addition of solutes can do little to raise it but can lower it considerably. Other substances that also consist of polar molecules tend to attract water molecules electrically and can be thought of as being pulled into the body of

Table 4-4. Cohesion forces between organic groups; the larger the forces the more polar the group*

Groups		Molar cohesion energy (kcal/mole)
Aliphatic C:		
Methyl and methylene groups	$-CH_3$ and $=CH_2$	1.78
	$=CH_2$ $=CH-$	0.99
Ether bridge	$-O-$	1.63
Amino group	$-NH_2$	3.53
Carbonyl group	$=CO$	4.27
Aldehyde group	$-CHO$	4.70
Hydroxyl group	$-OH$	7.25
Carboxyl group	$-COOH$	8.97

*From Frey-Wyssling, A., and K. Mühlethaler. 1965. Ultrastructural plant cytology. American Elsevier Publishing Co., Inc., New York.

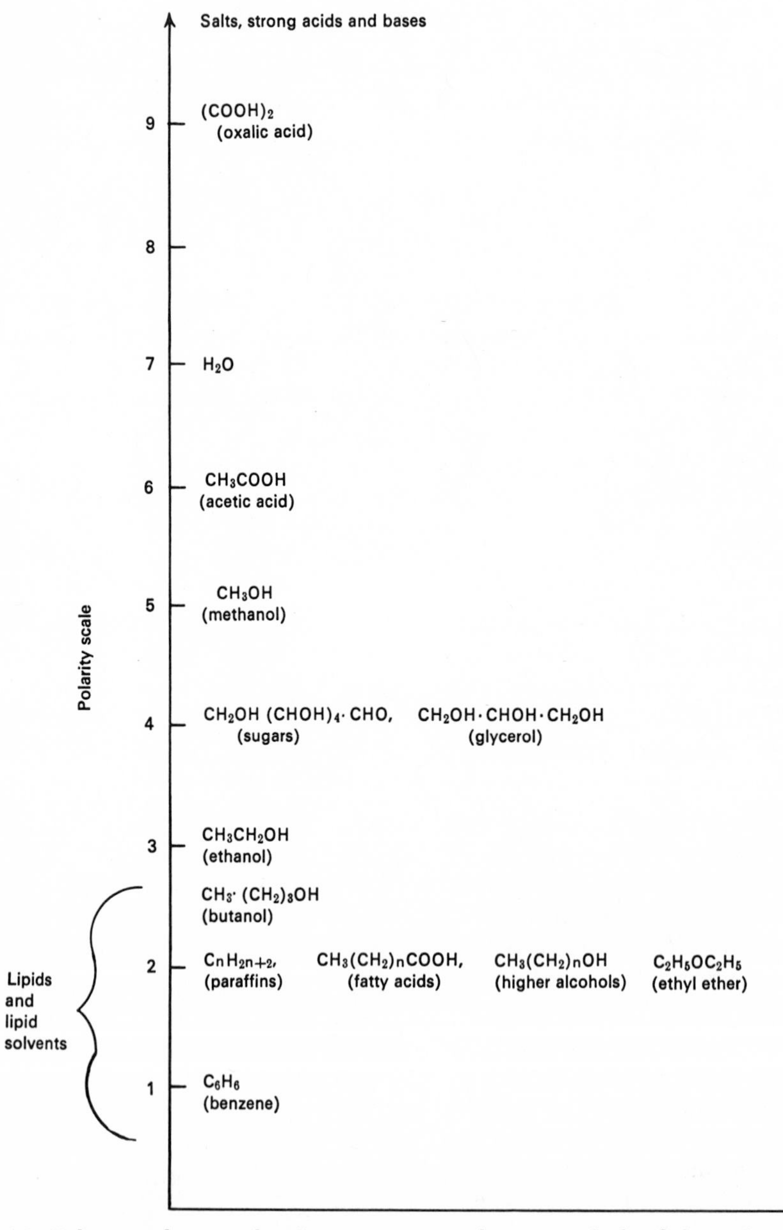

Fig. 4-2. Relative polarity scale of some organic substances calculated from the theoretical values in Table 4-4.

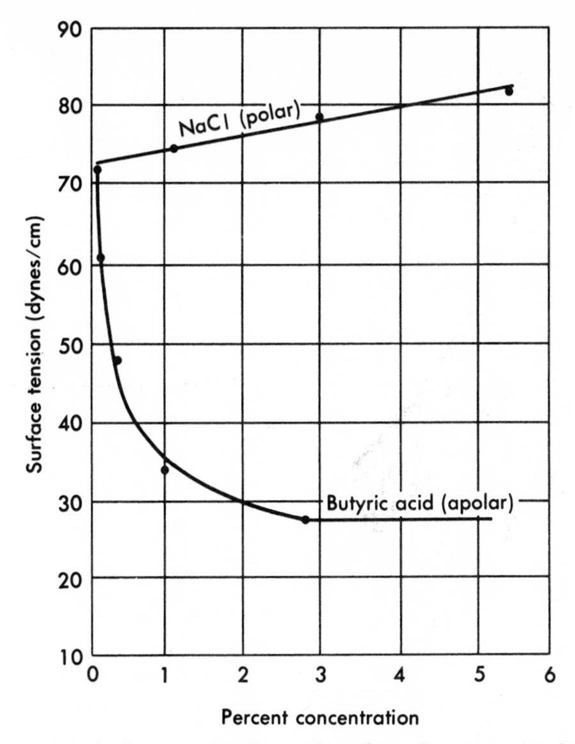

Fig. 4-3. Surface tension of solutions of polar and apolar substances. (Redrawn from Freundlich, H. 1922. Colloid and capillary chemistry. E. P. Dutton & Co., Inc., New York.)

an aqueous solution (negative adsorption). Salts and sugars are polar substances and therefore do not concentrate at the surface. Consequently, they do not reduce the surface tension of water and are said to be surface inactive, although they may actually slightly increase surface tension (Fig. 4-3). Most organic substances are more or less apolar and therefore are not so strongly attracted to the water molecules as the water molecules are to each other. As a result they can be thought of as being pushed to the surface, where they reduce the surface tension because of the small molecular forces of attraction between them. They can even form a monomolecular layer on the water, with their polar groups extending down into the polar water layer and their apolar groups away from the water. Such substances are said to be surface active. Proteins may be surface active; they consequently form an adsorbed layer around latex particles in rubber-producing plants. But proteins do possess definite electrical properties and therefore are not nearly as surface active with respect to water as are fatty substances. Consequently, in the case of living protoplasm there is a far greater concentration of lipids than of proteins at the surface. As a result, the surface tension of protoplasm is lowered from 72 dynes/cm (for pure water) to a range of 0.1 to 2 dynes/cm. According to the calculations of Danielli (1966), a bimolecular lipid membrane associated with proteins has less surface-free energy than that of any other thickness. He therefore concludes (on the basis of the second law of thermodynamics) that biological membranes normally must be bimolecular.

2. Neutralization of surface charge. The phenomenon of staining is caused by this kind of adsorption. Dyes are weak organic (aromatic) acids or bases, consisting mainly

of apolar groups. The dissociation of these weak acids and bases is extremely low, and because of their high degree of apolarity they are essentially insoluble in water. But the salts of even weak acids and bases may have a high degree of dissociation (e.g., sodium salts of the acids and chlorides of the bases). Consequently, the salts of these apolar dye acids and bases possess a charge and are therefore polar. As one result, the salts are highly soluble in water (a polar solvent); as a second result, these dye ions are strongly adsorbed to oppositely charged surfaces. The salt of a dye base is called a basic dye, the salt of a dye acid, an acidic dye:

1. $R_1OH + HCl \rightleftharpoons R_1Cl + H_2O$
   Dye base — Dye salt (basic dye)

   $$R_1Cl \rightleftharpoons R_1^+ + Cl^-$$

2. $R_2H + NaOH \rightleftharpoons R_2Na + H_2O$
   Dye acid — Dye salt (acidic dye)

   $$R_2Na \rightleftharpoons R_2^- + Na^+$$

Vital staining is caused by both the polar and apolar properties of dyes. Neutral red is a basic dye (RCl) and therefore dissociates as the red (R^+) dye ion at low (acid) pHs. At high pHs (about the neutral point) it is converted to the yellow, undissociated base (ROH), since the dye ion (R^+) is also in equilibrium with the OH^- of the water and must combine with it when the OH^- concentration becomes relatively high. Because of this color change it can be used as a pH indicator. The red dye ion is polar and highly soluble in water and readily stains (i.e., adsorbs to) oppositely charged negative surfaces. The cell wall is negatively charged, partly because of dissociation of pectic acids to negative pectate ions. Consequently, the cell wall stains in red (acid) solutions of neutral red. In the presence of Ca^{++} this staining does not occur because the doubly charged calcium ions are preferentially adsorbed to the pectate or other negative surfaces.

The vacuole of most plant cells also contains negatively charged particles, primarily tannins. Yet the vacuole does not stain when the living cell is immersed in a red solution of neutral red. This is because the surface of the living cell consists of apolar lipids, in which the polar NR^+ is insoluble. However, if the pH of the neutral red solution is raised to 7 or 8, it becomes orange in color because of the presence of both the red (NR^+) and yellow (NROH) forms. The latter has no charge and is apolar. Therefore it is soluble in the plasma membranes and diffuses through both. On reaching the vacuole, with its acid pH, the apolar molecule (NROH) is converted to the ion (NR^+) and is quickly adsorbed by the tannin particles (T^-), thus staining the vacuole vitally. It is evident, then, that vital staining of the cell sap is possible only (1) with basic dyes; (2) if the dye is applied at a high enough pH to produce a sufficient concentration of the apolar, undissociated molecule; and finally (3) strong staining requires the presence of a high concentration of negatively charged (tannin) particles in the vacuole.

Adsorption has proved to be important in the case of uptake of nutrients from the surrounding medium by plant roots. There is strong evidence that the first step in the uptake is adsorption at a specific surface in the roots. It has been found that in weak solutions the uptake is more or less proportional to the external concentration, but that in more concentrated solutions it is almost independent of the external concentration. This is believed to result from saturation of the surface by the adsorbed nutrient at relatively low concentrations (Robertson, 1941). Due largely to the charge on proteins, some of the ions in the protoplasm are in the adsorbed rather than the free ionic form. Ling and Cope (1969), in fact, have proposed that the bulk of the K^+ in protoplasm is adsorbed. This, however, is an extreme view and does not agree

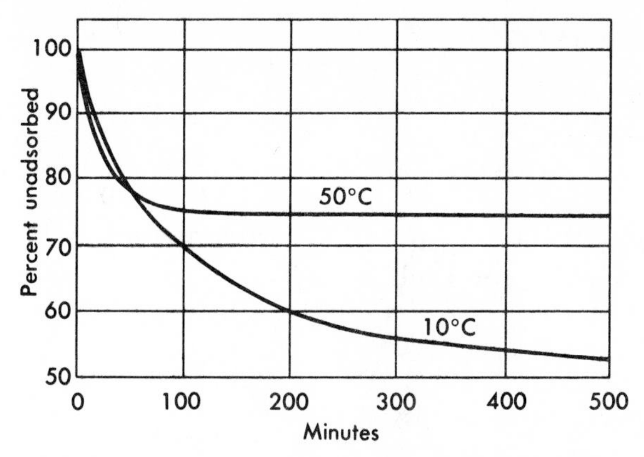

Fig. 4-4. Effect of temperature on the rate of adsorption of Congo red by paper. (Redrawn from Belehradek, J. 1935. Temperature and living matter. Protoplasma-Monographien 8. Gebrüder Borntraeger, Berlin.)

with measurements of electrical potential made on plant cells (Chapter 8). These measurements show that K^+ activity in protoplasm is sufficiently close to the measured concentration to prove that practically all is unbound (Spanswick, 1968).

Temperature has a pronounced effect on adsorption. The amount adsorbed greatly decreases with a rise in temperature. This can readily be understood, since the higher the temperature, the greater the kinetic energy of the adsorbed molecules or ions and the more easily they can escape from the surface. However, temperature has the opposite effect on the time required for adsorption equilibrium to be attained; this decreases with a rise in temperature (Fig. 4-4).

QUESTIONS

1. What is meant by specific surface?
2. How does it vary with particle size?
3. What is meant by surface energy?
4. In what two forms does it exist in plants?
5. What relation is there between the extent of plant surface and the activity of the plant?
6. With respect to what is a polar substance polarized?
7. On what property of the elements does the polarity of a substance depend?
8. Is water polar?
9. Are fatty substances polar?
10. Are salts polar?
11. Is there a sharp line of separation between the two kinds?
12. Do polar substances attract or repel each other?
13. Do apolar substances attract or repel each other?
14. What do we know about solubilities of polar and apolar substances in each other?
15. How can the free energy at a surface be decreased?
16. How can the polarity of chemical groups affect a single isolated large molecule?
17. What is the relation between the polarity of a liquid and its surface tension?
18. How can added substances affect the surface tension of a liquid?
19. What are "surface active" substances?
20. What is the surface tension of water? Of protoplasm?
21. What substances can be expected to concentrate at the protoplasm surface?
22. What is a basic dye? An acidic dye?
23. Which dye is suitable for vital staining? Why?
24. Which form of a dye is readily adsorbed?

25. Which form readily penetrates a living cell? A dead cell?
26. Which form concentrates in the vacuole of living cells? Why?
27. Which form concentrates in cell walls? Why?
28. Which cell substances are responsible for each of the preceding two kinds of staining?
29. At what pH will acidic dyes penetrate living cells readily?
30. What relation is there between adsorption and temperature?

SPECIFIC REFERENCES

Abdul-Baki, A., and J. E. Baker. 1969. Some properties of mitochondria from developing mature and dry barley (*Hordeum vulgare.* L.) seeds. Abs. XI. Internat. Bot. Congr., Seattle: 1.

Belehradek, J. 1935. Temperature and living matter. Protoplasma-Monographien 8. Gebrüder Borntraeger, Berlin.

Danielli, J. F. 1966. On the thickness of lipid membranes. J. Theor. Biol. **12**:439-441.

Ling, G. N. 1962. A physical theory of the living state. Blaisdell Publishing Co., Division of Ginn & Co., Waltham, Mass.

Ling, G. N., and F. W. Cope. 1969. Potassium ion: is the bulk of intracellular K^+ adsorbed? Science **163**:1335-1336.

Puckridge, D. W. 1971. Photosynthesis of wheat under field conditions. Aust. J. Agr. Res. **22**:1-9.

Robertson, R. N. 1941. Studies in the metabolism of plant cells. I. Accumulation of chlorides by plant cells and its relation to respiration. Aust. J. Exp. Biol. Med. Sci. **19**:265-278.

Spanswick, R. M. 1968. Measurements of potassium ion activity in the cytoplasm of the Characeae as a test of the sorption theory. Nature **218**:357.

Stoy, V. 1969. Interrelationships among photosynthesis, respiration, and movement of carbon in developing crops, p. 185-206. *In* Estin, J. D., F. A. Haskins, C. Y. Sullivan, and C. H. M. van Bavel (Eds.). Physiological aspects of crop yield. Amer. Soc. Agron. and Crop Sci. Soc. Amer., Madison, Wis.

Verleur, J. D. 1969. Observations on the induction of mitochondrial particles in potato tuber tissue after wounding. Z. Pflanzenphysiol. **61**:299-309.

GENERAL REFERENCES

Conn, H. J. 1961. Biological stains. The Williams & Wilkins Co., Baltimore.

Danielli, J. F., K. G. A. Parkhurst, and A. C. Riddiford (Eds.). 1964. Recent progress in surface science. Vol. 2. Academic Press, Inc., New York.

Freundlich, H. 1922. Colloid and capillary chemistry. [Transl. by George Barger.] E. P. Dutton & Co., Inc., New York.

Gortner, R. A. 1949. Outlines of biochemistry. John Wiley & Sons, Inc., New York.

Scarth, G. W., and F. E. Lloyd. 1930. Elementary course in general physiology. John Wiley & Sons, Inc., New York.

5
Colloids

SIGNIFICANCE IN PHYSIOLOGY

In science, as in any human field of activity, there are always fads. A topic will become popular, then later it will be forgotten. The role of colloids in the plant is an excellent example. For a time all of the properties of protoplasm were explained as being basically a result of its colloidal character. With the development of the electron microscope the organelles of protoplasm were discovered, proving that protoplasm is much more than a complex colloidal system. The pendulum has swung to the other extreme, and the trend now is to ignore the colloidal properties of protoplasm in particular and of the cell in general. The result is an unfortunate loss of undertsanding of many cell properties that are still definitely colloidal and therefore explainable in no other way.

For, although the organelles explain many of the most important properties of protoplasm, they fail to explain the properties of the ground protoplasm or hyaloplasm. Thus despite the high water content of protoplasm, it is much more viscous than most ordinary aqueous solutions of the same water content. Furthermore, unlike pure solutions, it possesses some of the properties of solids, for example, elasticity. These properties cannot be understood without a consideration of the colloidal state of protoplasm. Even some of the organelles themselves are in the colloidal size range (see later discussion), and both the cell wall and the vacuole have colloidal properties. It is therefore impossible to have a complete understanding of the properties of the cell without an understanding of the properties of colloids.

The most obvious factor in adsorption is the extent of surface—the greater the surface area per unit volume (i.e., the specific surface), the greater is the amount adsorbed per unit volume. Adsorption therefore becomes more pronounced the more finely divided is the adsorbent. It reaches a maximum, in fact, in the *colloidal* range, where particle sizes are below visibility with the optical microscope but above ordinary molecular size (Table 5-1). It is for this reason that adsorption must play such an important role in the living cell and particularly in protoplasm, which consists largely of substances in the colloidal state.

Since the colloidal state depends simply on the particle size, any substance can (at least theoretically) be a colloid. There is, of course, no sharp line between colloids and noncolloids. Any particles that fall in

Table 5-1. Size limits (logarithmic scale) and properties of colloids relative to other dispersions*

Properties	Approximate ranges of diameters (in nm†) 2000 — 200	200 — 20 — 2	2 — 0.2
Dispersion	Suspension (2 μm — 200 nm)	Colloidal dispersion (200 — 5 nm)	Molecular dispersion (5 — 0.1 nm)
Visibility	Optical microscope	Ultramicroscope and electron microscope	Near or beyond limits of detection with electron microscope
Separability	Filtrable	Dialyzable	Nondialyzable except with semipermeable membranes
Stability	Unstable	Relatively stable	Stable
Example	Starch suspension	Proteins with mol wt 125,000 (diameter 6.6 nm‡) and larger	Proteins with mol wt 15,000 (diameter 3.3 nm‡) and smaller

*Adapted from Scarth, G. W., and F. E. Lloyd. 1930. Elementary course in general physiology. John Wiley & Sons, Inc., New York.
†1 mm = 1000 μ (microns) or 1000 μm (micrometer); 1 μ = 1000 mμ or 1 μm = 1000 nm (nanometer).
‡See Frey-Wyssling, A., and K. Mühlethaler. 1965. Ultrastructural plant cytology. American Elsevier Publishing Co., Inc., New York.

the range of about 5 nm to 0.2 μm are usually considered to be colloidal. Sometimes the upper limit may be extended to 0.5 μm, since certain systems with particles of this size are stable. The lower limit includes some molecules, since these may be large enough to take on colloidal properties. This is particularly true of proteins.

Colloidal systems really consist of two components, the *disperse phase* (particles) and the *dispersion medium*. Since both of these may be in any of the three states of matter (solid, liquid, and gas), there are nine possible combinations. Of these only one cannot represent a colloid, viz., gas disperse phase in gaseous medium, since all gases are molecularly miscible in all proportions. Of the eight remaining possible combinations, only two, the solid disperse phase in a liquid medium and the liquid disperse phase in a liquid medium, are important in physiology; and the only liquid medium of physiological significance is water. Consequently, the following discussion applies primarily to aqueous colloidal dispersions.

PROPERTIES OF COLLOIDS

There are many ways in which colloidal dispersions differ from molecular dispersions or *true solutions*. All these differences depend on the particle size. Thus the colloidal particles are large enough to scatter light. As a result, colloidal dispersions fre-

Table 5-2. Relative diffusion rates (or diffusion coefficients) of molecularly and colloidally dispersed substances*

Dispersion	Substance	Diffusion coefficient
Molecular	Nitric acid	2.1
	Sucrose	0.31
Colloidal	Nuclear gold (1.7 nm)	0.27
	Egg albumin	0.059
	Antitetanus serum	0.0021

*Adapted from Gortner, R. A., 1938. Outlines of biochemistry. John Wiley & Sons, Inc., New York.

Table 5-3. Osmotic pressures (atmospheres) of a molecularly dispersed (dextrose) and a colloidally dispersed (serum albumin) substance

Solution (%)	Dextrose	Bovine serum albumin
5	6.7	0.013
10	13.5	0.032
20	27.0	0.12

quently show a *Tyndall effect;* if a beam of light is shone on the dispersion, the path of the beam can be seen from either side at right angles to the direction of the beam. But such light scattering is not possible when the colloidal particles have a refractive index that does not differ greatly from that of the dispersion medium, for instance, when the particles are highly hydrated as in the case of many proteins.

Another result of their relatively large particle size is the slow rate of diffusion of colloids: as little as 0.001 the rate of small molecules (Table 5-2).

The osmotic effects (Chapter 6) of colloidal solutions are small, again because of the large particle size (Table 5-3).

The colloidal particles are too large to pass through the pores of many membranes that are easily penetrated by molecularly dispersed salts and sugars. This fact is made use of to free colloidal systems of crystalloids. When confined in such a membrane that is suspended in water, the crystalloids diffuse out, leaving the colloids behind. This process is known as *dialysis*. A similar process including the application of pressure is known as *ultrafiltration.*

Colloids may differ from each other in other ways besides size. Some may be nonspherical, and this may lead to other properties such as gel formation. Despite their large particle size, some may be electrolytes and, because of their dissociation, become charged. Proteins supply examples of both of these properties.

STABILITY OF COLLOIDAL DISPERSIONS

Molecular dispersions are stable because of the high kinetic energy of the molecules. When this energy is reduced, for instance when the molecules crystallize, they precipitate. In the same way, colloidal particles aggregate to form larger particles, which will then precipitate. If the colloid does not precipitate, it is stable; and since the particles do not possess the high kinetic energy of molecules, the colloid must owe its stability to some other property that prevents aggregation. There are two main factors that contribute to its stability: *charge* and *hydration.*

The charge of a colloidal particle results either from the capture of an ion or from ionization (dissociation) of the colloid. Colloidal bases (e.g., alkaloids, basic dyes, and hydroxides of metals) become positively charged in water. All other colloids are negative in water. Proteins are *amphoteric* (either positively or negatively charged) depending on whether they dissociate as bases or as acids. At some point between, the *isoelectric point,* their net charge is zero, and they are therefore least stable. Some proteins (the albumins) are *isostable;* that is, their own ionization supplies a sufficient net charge. Others (globulins) need other ions and therefore are stable only in salt solutions. An unstable colloid may therefore be dispersed by the addition of sufficient ions (i.e., by salts). This is called *peptization.* But if too high a salt concentration is used, the colloid may be discharged and precipitated. This is called *flocculation.* The higher the valence of the oppositely charged ion, the smaller is the concentration of salt required to flocculate a colloid (Table 5-4). Furthermore, the concentration of a divalent ion that is just sufficient to flocculate the colloid will be much less than half the concentration of the monovalent ion because the fraction adsorbed is inversely related to the concentration supplied.

Table 5-4. Concentration of salts required to flocculate (precipitate) sols*

Negative sol		Positive sol	
Salt	Millimoles	Salt	Millimoles
KCl	50	KCl	80
$BaCl_2$	0.70	K_2SO_4	0.28
$AlCl_3$	0.09	$K_4Fe(CN)_6$	0.08

*Adapted from Scarth, G. W., and F. E. Lloyd. 1930. Elementary course in general physiology. John Wiley & Sons, Inc., New York.

The second cause of stability is hydration. Aqueous colloids may be roughly classified into *hydrophilic* and *hydrophobic* colloids, although there is some gradation between the two groups. The hydrophobic colloids are less stable and can be flocculated by low concentrations of electrolytes (Table 5-4). The hydrophilic colloids cannot be flocculated but may be *salted out* by high salt concentrations. The effectiveness of salts varies with the different ions in a definite order, which is known as the lyotropic (or Hofmeister) series (Fig. 5-1). The order of the ions in this series is directly related to their hydration. They may possess both tightly held (A) and loosely held (B) water. When both of these shells of water are included, the hydrated ions are found to have radii that decrease with their order in the lyotropic series, even though the radius of the unhydrated ion increases (Table 5-5).

Proteins are typically hydrophilic colloids. They cannot be flocculated by low concentrations of salts but can be precipitated by high concentrations, for example, by saturation or half saturation with ammonium sulfate. Some hydrophilic colloids form aqueous sols (i.e., have the physical

Cations: $Li^+ > Na^+ > K^+ > NH_4^+ > Rb^+ > Cs^+ > Mg^{++}$
Anions: $SO_4^= > HPO_4^= > Cl^- > Br^- > NO_3^- > I^- > CNS^-$

Fig. 5-1. Order of effectiveness of ions in salting out hydrophilic sols: the lyotropic series. (Redrawn from Scarth, G. W., and F. E. Lloyd. 1930. Elementary course in general physiology. John Wiley & Sons, Inc., New York.)

properties of a liquid); gels are formed when a more or less viscous solution is transformed into a highly deformable solid without separation of fluid. Protoplasm itself is intermediate and exhibits the properties of both sols and gels due to its high concentration of protein; it shows streaming and brownian movement, as would be expected in a sol, but both of these movements may be stopped by removal of some of its water, converting it into a rigid gel.

It must be emphasized that although proteins are hydrophilic, they also possess hydrophobic groups that are normally repelled by the water and that therefore occur mainly within the three-dimensional protein molecule. The hydrophilic groups on the outside of the molecule are responsible for the solubility of the proteins. The proteins associated with membranes, however, are so hydrophobic as to be insoluble in water. The anions that are least effective in salting out hydrophilic sols (SCN^-, ClO_4^-, NO_3^-) may actually have the reverse effect on such hydrophobic proteins, increasing their solubility. They have been called *chaotropic* ions (Hatefi and Hanstein, 1969) because they are thought to break down (or to convert to a more chaotic form) the water structure. Because of this greater disorder, the water molecules no longer repel the apolar or hydrophobic components of the protein molecule and may actually dissolve these normally insoluble molecules. Only anions can be chaotropic because, unlike cations, they decrease the polarity of the surrounding water. In the hydration shell of the cation, the H atoms are directed outward and are therefore available for further bonding; in the hydration shell of the anion, they are directed inward and are therefore unavailable for H-bonding of water molecules.

The preceding classification of colloids into hydrophilic and hydrophobic systems, although useful, is an oversimplification. Booij and de Jong (1956) divide them into (1) nonequilibrium state and (2) equilib-

Table 5-5. Relative order of hydration of ions of lyotropic series as illustrated by ionic radii (A = tightly held water; B = loosely held water)*

Element	Kind of hydration	Unhydrated ionic radius (nm)	Hydrated ionic radius (nm)	Difference (hydration shell)
Li	A + B	6.0	38.2	32.2
Na	A	9.5	35.8	26.3
K	B	13.3	33.1	19.8
Rb	B	14.8	32.9	18.1
Cs	B	16.9	32.9	16.0
Mg	A + B	6.5	42.8	36.3
Ca	A + B	9.9	41.2	31.3
Ba	A + B	13.5	40.4	26.9
F	A	13.6	35.2	21.6
Cl	B	18.1	33.2	15.1
I	B	19.5	33.1	13.6

*Adapted from Erlander, S. R. 1969. The structure of water. Science J. **5A**(5):60-65.

rium state colloidal sols. Hydrophobic sols, as a rule, are in the nonequilibrium state; and their stability is more apparent than real. Although it may take several years, every sol of this type flocculates spontaneously because a small fraction of the particles will have sufficient kinetic energy to overcome their mutual electric repulsion. Protein molecules, on the other hand, disperse spontaneously and are, therefore, in true solution. Their sol state is a true equilibrium state. Because of their polar nature, proteins attract water and have, therefore, been called hydrophilic colloids, whether sols or gels. Booij and de Jong (1956) have labeled them macromolecular colloids. The hydrophobic sols are of little importance to the biologist. Equilibrium-state colloids are of fundamental importance, since many colloids of this kind occur in the cell.

However, the macromolecular colloids are not the only ones in the equilibrium state. A second group (according to Booij and de Jong) are the association colloids. They may consist of double films with a hydrophobic interior, such as those forming the cell membrane (Chapter 7). These association colloids may be in the sol, or solution form (analogous to gases); in the coacervate, or isotropic, form (analogous to liquids); or in the colloid crystal (solid) state. According to Booij and de Jong, the coacervate state resembles protoplasm the most. Their state is closely related to flocculation, since they arise as a result of a decrease in solubility. The amount of water associated with a protein, for instance, will decrease if it is part of a coacervate, because of the protection of its polar groups by the micromolecule with which it is associated and which may actually form a salt bond with it.

PROPERTIES OF GELS

Since protoplasm has the properties of gels, it is important to know what these are. In spite of the more or less rigid nature of aqueous gels, their water content is usually high, and this ensures just about as rapid diffusion of water-soluble substances through them as through water itself. The following are their main properties.

Swelling or imbibition pressure

Most aqueous gels are highly hydrophilic. Because of the large adsorptive forces they may have imbibition pressures as high as 1000 atm; that is, they may absorb water against water-removing forces of this value, or they may hold some water against pressures of 1000 atm in a hydraulic press. This is true of seeds because of their gel nature. The first phase of water uptake in the cotyledons and germ axes of pea seeds consists of pure imbibition (Kühne and Kausch, 1965). The maximal degree of swelling of the seed is almost independent of temperature. When this point is reached, the protoplasmic gel is prevented by the cell walls from attaining its own maximum possible degree of swelling.

Proteins may retain as much as 10 g water per 100 g protein at 100° C. At –35°

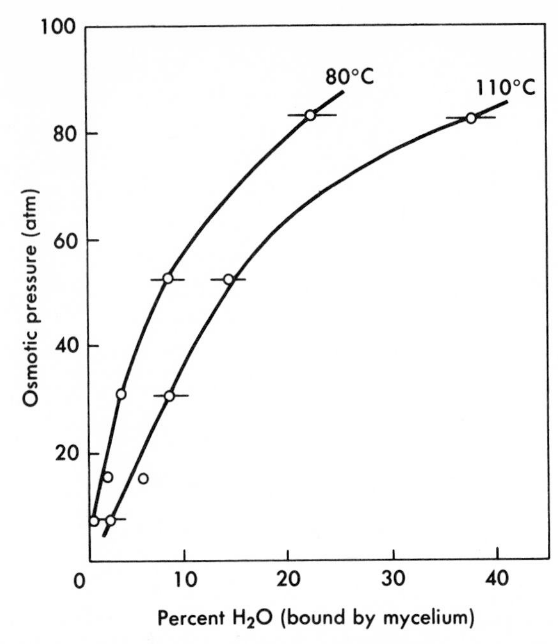

Fig. 5-2. Percent of bound water held by the mold *Aspergillus niger* after growth in media containing 0% to 50% sugar and therefore having osmotic pressures (Chapter 6) of 0 to nearly 100 atm. The bound water is that held at room temperature in an evacuated desiccator with a relative humidity of almost zero. The values obtained depend on the temperature used to drive off the bound water—less if 80° C than if 110° C. (From Todd, G. W., and J. Levitt. 1951. Plant Physiol. **26**:331-336.)

C. they bind 30 to 50 g water per 100 g protein (Kuntz et al., 1969), and nucleic acids hold 3 to 5 times as much water. This difference is at least partly because about half the surface of proteins consists of hydrophobic groups that do not bind water (Bull and Breese, 1968). Water that is adsorbed strongly by colloids is sometimes called *bound water.* The amount of such bound water held by tissues may vary with the conditions of growth (Fig. 5-2). The water forming a monomolecular layer on the adsorbent is most strongly bound. As more is added (e.g., by an increase in environmental relative humidity), the adsorbed water forms multimolecular layers that are held less and less strongly (Fig. 5-3). Finally, no more water can be bound, and any further uptake is simply due to condensation of the water in the capillary spaces between the adsorbent particles (Acker, 1969). In the case of plant material, the charge distribution is not uniform; therefore instead of forming a monomolecular layer, the most strongly bound water occurs around polar sites of high molecular weight proteins, starch, etc.

It was formerly thought that bound water is nonsolvent water, and that solutes could dissolve only in the unbound or "free" water. Experiments on equilibrium dialysis have demonstrated that the protein-bound water does not exclude solutes such as L-xylose and dimethylsulfoxide (De Bruijne and Van Steveninck, 1970).

Hysteresis

If several gelatin gels are made up of different concentrations and then dried, when allowed to reimbibe water, each will tend to regain its original concentration (Fig. 5-4). This is because of the structure of the gel. The colloidal particles are needle shaped and cross each other at definite points, forming a "brush-heap" type of structure. The water is held in the inter-

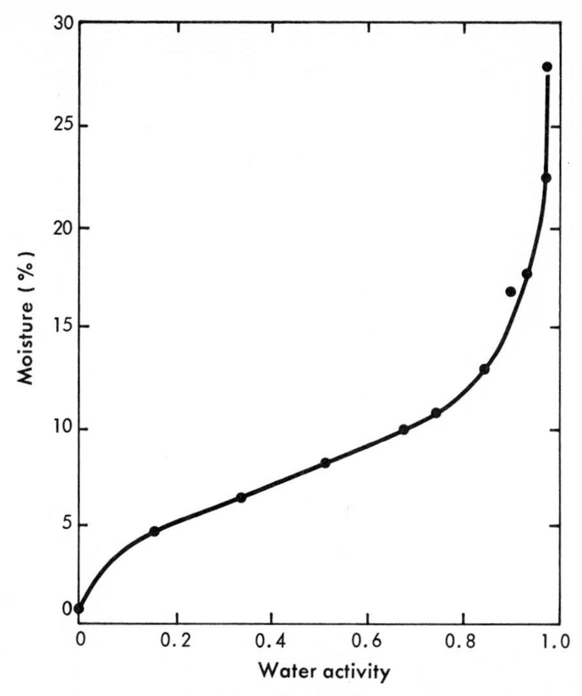

Fig. 5-3. Decrease in strength of binding of water by bean leaf protein (demonstrated by increase in activity) with increase in hydration (from Arkcoll, 1969). Water activity $= \frac{P}{P_o}$ where P = vapor pressure and P_o = saturated vapor pressure.

stices between the *micelles* (particles). Since these micelles tend to maintain a constant position with reference to each other (i.e., they act as though fastened together at the points of contact), the dilute gels will have larger interstices for holding water than will the concentrated gels. This phenomenon may be of fundamental importance in the case of protoplasm. If a plant has grown under low moisture conditions, its protoplasm may conceivably have this tight kind of structure and perhaps be unable to imbibe as much water as if it had grown under conditions of high moisture. This may possibly be one reason why such plants are sometimes permanently set

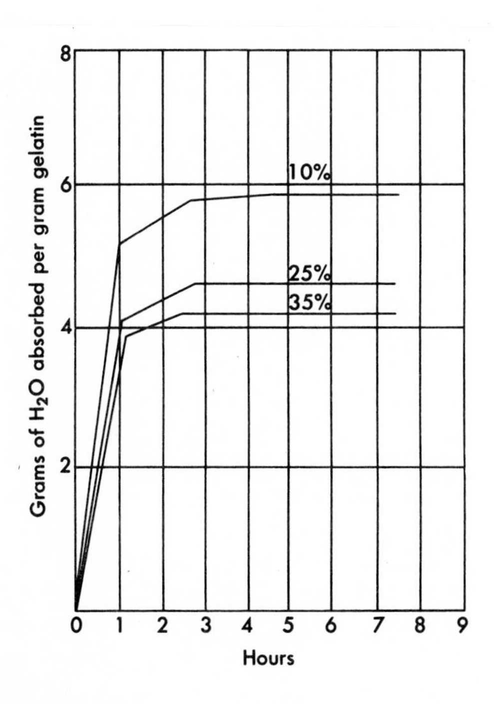

Fig. 5-4. Imbibition curves of gelatin made up of 10%, 25%, and 35% gels, dried to 3% moisture, then allowed to reimbibe water. (From Gortner, R. A. 1938. Outlines of biochemistry. John Wiley & Sons, Inc., New York.)

back by early droughts, even though not obviously injured.

Syneresis

Sometimes the ability of a gel to hold water decreases with age, and free water (or solution) is then liberated. This is called syneresis. Syneresis may also conceivably happen in the case of senescent cells. The vacuole would then enlarge at the expense of the protoplasm.

Thixotropy

Some gels may act as sols if stirred or shaken. After this treatment they may pour readily. On standing they become gels again and will not pour. This reversible sol $\rightleftharpoons$ gel transformation is called thixotropy. By use of microneedles it is possible to show that protoplasm is thixotropic.

QUESTIONS

1. Is a colloid a particular substance or a state in which the substance occurs?
2. Are the colloidal particles visible?
3. What is the approximate upper size limit?
4. What is the approximate lower size limit?
5. Is there a sharp line between colloids and noncolloids?
6. Can the colloidal particle be a single molecule?
7. Is the specific surface of a colloid large or small?
8. Does it show more or less adsorption than noncolloids?
9. How can one distinguish a colloidal dispersion from a molecular solution?
10. How do the two compare with regard to rate of diffusion?
11. How do the two compare with regard to osmotic pressure?
12. What colloids occur in the cell?
13. Name one property that may give the colloid stability.
14. Name another property that may give the colloid stability.
15. What is flocculation?
16. What is peptization?
17. What is salting out?
18. What is the lyotropic series?
19. Are proteins hydrophilic or hydrophobic (in general)?
20. What causes their charge?
21. Are they acids or bases?
22. What is an amphoteric substance?
23. What is the isoelectric point? How does it compare with the neutral point?
24. Is protoplasm a liquid or a gel?
25. How does the rate of diffusion through gels compare with that through sols?
26. Name three important properties of gels.

SPECIFIC REFERENCES

Acker, L. W. 1969. Water activity and enzyme activity. Food Technol. **23**:27-40.

Arkcoll, D. B. 1969. Preservation of leaf protein preparations by air drying. J. Sci. Food Agric. **20**:600-602.

Booij, H. L., and H. G. B. de Jong. 1956. Biocolloids and their interactions. Protoplasmatologia I(2). Springer-Verlag, Vienna.

Bull, H. B., and K. Breese. 1968. Protein hydration. I. Binding sites. Arch. Biochem. Biophys. **128**:488-496.

De Bruijne, A. W., and J. Van Steveninck. 1970. Apparent non-solvent water and osmotic behavior of yeast cells. Biochim. Biophys. Acta **196**:45-52.

Erlander, S. R. 1969. The structure of water. Science J. **5A**(5):60-65.

Hatefi, Y., and W. G. Hanstein. 1969. Solubilization of particulate proteins and nonelectrolytes by chaotropic agents. Proc. Nat. Acad. Sci. **62**: 1129-1136.

Kuhne, L., and W. Kausch. 1965. Über das Quellungsmaximum der Kotyledonen und Keimachsen von *Pisum sativum* L. Planta **65**:27-41.

Kuntz, I. D., Jr., T. S. Bradfield, G. D. Law, and G. V. Purcell. 1969. Hydration of macromolecules. Science **163**:1329-1331.

Todd, G. W., and J. Levitt. 1951. Bound water in *Aspergillus niger*. Plant Physiol. **26**:331-336.

GENERAL REFERENCES

Gortner, R. A. 1938. Outlines of biochemistry. John Wiley & Sons, Inc., New York.

Scarth, G. W., and F. E. Lloyd. 1930. Elementary course in general physiology. John Wiley & Sons, Inc., New York.

6

Diffusion and osmosis

DIFFUSION

Although many of the properties of the cell are dependent on the colloidal nature of its components, others are caused by the free molecular state of the water and the substances dissolved in it. Since these molecules are in constant motion (thermal agitation), there will be a continuous transfer of molecules from one region to another even without the application of external forces. At equilibrium, when the average free energy of each species of molecules is the same throughout the medium, as many molecules of one kind are transferred in one direction as in the other, and the net effect is zero. If the system is not in equilibrium, more molecules will move in one direction than in the other, and there will be a net transfer of the substance, called *diffusion*. An important question in the physiology of the plant is to what extent diffusion accounts for the continuous transfer of substances into, out of, and within the plant. This question can be answered only if the principles of diffusion are known.

If f is the frictional resistance to molecular movement:

$$\text{Diffusion rate} \propto \frac{1}{f}$$

This explains why the diffusion rate of water vapor is about 1500 times as rapid as that of liquid water. The molecules are so much farther apart in the vapor than in the liquid that the frictional resistance is extremely low. Thus gaseous diffusion is so rapid that no mechanism occurs in the plant for speeding up its movement. Liquid water, on the other hand, diffuses too slowly to keep up with the movement of the water vapor. To prevent its tissues from drying up, the land plant has therefore had to develop mechanisms for speeding up the movement of its liquid water and slowing down the movement of its water vapor.

The same principle of diffusion applies to the solutes dissolved in water. This is illustrated by plasmolysis. Hypertonic sucrose solutions take much longer to induce plasmolysis than do calcium chloride solutions of the same tonicity, partly because of their much greater viscosity, partly because of the slower diffusion rate of sucrose than of calcium chloride. But the aqueous solutions in the plant are usually so dilute that their viscosity is nearly that of pure water, and all the solutes are subjected to essentially the same frictional force. Consequently, when considering their movement within the plant, the frictional resistance can usually be neglected, and the diffusion rate will depend on any factor that affects the free energy of the diffusing substance. This is because, from the second law of thermodynamics, molecular movement will always be from a region of higher free energy of the moleecules to a region of lower free energy.

Since diffusion depends on molecular movement, it must be quantitatively related to (1) the rate of motion of the average molecule (its thermal agitation) and (2) the number of molecules per unit volume. The free energy of a substance that depends on these two components is called the chemical potential of the substance. Therefore the rate of diffusion increases with increase in μ, where μ = chemical potential of the diffusing substance. The chemical potential is given by the following equation:

$$\mu = RT \ln a \text{ joules/g}$$

where R = the gas constant (8.3 joules per degree)
T = the temperature (degrees absolute)
a = the activity or "effective" concentration of the substance
ln = the natural logarithm

The two variables, T and a, are measurements of the aforementioned two components. If the temperature is constant, the chemical potential is therefore proportional to the logarithm of the activity of the substance. But the activity of a nondissociating substance in the plant is, for all practical purposes, proportional to its concentration (for its dissociating substances, see Chapter 8). Therefore at constant temperature, it will diffuse from a region of higher concentration to one of lower concentration (Fig. 6-1). The greater this difference the more rapid is the diffusion. However, the rate of diffusion must also depend on two spatial factors. The greater the area across which the substance is diffusing the greater is the number of molecules diffusing per unit time. Conversely, the smaller the distance between the two concentrations the greater

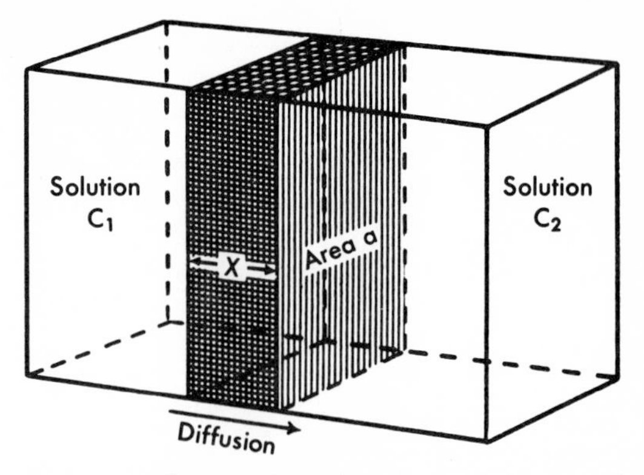

Fig. 6-1. Diffusion of a substance from a region of higher to one of lower concentration. (Redrawn from Scarth, G. W., and F. E. Lloyd. 1930. Elementary course in general physiology. John Wiley & Sons, Inc., New York.)

is the number crossing it per unit time. Two of the three factors—the concentration difference and the distance between them—are combined in what is called the *concentration gradient.* This is the concentration difference per unit distance.

The quantitative relation is given by Fick's law of diffusion:

$$\frac{s}{t} = Da \frac{C_1 - C_2}{x}$$

This use of the average change in concentration per unit distance is accurate only for short distances, such as across a membrane. For longer distances, the differential equation must be used:

$$\frac{ds}{dt} = Da \frac{dc}{dx}$$

where s = amount of substance diffusing (moles)
t = time (sec)
D = coefficient of diffusion, or specific diffusion rate
a = area of membrane (cm^2)
C_1 = higher concentration (moles per liter)
C_2 = lower concentration
x = membrane thickness (cm) or distance between C_1 and C_2

$$\frac{C_1 - C_2}{x} \text{ or } \frac{dc}{dx} = \text{concentration gradient}$$

D is a constant for a specific substance

diffusing through a specific medium at a standard temperature. When different substances are compared, small molecules are found to diffuse more rapidly than large molecules, that is:

$$D \propto \frac{1}{\text{molecular size}}$$

For small molecules:

$$D \propto \frac{1}{\text{mass}^{1/2}}$$

For large molecules and colloidal particles:

$$D \propto \frac{1}{\text{radius}} \text{ or } \frac{1}{\text{mass}^{1/3}}$$

It is now obvious why the larger colloidal particles diffuse so much more slowly than the smaller molecularly dispersed substances. In dilute solutions, such as are found in the plant, each solute diffuses independently of the other solutes.

If the partition x (Fig. 6-1) is a structure immersed in the aqueous medium, the concentration gradient in it will depend not only on the two concentrations in the aqueous media but also on the solubility of the diffusing substance in this structure. This solubility in the structure is called the *permeability* of the structure to the diffusing substance. If the structure is thin, it is called a membrane. The diffusion rate of a substance through a unit area of a membrane is therefore controlled not only by (1) the temperature, (2) the concentration gradient in the aqueous phase, and (3) the size of the diffusing molecule or particle, but also by (4) the permeability of the membrane to the substance. This will be considered in more detail in Chapter 7.

OSMOSIS AND OSMOTIC PRESSURE

Pressure caused by osmosis

The same laws of diffusion apply to solvent as to the solute. The basic difference is that the activity and therefore the free energy of the solute is most readily measured by its concentration, whereas the activity of the solvent (water) is most readily measured by its vapor pressure. If some standard concentration (C_0) or vapor pressure (p_0) is chosen, then the relative activity at any other concentration (C) or vapor pressure (p) is:

$$a_{solute} = \frac{C}{C_0}$$

$$a_{H_2O} = \frac{p}{p_0}$$

Because of the much larger quantity of the solvent and because of its liquid state, its diffusion through semipermeable membranes gives rise to hydrostatic pressures. The importance of these pressures has led to the use of the specific term *osmosis,* for diffusion of the solvent, and *osmotic pressure,* for the hydrostatic pressure produced by osmosis (for a stricter definition see later). Osmosis is sometimes defined as diffusion of solvent across a semipermeable membrane, but the preceding simpler definition is broader and therefore more useful. It permits use of the term, for instance, for the diffusion of water from a solution into a hydrophilic gel, although no semipermeable membrane separates the two systems.

Since pressure is force per unit area ($P = \frac{F}{A}$), the osmotic pressure that develops in the cell, because of osmosis, must produce forces of fundamental importance to the plant. These forces give rise to plant *turgor*—the stiffness or rigidity of the plant resulting from the hydrostatic pressure exerted on the cell walls. Plant turgor also results in growth, plant movements, and other plant responses. In order to study these phenomena, osmotic pressure must be measured and calculated quantitatively. The qualitative concept of osmotic pressure as the hydrostatic pressure produced by osmosis is therefore inadequate. In the strict quantitative sense, the term *osmotic pressure* is not used for nonequilibrium (i.e., temporary or even steady-state) pressures but only *for the final, equilibrium pressure (above atmospheric) attained as a result of osmosis under specific, standard conditions.* Since forces occur in pairs, osmotic pressure is determined by applying an increasing pressure until the one is found which is just sufficient to stop osmosis. The standard conditions are (1) separation of the solution from pure solvent at atmospheric pressure by a semipermeable membrane and (2) maintenance of the solution and solvent at the same standard temperature (20° C). The osmotic pressure of a solution is therefore defined as the extra pressure that must be applied to it in order to stop osmosis from pure water separated from the solution by a semipermeable membrane, when both are at 20° C.

Unfortunately, however, the cell is usually neither under the preceding standard conditions nor in a state of equilibrium. Therefore the actual forces prevailing in the cell are not those that would result from the defined (equilibrium) osmotic pressure. Furthermore, from Newton's third law of motion, the real forces must exist in pairs, equal in magnitude and opposite in direction. These forces are caused by (1) the pressure exerted by the piston and walls of the osmometer (Fig. 6-2) on the solution, which the physiologist calls wall pressure (P_w); and (2) the hydrostatic pressure exerted by the water on the walls, which the physiologist calls turgor pressure (P_t), since it can convert flexible walls into stiff walls and a flaccid cell into a turgid cell. The important difference between these three pressures is that osmotic pressure (P_0—the pressure at zero diffusion of

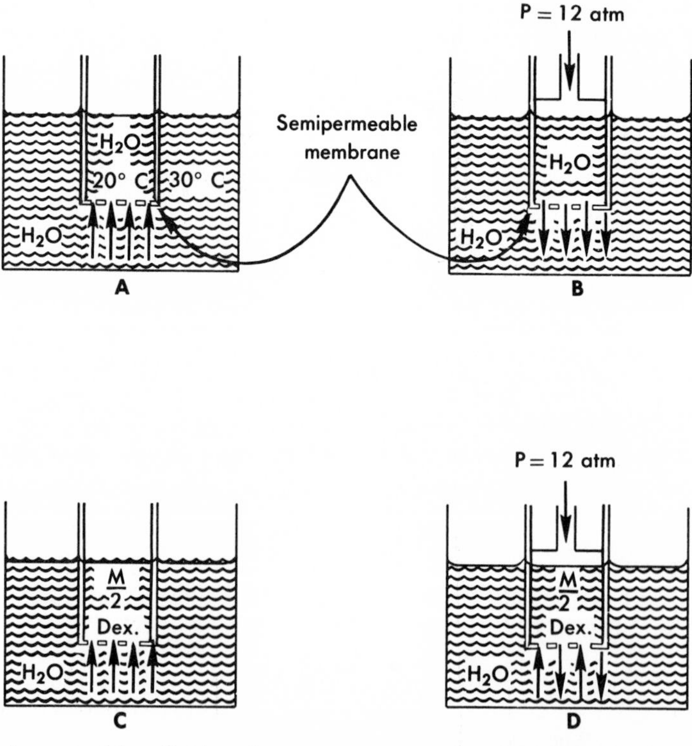

Fig. 6-2. Endosmosis (**A** and **C**) and exosmosis (**B**) produced by differences in **A**, temperature, **B**, pressure, and **C**, solute concentration. There is no osmosis in **D** because the solute effect is exactly counterbalanced by the pressure. The pores in the bottom of the container (or osmometer) indicate a semipermeable membrane.

Table 6-1. Colligative properties of a solution; effect of dissolving 186 g (1 gram molecular weight) of dextrose in 1000 g water, yielding a solution of 1.0 molal dextrose

Colligative property	Water	Value for molar dextrose	Effect of dextrose on property
Vapor pressure	17.5 mm Hg at 20° C	17.2 mm Hg	Lowered
Freezing point	0° C	−1.86° C	Lowered
Boiling point	100° C	100.51° C	Raised
Osmotic pressure	0 atm	24 atm	Raised
Water potential	0 joules/g	−2.4 joules/g	Lowered

water under standard conditions) refers only to the equilibrium pressure. Wall pressure (P_w) and turgor pressure (P_t) refer to the actual pressure under the particular conditions to which the cell is subjected, even though equilibrium has not been attained. At equilibrium under standard conditions, the following relation exists:

$$P_0 = P_w = P_t$$

In any nonequilibrium or nonstandard state (even in the steady state)

$$P_w = P_t$$

but both differ from P_0.

Energy relations

As in the case of any diffusion process, osmosis will always take place from a region of higher chemical potential of the solvent to one of lower chemical potential. *Endosmosis* (osmosis into a system) occurs if the *external* water is at a higher chemical potential; *exomosis* (osmosis out of a system) occurs if the *internal* water is at a higher chemical potential. Endosmosis therefore will be induced by any factor that decreases the free energy or chemical potential of the solvent within the system or that increases it outside the system. Exosmosis is induced by either of the reverse changes. There are three such factors (Fig. 6-2). The chemical potential (1) increases with the temperature, since chemical potential is a measure of energy resulting from molecular motion and temperature is simply a measure of the kinetic energy of the molecules; (2) increases with the pressure, since the piston does work on the liquid, raising its free energy; and (3) decreases with increased concentration of solute, since the force of attraction between solute and solvent lowers the free energy of the solvent molecules. The effect of the solute on the free energy of the solvent is proved by changes in the colligative properties of solutions (Table 6-1), since these all depend on the free energy of the solvent molecules. For instance, because of the lower free energy of the water molecules in a solution than in the pure solvent, the vapor pressure of the water is lowered and a higher temperature is required to boil the solution than the pure solvent. Similarly, the lower vapor pressure results in a lower diffusion rate (or osmosis) of the water. Theoretically, then, the ideal method for predicting the direction of osmosis between two points is by measuring the chemical potential of the solvent at these two points because the chemical potential measures the net effect of all three factors. Unfortunately, the *actual* chemical potential cannot be measured, but the *change* in relation to pure solvent under standard conditions can be determined.

The experiment illustrated in Fig. 6-2 reveals that a pressure of 12 atm on the 0.5 M sugar solution was required to prevent osmosis from pure water. This must mean that the pressure exactly counteracts the effect of the solute on the free energy of the solvent. Since the solute lowers the free energy of the water molecules (Table

6-1), the pressure must raise it by an equal amount. Therefore 0.5 mole of sugar per liter of water must lower the free energy of the water by 12 atm. This change in free energy of the water, produced by the solute is called the *osmotic potential* (π) of the solution. From the foregoing opposing effects of solute and hydrostatic pressure on the free energy of the solvent:

$$\pi = -P_0$$

This means that only the standard equilibrium pressure, which is called the osmotic pressure, is numerically equal to the osmotic potential (although opposite in sign). In the absence of any pressure, other than atmospheric, the negative value of this theoretical pressure yields the decrease in free energy of the water molecules resulting from the solute. Since solutes always lower the free energy of the solvent, π will always be negative in the plant. Dilution of the solution with solvent will increase the value of π (i.e., will decrease the negative numerical value) until, at infinite dilution, it reaches the maximum value of 0 atm.

Under normal conditions, however, the cell solution is neither at atmospheric pressure nor is it subjected to the full theoretical osmotic pressure, since it is turgid but not in equilibrium with pure water. Because of its turgor, the cell contents are subjected to a *wall pressure* (P_w) somewhere between these two extremes. This wall pressure increases the free energy of the cell's water. The algebraic sum of the preceding two factors (π and P_w) is therefore a measure of the net change in chemical potential of the water relative to that of pure water at atmospheric pressure and at the same temperature. This net change in the chemical potential of the water is called the water potential (Ψ), and is calculated from:

$$\Psi = \pi + P_w$$

The importance of Ψ lies in its control of the direction (and relative rate) of diffusion. Osmosis will always occur from a region of higher Ψ (smaller negative value) to a region of lower Ψ (larger negative value). Thus it will occur from cell 1 with a Ψ of –3 atm to cell 2 with a Ψ of –4 atm. If the solution is at atmospheric pressure, $P_w = 0$ and $\Psi = \pi$.

The originators of the term *water potential* (Slatyer and Taylor, 1960) define it as "the difference between the partial specific Gibbs' free energy of water in the system under consideration and of free, pure water at the same temperature."

Since π is always negative (the maximum value being zero for pure water) and since P_w is positive in a turgid cell (Fig. 6-3, *A*) and cannot exceed the numerical value of π, Ψ will be negative in a turgid cell or will have a maximum value of zero. But if a cell is allowed to lose water until it is flaccid (Fig. 6-3, *B*), the whole cell will collapse (unlike the separation of protoplast from wall, which occurs on plasmolysis, Fig. 2-2), and the positive osmotic force will be converted to a negative force or tension. Therefore P_w will have a negative energy value. Since both π and P_w are negative in the flaccid cell, the water potential must be lower (i.e., more negative) than π alone.

Osmotic quantities have, in the past, been measured in atmospheres. The recent trend, however, is to measure them in *bars* (a unit of the metric system). Fortunately, the values are interchangeable since 1 atm (approximately 15 lb/inch^2) = 1.01 bars. Before the recent adoption of water potential terminology, various other names were given for the water potential: suction pressure (SP), diffusion pressure deficit (DPD), osmotic equivalent (E), etc. These terms were used as positive values for what are now known to be negative water potentials.

In the preceding equations the free energy of the water is measured in terms of a unit of pressure. This is possible because of the twofold significance of pressure:

1. $P = \frac{F}{A}$ (pressure = force per unit area)

2. $P = \frac{W}{\Delta V}$ (pressure = work done per unit change in volume)

From the two equations for P, it can be expressed either in units of pressure (bars) or of work per unit volume (joules/cm³). Similarly, Ψ can also be measured in bars (Ψ bars) or in joules/cm³ (Ψ joules/cm³). But, by definition:

$$1 \text{ joule/cm}^3 = 10 \text{ bars}$$

Therefore when Ψ is measured in joules/cm³, it is numerically equal to 1/10 the number of bars, or:

$$\Psi_{\text{joules/cm}^3} = \frac{\Psi_{\text{bars}}}{10}$$

Also, since 1 cm³ water = 1 g, the energy relation is sometimes given as joules/g.

Another quantity, the *matric potential*, is sometimes included as a theoretical value for the effect of hydrophilic colloids (as opposed to the osmotic potential caused by solutes) on the water potential. It must be realized, however, that there is no difference thermodynamically as to the nature

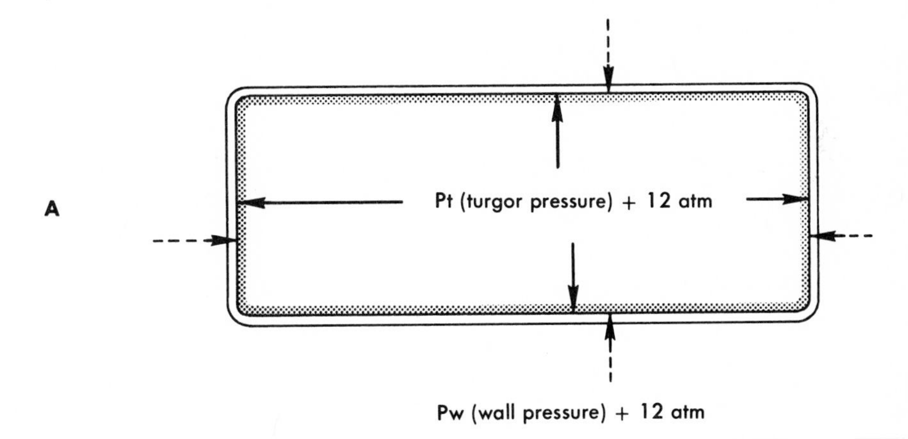

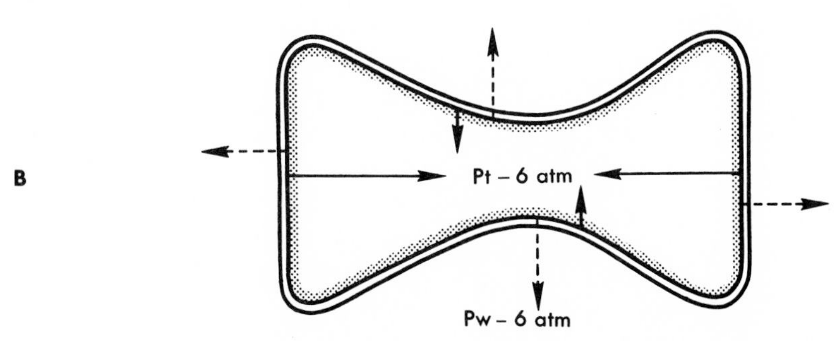

Fig. 6-3. Pressures in **A**, a turgid, and **B**, a flaccid cell. The + sign means (1) an increase in free energy of the water and (2) the direction of the forces in a turgid cell. The – sign means (1) a decrease in free energy of the water and (2) the direction of the forces in a flaccid cell. The extracted sap from the turgid cell would have an osmotic potential of –12 bars if the cell is at full turgor (i.e., in equilibrium with pure water).

of the effect of the colloid and the solute. Both lower the water potential by virtue of their attraction for the water molecules. Furthermore, the methods normally used to measure the osmotic potential include any effect resulting from the hydrophilic colloids present. Therefore in practice, what is called the osmotic potential is really a combination of the two quantities and no distinction between the two can be made. Attempts have been made to measure this matric potential separately (Wiebe, 1966; Boyer, 1967). But as long as the method for measuring osmotic potential combines the effects of both solute and colloid, there is no need to include matric potential as a separate quantity in the equation for Ψ.

From Table 6-2 it is obvious that large osmotic pressures have been found in plant cells, and since these theoretical pressures may be converted into actual turgor pressures of nearly the same numerical values, the question arises as to the ability of cell walls to withstand pressures of the order of 100 atm (i.e., 1500 lb/inch^2). It must be realized, however, that the dimensions of a cell are small, and therefore the force acting on the two opposite wall faces of a cell is normally far too small to exceed the elastic limit of the mature cell wall. In the case of growing cells, on the other hand, the walls are so soft that this elastic limit may be readily exceeded and the result is cell enlargement (Chapter 17).

Table 6-2. Approximate ranges of calculated osmotic pressures of plant saps of different environmental groups*

Plant group	Osmotic pressure (atm or bars)
Summer ephemerals	8-42
Succulents and winter ephemerals	4-24
Xerophytes	14-57
Hydrophytes	8-13
Air leaves	18-21
Water leaves	8-9
Epiphytes	3-6
Halophytes	30-115
Parasites	14-17
Hosts	11-14

*From Altman, P. L., and D. S. Dittmer (Eds.). 1964. Biology data book. Federation of American Societies for Experimental Biology, Washington, D. C.

It is necessary to distinguish between two pressures, both of which are included in the colligative properties of aqueous solutions, and both of which therefore depend on the water potential (Ψ). A decrease in Ψ is accompanied by a decrease in vapor pressure (Table 6-1). However, vapor pressures are too small (usually 0.02 to 0.04 atm) to account for appreciable forces in plant cells. In the liquid state, on the contrary, the water molecules are about 1500 times as concentrated as in the vapor state when both are at the same vapor pressure. The *hydrostatic* pressures produced by a decrease in Ψ (followed by osmosis) therefore may be expected to attain values 1500 times the change in vapor pressure. Thus 0.5 M dextrose solution has a vapor pressure only 0.0002 atm less than that of pure water, yet if it is surrounded by a semipermeable membrane and immersed in water, osmosis may lead to a maximum pressure of 12 atm (Fig. 6-2).

It is therefore of fundamental importance not to confuse the small vapor pressure (which may also be called the "diffusion pressure") of water with the much larger hydrostatic pressures, although the latter result from the former. This same confusion gave rise in the earlier literature to terms such as *diffusion pressure deficit* (DPD), which was a measure of potential *hydrostatic* pressure and not the much smaller vapor (or true diffusion) pressure difference.

Significance of osmotic quantities

The water potential of a cell enables the prediction of the direction of osmosis—from a high potential (or low negative value) to a low potential (or high negative

Table 6-3. Water potential of inner cortex in different organs of ivy*

Organ	Water potential (atm)
Petiole	–8.4
Stem (225 cm high)	–5.0
Stem (35 cm high)	–2.9
Root (older portion)	–2.4
Root (absorption zone)	–1.6

*Adapted from Molz, F. J. 1926. Amer. J. Bot. 13:433-501.

value). There is therefore a gradient from the highest value in the water-absorbing zone of the plant, the root, to the lowest value in the water evaporating part of the plant, the leaves (Table 6-3). Besides the direction of diffusion of water between any two systems, the water potential difference between them also indicates the relative rate of diffusion. This is true as long as relatively small differences are involved (e.g., less than either 150 bars or 15 joules/g). Unfortunately, water potential is proportional to the log of its vapor pressure, which controls its diffusion rate. Therefore for large Ψ differences the vapor pressure difference must be used, for example, for the diffusion of water from the leaves into the normally dry air (the water potential of the air at 60% relative humidity at normal temperatures is –680 atm).

The three pressures play specific roles in the physiology of the cell (Table 6-4). The osmotic pressure of a solution depends on the presence of solutes. Measured or calculated values for plant or cell sap are indirect measures of cell sap concentration. Cell sap concentration, measured in molarity, will not parallel osmotic pressure values exactly, since solutes differ in their osmotic pressures per unit molarity. It is therefore sometimes expressed in *osmolarity*—the molarity of a nearly ideal solution (e.g., dextrose) having the same osmotic pressure as the unknown solution. Osmotic potentials of cell saps, on the other hand, may reveal the relative degree of hydration of the protoplasm. Since both the vacuole and protoplasm are subjected to the same wall

Table 6-4. Significance of osmotic quantities

Quantity	Symbol	Sign in turgid cells	Meaning	Significance
Osmotic pressure	P	+	Calculated force per unit area required to stop osmosis	1. Theoretical maximum wall and turgor pressure 2. Measure of solute concentration
Osmotic potential	π	–	Water potential resulting from solutes (and colloids if matric potential included)	Degree of hydration of protoplasm
Turgor pressure	P_t	+	Instantaneous actual hydrostatic pressure exerted on cell wall	Stretches cell wall and may lead to cell enlargement if walls are sufficiently plastic
Wall pressure	P_w	+	Instantaneous back pressure by wall on protoplast	Increases free energy of water molecules
Water potential	Ψ	–	Chemical potential of water in system minus that of pure water under standard conditions	Indicates direction and relative rate of diffusion

pressure, their water potentials relative to each other are solely the result of their osmotic potentials (using the term as including matric potential, see earlier discussion). Consequently, if the solute concentration of the vacuole increases, the osmotic potential must decrease. Water must then move into the vacuole from the protoplasm, and the latter will be partially dehydrated. Rapidly metabolizing and growing cells are characterized by low cell sap concentrations and therefore high protoplasmic hydration. High cell sap concentrations and therefore low protoplasmic hydration are commonly found in inactive, dormant cells, for example, during the winter (Table 6-5). Osmotic pressure is also related to the ecology of the plant (Table 6-2), although there is, of course, considerable overlap between ecological groups. The highest values are definitely found among the halophytes, the lowest among the epiphytes, hydrophytes, and succulents. Osmotic pressure values may also reflect the balance between transpiration and absorption, as can be seen by the diurnal changes (Table 6-6), although changes in water potential are more striking.

Because it reflects the water status of the plant in all these cases, Walter (1963) has used the osmotic potential as an indirect measure of what he calls the "hydrature" of the plant, by analogy with temperature, which he describes as the "status" of the heat in a system. Hydrature, according to Walter, is the relative water vapor pressure (or relative humidity) of an aqueous system, independent of external factors, (i.e., at constant temperature and external pressure). It is therefore proportional to osmotic potential.

The wall pressure raises the water potential of the water in the cell. This increases the escaping tendency of the water molecules and may lead to diffusion out of the cell. But the resulting force is insufficient to squeeze the water through the extremely fine "pores" of the plasma membrane. Artificial pressures of 100 to 150 atm may be required to produce this flow. Nevertheless, wall pressure may be thought of as producing a flow in another way—the flow of water up the xylem (Chapter 8). Turgor pressure, on the other hand, may lead to wall stretching and therefore cell enlargement (Chapter 17).

The cells of land plants are not normally at equilibrium but are continuously losing and gaining water. The turgor and wall pressures therefore may be much less than

Table 6-5. Calculated osmotic pressures of plant saps in some Canadian evergreens in relation to the season*

	Osmotic pressures (atm)		
Month	*Linnaea borealis*	*Picea glauca*	*Pyrola rotundifolia*
Oct.	19.6	17.1	
Dec.	25.0	20.3	24.6
Jan.		20.0	
Feb.			23.9
March	25.6	24.9	
April		20.1	17.2
May	14.3	21.0	
June		19.7	12.6

*Adapted from Altman, P. L., and D. S. Dittmer (Eds.). 1964. Biology data book. Federation of American Societies for Experimental Biology, Washington, D. C.

Table 6-6. Diurnal changes in water potential and in calculated osmotic pressures of the cell sap*

Time	Water potential (atm)	Time	Osmotic pressure (atm)
5:15 A.M. (sunrise)	− 7.5	6:00 A.M.	12.5
7:00	− 9.5		
9:00	−12.0	10:00	15.3
11:00	−15.0		
1:00 P.M.	−17.5	2:00 P.M.	17.4
3:00	−17.5		
4:30	−15.0	5:00	16.5
7:30	−15.0	8:00	16.3

*Adapted from Herrick, E. M. 1933. Amer. J. Bot. **20**:18-34.

the osmotic pressure of the cell sap because of loss of water to the environment. Even when immersed in pure water (Fig. 6-4), the cell's turgor pressure can never fully attain the osmotic pressure of the cell sap at zero turgor pressure because any development of turgor results from endosmosis, which dilutes the cell sap and lowers its osmotic pressure (or raises its osmotic potential). As endosmosis occurs, all three values change, but the equation holds for any instantaneous set of values. These relations are shown in Fig. 6-4.

Starting with the protoplast of a sunflower leaf plasmolyzed to 0.9 of the cell wall volume, the osmotic potential is –16 bars and so is the water potential (Ψ), since wall pressure is zero. When the protoplast has expanded sufficiently just to fill the cell wall, it has an osmotic potential and water potential of –14.3 bars, and wall pressure is still zero. At a cell volume of 1.1, the osmotic potential has increased still further (to –13.0 bars) because of dilution of the cell contents by the entering water. The wall pressure has risen from zero to 3 bars and the Ψ is –10 bars (osmotic potential plus wall pressure). At full turgor (i.e., in equilibrium with pure water) the Ψ is zero and the osmotic potential and wall pressure are –11.7 bars and +11.7 bars, respectively (where +

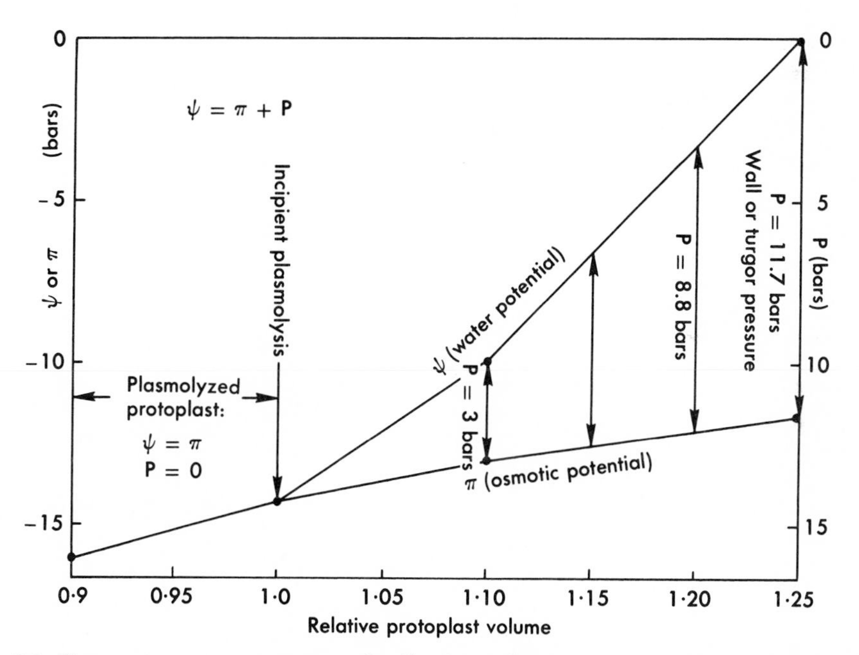

Fig. 6-4. Changes in osmotic quantities of cells of a sunflower petiole during deplasmolysis and final expansion in water. All three quantities increase, but the increase in Ψ for any one increase in cell volume always equals the sum of the increases in π and P. Therefore at any one point $\Psi = \pi + P$; for example:

Protoplast volume	Ψ	π	P
0.9	–16	–16	0
1.1	–10	–13	3
1.15	– 6.7	–12.7	6
1.25	0	–11.7	11.7

(Redrawn from Clark, J. A. 1956. Ph.D. Thesis. University of Missouri, Columbia, Mo.)

simply means above atmospheric pressure).

In all the above cases, the cell as a whole acts as an osmometer, due to the presence of a semipermeable plasmalemma. It follows that any organelle within the cell that is surrounded by its own semipermeable membrane may also act as an osmometer. Corn mitochondria, for instance, act as osmometers within the range of 1.8 to 8.4 bars (Lorimer and Miller, 1969). Even drops of cytoplasm, isolated from the large internodal cells of Characeae, are able to act as osmometers (Yoneda and Kamiya, 1969).

Determination of osmotic quantities

Osmotic potential. In most cases the osmotic quantities of cells are not determined directly. The values are much more easily obtained by comparing the cell with solutions of known osmotic values. It is therefore convenient to be able to calculate the osmotic potential of a solution from its concentration. This can be done by use of the following equation,

$$\pi = -CRT$$

where π = osmotic potential (bars)
C = concentration (moles per liter)
R = gas constant (0.082 atm/mole/°K)
T = temperature (degrees absolute or Kelvin)

If a solution is reasonably dilute and consists of monomolecular solute particles, this equation gives values that are sufficiently accurate for most physiological work. If, however, the molecules dissociate into smaller particles as in the case of electrolytes (salts, acids, bases), the osmotic potential will be numerically larger than calculated by an amount equal to the excess of particles over the expected number of molecules. In the case of sodium chloride, for instance, the right side of the equation has to be multiplied by 1.8 and in the case of calcium chloride, by 2.4. This factor is known as the *isotonic coefficient,* and the equation becomes

$$\pi = -iCRT$$

where i = isotonic coefficient (average number of particles per molecule). This equation is not accurate at high concentrations because of the hydration of the particles; and the error cannot be calculated for, because the degree of hydration is usually not known accurately enough and is not constant. Accurate osmotic potentials can, however, be calculated from two colligative properties of solutions: (1) the freezing point lowering and (2) the vapor pressure, both of which values may be obtained for many solutions from published tables (e.g., *Handbook of Chemistry and Physics* edited by C. D. Hodgman and others).

1. $\pi = -12.06\ \Delta$ bars*
where Δ = freezing point lowering of the solution (compared to pure water)

2. $$\pi = -\frac{RT}{V^1} \ln \frac{p_0}{p}$$

where ln = the natural logarithm
p_0 = vapor pressure of pure water
p = vapor pressure of the solution at the same temperature as p_0
V^1 = volume of one mole of solvent

The preceding calculations must be applied to determine the exact values for the solutions used in measuring the osmotic quantities of cells and tissues (Table 6-7). These solutions may be used for two different kinds of measurement.

Osmotic potential of the cell at zero turgor. The terms used earlier for plasmolytic effects of solutions (Chapter 2) can now be understood. A hypertonic solution is one with a higher concentration and therefore a lower osmotic potential than that of the cell sap. A hypotonic solution is one with a lower concentration and therefore a higher osmotic potential than that of the cell sap. An isotonic solution, theoretically, is one

*Strictly speaking, this gives the value for solutions at 0° C. At room temperature the value is nearly 10% higher, a negligible error for most plant measurements.

with the same osmotic potential as that of the cell sap. (In practice it is slightly lower.) There are two simple methods of determining osmotic potential by this relationship: (1) the minimum length method and (2) the incipient plasmolysis method. Since the stretching or contraction of living cells is caused by turgor pressure, once turgor pressure is zero, no further cell wall contraction can occur (when cells are in aqueous solutions). Consequently, the most dilute solution that is capable of reducing the cell (or tissue in the case of uniform strips) to minimum size has the same osmotic potential as the cell sap (i.e., is isotonic or isosmotic with it). In the second method the point is determined by going a little beyond it and observing for incipient (i.e., beginnings of) plasmolysis. If the merest trace of plasmolysis can be detected, the two methods give essentially identical results.

Osmotic potential of the cell at its momentary state. If the juice of the tissue or cell is extracted, its freezing point lowering can be measured and the osmotic potential calculated as just described. The value will, of course, vary depending on the degree of turgor or wilting of the plant. However, as long as the plant material is not wilted, the values obtained by this method should agree within about 10% with the values obtained for the cell at zero turgor.

Water potential. Since no diffusion of water can occur between systems at the same water potentials, the simplest method is to find the solution of known water potential in which the living cells show neither endosmosis nor exosmosis, that is, in which they neither swell nor contract. This can be done by measurements of cell or tissue strip length.

Turgor and wall pressure. In contrast to osmotic potential and water potential, turgor and wall pressure are usually not measured directly but are simply calculated from the other two measured values using the fundamental equation:

$$P_w \text{ (or } P_t) = \Psi - \pi$$

An ingenious direct method has been developed for measuring turgor pressure in higher plants (Virgin, 1955) but it has not been adopted by other investigators, presumably because the relation between the measured value and the turgor pressure

Table 6-7. Osmotic pressures of solutions of several substances, calculated from freezing point lowerings of the solutions listed in International Critical Tables; calculations from Lewis's formula

Molarity (weight molar)	Osmotic pressure (atm or bars)				
	$CaCl_2$	NaCl	KNO_3	Sucrose	Glucose
0.01	0.7	0.5	0.5		
0.10	6.2	4.5	4.3	2.6	2.4
0.20	12.3	8.9	8.2	5.1	4.8
0.30	18.5	13.2	11.9	7.7	7.2
0.40	25.3	17.6	15.4	10.2	9.6
0.50	32.2	21.9	18.8	12.9	12.1
0.60	39.2	26.2	22.0	15.5	14.5
0.70	47.2	30.5	25.1	18.3	17.0
0.80	56.2	34.9	27.9	21.1	19.4
0.90	65.4	39.2	30.6	23.9	21.8
1.00	75.0	43.6	33.2	26.9	24.3
2.00	194	88.8		56.9	
3.00	346	138.2		92.9	
4.00				137	
5.00				187	

proved to be more complicated than was originally assumed (Burström, 1971). Direct measurement is easier in the case of large algal cells (Green, 1968). A new method developed for roots has yielded pressures from 8 to 15 bars for cotton and pea roots at soil matric potentials of –⅓ bar (Eavis et al., 1969).

Many ingenious methods for measuring water potentials and osmotic potentials have been developed (Levitt, 1964; Wiebe et al., 1971). The precision of the vapor pressure method has been greatly increased in the hands of many investigators. As a result, the thermocouple psychrometer has now come into use as almost the standard method of measuring both water potential and osmotic potential. There are many modifications of both the instrument and the method of using it (Wiebe et al., 1971). The pressure chamber method may be equally satisfactory (Boyer and Ghorashy, 1971; Duniway, 1971). It is advisable, however, to check these newer methods against the classical methods, for instance, by determining the freezing point lowering.

QUESTIONS

1. What kind of movement is diffusion caused by?
2. If diffusion stops (because of equilibrium), does molecular movement stop?
3. If different substances are compared under the same conditions, will all diffuse at the same rate?
4. Does diffusion depend in any way on the medium?
5. What difference must exist in order for diffusion to occur?
6. What effect does temperature have on diffusion?
7. What three factors are included in Fick's law of diffusion?
8. What property of a substance can always be related to diffusion, including concentration difference, temperature, etc.?
9. What effect does a solute have on the free energy of the solvent molecules?
10. What is meant by osmosis?
11. What is meant by osmotic pressure?
12. What is the effect of pressure on the free energy of water molecules?
13. If a solution is separated from water by a semipermeable membrane, what happens?
14. How long will this process continue?
15. What are the real pressures in a living cell?
16. What is the effect of each pressure?
17. How do these pressures compare quantitatively?
18. How can these pressures be equal when the cell enlarges?
19. What quantity gives a quantitative measure of the ability of the cell's water to diffuse?
20. Does this quantity vary directly or inversely with this ability to diffuse?
21. How is this quantity related to π and P?
22. When a cell is in equilibrium with pure water, what is its Ψ?
23. What is the osmotic potential of 0.33 M dextrose?
24. What is the osmotic potential of 0.33 M sodium chloride?
25. At what temperature would a cell with an osmotic potential of –24 bars freeze?
26. If a cell is at incipient plasmolysis, what cell quantity does the external solution reveal?
27. If a cell remains unchanged in size on transfer to a solution, what cell quantity does the solution give?
28. How do you measure P_w? How do you measure P_t?
29. What osmotic quantity can be calculated from the freezing point lowering of the plant juice?

SPECIFIC REFERENCES

Altman, P. L., and D. S. Dittmer (Eds.). 1964. Biology data book. Federation of American Societies for Experimental Biology, Washington, D. C.

Boyer, J. S. 1967. Matric potentials of leaves. Plant Physiol. **42**:213-217.

Boyer, J. S., and S. R. Ghorashy. 1971. Rapid field measurement of leaf water potential in soybean. Agron. J. **63**:344-345.

Burström, H. G. 1971. Resonance frequency measurements on plant tissues. Endeavour **30**:87-90.

Clark, A. J. 1956. An investigation of the drought hardening of the soybean plant. Ph.D. Thesis. University of Missouri, Columbia, Mo.

Duniway, J. M. 1971. Comparison of pressure chamber and thermocouple psychrometer determinations of leaf water status in tomato. Plant Physiol. **48**:106-107.

Eavis, B. W., L. F. Ratliff, and H. M. Taylor. 1969. Use of a dead-load technique to determine axial root growth pressure. Agron. J. **61**: 640-643.

Green, P. B. 1968. Growth physics in *Nitella*: a method for continuous in vivo analysis of extensibility, based on a micromanometer technique for turgor pressure. Plant Physiol. **43**: 1169-1184.

Herrick, E. M. 1933. Seasonal and diurnal variations in the osmotic values and suction tension values in the aerial portions of *Ambrosia trifida*. Amer. J. Bot. **20**:18-34.

Lorimer, G. H., and R. J. Miller. 1969. The osmotic behavior of corn mitochondria. Plant Physiol. **44**:839-844.

Molz, F. J. 1926. A study of suction force by the simplified method. Amer. J. Bot. **13**:433-501.

Scarth, G. W., and F. E. Lloyd. 1930. Elementary course in general physiology. John Wiley & Sons, Inc., New York.

Slatyer, R. O., and S. A. Taylor. 1960. Terminology in plant and soil-water relations. Nature (London) **187**:922-924.

Virgin, H. 1955. A new method for the determination of the turgor of plant tissues. Physiol. Plant. **8**:954-962.

Walter, H. 1963. Zur Klärung des spezifischen Wasserzustandes im Plasma und in der Zellwand bei der höheren Pflanze und seine Bestimmung. Ber. Deutsch. Bot. Ges. **76**:40-71.

Wiebe, H. H. 1966. Matric potential of several plant tissues and biocolloids. Plant Physiol. **41**: 1439-1442.

Wiebe, H. H., G. S. Campbell, W. H. Gardner, S. L. Rawlins, J. W. Cary, and R. W. Brown. 1971. Measurement of plant and soil water status. Bull. 484. Utah Agr. Exp. Sta.

Yoneda, M., and N. Kamiya. 1969. Osmotic volume change of cytoplasmic drops isolated in vitro from internodal cells of Characeae. Plant Cell Physiol. **10**:821-826.

GENERAL REFERENCES

Crafts, A. S., H. B. Currier, and C. R. Stocking. 1949. Water in the physiology of the plant. Chronica Botanica Co., Waltham, Mass.

Dainty, J. 1969. The water relations of plants. p. 421-452. *In* M. B. Wilkins (Ed.). The physiology of plant growth and development. McGraw-Hill Book Co., New York.

Harris, J. A. 1934. Physico-chemical properties of plant saps in relation to phytogeography. University of Minnesota Press, Minneapolis.

Kramer, P. J. 1969. Plant and soil water relationships: a modern synthesis. McGraw-Hill Book Co., New York.

Levitt, J. 1964. Osmotic pressure measuring devices. *In* D. W. Newman. Instrumental methods of experimental biology. The Macmillan Co., New York.

Slatyer, R. O. 1967. Plant-water relationships. Academic Press, Inc., New York.

Stadelmann, E. J. 1966. Evaluation of turgidity, plasmolysis, and deplasmolysis of plant cells. p. 143-215. *In* D. M. Prescott. Methods in cell physiology. Vol. 2. Academic Press, Inc., New York.

7

Permeability

Diffusion through cell membranes
Quantitative relations
Mechanism of control by membranes

DIFFUSION THROUGH CELL MEMBRANES

The continued growth and survival of a plant depends on the maintenance of its cellular *semipermeability*—a nearly free permeability of the cells to water and nearly complete impermeability to the cell solutes. If, for any reason, a cell becomes more permeable to its solutes, the protoplasmic solutes begin to leak out, and the vacuole solutes begin to leak into the protoplasm. Both of these processes, if they proceed too far, lead to injury or death of the cell. As an example, the permeability to the cell solutes may be increased artificially by the application of octanoic acid (Lode and Pedersen, 1970) or of basic macromolecules such as histones (Drew and McLaren, 1970). This leads successively to leakage of cell solutes, inhibition of growth, and finally to death of the cells. Nevertheless, solutes are constantly entering and leaving the living plant cells. The quantitative aspects of permeability must be understood in order to explain this apparent paradox. It should be emphasized that the *membrane* is *permeable* or *impermeable* to a substance. The *substance* to which the membrane is permeable or impermeable may be called a *permeating* or *nonpermeating* substance.

The rate at which solutes diffuse into a cell will depend not only on their concentrations inside and outside the cell but also on the permeability of the cell to them. Thus from Fick's law of diffusion it follows that if the concentration difference, area, and distance between the two measured concentrations are kept constant and at a standard temperature, D (Fick's constant) will be a measure of the rate of movement specific for each substance investigated. D is called the *specific diffusivity* of the substance. However, D will be constant for a specific substance only when it diffuses through a specific medium, for example, water. Yet this specific diffusivity of a substance in water will not give the rate of its diffusion into a living cell, even though the water content of the cell may be 85% to 95%. This is because the protoplast is surrounded by an apolar structure of low water content called a membrane. Diffusion through membranes depends on their permeability. In its simplest sense, *the permeability of a structure to a substance may be defined as the specific diffusivity of the substance in the structure.* Permeability therefore varies directly with the solubility of the substance in the structure (as well as inversely with its molecular size as in the case of diffusion in general). Most water-soluble substances have low solubilities in these apolar membranes, and therefore their rates of diffusion into or out of the cell are usually low. When the rate of diffusion of a substance into a cell is measured, the specific diffusivity (D) then becomes a measurement of the permeability of the cell or of its membranes to the substances. It is, then, commonly referred to as the permeability constant of the cell. Measurements are available for the permeability of the cell to many substances (Stadelmann, 1962).

The same principles apply to the solvent as to the solute, and the membrane therefore may have its permeability measured to water as well as to the solute dissolved in it. Yet the statement that the cell is semipermeable may seem to imply that it offers no resistance to the movement of water but completely prevents the movement of solutes. Both of these conclusions are incorrect. Thus if a collapsed but still living cell is allowed to expand in water, the cell wall expands ahead of the protoplasm (Fig. 7-1), showing that it is more permeable to the water than is the protoplast. Direct measurements on *Nitella* cells have indicated that the wall is 3.6 times as permeable to water as is the protoplasm (Kamiya et al., 1962).

That the living cells are nearly ideally semipermeable with respect to their solutes has been shown by Scholander et al. (1965, see Chapter 9). By placing all except the cut ends of shoots in a bomb and leading in nitrogen gas under pressure, they were able to force as much as 90% of the cell water out of the cut end. This exuded water was almost completely free of solutes because the freezing point was only –0.01° C. Since the freezing point of the cell sap is probably at least as low as –1° C, this means that less than 1% of the cell solutes were forced through the plasma membrane, and even some of these solutes must have come from the cell walls. The living cell is therefore more than 99% perfect as a semipermeable system for its own solution. A still more extreme example is provided by mangroves that normally grow in seawater. When a pressure of 40 atm was applied to the root system of a decapitated seedling potted in seawater, nearly fresh water flowed out of the cut stem surface (Scholander, 1968). Yet there are many substances to which the living cell is highly permeable, although most of these substances do not normally occur in cells. With respect to such substances the cell is differentially permeable rather than semipermeable.

There is direct experimental evidence for the conclusion that the cell's permeability is controlled at the protoplasmic surface. Since the cell wall is a hydrophilic gel, diffusion of solutes through it is relatively unimpeded. This is evident from the fact that hypertonic aqueous solutions of all nontoxic solutes cause a rapid separation of the protoplast from the cell wall in the process of plasmolysis of a living cell, as

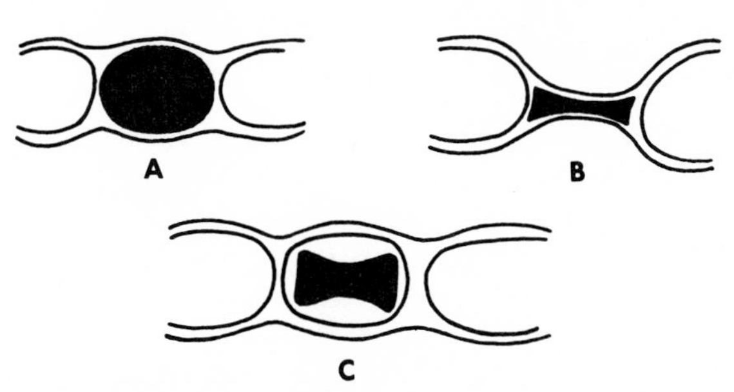

Fig. 7-1. Expansion of a collapsed (i.e., dried) living cell: **A,** in water; **B,** in air; **C,** immediately after transfer from air to water. Final stage as in **A.** (Redrawn from Iljin, W. S. 1930. Protoplasma **10**:379-414.)

long as the protoplast is less permeable to the solute than to the water. Similarly, if dyes to which the protoplasm is normally impermeable (or nearly so) are injected into protoplasm, they diffuse freely and rapidly throughout the layer. Finally, there is nothing to prevent free diffusion of solutes throughout the cell sap once they have entered the vacuole. But the same dyes that diffuse readily throughout the protoplasm fail to leave it either by diffusing out through the wall or into the vacuole. Thus the only impediments to free diffusion throughout the cell are the two protoplasmic surfaces, the so-called plasma membranes. The inner one is sometimes called the *tonoplast,* and the outer one, the *plasmalemma.* Although the membranes are too thin to be seen under the optical microscope, they are both readily observed in electron micrographs (Fig. 2-6) and are about 7.5 nm thick (Sitte, 1958, see Chapter 2). These surface membranes differ from the main body of the protoplasm in their highly lipid (fatty) nature. As a result, they slow up and in many cases completely prevent the passage of solutes in either direction. It is thus the lipids of plasma membranes that are responsible for the semipermeability of the living cells. Differences between the permeabilities of the tonoplast and the plasmalemma have often been sought without success. When differences have been reported, the results are contradictory: a 30 times higher permeability of the plasmalemma to urea (Dainty and Ginzburg, 1964*b*) and a 100 times higher permeability of the tonoplast to ions (Dainty, 1968, see Chapter 2). The latter result is supported by Ozerskii (1970).

It has been suggested that the main body of the cytoplasm layer is also responsible for the semipermeability of the cell, but all the evidence is against this. A direct test of this suggestion has been made in the case of cells having all the mesoplasm in one small part of the cell, and only the plasmalemma and tonoplast in the remaining greater part. Cells of this kind had no measurable difference in permeability from normal cells (Url, 1971), showing that the permeability is controlled by the membranes and not by the mesoplasm.

Although the endoplasmic reticulum (ER) forms a system of internal membranes that traverse the cytoplasm, they do not actually compartmentalize the protoplasm. On the other hand, the organelles are themselves surrounded by differentially permeable membranes, but this will not affect movement around them through the main body of the cytoplasm (the groundplasm). Even bodies within an organelle may possess osmotic properties because of the possession of semipermeable membranes, for example, the grana of chloroplasts (Gross and Packer, 1967).

Electron micrographs of the plasma membranes of cells have led to the suggestion that the whole concept of cell permeability must be changed, mainly because these micrographs indicate the presence of visible "pores" in the membranes or connections from the ER to the membrane. It is suggested that substances may therefore enter the cell freely through these pores or through the ER. More recent observations, however, indicate that this earlier interpretation of the electron micrographs was erroneous, and that the ER does not connect with the plasma membrane (Frey-Wyssling and Mühlethaler, 1965, see Chapter 2). As to the structure of the membrane, this will be considered later.

QUANTITATIVE RELATIONS

In most cases only relative permeability is measured. This is done by using a single cell or kind of cell and a standard concentration gradient. In this way the area, concentration gradient, and membrane thickness in Fick's law are kept constant. Therefore:

$$D = K \frac{s}{t} \text{ or } D \propto \frac{s}{t}$$

where $K = \dfrac{x}{a(C_1 - C_2)}$

D = specific diffusivity of the substance in the cell, or the cell's permeability to the substance

s = amount of substance diffusing (moles)

t = time (sec)

Relative permeability is then measured simply by $\frac{s}{t}$, the amount of substance entering the cell per unit time (moles/sec). Although such relative values are adequate for some purposes, absolute measurements of cell permeability are becoming more common. This requires measurements of the surface area of the cell. When this is known, the absolute permeability (cm/sec) can be obtained and becomes the diffusion coefficient of the cell membrane. The permeability coefficient of a membrane has been defined as the diffusion coefficient per unit membrane thickness (Stein, 1957); but it is not expressed on this basis, since the membrane thickness is not normally measured.

There are three main methods in common use for measuring cell permeability:

1. The deplasmolysis method. The cells are plasmolyzed in a hypertonic solution of the substance. If the solute penetrates, deplasmolysis occurs at a rate proportional to the permeability of the cell to the substance, provided that the other diffusion factors in Fick's equation are kept constant. This method is valid only for substances to which the cell is relatively impermeable (less than 1/100 the permeability of the cell to water). Other sources of error in this method are said to lead to an overestimate of permeability by 1½ to 2 times (Stein, 1967).
2. The chemical analysis method. The cells are immersed in a solution of the substance, and after a standard time the cell sap is removed and analyzed for the solute. This method is most easily used in the case of large-celled algae, (e.g., species of *Chara, Nitella,* and *Valonia*), which contain as much as a milliliter or more of sap per cell. This sap can be removed for analysis by means of a hypodermic needle or syringe.
3. The isotope method (Brooks, 1951). If the substance used in method 2 is radioactive, the sensitivity of the method is increased by several orders of magnitude, and it has even been possible to determine the rates of penetration into the protoplasm and vacuole separately. Deuterated and tritiated water (DHO and THO) have been used to measure water permeability.

Other methods have been used for specific cells and specific substances, for example, the betacyanin efflux method for the cells of red beetroots. It is still applicable in some cases (Siegel, 1970).

In the case of water a distinction has been made between the values obtained by methods 1 and 3. Method 1 measures hydraulic conductivity or osmotic permeability; method 3 measures diffusional permeability. Theoretically, the two should be identical, and direct measurements with artificial phospholipid membranes show that they are (Cass and Finkelstein, 1967). In the case of cells or tissues, however, differences between the two values have been obtained. However, the difference probably results from experimental error, and diffusional permeability values are commonly a gross underestimate (Kohn and Dainty, 1966). When the complicating factors are taken into account, measurements have, indeed, shown that the osmotic and diffusional permeability constants of the cell are identical (Gutknecht, 1967).

By means of such methods it can be shown that gases penetrate living cells freely, and all small molecules penetrate rapidly. This holds true as long as the molecular weight is not greater than 50 to 60 (e.g., water, ethyl alcohol, ethylene glycol). Electrolytes appear to be exceptions. Although in many cases their molecules are small, they penetrate slowly or not at all. But it must be remembered that electrolytes dissociate into ions that are highly hydrated and therefore form rather large particles (larger than the undissociated molecules), since each ion must diffuse together with its shell of water molecules. The charge on the ions may perhaps also impede penetration. As a result, weak acids and bases penetrate more rapidly than strong acids and bases because the former consist mainly of undissociated and therefore unhydrated (and uncharged) molecules. The solubility of the undissociated electrolytes in the lipid membrane is therefore greater than that of the ions.

In the case of nonelectrolytes each substance penetrates at a rate essentially independent of the presence of other substances. In the case of electrolytes this is far from true. Monovalent cations penetrate much more rapidly from single salt solutions than when a salt of a divalent cation is also present. In fact, single salt solutions may be toxic, although the two-salt solutions are harmless. Thus a solution of potassium chloride or sodium chloride (or other salts of potassium and sodium) may cause swelling of the cytoplasm after rapid penetration, accompanied by vacuole contraction because of the transfer of water from the vacuole to the swelling cytoplasm (Fig. 7-2). This may eventually lead to injury or death. However, if one tenth as much calcium chloride (on a molar basis) is added to the potassium chloride or sodium chloride solution, no such extreme swelling occurs, and the cells may remain alive for days. This is because the Ca^{++} reduces or prevents the penetration of the K^+ or Na^+. Such effects of ions on each other are known as *antagonism.* The nontoxic solution of the two salts is called a *balanced solution.* It is interesting to note that potassium chloride and sodium chloride penetrate the cytoplasm rather readily in single salt solutions but do not seem to enter the vacuole as readily. This is sometimes taken as evidence of a difference in permeability between the outer and inner plasma membranes, but it may simply be caused by the great increase in water imbibed by the protoplasmic colloids. Conversely, it has also been suggested (Collander, 1959) that the plasma membrane is completely impermeable to the ions. From this view, ion antagonism is not a lowering of the normal permeability of the cell but is a protection of the membrane from injury by the unbalanced single salt solution, which penetrates only after injuring the membrane.

Ion antagonism can be explained on the basis of the colloidal properties of the membrane (Booij and de Jong, 1954, see Chapter 5). Ion antagonism occurs between two cations if the concentration necessary to reverse the negative charge of the colloid is more than 10 times as high for one ion as for the other. This antagonism between two ions may occur for one biological colloid and not for another. Phospholipids show the greatest spread of activity between ions, and therefore antagonism is well developed in the case of these substances. They are the only biological colloids in which antagonism is readily observed between Na^+ and Ca^{++}. The ex-

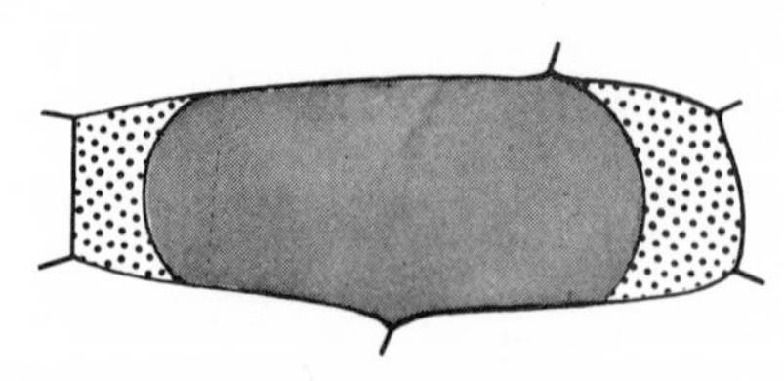

Fig. 7-2. Penetration of potassium into cytoplasm indicated by *vacuole contraction.* This is caused by the increased hydrophily of the cytoplasm, causing movement of water from the vacuole (normally almost filling the cell, see Fig. 2-1) to the cytoplasm (normally an almost invisibly thin layer).

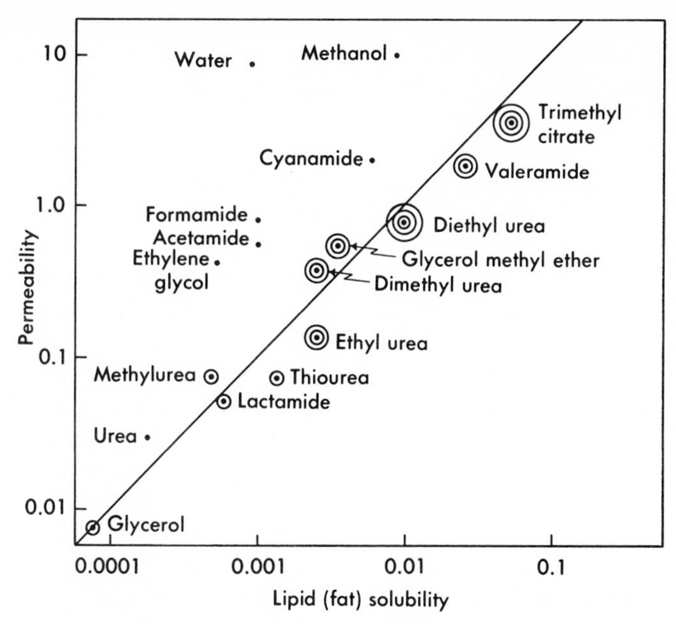

Fig. 7-3. Permeability of *Chara* cells to various organic nonelectrolytes. Molecular size is indicated by the number of circles. (From Collander, R. 1949. Physiol. Plant. **2**:300-311.)

istence of such antagonism is evidence that cell permeability is controlled by phospholipids. There are also other reasons why phospholipids are ideal compounds for biological membranes. They consist of both apolar (lipid-soluble) and polar (water-soluble) groups in the same molecule. Surface-active molecules of this type can form structures of colloidal size called micelles, in which the lipid groups form a lipid phase, separated from the surrounding aqueous medium by the water-soluble groups. This applies to both plasma membranes.

The permeability of cells to large molecules (i.e., with molecular weights above 50 to 60) varies from the highest permeability (nearly instantaneous penetration) to zero permeability, depending on the substance. Those large molecules that are lipid (fat) soluble penetrate rapidly, and the more lipid soluble they are the more rapid is the penetration (Fig. 7-3). This relation explains the accumulation of acids in the vacuole. Those commonly found in the vacuole (oxalic, malic, and citric acids) are lipid insoluble and therefore cannot leak back into the cytoplasm (Collander, 1959). Lipid-soluble acids (acetic, lactic, pyruvic acids) are never found in the vacuole. Even the same substance will penetrate more rapidly under conditions that increase its lipid solubility. The permeability of cells to organic acids, for instance, increases with a decrease in pH because of the decrease in dissociation of the acid and the consequent increase in concentration of the lipid-soluble molecular form.

MECHANISM OF CONTROL BY MEMBRANES

These results with large molecules, taken in conjunction with the aforementioned rapid penetration of small molecules that are not lipid-soluble, have led to the suggestion that the semipermeable membrane is a lipid sieve; that is, it consists of relatively large fatty particles separated from each other by much smaller aqueous pores (Fig. 7-4). According to the *lipid-sieve theory* of permeability, lipid-soluble molecules of any size pass readily through the

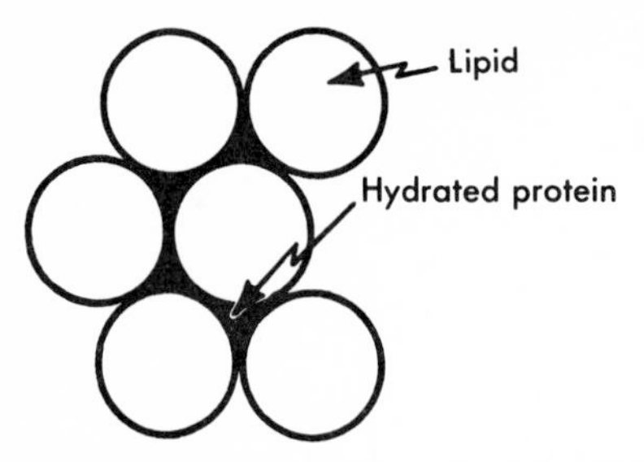

Fig. 7-4. Simplified diagram of the possible submicroscopic structure of the plasma membrane. (Redrawn from Scarth, G. W., and F. E. Lloyd. 1930. Elementary course in general physiology. John Wiley & Sons, Inc., New York.)

colloidal fatty particles, and only the smallest lipid-insoluble molecules pass through the small aqueous pores.

This concept is simplified and does not agree with the most popular diagram of the membrane, proposed by Davson and Danielli (1943). They proposed that the plasma membrane consists of a bimolecular leaflet of phospholipids. The nonpolar portions are pictured as oriented within the leaflet, perpendicular to the plane of the membrane; the polar portions must therefore point outward on both sides. The leaflet is supposed to be sandwiched between two layers of protein (Fig. 7-5). Although their concept was based primarily on permeability results, it has since been supported by electron micrographs and x-ray diffraction data. The membrane appears in electron micrographs as two dark (presumably protein) layers separated by a lighter (presumably lipid) layer, the whole structure being called the "unit membrane" and having a thickness of about 5 nm (Stein, 1967). In further support of their concept, Danielli (1966) has calculated that a bimolecular membrane has less surface-free energy than that of any other thickness and therefore is the most stable and is to be expected from the laws of thermodynamics.

Analyses of isolated membranes have been made in the case of animal cells (e.g., the "ghosts" of red blood cells) and these do agree with the concept of a lipid protein structure (Stein, 1967). The lipids consist largely of cholesterol and phospholipids in roughly equimolar amounts; the bulk of the phospholipids is lecithin (phosphatidylcholine) and cephalin (phosphatidylethanolamine and phosphatidylserine). The protein content is double the lipid by weight, and includes a number of different proteins. Unfortunately, the membranes of plant cells have not yet been isolated and analyzed. The available evidence indicates that they resemble those of animal cells, the lipids consisting of phosphatides and cholesterol.

The role of calcium in the membrane and the phenomenon of antagonism can be explained on the basis of the Davson-Danielli model. The membrane phospholipids resemble the proteins in possessing both + and − charges. The net charge is negative (in opposition to the + charge assigned to lipids in the Davson and Danielli model), leading (in vitro) to a strong electrostatic interaction only with positively charged proteins (e.g., the phospholipid

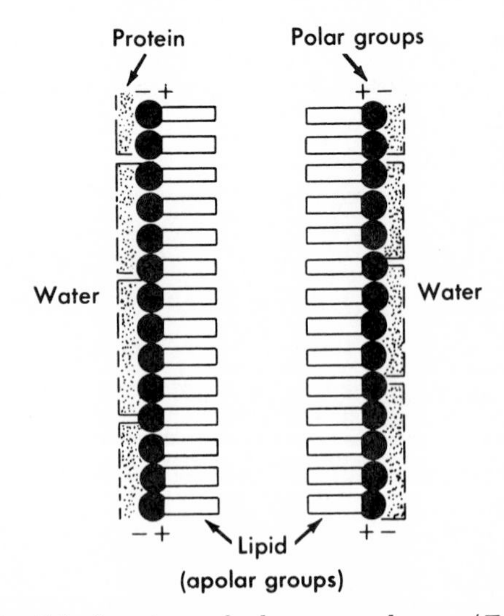

Fig. 7-5. Structure of plasma membrane. (From Davson, H., and J. F. Danielli. 1943. The permeability of natural membranes. Cambridge University Press, New York.)

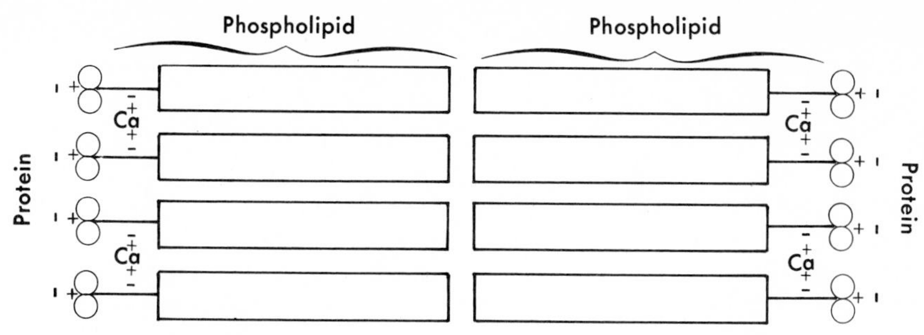

Fig. 7-6. Hypothetical role of Ca^{++} in the Davson and Danielli membrane.

cephalin and the protein prolamine, Joos and Carr, 1969). This interaction cannot, however, occur in the membrane, since the membrane proteins are not basic, and must, in fact, have a net negative charge at the pH of protoplasm which is undoubtedly above their isoelectric point (Chapters 5 and 8). The Ca^{++}, however, competes (in vitro) with the above basic proteins for the negatively charged phospholipid sites (Joos and Carr, 1969). Therefore in the normal membrane the negative charges of the phospholipids are presumably neutralized by Ca^{++} ions, leaving a net positive charge which is attracted by the net negative charge of the membrane proteins (Fig. 7-6). As mentioned above (Chapter 5), a small amount of Ca^{++} is sufficient to neutralize the negative charges of the phospholipids. It would require about 10 times as high a concentration of monovalent cations to neutralize the same amount of negative charge. Such a large number of cations, with their large hydration shells, would increase the water content of the lipid layer sufficiently to destroy its semipermeability. The Mg^{++} is perhaps able to replace the Ca^{++} in the case of some microorganisms (e.g., *Mycoplasma laidlawii,* Tillack et al., 1970), but due presumably to its larger hydration shell, is unable to replace Ca^{++} in the case of plant cells (Chapter 12).

It has even been possible to produce artificial membranes having the properties of the plasma membrane. Surprising success has been achieved in some cases. Thus a bimolecular lipid membrane consisting of phosphatidylcholine, cholesterol, and n-decane has been found to possess a permeability to water well within the range of most living cells (Hanoi and Haydon, 1966). Similar results have been obtained with artificial phospholipid membranes (Cass and Finkelstein, 1967). Despite this similarity to the plasma membrane of cells, no evidence could be obtained for aqueous pores in them. All this evidence in support of the Davson and Danielli concept of membrane structure appears to exclude the lipid-sieve theory. In further support of this conclusion, some evidence (Gutknecht, 1967), indicates that no actual "water-filled pores" occur in the membrane, as has been postulated on the basis of erroneous differences between osmotic and diffusional permeability to water (see earlier discussion).

It is difficult, however, to obtain direct evidence of the existence or nonexistence of pores in the protoplasmic membranes. Electron micrographs cannot detect pores any less than 20 Å in diameter, yet measurements of electroosmosis in Nitella are explainable by membrane pores 3 to 12 Å in diameter (Fensom and Wanless, 1967). Furthermore, there are objections to the Davson and Danielli concept and many facts with which it does not agree (Korn, 1966). One investigator, for instance, favors a single layer of lipoproteins, in which the bond between the lipid and protein molecules is hydrophobic rather than electrical (Benson, 1968). This model, however, is too extreme and would actually destroy the semipermeability of the membrane (Tanford, 1972). Nevertheless, it is now generally agreed that the Davson and Danielli concept is an oversimplification. The best evidence for this conclusion is provided by electron micrographs. They have actually revealed a surface sculpturing of the membrane (Frey-Wyssling and Mühlethaler, 1965, see Chapter 2), which may be interpreted according to the lipid-sieve theory. In fact, all biological membranes so far studied by means of the "freeze-etch" technique seem to possess particles (Staehlen, 1968).

Some attempts have been made to reconcile these two opposing concepts of the plasma membrane. Branton (1969) proposes a basic Davson-Danielli model with granules contained inside of an essentially continuous bilayer sandwich. Hendler (1971) concludes that the bulk of the new information favors the lipid bilayer. He also concludes, however, that it can be interrupted by protein pores and occasional cross bridges, and that it may contain various granular or micellar inclusions, although still providing an essentially continuous core. Although the lipid-protein interactions are primarily polar, he admits the possibility of limited areas of hydrophobic interactions and of transient modifications to accomplish certain specific functions.

On the basis of these analyses, a picture of the plasma membrane has begun to emerge as a composite of the two extreme concepts or as a reversible transition between the two. According to current concepts, the bilayer model of Davson and Danielli still serves as the best basis, although it must be modified (Stoeckenius and Engelman, 1969). The continuous bilayer of lipids probably has a disordered central hydrocarbon region. The protein is predominantly arranged on the surface of the lipid, but the amount and the arrangement may vary, and the membrane may be stabilized by both ionic and hydrophilic bonding between protein and lipid. Protein may or may not extend into and even through the lipid (Stein, 1967; Wright and Diamond, 1969). Such proteins would presumably form a bimolecular strand or "plug," with the hydrophobic sides of the protein molecules in contact with the adjacent membrane lipids and the hydrophilic sides adjacent to each other (Stein, 1967). These protein plugs would form a microgel of hydrated filaments, allowing water and small polar molecules to diffuse through the aqueous gel phase, but not permitting bulk flow. Lipid micelles might exist, stabilized by a layer of protein adsorbed to their surface, as is known to occur in the case of latex droplets. The adsorbed protein layer would be continuous with the protein on each side of the membrane. A good analogy to the cell membrane is the leaf surface, only 1% of which is occupied by stomata, yet these stomata may account for essentially all the gas flow into or out of the leaf (Chapter 11). Similarly, only a small part of the membrane need consist of protein passageways in order to account for the total membrane permeability to polar solutes.

In this way, the membrane as a whole could still conform to the Davson-Danielli model. Furthermore, this bilayer model may be an extreme form which is approached most closely in fully quiescent cells. The greater the activity of the cell, the nearer the membrane would be to a completely micellar form. All transitions between these two extremes may conceivably occur, including the three conformational states in the mitochondrial membrane described by Green and MacLennan (1969).

It must therefore be concluded that modern evidence and concepts of membrane structure do not rule out the lipid-sieve theory. This theory has, indeed, proved to be a useful explanation of permeability phenomena, and the relative rates of penetration of different substances can usually be explained in this way. But there are exceptions. Urea (molecular weight 60, lipid insoluble) penetrates some cells more rapidly than does glycerol (molecular weight 92, lipid insoluble) in accordance with theory. Yet there are cells in which the relation is reversed. Such cells usually have low permeabilities to other amides beside urea and are therefore called amidophobic. Differences in relative permeabilities to these two substances may even occur in the same cell at different times (Mayer, 1965).

Modern workers have ascribed many other functions to the plasma membrane beside the control of cell permeability. It is believed to function in protein synthesis, energy transfer, active transport of substances (Chapter 8), and sometimes in other ways (Korn, 1966). Such properties are possible only if proteins and other substances are included in the term *membrane*. The lipids alone could not account for these properties. Specific enzymes have, indeed, been found in isolated membranes—e.g., a Mg^{++}-dependent adenosine triphosphatase enzyme in the plasma membrane of yeast cells (Nurminen et al., 1970).

Many substances when applied to cells can greatly alter permeability. Some of these effects are readily explainable by the lipid-sieve theory. Anesthetics such as ether and chloroform are lipid solvents and therefore would be expected to alter cell permeability because of their solubility in the lipid plasma membrane. This would increase the lipid phase and presumably decrease the cell's permeability to polar substances. If sufficient anesthetic is available to dissolve away some of the lipid membrane, the effect would be the opposite. Thus such substances may be expected to produce opposite effects depending on the concentrations used. On the other hand, azide (a respiratory inhibitor) would not be expected to affect permeability, yet it decreases cell permeability to glycerol and sucrose (Burström, 1962).

Cells of different plants differ in permeability to a substance. Thus the values for permeability to water usually range from 1×10^{-4} to 15×10^{-4} cm/sec in different cells (Stadelmann, 1963). Even the permeability of a specific cell to a specific substance may not be constant. Besides changes induced by the ions themselves (antagonism, see earlier discussion), changes may occur, for instance, with the season. The permeability to water increases during fall and winter in many plants (Levitt, 1956). However, this is true only of polar substances, which, according to the lipid-sieve theory, "leak" through the smaller aqueous phase of the membrane, the proportion of which is presumably readily altered. Since the plasma membrane is made up largely of lipids, even doubling or trebling the aqueous phase will not change the lipid proportion appreciably, and the cell will be highly permeable to lipid-soluble substances under any conditions.

The fact that the cell is differentially rather than perfectly semipermeable has been clearly shown by the surprisingly large amounts of substances that can be leached from living leaves, both naturally by rain and artificially. Some of this material, of course, comes from dead cells, and the amount leached from the living cells is probably less than 1% of the total. As mentioned earlier, pressure can also remove small amounts of substances.

QUESTIONS

1. What is the relation between diffusion and permeability?
2. Is the cell perfectly semipermeable?
3. Is the cell wall differentially permeable?

4. Is the protoplasm as freely permeable to water as is the wall?
5. Where is the property of semipermeability located?
6. What is the name of the structure involved?
7. What is the structure's chemical nature?
8. How permeable is the cell to gases?
9. How permeable is the cell to small molecules?
10. How can you prove that a solute does enter a cell?
11. Is the cell equally permeable to all substances of small molecular size?
12. How would the permeability to ions relate to the lyotropic series?
13. Does the rate of penetration of a substances depend on the presence or absence of other substances ?
14. What is antagonism?
15. What is a balanced solution?
16. Is the cell impermeable to all large molecules?
17. What is the name of the theory that attempts to explain permeability to any substance?
18. Is the permeability of a specific cell to a specific substance constant? Explain.

SPECIFIC REFERENCES

Benson, A. A. 1968. The cell membrane: a lipoprotein monolayer. p. 190-202. *In* L. Bolis and B. A. Pethica. Membrane models and the formation of biological membranes. North-Holland Publishing Co., Amsterdam.

Branton, D. 1969. Membrane structure. Ann. Rev. Plant Physiol. **20**:209-238.

Brooks, S. C. 1951. Penetration of radioactive isotopes, P^{32}, Na^{24}, and K^{42} into *Nitella.* J. Cell. Comp. Physiol. **38**:83-93.

Burström, H. 1962. Influence of azide on the permeability of *Rhoeo* cells. Indian J. Plant Physiol. **5**:88-96.

Cass, A., and A. Finkelstein. 1967. Water permeability of thin lipid membranes. J. Gen. Physiol. **50**:1765-1784.

Dainty, J., and B. Z. Ginzburg. 1964*a*. The measurement of hydraulic conductivity (osmotic permeability to water) of internodal characean cells by means of transcellular osmosis. Biochim. Biophys. Acta **79**:102-111.

Dainty, J., and B. Z. Ginzburg. 1964*b*. The permeability of the cell membranes of *Nitella translucens* to urea, and the effect of high concentrations of sucrose on this permeability. Biochim. Biophys. Acta **79**:112-121.

Danielli, J. F. 1966. On the thickness of lipid membranes. J. Theor. Biol. **12**:439-441.

Drew, M. C., and A. D. McLaren. 1970. The effect of histones and other basic macromolecules on cell permeability and elongation of barley roots. Physiol. Plant. **23**:544-560.

Fensom, D. S., and I. R. Wanless. 1967. Further studies of electroosmosis in *Nitella* in relation to pores in membranes. J. Exp. Bot. **18**:563-577.

Green, D. E. and D. H. MacLennan. 1969. Structure and function of the mitochondrial cristal membrane. Bio Science **19**:213-222.

Gross, E. L., and L. Packer, 1967. Ion transport and conformational changes in spinach chloroplast grana. Arch. Biochem. **121**:779-789.

Gutknecht, J. 1967. Membranes of *Valonia ventricosa;* apparent absence of water-filled pores. Science **158**:787-788.

Hanoi, T., and D. H. Haydon. 1966. The permeability to water of bimolecular lipid membranes. J. Theor. Biol. **11**:370-382.

Hendler, R. W. 1971. Biological membrane ultrastructure. Physiol. Rev. **51**:66-97.

Iljin, W. S. 1930. Die Ursachen der Resistenz von Pflanzenzellen gegen Austrocknen. Protoplasma **10**:379-414.

Joos, R. W., and C. W. Carr. 1969. The binding of calcium to phospholipid-protein complexes. Proc. Soc. Exp. Biol. Med. **132**:865-870.

Kamiya, N., M. Tazawa, and T. Takata. 1962. Water permeability of the cell wall in *Nitella.* Plant Cell Physiol. **3**:285-292.

Kohn, P. G., and J. Dainty. 1966. The measurement of permeability to water in disks of storage tissue. J. Exp. Bot. **17**:809-821.

Korn, E. D. 1966. Structure of biological membranes. Science **153**:1491-1498.

Levitt, J. 1956. The hardiness of plants. Academic Press, Inc., New York.

Levy, M., and M. T. Sauner. 1967. Composition en phospholipides des membranes interne et externe des mitochondries. Compt. Rend. Séances Soc. Biol. (Paris) **161**:277-279.

Lode, A., and T. A. Pedersen. 1970. Fatty acid induced leaking of organic compounds from *Boletus variegatus.* Physiol. Plant **23**:715-727.

Mayer, E. 1965. On pore permeability of protoplasm. Protoplasma **60**:159-161.

Nurminen, T., E. Oura, and H. Suomalainen. 1970. The enzymic composition of the isolated cell wall and plasma membrane of baker's yeast. Biochem. J. **116**:61-69.

Ozerskii, M. I. 1970. Correlation between the

values of ionic currents through plasmalemma and tonoplast of Charophyta. Biofizika **15**:466-472.

Scholander, P. F. 1968. How mangroves desalinate seawater. Physiol. Plant. **21**:251-261.

Siegel, S. M. 1970. Further studies on regulation of betacyanin efflux from beetroot tissue: Caion-reversible effects of hydrochloric acid and ammonia water. Physiol. Plant. **23**:251-257.

Stadelmann, E. J. 1963. Comparison and calculation of permeability constant of water. [Transl. title.] Protoplasma **57**:660-718.

Staehlen, L. A. 1968. The interpretation of freeze-etched artificial and biological membranes. J. Ultrastruct. Res. **22**:326-347.

Stein, W. D. 1967. The movement of molecules across cell membranes. Academic Press, Inc., New York.

Stoeckenius, W. and D. M. Engelman. 1969. Current models for the structure of biological membranes. Cell Biol. **42**:613-646.

Tanford, C. 1972. Hydrophobic free energy, micelle formation, and the association of proteins with amphiphiles. J. Mol. Biol. **67**:59-74.

Tillack, T. W., R. Carter, and S. Razin. 1970. Native and reformed *Mycoplasma laidlawii* membranes compared by freeze-etching. Biochim. Biophys. Acta **219**:123-130.

Url, W. G. 1971. The site of penetration resistance to water in plant protoplasts. Protoplasma **72**: 427-447.

Wright, E. M., and J. M. Diamond. 1969. Patterns of non-electrolyte permeability. Proc. Roy. Soc. (London) **B172**:227-271.

GENERAL REFERENCES

Collander, R. 1949. The permeability of plant protoplasts to small molecules. Physiol. Plant. **2**: 300-311.

Collander, R. 1959. Cell membranes; their resistance to penetration and their capacity for transport, p. 3-104. *In* F. C. Steward. Plant physiology. Vol. II. Academic Press, Inc., New York.

Davson, H., and J. F. Danielli. 1943. The permeability of natural membranes. Cambridge University Press, New York.

Kavanau, J. L. 1965. Structure and function in biological membranes. Holden-Day, Inc., San Francisco.

Scarth, G. W., and F. E. Lloyd. 1930. Elementary course in general physiology. John Wiley & Sons, Inc., New York.

Stadelmann, E. J. 1962. Permeability, p. 493-528. *In* R. A. Lewin. Physiology and biochemistry of algae. Academic Press, Inc., New York.

Strugger, S. 1949. Praktikum der Zell- und Gewebsphysiologie der Pflanze. Springer-Verlag, Berlin.

8 Absorption

Solute absorption
Passive absorption
Active absorption
Water absorption
Elimination
Recretion
Secretion and excretion

With the help of the physical principles of diffusion, it should now be possible to attack the transfer of substances in the plant. There are three links in the chain of movement:

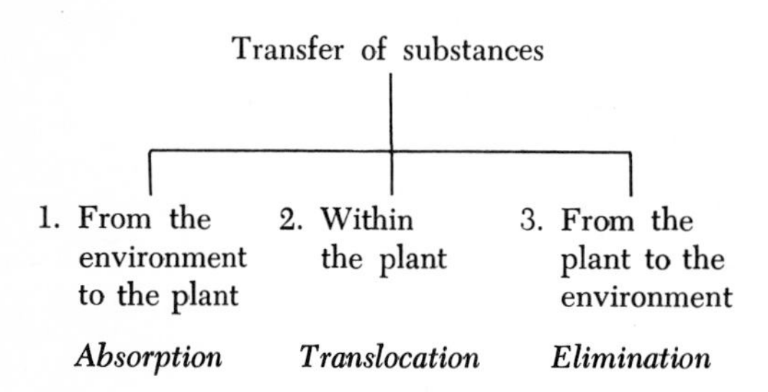

The first of these, absorption, is the entrance of substances into the plant. It is convenient to exclude gas absorption since the mechanism is the same as for gas elimination (Chapter 11).

SOLUTE ABSORPTION

Permeability, by definition, controls entrance into any system only if it is caused by diffusion. Thus differential permeability of the cell completely prevents some substances from diffusing in, whereas it permits others to enter at rates that depend on the permeability of the cell to them. This is true not only in the laboratory but under natural conditions as well. Weed killers, for instance, when sprayed onto plants will enter the cells and produce the damage much more effectively if they are lipid soluble. This is why the lipid-soluble ester is often more effective than the lipid-insoluble salt of the weed killer.

However, substances that cannot diffuse into the cell may also enter. For example, as seen above, when sections of plant tissue are immersed in solutions of sugar or $CaCl_2$, these solutes are unable to enter the cells. Yet both of these substances may be absorbed if supplied to the root system of a plant. Some large protein molecules (e.g., ribonuclease) have been shown to enter living cells, perhaps due to a preliminary conversion to lipoprotein. Even in such cases, size may still be important. The protein lysozyme, when labeled with fluorescein, can be seen to penetrate root cells, complexing with nucleoli; but the much larger molecule ferritin is confined to the cell walls (Seear et al., 1968). In some cases, however, microscopically visible particles that are too large to diffuse and are lipid insoluble may be coated by lipid, and if brought in contact with the protoplast surface of some plant cells, may then pass through it into the protoplasm. Such plant cells resemble animal cells that ingest (eat) particles by *phagocytosis*. Recent evidence indicates that certain plant cells may ingest solution by a similar method called *pinocytosis* ("drinking")—an infolding of the membrane until it encircles the liquid by pinching off the fold and reconstituting the membrane (Frey-Wyssling and Mühlethaler, 1965, see Chapter 2).

When absorption is simply caused by diffusion, the cell does not participate actively in the process, which is then called *passive absorption.* Phagocytosis and pinocytosis differ from diffusion by the active participation of the cell in the absorption process; that is, the cell uses up energy and does work. Under normal conditions, however, the plant protoplast is surrounded by a cell wall that does not permit visible particles to come in contact with the protoplast surface; nor would it favor pinocytosis, which has been shown only in the case of free tapetal cells suspended in natural fluid and free protoplasts devoid of cell walls (Cocking, 1966). Polystyrene latex particles can be taken up pinocytotically by free protoplasts of tomato fruit (Mayo and Cocking, 1969); but they undoubtedly cannot be taken up through the polar, water-saturated cell wall. Nevertheless, it has been suggested that ATP, adenosine, and ribonuclease enter radish root hair cells as a result of pinocytosis, and are therefore located in lysosome-like bodies (Salyaev and Romanenko, 1970). In support of such proposals, the plasma membrane in immature cells may be irregular in contour, some of the irregularities becoming conspicuous folds that continue to enlarge into the protoplasm (Mahlberg et al., 1971). Despite these intriguing observations, the importance of pinocytosis in absorption by plant cells remains to be proved. There are, however, other methods by means of which the cell can actively control the absorption process. The main problem, then, is to find out what portion of the absorption of free molecules and ions is passive and what portion is active.

Passive absorption

If passive absorption is a simple diffusion process, the maximum absorption of a substance will be reached when its concentrations inside and outside the cell are equal. Yet substances are commonly "accumulated" by plants as a result of passive absorption; that is, they attain a higher concentration inside than outside the cell. Furthermore, some substances are accumulated in the sense of attaining a larger

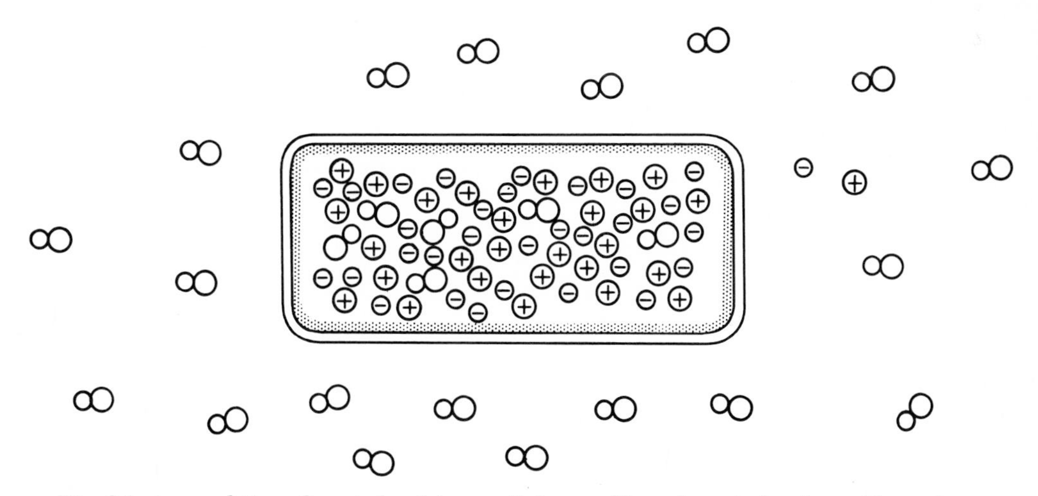

Fig. 8-1. Accumulation of neutral red by a cell from a dilute dye solution (e.g., 25 ppm) resulting from passive absorption in the molecular (⊖⊕-shaped ○○) form followed by dissociation (⊖⊕). Accumulation continues until the concentration of *undissociated molecules* inside the cell equals that outside it. At this point, total dye (dissociated plus undissociated) is many times as concentrated inside as outside the cell.

quantity per unit cell volume, even if not accumulated in the dissolved state. Passive absorption may result in both of these kinds of accumulation by one of the following three methods.

Entering molecules ionize. The gradient of molecular concentration can be maintained by the ionization of entering molecules. If the membrane is impermeable to the ions, absorption of the molecules will continue without loss of the ions, which will then accumulate. The end result will be an accumulation of the substance inside the cell, as in vital staining (Fig. 8-1). The cell is permeable to the neutral red (NROH) molecule because it is a lipid-soluble substance, but is impermeable to the ion (NR^+), which is lipid insoluble and large (molecular weight 289). The dye molecule is essentially undissociated at pH 8 and above, mainly dissociated at pH 5.5 and below. The two forms are in approximately equal concentrations at pH 7. If the pH of the staining solution is about 7 to 8, the dye molecules continue to enter the vacuole and immediately on entering, dissociate because of the acidity of the vacuole. This maintains the molecular concentration gradient, and the dye accumulates as the ion, which cannot leak out. Accumulation stops when the concentration of undissociated molecules is the same inside and outside the vacuole. In the case of the dye fluorescein diacetate, the cell apparently participates in the process metabolically. Again, the nonpolar dye ester enters the living cell by dissolving in the plasma membrane. Once in the protoplasm, however, it is hydrolyzed by the enzyme esterase to give the polar product fluorescein, which is retained in the protoplasm by the apolar plasma membrane (Heslop-Harrison and Heslop-Harrison, 1970, see Chapter 2).

Entering solute particles become insoluble either by precipitation or by adsorption on colloidal particles. The concentration gradient in this way will be maintained, and the substance will accumulate in the undissolved form. This occurs when cells containing colloidal tannin accumulate a vital stain in the vacuole; accumulation is then more intense than by the first method.

Entering ions are accumulated electrostatically. As just indicated, absorption is passive if the cell does not participate actively in the process. The substance must therefore move in passively if (1) it is at a higher level of free energy outside than inside the cell, and (2) the cell is permeable to it. The rate of absorption will be quantitatively related to these two factors. If the cell is sufficiently permeable to the substance, only the free energy levels need be considered. In the case of a substance in the molecular state, its free energy is caused by the thermal agitation of the molecules and is therefore given quantitatively by the chemical potential of the substance. The rate of movement (or diffusion) is proportional to the chemical potential gradient. But ions possess electrical as well as thermal energy. As a result, there is an electrical potential difference between the two sides of a cell membrane, usually of 10 to 100 mv, with the inside of the cell negative to the outside (Curran, 1963). This is caused by the high electrical resistance of the lipid plasma membrane. If the membrane is 100 Å thick, the potential difference across it must be 10^4 to 10^5 v/cm, an extremely high value. Thus there are two "forces" tending to move an ion (Dainty, 1969): (1) that resulting from its chemical potential gradient and (2) that resulting from its electrical potential gradient. Their movement will therefore depend on the algebraic sum of these two quantities, just as the movement of water molecules was seen to depend on the algebraic sum of osmotic potential and pressure. The algebraic sum of the forces acting on ions is called the electrochemical potential gradient:

Electrochemical potential gradient = Chemical potential gradient + Electrical potential gradient

Water	Ions
Water potential inside cell = water potential outside cell	Electrochemical potential inside cell = electrochemical potential outside cell
$\psi_i = \psi_o$	$\bar{\mu}_i = \bar{\mu}_o$
π_i may be greater or less than π_o	μ_i may be greater or less than μ_o
if P_i is less or greater than P_o	if E_i is less or greater than E_o
where ψ = water potential π = osmotic potential P = wall pressure	$\bar{\mu}$ = electrochemical potential μ = chemical potential E = electric potential

Fig. 8-2. Conditions for diffusion equilibrium.

It is therefore possible for a cell to be in equilibrium with its surrounding solution, although the ion concentrations (or chemical potentials) differ. A species of cation, for instance, can be at equilibrium, although it is at a higher concentration inside than outside the cell because the cell contents have a negative electric potential relative to the external solution. But the sum total of these two potentials for the ion—its electrochemical potential—must be the same inside and outside the cell at equilibrium. In other words, the outwardly directed concentration gradient must be equal to the inwardly directed electrical gradient if electrochemical equilibrium prevails. This is analogous to the condition for the cell's water (Fig. 8-2). At equilibrium the cell's osmotic potential may be lower (or its concentration may be greater) than that of the surrounding solution, provided that it is counterbalanced by wall pressure; but the water potential must be the same inside and outside the cell.

A chemical potential gradient between a cell and its surrounding solution is readily produced by a concentration difference between the two. The electrical potential gradient is somewhat more complicated (Dainty, 1969), and only the simplest aspects will be considered here. When a cell is immersed in a solution of electrolyte, each ion tends to move according to its own concentration gradient. However, because of the electrical attraction between oppositely charged ions, they must enter either by (1) exchanging for others of the same sign or else (2) moving in pairs (Fig. 8-3). The large, negatively charged protein ions (anions) cannot pass through the plasma membrane, whereas the inorganic ions, both of the same sign (anions) and of the opposite sign (cations), can pass, although slowly. The protein anions inside the cell must be electrically balanced by enough cations to supply an equal number of opposite charges. If now the cell is bathed in a solution containing the same kind and concentration of cation as is inside the cell, plus an anion not found inside the cell (or present there in lower concentration), the new anion will tend to diffuse in. But electrical balance must be maintained because although the more mobile ion of a pair may move slightly ahead of the other, thus converting its thermal energy into a small electric potential *(diffusion potential),* it cannot move completely out of the sphere of influence of the other ion unless work is done on it. Such a separation would result in

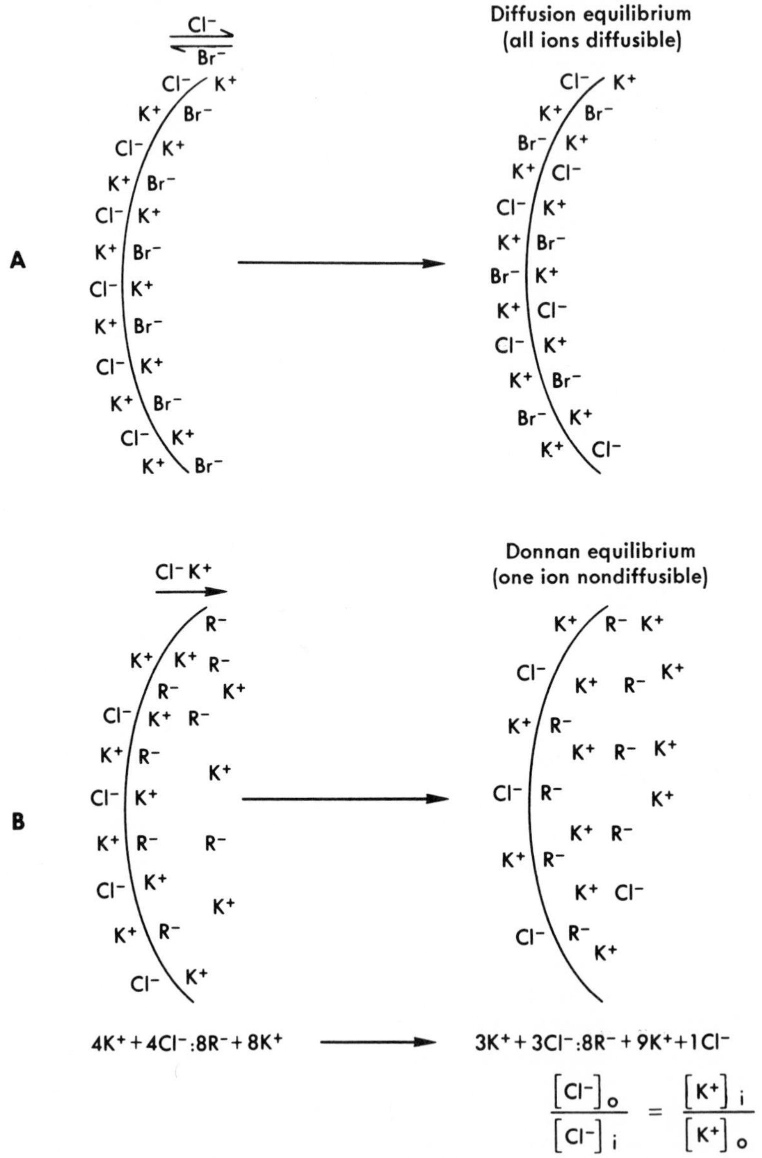

Fig. 8-3. Passive absorption of ions and Donnan equilibrium. **A,** Ion exchange. There is diffusion of Cl^- into the cell without disturbing the electrical neutrality because of a simultaneous and equal diffusion of Br^- out of the cell. **B,** Ion pair diffusion. Ion exchange is prevented because of the presence of nondiffusible ions (R^-) inside the cell. At equilibrium the ratio of Cl^- concentration outside to that inside the cell $\frac{[Cl^-]_o}{[Cl^-]_i}$ equals the ratio of K^+ concentration inside to that outside the cell $\frac{[K^+]_i}{[K^+]_o}$.

a net increase in free energy, which is an impossibility according to the second law of thermodynamics. Consequently, for each anion that diffuses in, a cation of equal charge tends to be dragged in against its concentration gradient. Simple diffusion equilibrium therefore cannot prevail, and the cation will reach a concentration inside the cell greater than that outside the cell. On the other hand, the anion diffusing into the cell will reach a concentration less than that outside the cell because of the equal number of ion pairs moving in and out of the cell. This electrochemical equilibrium, which is controlled by electrical as well as diffusion phenomena, is known as the *Donnan equilibrium.* If the cell is in a solution of a salt MA that dissociates into M^+ and A^-, the Donnan equilibrium can be represented quantitatively in simple form:

$$\frac{[M^+]_i}{[M^+]_o} = \frac{[A^-]_o}{[A^-]_i}$$

where $[M^+]_i$ = concentration of diffusible cation inside the cell
$[M^+]_o$ = concentration of diffusible cation outside the cell
$[A^-]$ refers similarly to the diffusible anion

An accumulation of M^+ inside the cell can occur only because of the presence of a nondiffusible ion of opposite charge (R^- in Fig. 8-3), and it must be accompanied by a corresponding negative accumulation (lower concentration) of anion. When the Donnan equilibrium prevails, complete electroneutrality on each side of the membrane is prevented by the indiffusible R^- which produces an almost infinitesimal excess of anions within the cell and of cations outside the cell. This produces a membrane (or Donnan) potential, the inside of the cell being negative to the outside. There are, however, a number of factors involved in this membrane potential.

The electrical potential gradient associated with accumulation by the Donnan equilibrium can be calculated from the Nernst equation, which in its simplest form is given as

$$E = -58 \log \frac{[M^+]_i}{[M^+]_o}$$

where E = the electric potential difference (mv) between the cell and its external solution. This equation states that when the cell is at electrochemical (or Donnan) equilibrium, an electric potential difference E exists between the inside and outside of the cell, tending to drive the cations in, which is just balanced by the chemical potential difference. If measurements of the actual electric potential difference E yield values that differ from the calculated value ($-58 \log \frac{[M^+]_i}{[M^+]_o}$), based on measurements of $[M^+]_i$ and $[M^+]_o$, then the cell is not at electrochemical equilibrium with respect to the specific ion M^+. Other ions present in the same solution can be treated in the same way (Higinbotham et al., 1967). It is therefore possible to determine from the preceding equation whether a specific ion is at electrochemical equilibrium inside and outside the cell.

Because of this Donnan equilibrium, K^+ ions may accumulate in algal cells to 30 times their concentration in the external solution, although at electrochemical equilibrium inside and outside the cells. (Dainty, 1969). The actual accumulation will, of course, vary with the cell and the external concentration.

Active absorption

Passive absorption is therefore unable to lead to an accumulation of solutes in the molecular state, since molecular diffusion can lead to no more than an equal concentration inside and outside the cell. As a result of the additional electrical energy of ions, it can, however, lead to their accumulation because of the presence of nondiffusible protein ions inside the cell. However, if the accumulation is greater than can be accounted for by the Nernst equation, the excess must be caused by an additional source of energy—specifically the metabolic energy of the cell. When this condition prevails, the absorption that occurs at the expense of metabolic energy is called active. Unfortunately, it is difficult (but not impossible—see following discussion) to apply the Nernst equation to higher plant cells, since this requires measurement of the electrical potential between the inside and the outside of a cell (Higinbotham et al., 1967). In practice the existence of active absorption for any one substance can be demonstrated by any one of these four criteria:

1. If the rate of absorption is too rapid to be explained by passive absorption, that is, by the permeability of the cell and the chemical or electrochemical potential gradient of the substance
2. If the steady state electrochemical (or in the case of uncharged particles, the chemical) potential is higher inside than outside the cell
3. If the absorption is quantitatively related to the expenditure of metabolic energy
4. If the *mechanism* of absorption can be demonstrated to depend on cell activity

The following discussion gives some of the evidence for all four of these criteria.

1. Since this kind of active absorption is not against a diffusion gradient but simply involves a more rapid rate of absorption than can be accounted for by diffusion, it has been called "facilitated diffusion" (Hogben and Adrian, 1963). Many substances are readily absorbed by living cells that are almost impermeable to them, according to standard tests—substances with molecular (or hydrated) weights greater than 50 to 60 but insoluble in lipids. This includes many of the substances most important to the plant, for example, sugars and many inorganic salts. That these substances do, indeed, fail to penetrate cells in sections of tissue is shown by the fact that one can plasmolyze the cells in solutions (e.g., of dextrose or calcium chloride), and no deplasmolysis will occur even over a period of several days. On the other hand, that they are absorbed by the living cells of the unsectioned plant can be shown by feeding them to the roots or even by immersing tissue slices (e.g., potato tuber, carrot root, etc.) in these solutions and bubbling air through the solutions. In periods of 24 to 48 hr considerable amounts will go into the living cells. This absorption is therefore something more than the result of diffusion and cell permeability.

2. Free ions are usually accumulated by living (and particularly by actively growing) cells until they are present in much higher concentrations in the cell sap than in the surrounding medium (Table 8-1). Although electrochemical potentials were not measured in these cases, the K^+ concentration inside the cell is much higher than the accumulation found by modern methods at electrochemical equilibrium, at least in the case of *Nitella* in pond water. Furthermore, both anions and cations are accumulated, whereas the nondiffusible proteins anions in the cell can account only for the accumulation of the oppositely charged cations. Finally, the Donnan equilibrium would lead to a greater accumulation of divalent and trivalent ions than of monovalent ions, whereas the opposite occurs. Consequently, at least some of the ions must be at higher electrochemical potentials inside than outside the cell. Direct measurements of electrochemical potential have shown that this does not necessarily hold for all ions and under all conditions. The general pattern for large algal cells is a lower electrochemical potential (e.c.p.) for Na^+ in the cytoplasm than in the external solution, and a higher e.c.p. for K^+ and Cl^- in the cytoplasm than outside. The relation between the cytoplasm and vacuole varies (Dainty, 1969). The differences across the tonoplast are generally, although not always, smaller than across the plasmalemma. It has therefore been

Table 8-1. Ratio of sap concentration to concentration in the surrounding medium*

Ion	*Valonia macrophysa* (in sea water)	*Nitella clavata* (in pond water)
Cl	1.03	100.5
Na	0.18	46.1
K	41.6	1065.1
Ca	Very small	13.2
Mg	Very small	10.5
SO_4	0	25.8

*Adapted from Osterhout, W. J. V. 1936. Bot. Rev. 2:283-315.

concluded that the cell has inwardly directed "pumps" for some ions and outwardly directed pumps for others. The latter would then lead to negative active absorption (active elimination). In the case of pea and oat roots, none of the measured ions (K^+, Na^+, Mg^{++}, Ca^{++}, NO_3^-, $H_2PO_4^-$, $SO_4^=$) was in equilibrium, and none appeared to move passively (Higinbotham et al., 1967; Higinbotham, 1970). As in the alga *Nitella*, both inwardly (for the anions) and outwardly (for Na^+, Mg^{++}, and Ca^{++}) directed pumps appeared to occur, although the possibility of passive movement by the divalent cations was not completely excluded. In the case of sunflower, the anions were found to be actively absorbed, the cations passively absorbed (Bowling, 1966). Similarly, on the basis of exudation experiments, Cl^-, NO_3^-, and K^+ were absorbed actively by sunflower roots; Na^+, Ca^{++}, and Mg^{++} passively (Sobey et al., 1970). Results support the concept that ATP (Chapter 13) is the energy source for ion transport in roots, and that an adenosine triphosphatase participates in the energy transfer (Fisher et al., 1970).

Animal physiologists use the term *active transport* for those cases where the ion is moved in against an electrochemical potential gradient (Hogben and Adrian, 1963). The plant physiologist has only recently (Etherton and Higinbotham, 1960) begun to measure this gradient because it is so difficult to do in cells of higher plants. In any case the preceding definition of active transport is too narrow for the plant physiologist, since the plant is normally not at equilibrium with its environment, nor is the *steady state* between active absorption and passive loss usually achieved. Under normal conditions of plant growth, ions can therefore be actively absorbed without necessarily attaining a higher electrochemical potential inside than outside the cells. Thus NO_3^- and NH_4^+ ions are metabolized inside the cell and cannot be expected to attain either equilibrium or the steady state with the external solution, even though absorbed actively. Ions that are not metabolized are translocated from the root to other parts of the plant, and therefore they, too, cannot possibly attain equilibrium or the steady state with the external medium. The plant physiologist must therefore usually determine whether absorption by the intact plant is active or passive by use of the remaining two criteria.

3. The main source of metabolic energy is the respiratory process. It has been shown conclusively that the rate of absorption is frequently proportional to the respiratory rate (Table 8-2). Respiratory inhibitors prevent this kind of absorption. In fact, if the cells are deprived of oxygen, not only are they unable to accumulate ions but they will actually lose much of what they had previously accumulated. This shows

Table 8-2. Absorption of potassium bromide from 0.00075 N solution by carrot discs*

Percent O_2	Relative respiration rate	Relative K absorption	Relative Br absorption
2.7	43	22	42
12.2	78	96	86
20.8	100	100	100
43.4	106	117	118

*Adapted from Steward, F. C., W. E. Berry, and T. C. Broyer. 1936. Ann. Bot. 50:345-366.

that respiratory energy is needed not only to accumulate the ions but also to maintain them in the cells. Thus the plasma membrane may be visualized as a microsieve through which the accumulated ions are constantly leaking out. Only if the living cell has a pumping system, capable of pumping the ions back into itself as rapidly as they are leaking out, can these high concentrations be maintained.

4. The final method depends on a knowledge of the mechanism of active absorption. If this mechanism could be shown to depend on metabolic activity, this would be final and definitive proof of active absorption. Many suggestions have been made of a "biological pump" that is able to force the ions into or out of the cell. Lundegårdh proposed the first such mechanism, but attempts to support it by direct experimental evidence have failed. It was then proposed that the ions must first be captured by "carriers" with different "sites," each of which can specifically capture a particular ion or group of ions. This would account for the selectivity of the process. By selectivity is meant the ability of the cell to absorb one ion in preference to another that is chemically closely related (e.g., K^+ as opposed to Na^+). Such selectivity cannot possibly be accounted for by passive processes.

Until recently, the carrier concept was purely hypothetical, since no substances with this property had been isolated. Several antibiotics (valinomycin, monactin, enniatin B, nigericin, etc.) have now been shown to act as ion carriers (Pressman et al., 1967; Läuger, 1972). They are depsipeptides (α-amino acids and α-hydroxy acids in alternating sequence). They all share a common property: they are macrocyclic molecules in which one side of the ring is hydrophilic, the other is strongly hydrophobic. They form complexes with alkali ions in organic solvents with a high degree of specificity. For instance, the stability constant of the K-complex of valinomycin is 10^4 times that of the Na-complex. When they contact an ion of suitable size and charge, their polar side becomes wrapped around the polar ion, and the antibiotic-ion complex then has a lipid-soluble apolar exterior which permits it to move through the membrane. At the inner surface of the membrane, the apolar exterior presumably remains in the lipid layer of the membrane, the polar interior unfolds due to attraction for the polar protoplasm, and the ion is released into the protoplasm. The whole process is rapid, and the carrier can shuttle back and forth between the external and internal surfaces of the membranes, carrying across the membrane ions to which the membrane is impermeable. Valinomycin was first shown to increase the permeability of mitochondria and erythrocytes to K^+. A single valinomycin molecule has been calculated to transport 10^4 K^+/sec across a membrane.

The above antibiotics, of course, have been obtained from fungi. No such carriers have yet been isolated from higher plants. The only possible candidate to date is an adenosine triphosphatase enzyme known to be associated with the membrane of root cells (Hansson and Kylin, 1969; Fisher et al., 1970). Since it is a protein, it resembles the above antibiotics in containing both hydrophilic and hydrophobic groups. Since it occurs in the lipid membrane, it must also be folded, like the antibiotics with its hydrophobic groups external to it and therefore in contact with the membrane lipids. Furthermore, this membrane-associated adenosine triphosphatase has been shown to act as a Na^+ efflux, K^+ influx (or a H^+ efflux, K^+ influx) pump in the presence of ATP, which supplies the energy for the process.

All the preceding four methods of investigation have pointed to the requirement of metabolic energy for the normal absorption of ions by cells. Furthermore, it has become obvious that active ion absorption is associated with protoplasmic membranes in general. Thus the plasma membranes (specifically the plasmalemma) lead to ion accumulation in the protoplasm as a whole; the mitochondrial membranes lead to ion accumulation in the mitochondrion; and the chloroplast membranes lead to ion accumulation in the chloroplast.

It follows that for cells to grow and multiply (and in many cases even to stay alive), they must possess the ability to absorb actively. Although most of the experimental work has used electrolytes, there is some evidence that this is also true of nonelectrolytes such as sugars (Street and Lowe, 1950). In the case of amino acids, the uptake of arginine by the mycelium of *Aspergillus nidulans* is severely restricted by the presence of lysine (Cybis and Weglenski, 1969). This has been interpreted as competition for the carrier ("permease"). On the other hand, the living cell cannot actively absorb substances that are highly lipid soluble, since it would require too much energy to do so; these substances would "leak" out through the plasma membrane so rapidly that it would be like keeping a large-pored sieve full by pumping the water back as fast as it leaked out (Fig. 8-4).

Attempts have been made to determine whether the foliar absorption of substances sprayed onto leaves may also involve active absorption. Some respiratory inhibitors apparently inhibit the process, though others do not (Sargent and Blackman, 1969). Finally, it must be remembered that the roots of many plants in their natural habitats are characteristically associated with mycorrhizae. These fungal growths must have pronounced effects on the absorption by roots of the host plant (Hacskaylo, 1972).

Although ions are normally absorbed actively, this does not eliminate passive processes. Cations are adsorbed passively on the root surface and may subsequently be transported actively across the plasma membrane. Similarly, the root contains "free space" into which solutes can diffuse readily. This space accounts for about 10% to 15% of the root volume (Ingelsten and Hylmö, 1961) and consists essentially of the cell walls (Pitman, 1965). Diffusion into this free space must precede active absorption at the protoplasmic surface. It is also probable that beside the active ab-

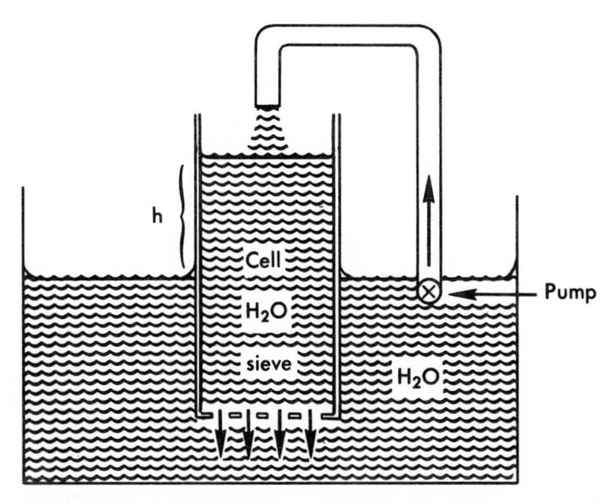

Fig. 8-4. Steady state maintenance of water at a higher free energy level (height **h**) by continuous pumping. The rate of energy expenditure varies directly with the permeability of the sieve.

sorption of ions, some passive absorption by the protoplasm and vacuole may also occur, particularly from high external concentrations. This question will be considered later (Chapter 11).

WATER ABSORPTION

Unlike solutes, absorption of water is purely a passive process, and the cell does not control it by direct expenditure of energy. This conclusion is based on both experimental results and theoretical considerations. Earlier experimental results led some investigators to conclude that water, like inorganic ions in general, may be actively absorbed. A reevaluation of these results, however, showed that they had been misinterpreted (Levitt, 1953), and later experiments by two of the earlier proponents of active water absorption were completely negative (Thimann and Samuel, 1955; Ordin and Bonner, 1956). In fact, there is no substantial evidence for active absorption of water in any plant or animal cell system other than in the case of the insect cuticle, which has a much lower permeability than plasma membranes. This conclusion has been supported by recent investigations which seemed, at first, to oppose it. Evidence for active water transport in corn roots (Ginsburg and Ginzburg, 1971*a*, 1971*b*) was followed by an analysis based on the thermodynamics of irreversible processes (Ginsburg, 1971), which concluded that the results do not necessitate a water pump.

The reason why the plant cell cannot absorb actively becomes clear from Fig. 8-4. The water potential in the cell can be kept above that of its environment only by the cell continuously pumping water into itself as rapidly as it leaks out through the semipermeable membrane. Since this membrane is highly permeable to water and the specific surface of the cell is considerable, large amounts of energy would have to be expended continuously by the cell. Calculations (Levitt, 1953) reveal that the cell does not release enough energy to maintain an appreciable increase in this water potential. A more sophisticated calculation on the basis of irreversible thermodynamics leads to the same conclusion (Simons, 1969).

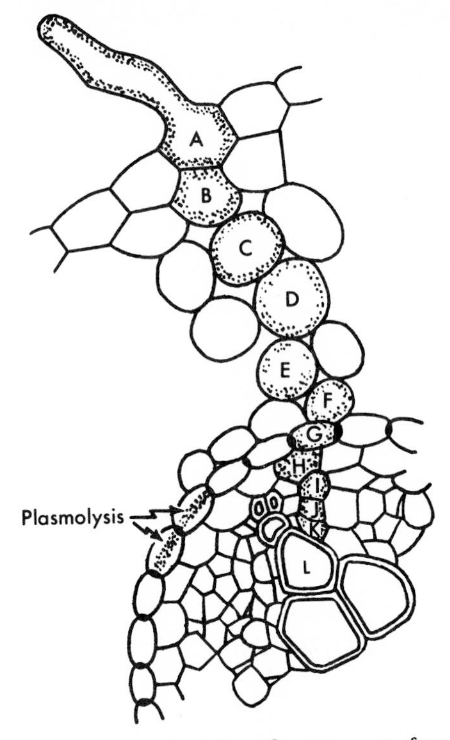

Fig. 8-5. Water potential gradient in roots from a root hair (cell **A**) to a vessel (cell **L**). Water enters by moving along this gradient. Plasmolysis of two endodermal cells is shown to illustrate the strong adhesion of the protoplasts to the Casparian strips. (Redrawn from Priestley, J. H. 1920. New Phytol. **19**:189-200.)

Unfortunately, there is some difference in terminology among investigators. "Active" and "passive" water absorption have been used in the past for water uptake resulting from root pressure and transpirational pull, respectively. These terms will not be adopted in this text since (1) it is believed that the same terminology should be used for solutes as for solvents and (2) "active" implies direct participation of metabolic energy.

As in the case of solutes, according to

the laws of diffusion, water should enter a cell from any medium in which the chemical potential of the water is higher than that in the cell. Thus in order for water to diffuse from the root medium through the root epidermis, cortex, endodermis, pericycle, and finally into the vessels, it must follow a water potential gradient (Fig. 8-5). The living cells would affect such water movement by maintaining this gradient. The net water potential gradient would be that between the vessels and the external medium. A hydrostatic (osmotically induced) pressure develops as a result of this difference and is known as the *root pressure* because all the living cells between the xylem vessels of the root and the soil solution act as a single semipermeable membrane. The root xylem is therefore an osmometer immersed in the soil solution, and the maximum osmotic pressure it can develop depends on the concentration of the vessel sap as well as that of the soil solution:

$$P_r = \pi_s - \pi_v$$

where P_r = maximum root pressure
π_s = osmotic potential of soil solution
π_v = osmotic potential of vessel sap

In most cases the vessel sap is dilute, so that even if the soil solution is pure water, pressures of 1 atm or less are common (Table 8-3). Values as high as 5 to 6 atm have, however, been recorded.

But how can this root pressure be maintained? Since the vessels are nonliving cells, they do not possess semipermeable membranes. It might therefore be expected that the solutes in the vessels would diffuse through the walls into the walls of adjacent cells, and from cell wall to cell wall until they leaked out of the roots into the surrounding medium. This leakage is believed to be prevented by a hollow cylinder of cells, one row thick, surrounding the central stele (in which the vessels are located) and known as the *endodermis* (Fig. 8-5, *G*). The endodermal cells have lateral walls that are impermeable to water because of impregnation with fatty substances. Sometimes the fatty substances form a ribbon around the cell wall known as the *Casparian strip.* These endodermal cells are alive. Therefore in order to pass through them, sap solutes would have to penetrate the semipermeable protoplasm, since they cannot leak through the lateral fatty cell wall unless they are fat soluble. Even when these cells are plasmolyzed, the protoplasm remains attached to the lateral walls (Fig. 8-5).

A second factor that would oppose maintenance of root pressure is dilution of the vessel sap by the water entering the ves-

Table 8-3. Osmotic pressure of tracheal (i.e., xylem) sap*

Species	Origin of sap	Osmotic pressure (atm)
Cotoneaster frigida	Centrifuged from stems	0.48-1.08
Cucurbita pepo	Stump exudate	1.9
Fagus sylvatica	Centrifuged from stems	0.26-1.23
Gossypium sp.	Stump exudate (low-salt plants)	0.92
	Stump exudate (high-salt plants)	3.00
Impatiens balsamina	Stump exudate	0.36
Lycopersicon esculentum	Stump exudate	1.5 -2.4
Salix babylonica	Centrifuged from stems	0.41-1.14
Ulmus procera (syn. *campestris*)	Centrifuged from stems	1.13-3.52
Xanthium strumarium	Stump exudate	0.67
Zea mays	Stump exudate	1.46

*From Altman, P. L., and D. S. Dittmer (Eds.). 1966. Federation of American Societies for Experimental Biology, Washington, D. C.

sels. This would have to be opposed by more rapid absorption of solutes than of water to ensure a higher concentration in the vessels than in the root medium; or there might also be an active movement of solutes (e.g., a secretion of salts or sugars) into the vessels from the adjacent living cells. Thus active absorption or active secretion of *solutes* must accompany absorption of water if root pressure is to be maintained. Experiments clearly demonstrate the need of oxygen for maximum water absorption by roots (Rosene, 1950). This is not surprising in view of the need for active absorption of solutes to maintain root pressure.

The movement of water into a cell may therefore be said to involve *osmotic work*. But it is important to understand the true meaning of this term in order to avoid the misconception that this implies work done on the water molecules by the plant. Osmotic work does not involve a "water pump." The work is performed by the plant in pumping *solute* molecules into itself. This lowers the free energy of the water molecules in the plant, leading to a passive diffusion, or osmosis, of water into the plant. The osmotic work is therefore equal to the work done by the water molecules in diffusing into the plant, and it cannot be greater than the work done by the plant in actively absorbing its solutes. The amount of energy involved in osmotic work is small—of the order of 1% of the total expended by the living cell (Robertson, 1941.)

Some aspects of root pressure remain to be explained. Although a water potential gradient does normally exist between the external medium and the vessel sap (i.e., the vessel sap does have a lower osmotic potential than the soil solution) and although a cell-to-cell gradient has been found from the epidermis to the endodermis, yet the gradient apparently reverses itself from the endodermis to the vessel sap. It must be realized, however, that all measurements of these quantities are probably meaningless, since the steady state between water absorption by roots and water loss by the plant as a whole is maintained in the intact plant under normal conditions. Any sectioning of the plant for experimental determinations destroys this steady state and therefore changes the conditions of the cells.

ELIMINATION

Although elimination is one of the three basic components of the overall transfer of substances, only the elimination of gases (Chapter 11) by the plant has received as much attention by the plant physiologist as the other two components. In an attempt to correct this neglect, Frey-Wyssling (1970) has distinguished three basic kinds of elimination: (1) *recretion*—the elimination of absorbed substances; (2) *secretion*—the elimination of assimilated substances, and (3) *excretion*—the elimination of dissimilated (breakdown) substances.

Recretion

It has long been known that roots not only absorb substances from the soil but that they also give off substances to the soil. This follows from the mechanism of absorption. If passive absorption occurs, it must be from a high electrochemical or chemical potential to a lower one. Consequently, those substances at a higher potential outside the cell will tend to move in; those at a higher potential inside the cell will tend to move out. If active absorption occurs, there must again be a two-way movement. If the movement is carrier mediated, absorption of cations from the soil must be at the expense of H^+ given off into the soil, and the soil may be expected to become more acid. Similarly, if anions are absorbed via carriers, they must be exchanged for other anions (e.g., HCO_3^-), and the soil may be expected to increase in bicarbonates.

Secretion and excretion

Many other substances may also be given off into the root medium, including plant metabolites: sugars, amides, organic acids (Slankis et al., 1964). The quantities, however, are small, requiring labeling with ^{14}C for detection. Thus as many as thirty-five radioactive substances have been detected in the medium in which *Pinus strobus* seedlings were grown aseptically (Street, 1966), and forty-five have been detected in wheat root exudate (Rovira, 1969). Both qualitative and quantitative differences exist between species (Smith, 1969). The substances that were given off by the roots included a wide variety of metabolites: alkaloids, vitamins, nucleotides, flavones, sugars, auxins, amino acids, and organic acids. The root cap may also produce polysaccharide slime substances (Kadej and Kadej, 1970). This occurs by a "reversed pinocytosis" of polysaccharides synthesized in the Golgi apparatus (Rougier, 1971). On the basis of Frey-Wyssling's terminology, elimination of most of the above substances is secretion, although it is difficult to distinguish between the secretion and the excretion of substances such as amino acids.

According to Street, the loss from roots is probably of minor significance in the physiology of higher plants. Thus less than 3 μg of amino N was released from a culture of fifteen wheat plants after 14 days of growth, though up to thirteen amino compounds were released (Ayers and Thornton, 1968). Nevertheless, it has been found that some substances secreted into the soil may prevent competition by other plants, by the process of *allelopathy* (Tukey, 1969). Two such substances are chlorogenic acid and scopolin leached from sycamore leaves (Al-Naib and Rice, 1971). Substances may also be given off from the top of the plant in the form of liquid drops. This kind of elimination is called *exudation,* and may involve recretion, secretion, or excretion. The exudation of droplets of dilute solution from a highly turgid plant into a saturated atmosphere is called *guttation.* Nectaries secrete sugar solutions. Insectivorous plants secrete solutions of digestive enzymes. Some halophytes recrete salt from salt glands in exudates with concentrations as high as 0.5 N, or 50 times the concentration of the nutrient solution in which the plant is grown (Berry, 1970). Eight inorganic ions accounted for 99% of the solutes.

Considerable quantities of both organic and inorganic metabolites may be leached from plants by rain, dew, and mist (Tukey, 1966). As in the case of roots, only a small fraction of the plant's solutes can be removed in this way. Conversely, dew can be absorbed by leaves, but only in relatively small quantities. The term *secretion* has also been used for the active movement of solutes from the protoplasm into the vacuole of the same cell, and it therefore may imply either absorption or elimination of substances.

QUESTIONS

1. Can a cell absorb substances to which it is impermeable?
2. Give some examples.
3. If absorption is caused by diffusion, what will the concentration inside the cell be at equilibrium?
4. If there is more of a substance per unit volume of cell than per unit volume of external solution, can this still be caused by diffusion?
5. How can this be explained?
6. When can you be sure that the substances inside the cell are not there solely as a result of diffusion?
7. What is meant by passive absorption?
8. Can an ion diffuse across a membrane from a region of higher concentration to one of lower concentration (for the same ion) without any other ions moving?
9. What two possibilities are there for such movement?

10. In a nonliving system at equilibrium can a higher concentration of an ion occur on one side of a membrane?
11. What is the name of this phenomenon?
12. What must be present inside the membrane to give rise to this phenomenon?
13. Can it explain the higher concentration of ions inside the cell than outside?
14. If it can, explain why. If it cannot, explain why not.
15. What is meant by active absorption?
16. With what process can we associate absorption of ions by cells against an electrochemical potential gradient?
17. Which ions are absorbed actively—cations or anions?
18. What is the source of the energy for the process?
19. What evidence is there that this is the source of energy?
20. Can lipid soluble substances be actively absorbed?
21. Can water be actively absorbed?
22. What is root pressure?
23. How can it be measured?
24. How is the concentration of the vessel sap maintained?
25. What is the relation of water absorption to oxygen?

SPECIFIC REFERENCES

Al-Naib, F. A.-G., and E. L. Rice. 1971. Allelopathic effects of *Platanus occidentalis*. Bull. Torrey Bot. Club **98**:75-82.

Altman, P. L., and D. S. Dittmer (Eds.). 1966. Environmental biology. Federation of American Societies for Experimental Biology, Washington, D. C.

Ayers, W. A., and R. H. Thornton. 1968. Exudation of amino acids by intact and damaged roots of wheat and peas. Plant Soil **18**:193-207.

Berry, W. L. 1970. Characteristics of salts secreated by *Tamarix aphylla*. Amer. J. Bot. **57**:1226-1230.

Bowling, D. J. F. 1966. Active transport of ions across sunflower roots. Planta **69**:377-382.

Cocking, E. C. 1966. Electron microscopic studies on isolated plant protoplasts. Z. Naturforsch. [B] **21b**:581-584.

Curran, P. F. 1963. The biophysical nature of biological membranes. *In* R. H. Wasserman. The transfer of calcium and strontium across biological membranes. Academic Press, Inc., New York.

Cybis, J., and P. Weglenski. 1969. Effects of lysine on arginine uptake and metabolism in *Aspergillus nidulans*. Mol. Gen. Genet. **104**:282-287.

Dainty, J. 1969. The ionic relations of plants, p. 455-485. *In* M. B. Wilkins (Ed.). The physiology of plant growth and development. McGraw-Hill Book Co., New York.

Etherton, B., and N. Higinbotham. 1960. Transmembrane potential measurements of cells of higher plants as related to salt uptake. Science **131**:409-410.

Fisher, J. D., D. Hansen, and T. K. Hodges. 1970. Correlation between ion fluxes and ion-stimulated adenosine triphosphatase activity of plant roots. Plant Physiol. **46**:812-814.

Frey-Wyssling, A. 1970. Betrachtungen über die pflanzliche Stoffelimination. Ber. Schweiz. Bot. Ges. **80**:454-466.

Ginsburg, H. 1971. Model for iso-osmotic water flow in plant roots. J. Theor. Biol. **32**:147-158.

Ginsburg, H., and B. Z. Ginzburg. 1971*a*. Evidence for active water transport in a corn root preparation. J. Membrane Biol. **4**:29-41.

Ginsburg, H., and B. Z. Ginzburg. 1971*b*. Radial water and solute flow in roots of *Zea mays*. III. Effect of temperature on THO and ion transport. J. Exp. Bot. **22**:337-341.

Hackskaylo, E. 1972. Mycorrhiza: the ultimate in reciprocal parasitism. Bio Science **22**:577-583.

Hansson, G. and A. Kylin. 1969. ATPase activities in homogenates from sugar-beet roots, relation to Mg^{2+} and $(Na^+ + K^+)$– stimulation. Z. Pflanzenphysiol. **60**:270-275.

Higinbotham, N. 1970. Movement of ions and electrogenesis in higher plant cells. Am. Zoologist **10**:393-403.

Higinbotham, N., B. Etherton, and R. J. Foster. 1967. Mineral ion contents and cell transmembrane electropotentials of pea and oat seedling tissue. Plant Physiol. **42**:37-46.

Higinbotham, N., J. S. Graves, and R. F. Davis. 1970. Evidence for an electrogenic ion transport pump in cells of higher plants. J. Membrane Biol. **3**:210-222.

Hogben, C., and M. Adrian. 1963. General aspects of ion transport. *In* R. H. Wasserman. The transfer of calcium and strontium across biological membranes. Academic Press, Inc., New York.

Ingelsten, B., and B. Hylmö. 1961. Apparent free space and surface film determined by a centrifugation method. Physiol. Plant. **14**:157-170.

Kadej, F., and A. Kadej. 1970. Ultrastructure of the rootcap in *Raphanus sativus*. Acta Soc. Bot. Pol. **39**:733-737.

Läuger, P. 1972. Carrier-mediated ion transport. Science **178**:24-30.

Levitt, J. 1953. Further remarks on the thermodynamics of active (non-osmotic) water absorption. Physiol. Plant. **6**:240-252.

Mahlberg, P., K. Olson, and C. Walkinshaw. 1971. Origin and development of plasma membrane derived invaginations in *Vinca rosea* (L). Amer. J. Bot. **58**:407-416.

Mayo, M. A., and E. C. Cocking. 1969. Pinocytic uptake of polystyrene latex particles by isolated tomato fruit protoplasts. Protoplasma **68**:223-230.

Ordin, L., and J. Bonner. 1956. Permeability of *Avena* coleoptile sections to water measured by diffusion of deuterium hydroxide. Plant Physiol. **31**:53-56.

Pitman, M. G. 1965. The location of Donnan free space in disks of beetroot tissue. Aust. J. Biol. Sci. **18**:547-553.

Pressman, B. C., E. J. Harris, W. S. Jagger, and J. H. Johnson. 1967. Antibiotic-mediated transport of alkali ions across lipid barriers. Proc. Nat. Acad. Sci. USA **58**:1949-1956.

Priestley, J. H. 1920. The mechanism of root pressure. New Phytol. **19**:189-200.

Robertson, R. W. 1941. Studies in the metabolism of plant cells. I. Accumulation of chlorides by plant cells and its relation to respiration. Aust. J. Exp. Biol. Med. Sci. **19**:265-278.

Rosene, H. F. 1950. The effect of anoxia on water exchange and oxygen consumption of onion root tissues. J. Cell. Comp. Physiol. **35**:179-193.

Rougier, Mireille, 1971. Cyto-chemical study of the secretion of plant polysaccharides with the aid of choice material: root cap cells from *Zea Mays*. J. Microsc. (Paris) **10**:67-82.

Rovira, A. D. 1969. Plant root exudates. Bot. Rev. **35**:35-58.

Salyaev, R. K., and A. S. Romanenko. 1970. Effect of pinocytosis inductors on the formation of lysosome-like particles in plant cells. Fiziol. Rast. **17**:425-430.

Sargent, J. A., and G. E. Blackman. 1969. Studies on foliar penetration. IV. Mechanisms controlling the rate of penetration of 2,4-dichlorophenoxyacetic acid (2,4-D) into leaves of *Phaseolus vulgaris*. J. Exp. Bot. **20**:542-555.

Seear, J., O. E. Bradfute, and A. D. McLaren. 1968. Uptake of proteins by plant roots. Physiol. Plant. **21**:979-989.

Simons, R. 1969. A thermodynamic analysis of particle flow through biological membranes. Biochim. Biophys. Acta **173**:34-50.

Slankis, V., V. C. Runeckles, and G. Krotkov. 1964. Metabolites liberated by roots of white pine (*Pinus strobus* L.) seedlings. Physiol. Plant. **17**:301-313.

Smith, W. H. 1969. Release of organic materials from the roots of tree seedlings. Forest Sci. **15**:138-143.

Sobey, D. G., L. B. MacLeod, and D. S. Fensom, 1970. The time-course of ion and water transport across decapitated sunflowers for 32 hr after detopping. Can. J. Bot. **48**:1625-1631.

Steward, F. C., W. E. Berry, and T. C. Broyer. 1936. The absorption and accumulation of solutes by living plant cells. Ann. Bot. **50**:345-366.

Street, H. E., and J. S. Lowe. 1950. The carbohydrate nutrition of tomato roots. Part II. The mechanism of sucrose absorption by excised roots. Ann. Bot. **14**:307-329.

Street, H. E. 1966. The physiology of root growth. Ann. Rev. Plant Physiol. **17**:315-344.

Thimann, K. V., and E. W. Samuel. 1955. The permeability of potato tissue to water. Proc. Nat. Acad. Sci. **41**:1029-1033.

Tukey, H. B., Jr. 1966. Leaching and metabolites from above-ground plant parts and its implications. Bull. Torrey Bot. Club **93**:385-401.

Tukey, H. B., Jr. 1969. Implications of allelopathy in agricultural plant science. Bot. Rev. **35**:1-16.

GENERAL REFERENCES

Epstein, E. 1972. Mineral nutrition of plants. John Wiley & Sons, Inc., New York.

Lundegårdh, H. 1945. Absorption, transport, and exudation of inorganic ions by the roots. Ark. Bot. **32A**(12): 1-139.

Osterhout, W. J. V. 1936. The absorption of electrolytes in large plant cells. I. Bot. Rev. **2**:283-315.

Sutcliffe, J. F. 1962. Mineral salts absorption in plants. Pergamon Press, Inc., New York.

9

Ascent of sap

BULK (OR MASS) FLOW

Diffusion can move water into a cell rapidly enough to account for absorption because the distances are small. After the water has entered the roots it rises to all parts of the plant. This process is known as the ascent of sap. Flow rates as high as 75 cm/min have been recorded (Huber, 1932). Calculations from Fick's law reveal that diffusion can account for only an infinitesimal fraction of this movement (Levitt, 1956). Consequently, another kind of movement must be involved.

When a liquid diffuses, it moves as a result of the thermal agitation of the individual molecules. Diffusion can therefore occur under conditions of constant and uniform hydrostatic pressure. However, when the liquid is exposed to a greater pressure in one region than in another, movement will occur from the region of higher to that of lower pressure. This movement is called *bulk flow* by physicists, or mass flow by many physiologists.

The bulk flow of a liquid can be represented by:

$$F = kp$$

where F = flux or flow (cm^3/sec for volume flow, cm/sec for linear flow)

p = pressure difference between the two ends of the flowing system (bars)

k = conductivity of the system in which the liquid is flowing (cm^3 [or cm]/sec/bar)

As in the case of electricity, the conductivity factor in water flow (k) is inversely proportional to resistance to flow (R):

$$k \propto \frac{1}{R}$$

Both the similarity to and the difference from diffusion are obvious if the pressure is expressed as a gradient $\left(\frac{p}{l}\right)$:

$$F = \frac{kp}{l}$$

where l = length of the path of flow or the distance between the two points differing in pressure by p bars.

It is obvious from the preceding that bulk flow differs from diffusion flow in being proportional to the pressure gradient instead of the concentration gradient. Therefore for the sap to flow through the tree, the pressure gradient $\frac{p}{l}$ must be sufficient to overcome the frictional resistance to movement. But when water flows through any tube, the frictional resistance is that between the flowing water and a stationary film of water held at the inner surface of the tube. In other words, the frictional resistance is the viscosity of the water. The relation between flow or flux (F) through a horizontal tube and viscosity is given by Poiseuille's law:

$$F = \frac{pr^2}{8ln}$$

where F = flux (linear, not volume flow)
p = pressure difference at two ends of the tube
r = tube radius
l = tube length
n = viscosity (0.01 cgs unit for water)

Poiseuille's law is simply related to the first equation, since it can be rearranged as follows:

$$F = \frac{r^2}{8n} \times \frac{p}{l}$$

It is now obvious that $\frac{r^2}{8n}$ is simply the constant k in the first equation and is therefore a measure of the conductivity of the tube for a specific liquid. Therefore for any one pressure gradient, the rate of flow is directly proportional to the square of the radius of the tube through which the liquid flows, and inversely proportional to the viscosity of the flowing liquid.

Poiseuille's law can also be rearranged to permit calculation of the pressure necessary to cause a flow under any one set of conditions:

$$p = \frac{8Fln}{r^2}$$

But Poiseuille's law applies to flow through a horizontal tube. Flow up a tree requires an additional pressure to overcome the force of gravity, that is, 1 atm or bar for every 30 feet of vertical tube length. In the case of sap flowing at the maximum measured rate (75 cm/min) through a tree 30 m (i.e., nearly 100 feet) tall, the necessary pressures for commonly found vessel diameters can be readily calculated (Table 9-1).

How do these theoretical values compare with experimentally determined values for trees? Early investigators found that the pressure required to force a flow through a horizontal log is about equal to the pressure required to support a stationary column in the vertical log. This agrees with the calculated values for vessels with an average diameter of 0.2 mm (Table 9-1). Recent measured values have yielded a gradient of 0.1 atm/m (Scholander et al., 1965). This agrees with the calculated values for vessels with an average diameter of 1.0 mm (Table 9-1). Both of these calculations, however, are for maximum flow rates. Although the rates of flow and the diameters of the vessels through which the major flow occurred were not measured in these cases, the values certainly fall within the calculated range. Consequently, it can be concluded that the ascent of sap follows the laws of bulk flow.

The secret of success of vines is revealed by Poiseuille's law. Despite the small cross section of their stems, they are able to transport water rapidly to great heights.

Table 9-1. Pressures required to overcome gravity and frictional resistance to flow in vessels of different diameters (calculated from Poiseuille's law)*

	Pressure (bars)			
Vessel diameter (mm)	To overcome gravity	To overcome frictional resistance	Total	$\frac{p}{l}$ (bar/m)
1.0	3	0.12	3.12	0.1
0.2	3	3	6	0.2
0.1	3	12	15	0.5

*Maximal flow rate assumed (75 cm/min) in a tree 30 m tall. Gravitational pressure, 1 bar/10 m (app.).

This results from their unusually large vessel diameters, permitting maximum flow rates with minimum pressures. Tube diameter is particularly important to them, since the viscosity of liquids is directly proportional to flux velocity. In the case of trees with slow flux velocities, viscosity is so low that xylem elements of small diameter are fully adequate.

It must be emphasized that Poiseuille's law applies only to pores that are not too small and are nearly right cylinders. It is therefore applicable to the flow through the xylem vessels, although it is questionable whether it applies to the small, irregular, porous channels of the cell wall (Tyree, 1969). Yet the flux equation based on the thermodynamics of irreversible processes (using the Onsager theory, see Chapter 1) predicts that a significant fraction of water is transported over short distances via the cell walls. Electrical gradients are suggested to contribute significantly to such short-distance translocation.

Bulk flow as described by Poiseuille's law is similar in some respects to diffusion as described by Fick's law. Thus if we use the symbols for Poiseuille's law, Fick's law becomes:

$$F_d \propto r^2 \frac{C_1 - C_2}{l}$$

Poiseuille's law becomes:

$$F_b \propto r^2 \frac{p_1 - p_2}{l}$$

where F_d = flow caused by diffusion
F_b = bulk flow

In both cases the flux or movement is proportional to the gradient: pressure gradient in the case of Poiseuille's law, concentration gradient in the case of Fick's law. In both cases the flux or movement is proportional to the cross-sectional area and inversely proportional to the resistance to flow. In the case of diffusion the resistance factor is inversely related to the constant D, in bulk flow directly to the viscosity. Because of these similarities, it may sometimes be difficult to distinguish between the two. For this reason it has been suggested that when water diffuses into a cell, the portion moving through the "microtubes" of the semipermeable membrane is actually a bulk flow (Ray, 1960). More recent results, however, oppose this concept of water-filled pores in the membrane (Chapter 7).

In general, however, there are three differences between the two kinds of movement (Fig. 9-1):

1. Bulk flow is always associated with a measurable hydrostatic pressure gradient, and diffusion is not.
2. If solutes are present, they move by bulk flow at the same rate and in the same direction as the solvent, but in the case of diffusion they do not.
3. Bulk flow usually involves much more rapid movement than that caused by diffusion, since the whole column of water moves at once as a single body. The speed of movement in bulk flow is limited only by the frictional resistance or viscosity of the water and the impressed pressure gradient.

Exact equations for bulk flow have been developed on the basis of irreversible thermodynamics, using the 1958 formulation of Kedem and Katchalsky (Chapter 1).

ROOT PRESSURE AS OPPOSED TO TRANSPIRATIONAL PULL

The bulk flow may result either from (1) a push—a larger positive (hydrostatic) force at the base than at the top of the plant, or (2) a pull—a smaller negative force (or tension) at the base than at the top of the plant. The push is developed due to the diffusion of water into the root cells from an external root medium at a higher water potential than that within the root. The kinetic energy of the entering water molecules is converted into a hydrostatic pressure within the root, which is called *root pressure*. The pull is caused by transpiration leading to a fall in hydrostatic pressure at the top of the plant. It is therefore called the *transpirational pull*. In either case, water is flowing against the force of gravity, and 1 atm of pressure difference is required per 30 feet, just to support the column of water. Consequently, for a 300-foot tree a pressure of 10 atm is just sufficient to support the column of water; and in order to overcome frictional resistance, according to the available experimental evidence, about another 10 atm is needed if the water is flowing through the tree at a normal rate.

Since the maximum root pressure (P_r) is the excess of the osmotic potential of the soil solution over that of the root:

$$P_r = \pi_s - \pi_v$$

where P_r = pressure difference between base of column and top of column (which is at atmospheric pressure)
π_s = osmotic potential of soil solution
π_v = osmotic potential of vessel sap

This equation has been recently tested and proved to be correct by simultaneous measurement of all three quantities (Morizet, 1972).

The volume flow of water caused by root pressure (F_r) is

$$F_r = k\,(\pi_s - \pi_v)$$

where k is the water conductivity of the root (g/hr/bar, Brouwer, 1953).

Since measurements of the root pressure of trees usually yield values of only 1 to 2 atm, the ascent of sap in tall trees cannot be caused by root pressure. This does not, of course, detract from the importance of root pressure. In grapevines, for instance, the vessels are emptied during winter and

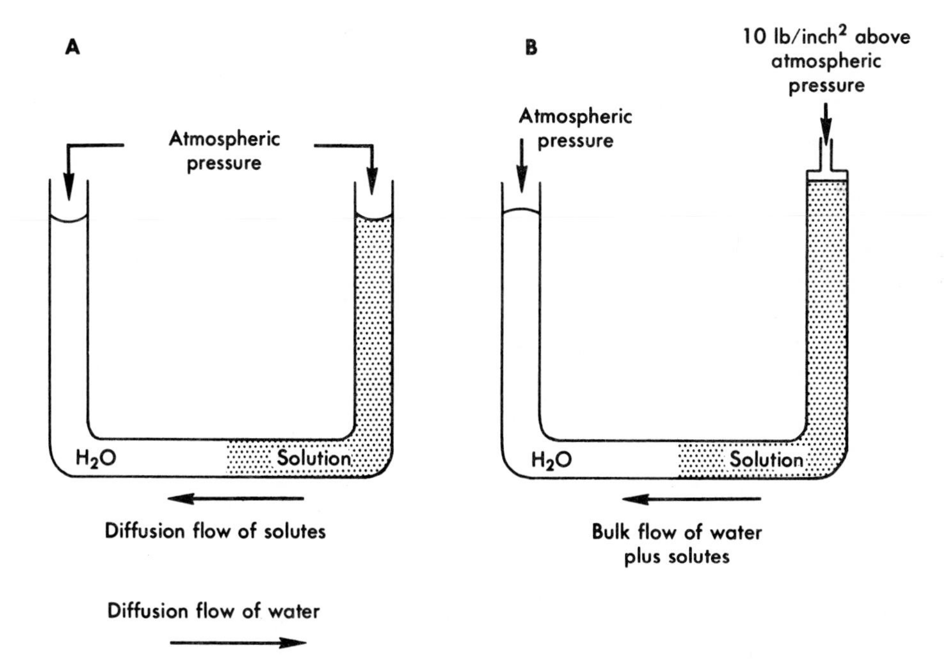

Fig. 9-1. Diffusion flow as opposed to bulk flow. **A**, Diffusion flow: $F \propto \frac{\Psi\ \text{solution}}{x}$, where x is distance between water and solution. Water and solutes move independently. **B**, Bulk flow: $F \propto p/l$, where l is distance between applied pressure (p) and liquid at atmospheric pressure. Water and solute move together and flow is independent of any diffusion within the tube.

are refilled in spring because of a root pressure of 2 to 5 atm (Scholander et al., 1955). At this time there is a gradient of about 0.1 bar/m in the stem. In opposition to the grapevine, however, most trees keep their vessels filled with water throughout the winter and, in fact, have their highest water contents at this time (Gibbs, 1935). This difference is undoubtedly related to the unusually wide vessels in the grapevine.

The most conclusive evidence that the ascent of sap is not commonly the result of root pressure is the fact that tensions or stresses instead of positive forces or pressures usually exist in the xylem (the water-transporting tissues) under normal growing conditions. This is shown by the absorption of water through bore holes in tree trunks. If such holes are connected by a syphon to an elevated bottle of water (Fig. 9-2), as much as 1 gallon can be absorbed over a 24 hr period by an average-sized apple tree on a warm summer day. On the other hand, when root pressure is operative, sap exudes from such bore holes. The only time such positive forces occur in the xylem is when the leaves are not on the plant (in the case of deciduous plants) or when the plant is in a saturated or near-saturated environment. In the latter case water may actually be exuded from specialized structures in the leaves called hydathodes. This exudation is known as *guttation,* and it may be stopped by reducing the root pressure, for example, by watering the soil with a sugar or salt solution, decreasing the osmotic potential of the root medium. The guttation fluid is not, however, pure water but may contain sugars, amino acids, inorganic elements, and a variety of vitamins (Goatley and Lewis, 1966).

That the stress or tension increases during the day can be shown by the use of a sensitive instrument known as a *dendrograph.* This measures the diameter of the tree trunk and reveals a definite contraction during the day and an expansion at night (Fig. 9-3). Even roots may show this contraction due to the tension resulting from water loss. The minimum diameter occurred about 3:00 P.M. in the stem, about 6:00 P.M. in the root (Klepper et al., 1970).

Thus the ascent of sap is usually associated with a pull from above rather than a push from below. If a capillary tube is held vertically in a beaker of water, a similar pull occurs, resulting in the rise of water in the tube to well above the surface of the water in the beaker. It is easy to calculate whether the rise in the plant can be

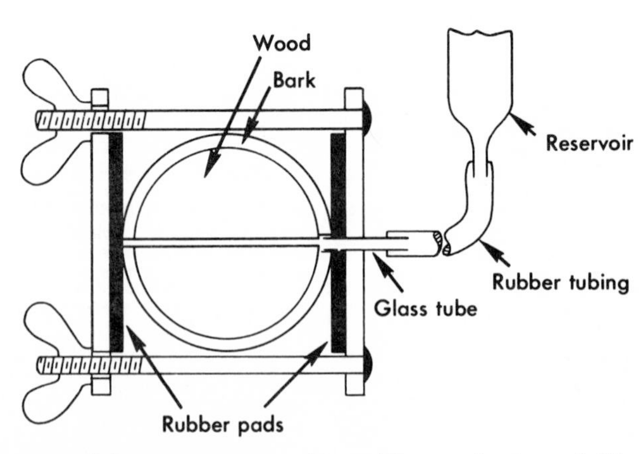

Fig. 9-2. Apparatus used for tree injection. (From Thomas, L. A., and W. A. Roach. 1934. J. Pomol. Hort. Sci. **12:**151-166.)

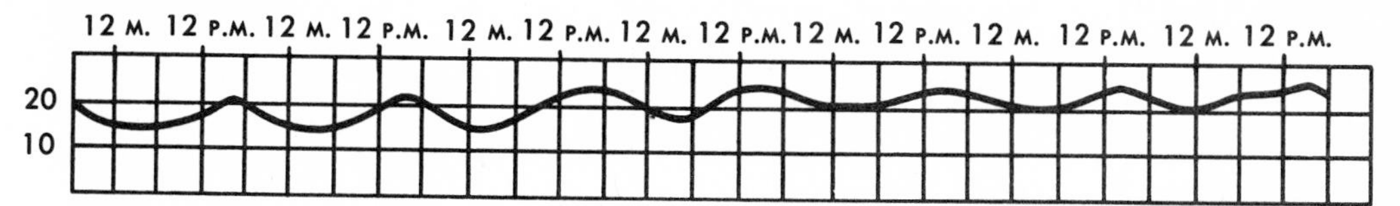

Fig. 9-3. Daily expansion and contraction of a redwood tree, amplified ten times. **12 M.**, noon, **12 P.M.**, midnight. (Redrawn from MacDougal, D. T., and F. Shreve. 1924. Carnegie Institution of Washington, Pub. No. 350. Washington, D. C.)

explained by a capillary pull in the vessels:

$$h = \frac{2S}{rdg}$$

where h = height (cm) supported in the capillary tube
S = surface tension of liquid (dynes/cm)
r = radius of tube (cm)
d = density of liquid (gm/cm^3)
g = acceleration caused by gravity (981 cm/sec^2)

A vessel diameter of 0.1 mm (0.01 cm) would account for only a 30 cm rise. This explains why plants like the grapevine, in which the vessels empty in the fall, cannot refill by capillarity and therefore must depend on root pressure to refill them. But in most plants the water system remains unbroken (at least in part) thoughout the life of the plant. Consequently, it is not in the vessels themselves that the capillary pull occurs. Since this is a surface tension phenomenon, the pull must occur at the water surface, and since the water system of a plant is continuous from the lowest roots to the highest leaves, the upper surface of the water column is in the leaves. Therefore the pull must be at the outer surface of the leaf (mesophyll) cells that are in contact with the intercellular spaces, that is, in the microcapillaries of their walls. Since these microcapillaries are so fine that they cannot be seen with the ordinary optical microscope, their diameters must be less than 0.1 μm. Using this value in the preceding equation for capillary height, we find that these microcapillaries are capable of exerting sufficient pull to support a water column 300 m high —three times the height of the tallest trees!

Unfortunately, physicists have not determined within what limits the preceding equation holds. Some go so far as to state that the rise is basically a push from below by atmospheric pressure and therefore cannot exceed 30 feet. However, this concept is easily shown to be incorrect, since the capillary rise occurs just as readily in a vacuum. In the case of microcapillaries 200 Å (i.e., 0.02 μm) in radius, measurements by Fedyakin (Derjaguin, 1966) have shown that the calculated value is double the real value. But this still leaves a capillary stress (or transpirational pull) of more than enough to account for the rise to the top of the tallest trees. The classical Askenasy experiment has shown that the rise in purely physical systems can be 2 to 3 times that caused by atmospheric pressure. However, such experiments are difficult to perform and have not succeeded in duplicating the rise of hundreds of feet in tall trees. The theoretical concept is shown in Fig. 9-4. The following three stages in the process may be distinguished: (1) transpiration (the loss of water vapor from the leaves), accompanied by evaporation from the surface of the microcapillaries; (2) capillary rise of water threads resulting from the force of adhesion between the water and the cell wall (i.e., the force of imbibition by the cell wall colloids); and (3) bulk flow—the whole column of water moves en masse because of the force of cohesion between the water molecules. These three stages, of course,

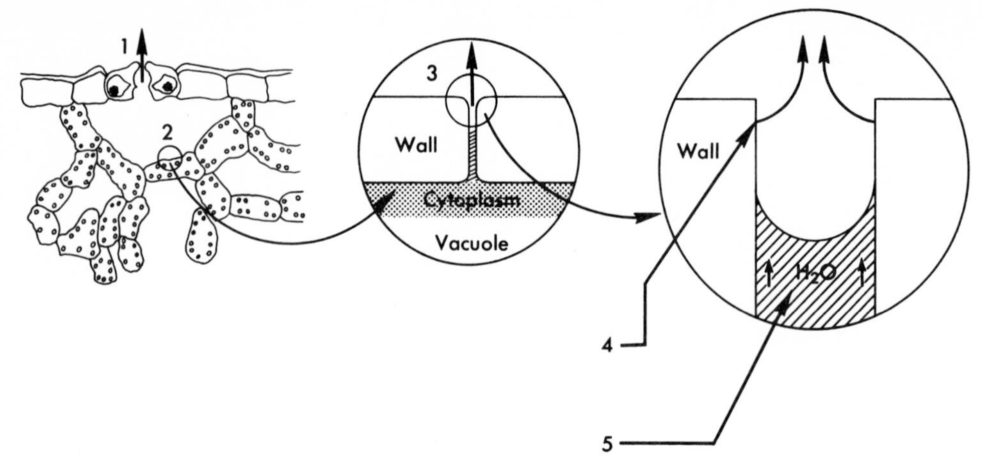

Fig. 9-4. The rise of sap as a result of transpirational pull. **1,** Transpiration (through stomatal aperture indicated by arrow); **2,** evaporation from microcapillaries into intercellular space; **3,** partial drying of wall surface; **4,** water from film attracted by imbibitional force of partially dried wall; **5,** whole water thread rising because of cohesive force of water.

occur almost simultaneously, although sap flow has been observed to lag behind transpiration by as much as 15 to 45 min (Hinckley, 1971). The so-called transpirational pull is therefore actually a capillary pull that results from transpiration, and the whole column is pulled en masse as though it were a solid wire. This concept is usually known as Dixon's Cohesion Theory of the ascent of sap, although it was proposed independently by Dixon and Joly in 1894 and by Askenasy in 1895.

Some points are difficult (Greenidge, 1958), although not impossible to explain by Dixon's Cohesion Theory. Even when it is stationary, a column of water 100 m high and 1 inch (2.5 cm) in diameter is subjected to a downward force of 150 lb because of the gravitational pressure of 10 atm, and if the preceding reasoning is correct, this is counteracted by an equal upward pull caused by surface tension. The column is therefore being subjected to a powerful force tending to break it. The question is whether a liquid such as water has sufficient tensile strength to resist the pull without rupture, and whether the force of attraction between the water and the vessel wall is sufficient to prevent separation from the wall. In other words, are the forces of cohesion and adhesion adequate?

Physical chemists have shown experimentally that pure water has a cohesive force capable of resisting a stress of as high as 275 bars (Hayward, 1971)—theoretically of the order of 1000 atm. Under such stresses, the water actually expands, to the same degree as it contracts when subjected to positive pressure (Winnick and Cho, 1971). But the water in the vessels is not pure. In addition to solids, gases are present in solution. The reduction in pressure because of the capillary pull may therefore reduce the solubility of these gases until they separate from the liquid, causing rupture of the column. Under conditions of excessive tension, vessels have in fact been found to become gas filled. By means of an apparatus that detects vibrations in the plant, it has recently been possible to demonstrate such "cavitation" in the xylem of many plants (Milburn and Johnson, 1966). But since there are many columns of vessels side by side, it is unnecessary for all of them to be continuous, and therefore the temporary filling of some vessels with gas may not be injurious. When the tension is relieved, by rain or simply at night, the gases go back into solution and the column becomes continuous again. The same mechanism operates during the thawing of frozen xylem elements. Air is present in the frozen wood of conifers; yet on thawing, there is no blockage of flow (Sucoff, 1969). This is explained by the assumption that only the largest bubble expands on thawing, releasing the tension and allowing all the bubbles in the adjacent tracheids to redissolve. It was calculated that these bubbles redissolved within 0.1 sec. When the diameter of the vessel is large (e.g., 1 mm), as in grapevine, the bubbles of gas are carried along with the stream and have little or no effect on the water movement (Scholander et al., 1955). Cavitation is actually prevented in the case of gymnosperms even when their xylem sap is frozen, presumably because of their bordered pits (Hammel, 1967).

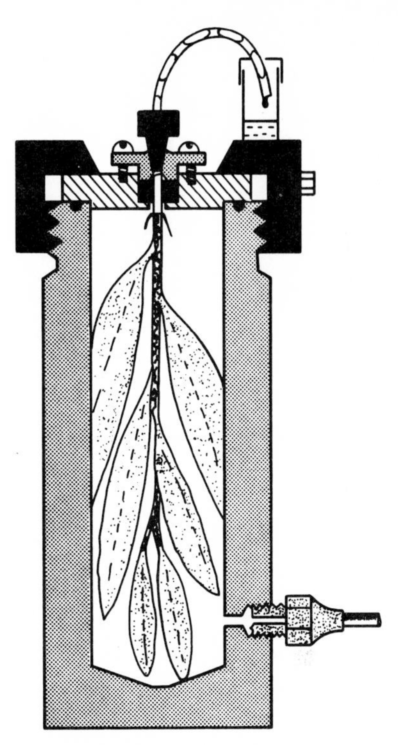

Fig. 9-5. Pressure bomb for the determination of the negative pressure of xylem sap. (From Scholander, P. F., et al. 1966. Plant Physiol. **41**:529-532.)

Many investigators have attempted to measure the stresses in the xylem using ingenious indirect methods, and a wide range of values has been reported (Loomis et al., 1960). Scholander and co-workers (1965) developed an elegant method called the "pressure-bomb" technique (Fig. 9-5). They severed a shoot, allowing the tension in the vessels to pull the liquid back, and immediately enclosed all except the cut surface in the bomb. They then measured the counterpressure (by means of nitrogen gas) that must be exerted on the shoot to force the liquid back to the cut surface. They discovered stresses as high as 80 atm. Even actively growing annuals and perennials were found by this method to have tensions greater than 70 atm (Hickman, 1970). However, there are still problems in the use of this method, and it sometimes yields impossibly low pressure gradients (less than the hydrostatic gradient—Tobiessen et al., 1971).

The transpirational pull, of course, would be greater than the forces needed simply to support the column, since the moving column must also overcome the frictional resistance to movement. Since the column of water moves en masse, this means that the transpirational pull is transmitted from the microcapillaries in the mesophyll cell walls to the protoplasm and vacuole, to the adjacent cells, and to the nearest vessel. It is transmitted down the vessels through the leaf blade, petiole, stem, and all the way to the roots with the possible exception of the plasma membrane, across which diffusion occurs according to the classical concept. As mentioned earlier it has been suggested, however, that bulk flow may be possible even across these membranes (Ray, 1960). The pull is also thought to move water along the cell walls, bypassing the protoplasts. Direct evidence of this has been produced for cell walls in the leaves (Gaff et al.,

1964), but indirect evidence (Weatherley, 1963) indicates that it cannot account for much of the flow through roots. From the base of the vessels in the roots the pull is transmitted through the adjacent cells to the epidermal cells and even to the medium surrounding these cells, that is, solution or soil particles. Thus the transpirational pull is responsible not only for the movement of water within the plant but also for absorption of water from the root medium. When the transpirational pull is pronounced, it follows that the absorption of water is greatly enhanced because even if no bulk flow occurs across the membranes, diffusion would be directly involved only over an infinitesimally small fraction of the total distance. Indirectly, of course, diffusion is involved since transpiration is itself a diffusion process (Chapter 11).

The mass movement of water under tension in the plant is rapid, as can be shown by cutting the stem of a wilted plant under water. Recovery occurs more rapidly than if the roots of the intact plant are watered. This illustrates the fact that although nearly the whole path of water movement is through dead vessels, the short path through the living cells in the root offers a greater resistance to movement than do the vessels. This is no doubt because of the retarding effect of the semipermeable plasma membranes, which are considerably less permeable to water than are the cell walls. The resistance to flow in the vessels is simply caused by the low frictional resistance or viscosity of the water itself. However, if the vessels are physically blocked, for example, by the growth of microorganisms in the vessels of diseased plants or by entrance of particles through cut surfaces, wilting and death may occur. These principles are made use of in prolonging the life of cut flowers. Vascular blockage is prevented by treatment with 8-hydroxyquinoline citrate (Marousky, 1969).

The high tensions in the vessels and tracheids point to the need of the thick walls always found in these cells. In the absence of such thick walls, the cells would collapse, closing the cavity and stopping the flow of water. Since water movement occurs only in a relatively narrow zone of the sapwood (the heartwood vessels are blocked by growths known as *tyloses* and by resins), it is here that the tension is localized. This means that when tensions are high, the vapor pressure of the water in the sapwood is greatly reduced below that in the heartwood. As a result, there is evidence that water distills from the heartwood to the sapwood on a hot summer day (Reynolds, 1939).

QUESTIONS

1. What is meant by bulk flow?
2. To what is the rate of bulk flow proportional?
3. How does bulk flow differ from diffusion flow?
4. Can the ascent of sap be caused by diffusion? Why?
5. Can it be generally caused by root pressure? At any time?
6. How tall a tree could be supplied with water by a root pressure of 5 atm?
7. What physiological process, other than ascent of sap, is caused by root pressure?
8. What would be the root pressure on a normal sunny day?
9. How does a tree trunk change during a summer day?
10. What is the instrument that measures this?
11. Is the water pushed up or pulled up a tree on a summer day?
12. What is the pull called?
13. What theory explains this pull?
14. What forces are primarily responsible for the rise?
15. Is the role of capillarity a result of the xylem capillaries?
16. What is the relation of water absorption to the ascent of sap?

17. Is the cohesion of water adequate to account for the rise of water to the top of trees?
18. Is there any complicating factor under natural conditions?
19. Where is the greatest resistance to the flow of liquid water in the plant?
20. Are the cohesive tensions on the water throughout the wood?
21. Do vessels collapse under tension? Why?

SPECIFIC REFERENCES

Brouwer, R. 1953. Water absorption by the roots of *Vicia faba* at various transpiration strengths. Proc. Kon. Nederl. Akad. Wet. (Biol. Med.) **56**:106-136.

Derjaguin, B. V. 1966. Recent research into the properties of water in thin films and in microcapillaries. Sympos. Soc. Expr. Biol. **19**:55-60.

Gaff, D. F., T. C. Chambers, and K. Markus. 1964. Studies of extrafascicular movement of water in the leaf. Aust. J. Biol. Sci. **17**:581-586.

Gibbs, R. D. 1935. Studies of wood. Canad. J. Res. **12**:715-787.

Goatley, J. L., and R. W. Lewis. 1966. Composition of guttation fluid from rye, wheat, and barley seedlings. Plant Physiol. **41**:373-375.

Greenidge, K. N. H. 1958. Moisture movement in plants. Scientia **93**:191-195.

Hammel, H. T. 1967. Freezing of xylem sap without cavitation. Plant Physiol. **42**:55-66.

Hayward, A. T. J. 1971. Negative pressure in liquids: Can it be harnessed to serve man? Amer. Sci. **59**:434-443.

Hickman, J. C. 1970. Seasonal course of xylem sap tension. Ecology **51**:1052-1056.

Hinckley, T. M. 1971. Estimate of water flow in Douglas-fir seedlings. Ecology **52**:525-528.

Huber, B. 1932. Beobachtung und Messung pflanzlicher Saftströme. Ber. Deutsch. Bot. Ges. **50**:89-109.

Klepper, Betty, M. G. Huck, and H. M. Taylor. 1970. Time-lapse cinematography of diurnal changes in lateral root diameter. Plant Physiol. **46**(Suppl.):21.

Levitt, J. 1956. The physical nature of transpirational pull. Plant Physiol. **31**:248-251.

Loomis, W. E., R. Santamaria, and R. S. Gage. 1960. Cohesion of water in plants. Plant Physiol. **35**:300-306.

MacDougal, D. T., and F. Shreve. 1924. Growth in trees and massive organs of plants. Carnegie Institution of Washington, Pub. No. 350. Washington, D. C.

Marousky, Francis J. 1969. Vascular blockage, water absorption, stomatal opening, and respiration of cut Better Times roses treated with 8-hydroxyquinoline citrate and sucrose. J. Amer. Soc. Hort. Sci. **94**:223-225.

Milburn, J. A., and R. P. C. Johnson. 1966. The conduction of sap. II. Detection of vibrations produced by sap cavitation in *Ricinus* xylem. Planta **69**:43-52.

Morizet, J. 1972. Étude expérimentale de l'absorption hydrique par les plantes. Compt. Rend. Acad. Sci. (Paris) **275**:1773-1775.

Ray, P. M. 1960. On the theory of osmotic water movement. Plant Physiol. **35**:783-795.

Reynolds, E. S. 1939. Tree temperatures and thermostasy. Ann. Mo. Bot. Gard. **26**:165-255.

Scholander, P. F., W. E. Love, and J. Kanwisher. 1955. The rise of sap in tall grapevines. Plant Physiol. **30**:93-104.

Scholander, P. F., H. T. Hammel, E. D. Bradstreet, and E. A. Hemmingsen. 1965. Sap pressure in vascular plants. Science **148**:339-346.

Scholander, P. F., E. D. Bradstreet, H. T. Hammel, and E. A. Hemmingsen. 1966. Sap concentrations in halophytes and some other plants. Plant Physiol. **41**:529-532.

Sucoff, E. 1969. Freezing of conifer xylem and the cohesion-tension theory. Physiol. Plant. **22**:424-431.

Thomas, L. A., and W. A. Roach. 1934. Injection of fruit trees; preliminary experiments with artificial manures. J. Pomol. Hort. Sci. **12**:151-166.

Tobiessen, P., P. W. Rundel, and R. E. Stecker. 1971. Water potential gradient in a tall *Sequoiadendron*. Plant Physiol. **48**:303-304.

Weatherley, P. E. 1963. The pathway of water movement across the root cortex and leaf mesophyll of transpiring plants. *In* A. J. Rutter and F. W. Whitehead. The water relations of plants. John Wiley & Sons, Inc., New York.

Winnick, J. and S. J. Cho. 1971. PVT behavior of water at negative pressures. J. Chem. Physics **55**:2092-2097.

GENERAL REFERENCES

Dixon, H. H. 1924. The transpiration stream. University of London Press, London.

Kramer, P. J. 1949. Plant and soil water relationships. McGraw-Hill Book Co., New York.

10

Translocation of solutes

Although the aim of physiology is to explain all living processes in terms of physics and chemistry, this goal in most cases has not yet been reached. As a result, it is frequently necessary to give a number of possible explanations (hypotheses) and to attempt to evaluate each on the basis of physical and chemical theory, together with the available physiological evidence. Sometimes it is possible in this way to exclude all the hypotheses except one. This has been done in the case of the ascent of sap. Although several explanations have been proposed from time to time for the phenomenon, all except the cohesion theory have been conclusively disproved. In an elementary course there is no time to discuss these disproved theories, and the cohesion theory is accepted, not because it is conclusively proved, but because it, and it alone, agrees with the known facts. In the case of the translocation of solutes, however, the evidence does not permit acceptance of any one hypothesis. This has led to more controversy than in almost any other field of plant physiology and requires reference to an unusually large number of original papers; for in science the number of publications in a specific area is often inversely related to the state of knowledge. Documentation of each statement with references presents two safeguards against error: it prevents the writer from making statements not established on evidence and it permits the reader to check the author's interpretation against the experimental evidence.

The development of many new techniques has produced a large body of new information and has led to the hope that the uncertainties will soon be resolved. But a clear understanding of the problem requires a consideration of both the old and the new results, both the old and the new hypotheses.

THE TRANSLOCATORY SYSTEMS

There are two principal translocatory systems in the plant, the xylem vessels and the phloem sieve-tubes. The individual cells are referred to as vessel elements and sieve-tube elements, respectively. These two systems are different, both morphologically and physiologically (Table 10-1).

The structure of the vessels is simple and has long been known, since they are dead when mature and consist only of cell walls, which can be adequately studied

Table 10-1. Comparison of the two translocatory systems

Vessels or tracheids	Sieve-tubes
Anatomical comparisons	
1. Occur in xylem tissue	1. Occur in phloem tissue
2. Elements have open ends	2. Sieve plates at ends (and sometimes at sides) of element
3. Long, narrow cells (usually wider than sieve-tubes)	3. Long, narrow cells
4. Thick, rigid walls	4. Thin, more extensible walls
Physiological comparisons	
1. Dead when mature and functional	1. Alive when mature and functional
2. Permeable to both solute and solvent	2. Semipermeable
3. Low sap concentration	3. High sap concentration (osmotic pressure of 15 to 34 atm)
4. No turgor pressure	4. High turgor pressures (in turgid plants)
5. Partially collapsed when functioning (except when root pressure is operating)	5. Distended by pressure when functioning
6. Absorb water or air when cut (except when root pressure is operating)	6. Exude when cut (if functioning)
7. Translocate both solvent and solutes	7. Translocate only solutes
8. Speeds up to 75 cm/min	8. Speeds up to 5 cm/min

under the optical microscope in dead, permanently mounted sections. The sieve-tubes, however, are much more difficult to study. Because they are so long and narrow, it is difficult to cut thin enough sections of living tissue without cutting across every sieve-tube and therefore killing them. Improved techniques, however, have succeeded in revealing some of the properties of the living sieve-tubes; for example, it is possible to plasmolyze them in the same way as other living plant cells (Esau, 1966). Since they are alive when functional, they must be carefully fixed, then observed under the electron microscope to determine the protoplasmic structure, including that of the organelles. This, too, gives rise to difficulties, and some investigators disagree as to the details of protoplasmic structure. No nucleus is found in the mature sieve-tube element, and many investigators have produced evidence that the tonoplast also disappears (Buvat, 1963). However, in opposition to the many negative results, nuclei have been found in all the mature sieve-tubes of three species of *Taxodiaceae*, and in some mature sieve-tubes in twelve of thirteen species of woody dicotyledons (Evert et al., 1970). Mitochondria remain, although having fewer internal membranes than in parenchyma cells. Contrary to earlier reports, plastids are also present (Evert and Deshpande, 1971), although as in the case of the mitochondria, they lack a well-developed membrane system (Zee, 1969). Ribosomes and dictyosomes are lacking (Murmanis and Evert, 1966). However, it is the sieve plates that have been most intensely studied, and the connecting strands across the pores have been clearly demonstrated (Fig. 10-1) by some although not by all observers (Esau, 1966). At least in some cases these appear to consist of endoplasmic reticulum (ER), which fills the pores. More recent observations indicate that beside the ER, strands derived from slime bodies traverse the pores, which are lined by the plasmalemma (Murmanis and Evert, 1966).

The slime bodies extend throughout the length of the sieve-tube (Fig. 10-1) as fibrils, strands, or so-called "P-protein bodies" connecting the sieve plates at both ends (Cronshaw and Esau, 1968). They occur first as tubules 200 Å in diameter in *Coleus* (Steer and Newcomb, 1969) and are later converted into fibrils 70 Å in diameter. They vary somewhat in size and structure from one species to another but have been found in all kinds of higher

Fig. 10-1. Cucurbit sieve element with slime-impregnated filaments running from sieve plate to sieve plate; the cut end is at the bottom. (From Crafts, A. S. 1961. Translocation in plants. Holt, Rinehart & Winston, Inc., New York.)

plants, with the exception of corn (Singh and Srivastava, 1972). The companion cells adjacent to the sieve-tubes are also believed to play some role in phloem transport. It has, for instance, been suggested that they, rather than the sieve-tubes, are the paths of translocation in the minor veins (Trip, 1969). By means of histoautoradiography, however, tritium-labeled sucrose has been localized only in the sieve-tubes (Schmitz and Willenbrink, 1968).

Despite the slime bodies that traverse them, the sieve plate pores are unplugged in vivo, even allowing small starch grains to pass through them and therefore permitting flow across them from sieve-tube to sieve-tube (Anderson and Cronshaw, 1969). At times of emergency, the plant must have a mechanism for rapidly sealing off these connecting holes between the sieve-tubes. This is accomplished by the formation of slime plugs, which seem to be due to the jelling of an alkaline protein (Walker, 1971) and of *callose* (composed of an insoluble carbohydrate), which seal off the holes. A number of treatments lead to one or both of these sealing mechanisms.

1. Unlike the xylem, which may conduct as well after killing as when alive, killing the phloem by scalding a thin portion or cooling it to near freezing stops the flow across it. This is to be expected, since the conducting xylem vessels are dead, whereas the conducting sieve-tubes are alive.

2. Stresses that do not kill (e.g., cooling) may also stop or retard the flow, but in this case the effect may be reversible. The effect of cooling on translocation in the phloem depends on the species. *Cucurbita* plants show the highest degree of sensitivity. Movement of ^{14}C-labeled substances from the leaf blade is restricted at any temperature from 15° C downward (Webb, 1970). At 5° C or lower, translocation is almost completely inhibited but recovers completely after 19 to 20 hr of continuous cold at 1° C (Webb, 1971). Bean plants are slightly less sensitive. When their petioles were cooled to 7° C, the translocation dropped to 10% of normal during the first 100 min but recovered to 40% of normal after 7 hr (Geiger and Sovonick, 1970). It failed to recover at 3° C. The sugar beet is relatively insensitive. Cooling of its petioles to 2° C lowered the translocation rate to 50% of the control in 20 min, but it fully recovered the original rate within 6 hr. In the case of the willow *(Salix viminalis),* translocation was stopped only by freezing (Weatherley and Watson, 1969). Heating at 45° C for 15 min along 4 cm of hypocotyl also brings about a deposition of callose in the sieve plates (McNairn, 1970). Similarly, a temporary massage during exudation stops the flow completely from the castor bean, presumably due to a sealing mechanism (Milburn, 1970). On the other hand, massaging the stem before tapping enhances the exudation, presumably by inhibiting the normal, pressure-sensitive sealing mechanism. Nat-

ural cooling in the fall produces the same reversible blockage of the sieve-tubes as artificial cooling, although dormancy may also be involved. In woody vines the sealing process begins in early or mid-October (Davis and Evert, 1970), and by late December all mature sieve-tubes are either dead or dormant due to sealing. With the break of dormancy in the spring, the seal is redissolved, and the pores are again open.

Pure vessel sap is much more easily obtained than sieve-tube sap. All that is necessary is to cut a length of stem (cut surfaces at both ends) and stand it on the shoulder of a specially constricted centrifuge tube. On centrifuging the stem, the vessel sap is thrown out of the vessels into the bottom of the centrifuge tube. Suction may also be used to obtain vessel sap. In the case of the pear and apple, the composition of the sap obtained by these two methods is identical (Jones and Rowe, 1968). In the case of the grapevine, however, the fluid obtained by suction contains considerable quantities of material that does not move in the vessels, such as sucrose (Hardy and Possingham, 1969).

A relatively simple method of obtaining sieve-tube sap is also used, although there may be complications. The sieve-tubes are normally near full turgor because of the close proximity of the vessels with their sap of high water potential. Thus in a transpiring plant, although the vessels are slightly collapsed by the tension, the sieve-tubes may still be distended by their turgor pressure. As a result, a cut across the stele causes exudation ("bleeding") from the sieve-tubes, and the exudate may be collected for analysis. Although sieve-tube turgor may be retained even in wilted plants, a high rate of transpiration and relatively low rate of absorption by the plant may result in a sufficiently severe tension on the vessels to reduce the sieve-tube turgor to zero. Under these conditions no exudation is possible. Exudation will also occur from the vessels after some time because of root pressure. Consequently, the phloem exudate must be collected immediately after the cut is made in order to prevent contamination with vessel sap. Despite such precautions, the phloem exudate may be contaminated with substances from other cells when obtained from a cut surface. A far more elegant method of obtaining pure sieve-tube exudate has therefore been adopted by recent investigators. The aphid is able to insert it proboscis between the cells of a plant until it reaches a sieve-tube, which it then penetrates. If the anesthetized insect is separated from its proboscis, this now acts as a micropipette from which pure sieve-tube sap continues to exude for days. With modern methods of chromatography this sieve-tube sap has been quantitatively analyzed.

PATHS OF MOVEMENT

Xylem vs. phloem

Although the ascent of water was long ago conclusively shown to occur in the xylem vessels and tracheids, the path of solute movement has been the subject of considerable controversy. According to the classical concept, mineral substances are carried up the xylem vessels by the ascending sap, but organic substances move downward in the phloem sieve-tubes. There is much indirect evidence for this view. Analysis of the xylem sap (obtained by centrifuging) reveals the presence of inorganic salts (Table 10-2). On the other hand, sieve-tube exudate consists mainly of sugars (Table 10-3). Most of the remaining solutes are accounted for by the amino acids, other organic acids, and inorganic cations and anions. The concentration of total solutes may be as high as 15% to 30%, of which 50% to 90% may be sucrose. Besides the sucrose and other solutes, the exudate contains fine filamentous structures that are protein in nature and are undoubtedly the "slime" or "P-protein" observed in intact sieve-

Table 10-2. Analysis of tracheal (vessel) sap of the pear*

Element or ion	Parts per million in sap Nov. 10	May 10
Ca	16.6	84.7
Mg	0.8	23.5
K	23.6	59.6
Fe	1.0	2.1
SO_4	8.3	31.8
Cl	3.2	4.5
PO_4	10.6	25.2

*Adapted from Anderssen, F. G. 1929. Plant Physiol. 4:459-476.

tubes (Kollmann et al., 1970). Only traces of nucleic acids and neither lipids nor polysaccharides were found in the exudate of *Cucurbita* and *Nicotiana*.

Many recent investigations have identified several larger moleculed sugars than the disaccharide (Chapter 14) sucrose in the phloem exudate, for example, raffinose (a trisaccharide), stachyose (a tetrasaccharide), verbascose (a pentasaccharide). But even these so-called oligosaccharides all consist of sucrose, combined with one, two, or three molecules of galactose, respectively; and sucrose is still the commonest sugar in the phloem of most plants and the one usually found in the largest quantity. In the case of corn, it is the only carbohydrate translocated (Hofstra

Table 10-3. Composition of the exudate obtained from incisions made in the bark of *Ricinus* plants (concentrations expressed in mg/ml and also in meq/l or mM where relevant*)

	mg/ml	
Dry matter	100-125	
Sucrose	80-106	
Reducing sugars	Absent	
Protein	1.45-2.20	
Amino acids	5.2 (as glutamic acid)	35.2 mM
Keto acids	2.0-3.2 (as malic acid)	30-47 meq/l
Phosphate	0.35-0.55	7.4-11.4 meq/l
Sulfate	0.024-0.048	0.5-1.0 meq/l
Chloride	0.355-0.675	10-19 meq/l
Nitrate	Absent	
Bicarbonate	0.010	1.7 meq/l
Potassium	2.3-4.4	60-112 meq/l
Sodium	0.046-0.276	2-12 meq/l
Calcium	0.020-0.092	1.0-4.6 meq/l
Magnesium	0.109-0.122	9-10 meq/l
Ammonium	0.029	1.6 meq/l
Auxin	10.5×10^{-6}	0.60×10^{-4} mM
Gibberellin	2.3×10^{-6}	0.67×10^{-5} mM
Cytokinin	10.8×10^{-6}	0.52×10^{-4} mM
ATP	0.24-0.36	0.40-0.60 mM
pH	8.0-8.2	
Conductance	13.2 micromhos/cm at 18° C	
Solute potential	−14.2 to −15.2 bars	
Viscosity	1.34 cP at 20° C	

*From Hall, S. M., and D. A. Baker. 1972. The chemical composition of *Ricinus* phloem exudate. Planta **106**:131-140.

and Nelson, 1969). In general, hexoses (e.g., glucose and fructose) do not occur in sieve-tubes. They cannot even be induced artificially to move in the phloem; for when ^{14}C-labeled fructose or glucose were fed to Yucca stalks, the phloem exudate from the other end always contained sucrose as the major labeled component, and no ^{14}C-labeled hexoses could be detected (Tammes et al., 1967). The same result was obtained with a soybean leaf (Trip and Gorham, 1968). Feeding experiments with ^{14}C-U-glucose and fructose led to the conclusion that the glucose and fructose that move from the mesophyll are utilized in the veins of the first and second order of branching for the synthesis of sucrose, which is then translocated by way of the midrib to the stem (Yamamoto et al., 1970). The suitability of sucrose for translocation has been explained by its nonreducing nature and by its easy hydrolyzability to glucose and fructose (Arnold, 1968). In contrast to all the above results, however, in apple trees the main translocation of carbohydrate occurs in the form of the hexose alcohol, sorbitol (Bieleski, 1969; Hansen, 1970), which is also a nonreducing substance.

Nevertheless, it cannot be concluded that the only translocatory system in which organic substances occur is the phloem. Sugars may reach a relatively high concentration in the xylem, at least during spring (Fig. 10-2). Furthermore, although it is true that organic nitrogen is translocated downward and even upward (Joy and Antcliff, 1966) in the sieve-tubes of many herbaceous plants, in the case of trees this occurs primarily, if not solely, upward in the vessels (Bollard, 1957; Barnes, 1963). The organic nitrogen compounds of major importance in the translocatory stream through the xylem of fifty-three species of trees are allantoic acid,

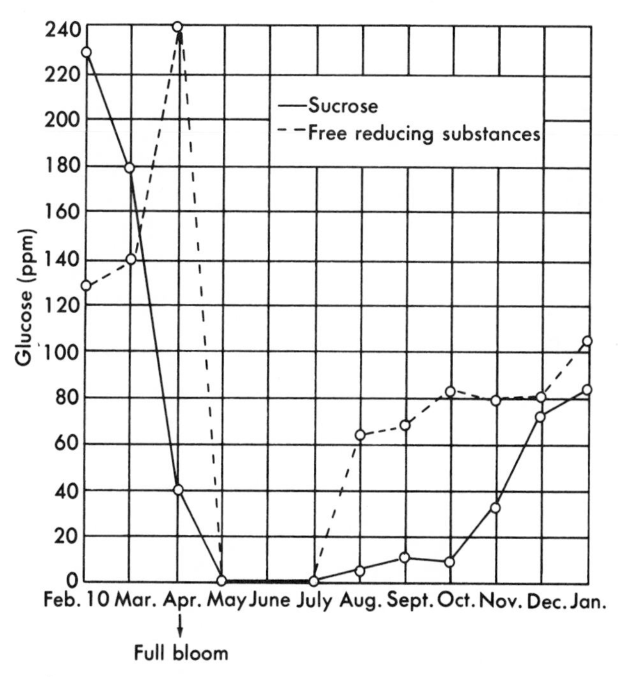

Fig. 10-2. Free reducing substances and sucrose in tracheal sap of pear branches. (From Anderssen, F. G. 1929. Plant Physiol. **4**:459-476.)

arginine, asparagine, aspartic acid, citrulline, glutamic acid, and glutamine. In seven species of *Pinus*, glutamine accounted for 73% to 88% of the nitrogen. In at least some cases this is apparently due to a selective binding of amino acids during their translocation in the xylem (Van Die and Vonk, 1967); for when ^{14}C-labeled amino acids were allowed to percolate through the xylem vessels of tomato stems, glycine and alanine were strongly retained by the stem and partly metabolized, and glutamic acid was not. In the case of *Zea mays* (Bolli, 1967) and kidney beans (Martin, 1971), translocation of amino acids appears to involve both paths of translocation. Even the inorganic ions, such as iron and manganese, may be translocated in the xylem as components of organic substances (Hoefner, 1970).

Even the early workers soon began to question the classical concept that organic substances move in the phloem, inorganic in the xylem. Many of the preceding analyses showed that this could not always be true because if sugars are found in the xylem sap, they must be translocated in this system, and if inorganic ions are found in the phloem sap, they must be translocated there. Some even suggested that all movement occurred only in the xylem, others that all occurred only in the phloem. To test these concepts it was not sufficient to analyze the two saps; it was necessary actually to show movement.

One method of determining the path of movement was to interrupt the flow by cutting across one or another of the two translocatory systems, leaving only the other intact. *Girdling* or *ringing*, for instance, is the removal of a ring of bark (containing phloem), leaving only the xylem intact. The general effects of ringing have long been known (e.g., the accumulation of carbohydrates above the ring), and such effects have been explained by the movement of sugars in the phloem. In other cases the two systems were separated from each other in order to supply substances only to one of the two systems. More recently, the movement of solutes has been followed mainly by the use of tagged atoms, for example, radioisotopes. The great advantage of this method, of course, is the ability to follow quickly and quantitatively the movement of even small amounts of introduced substances by measuring their radiation. With the aid of such substances Stout and Hoagland (1939) conclusively showed that inorganic substances move up the plant in the xylem (Fig. 10-3). These results also revealed one source of difficulties: Whenever bark and wood were left in normal contact, the substances moved readily into the phloem from the xylem. The substances in this way reached relatively high concentrations in the phloem both above and below the region of separation of bark from wood, sometimes even higher than in the adjacent xylem. Yet the extremely low concentrations in the separated phloem revealed that no movement occurred up or down the phloem other than that caused by simple diffusion.

Even these experiments must be accepted with some reservation. As in all such techniques, injury to the bark is difficult to preclude completely, and their results do not prove that movement of these substances cannot also occur in the phloem. In fact, later results conclusively showed that they can. Biddulph and Markle (1944) injected radiophosphorus into leaves of cotton plants having bark and wood separated as in Stout and Hoagland's experiment. In this case movement was downward instead of upward. They found it moved only via the phloem and at rates in excess of 21 cm/hr (Table 10-4). As in Stout and Hoagland's experiments, the radioisotope was able to move from one tissue to the other when the two were left in contact. These results also seem to disprove any suggestion of injury to the phloem in Stout and Hoagland's experiments because if separation of the

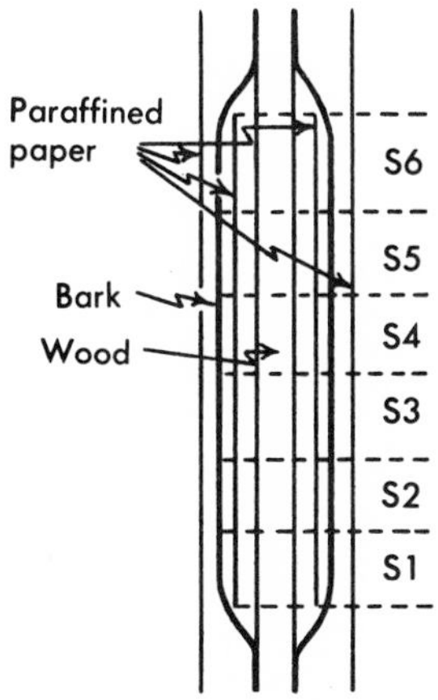

Paraffined paper

Bark

Wood

Willow segment	Willow (Gain of K, 5 hr after feeding) Bark	Wood
	53	47
S6	11.6	119
S5	0.9	122
S4	0.7	112
S3	<0.3	98
S2	<0.3	108
S1	20	113
	84	58

Geranium segment	Geranium (Gain of PO_4, 6 hr after feeding) Bark	Wood
	270	860
S8	9.0	112
S7	0.5	120
S6	0.6	132
S5	0.8	138
S4	<0.3	147
S3	<0.5	137
S2	<0.3	152
S1	11.1	131
	316	442

Fig. 10-3. Upward movement of radioisotopes fed to the roots of willow and geranium (5 meq per liter). Bark is stripped from the wood 1½ hr before feeding. (Redrawn from Stout, P. R., and D. R. Hoagland. 1939. Amer. J. Bot. **26**:320-324.)

Table 10-4. Downward movement of radiophosphorus injected into a leaf of a cotton plant; leaf strip immersed in solution for 5 min, then in water; bark stripped as in Fig. 10-3*

	Micrograms PO_4 in stripped stem after 3 hr			Micrograms PO_4 in unstripped stem after 1 hr		
	Bark		Wood	Bark		Wood
		0.904			0.444	
Bark separated from wood	0.684		0.004	0.160		0.055
	0.544		0.000	0.103		0.063
	0.120		0.019	0.055		0.018
	0.123		0.002	0.026		0.007
		0.160			0.152	

*Adapted from Biddulph, O., and J. Markle. 1944. Amer. J. Bot. **31**:65-70.

phloem from the xylem fails to prevent downward movement in the phloem, it is hardly likely to prevent upward movement. But it is not yet possible to generalize safely, since not all inorganic substances have been investigated. Manganese, for instance, has been reported to be translocated mainly in the phloem (Klimovitskaya, 1963).

Without the help of radioisotopes, Mason, Phillis, and Maskell (see Mason and Phillis, 1937) had earlier produced evidence that nitrogen, phosphorus, potassium, sulfur, magnesium, and chlorine are phloem mobile in a downward direction but that calcium is not (in opposition to exudate analyses, Table 10-3). Other results have confirmed this immobility of calcium in the phloem (Arisz, 1952). According to Ziegler (1961), calcium accumulates markedly in transpiring organs because it is translocated in the xylem, but potassium is translocated mainly in the phloem and its concentration in the plant is much less dependent on transpiration. Therefore plant parts such as fruit or nectaries that are supplied mainly by the phloem always contain much higher quantities of potassium than of calcium. But others have shown that at least in some plants there can be a slow movement of calcium in the phloem.

An elegant method of measuring ions quantitatively in undisturbed plants or their individual cells, the electron probe x-ray microanalyzer, has revealed long-distance transport of potassium, phosphorus, calcium, and strontium through the fruit stalks of pea pods, mainly by way of the sieve-tubes (Läuchli, 1968). The calcium supply to the sieve-tubes, and therefore its translocation through them, was small. In potato plants, ^{32}P moves both in the xylem and the phloem, but both calcium and strontium are relatively immobile in the phloem (Baker and Moorby, 1969). The movement of calcium in the phloem apparently requires high concentrations to saturate the fixation capacity of the plant and to leave some in the mobile form (Wiersum et al., 1971). In contrast to calcium, magnesium moves readily in the phloem. Fully 7% to 11% of the ^{28}Mg applied to bean and barley leaves was transported basipetally (Steucek and Koontz, 1970). Since it was not transported past a heat-killed section, translocation apparently occurred in the phloem. This conclusion was supported by autoradiograms.

Thus we are forced to admit that inorganic substances may move both in the xylem and the phloem. However, the available evidence favors the view that upward movement of these substances is primarily, if not solely, in the xylem; downward movement under normal conditions is probably in the phloem. Since by far the greatest part of the movement of these substances is in the upward direction, this would assign to the xylem the main path of movement of inorganic substances, whereas the phloem is normally almost the sole path of movement of organic substances (except for the upward movement in trees [1] of organic nitrogen and [2] of sugars during spring) and the path of downward movement of all solutes. The classical concept is therefore reasonably near the truth.

Sources and sinks

Since movement in the phloem may be either downward or upward, what controls the direction of flow at any one time? The commonly observed downward movement in the phloem is from the region of synthesis in the leaves to regions of utilization such as the roots and underground storage organs. These two regions are referred to as the *source* and the *sink*, respectively (Beevers, 1969). But the sink is not necessarily below the source. There is considerable movement upward in the phloem from the fully grown, photosynthesizing leaves to the growing leaves, stems, and fruit (particularly the seed). In orange trees, for instance, all the expanding leaves showed a strong import of ^{14}C assimilates but no export, whereas fully expanded leaves exported assimilates, principally to nearby fruits (Kriedermann, 1969). Mature leaves some distance from the sites of current growth activity transported principally to the roots. Tomato leaves that have attained half to two thirds of their full growth export and import ^{14}C simultaneously (Bonnemain, 1969).

Similarly, the source is not always the leaves. What was originally a sink during its development (e.g., underground storage organs, seeds, storage cells in the xylem) may later become a source, supplying the newly formed sinks, which are the sprouts, above it. It has long been believed that the source and sink relationship is dependent on the maintenance of a diffusion or concentration gradient, due to rapid utilization of the translocated substances for cell growth in the sink region. Some evidence, however, points to active absorption in the sink leaf as a factor augmenting the translocation (Geiger and Christy, 1971). Similarly, loading of the phloem in the source region must be a factor (Beevers, 1969), and this may involve active absorption by the phloem cells or secretion by the mesophyll or bundle sheath cells nearest to them. The relation is shown in Fig. 10-4.

Symplasmic movement

Only the translocation in the vascular tissue (xylem and phloem) can occur rapidly enough to supply the needs of the higher plant. Nevertheless, translocation over considerable distances can and must occur by another path in parts of the plant not adequately supplied by vascular tissue. Arisz (1952) suggested that this kind of translocation occurs by way of the *symplasm*—the protoplasm of the plant as a whole—through the well-known cell-to-cell interconnections called plasmodesmata. The interconnected regions in the plant that are outside the protoplasm he called the *apoplasm*. Many investigators have supported Arisz's concept. Investigations of xylem exudate of maize roots led to the conclusion that ion movement to the xylem is mainly symplasmic from the plasmalemma of the epidermal cells to the plasmalemma of the xylem parenchyma (Dunlop and Bowling, 1971; Läuchli and Epstein, 1971*a*). The rate is rapid, attaining values of 75 to 250 cm/hr (Läuchli and Epstein, 1971*b*), which are within the range of rates of movement in the phloem. Similar symplasmic movement must occur from the mesophyll cells in the leaf, where synthesis occurs, to the phloem, leading to "vein loading" of translocate (Geiger et al., 1971), which is also rapid, requiring only 2 to 3 min in rhubarb and 10 min in maize (Biddulph, 1969).

MECHANISMS OF MOVEMENT

The two classical theories

As to the mechanism of movement in the phloem, two principal theories have been proposed: (1) the mass, bulk, or

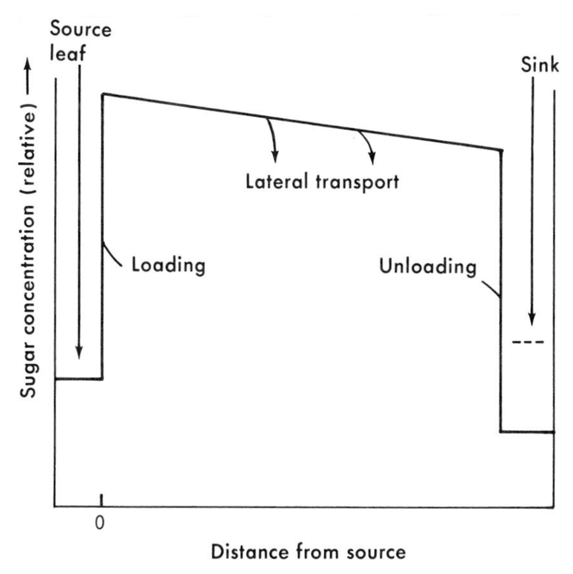

Fig. 10-4. Concentration gradients between source and sink (schematic). (From Beevers, H. 1969. *In* J. D. Eastin et al. [Eds.]. Physiological aspects of crop yield. Copyright 1969 by the American Society of Agronomy and reprinted by permission of the copyright owner.)

pressure flow theory (for arguments in its favor see Crafts, 1951, 1961, Crafts and Crisp, 1971, and Esau, 1966), and (2) the cytoplasmic streaming theory (for the earlier defense of this theory see Curtis and Clark, 1950). A third concept, "activated (or facilitated) diffusion," is simply giving a name to an unknown process. It is based on two facts: (1) the movement usually follows a diffusion gradient and (2) the rate is much too rapid (40,000 times in some cases) to be accounted for by diffusion. No attempt is made to explain the "activation" mechanism.

The first of these was proposed by Münch in 1926 (see Münch, 1931). He thinks of translocation in the plant as a kind of circulation analogous to that of the bloodstream, with the mesophyll cells in the leaves acting as a kind of heart. Water and salts move up the xylem, organic substances move down the phloem, and an osmotic mechanism controls the latter (Fig. 10-5). As a result of photosynthesis, the cell sap concentration of the mesophyll cells at the top of this circulatory system is maintained high, even though some solutes are being exported down the phloem. Because of the upward movement of water in the xylem, the water content is maintained at a high level. These two factors combine to yield a high turgor pressure in the mesophyll cells. Small pores known as pits occur in the cell wall through which cytoplasmic strands, known as plasmodesmata, connect these living cells with each other. These connections occur from mesophyll cell to mesophyll cell and ultimately to the sievetubes. Through these connections the turgor pressure succeeds in forcing some of the cell solution into and down the sievetubes. The loss of solutes from the mesophyll cells would then be compensated for by newly synthesized organic substances, the loss of water by movement from the xylem. Excess water forced into the phloem would be "squeezed" laterally through the

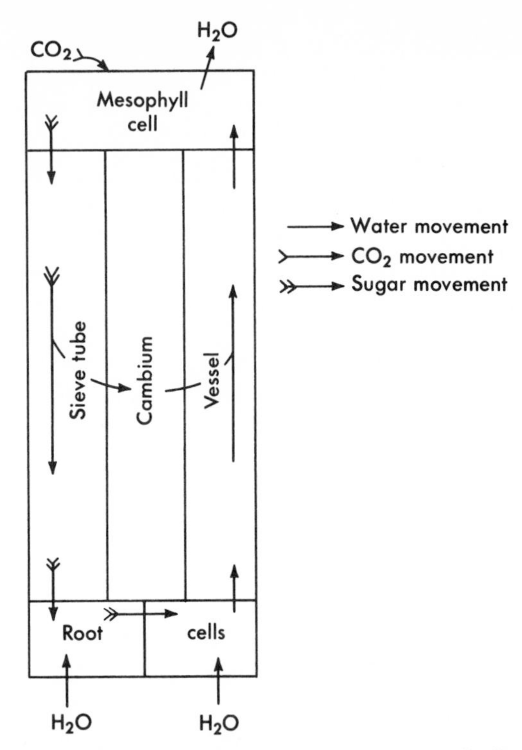

Fig. 10-5. Diagrammatic representation of the pressure flow theory. (Redrawn from Münch, E. 1931. Die Stoffbewegungen in der Pflanze. Gustav Fischer Verlag, Stuttgart.)

cambium into the xylem at the same time as the solutes are conducted downward. Movement in the phloem would then always have to be from a region of high to one of low turgor pressure. Therefore it could occur in an upward direction from storage tissues. But at any one time all substances would have to move in the phloem in the same direction. Movement would also depend on the existence of open plasmodesmata between all living cells. There are many other factors that have to be taken into account in evaluating the Münch "Druckström" theory. The strongest experimental evidence in its favor is the existence of exudation pressure in the phloem. Quantitatively, the theory also has experimental support:

1. Direct measurements (Hammel, 1968) indicate a turgor pressure from 0 to

3 atm higher at the upper levels than at the lower levels of oak trees, yielding a pressure gradient of approximately 0.5 atm/m.

2. The measured velocity is of the order of magnitude required by the assumption that the sieve-tube exudate is a moving solution (Zimmerman, 1969), and calculations reveal that each sieve-tube may have to be refilled as many as 3 to 10 times/sec.

The Münch mass-flow or pressure-flow theory therefore gives a complete, logical, and apparently reasonable concept of translocation in the phloem and has been accepted by many physiologists as "the mechanism."

Unfortunately, other direct attempts to prove the existence of a solution flow, at least in many cases, have yielded negative results. One of the central features of any mass flow concept is that the water must move along with the solutes. In some cases, water movement in the phloem has been conclusively demonstrated (Biddulph, 1969). Yet Aronoff and his group (see Peel et al., 1969) have consistently obtained negative results with labeled water. In agreement with them, Peel et al. (1969) have found substantial amounts of ^{14}C-sugars and ^{32}P-phosphates in the stylet exudate within 1 hr of application but little or no tritiated water even after 8 hr. Peel (1970) has therefore concluded that water is relatively immobile in the sieve-tubes. Furthermore, the pressure that is known to exist in the sieve-tubes does not behave as it would in externally operating flow mechanisms; it does not immediately cease with the severing of a phloem bundle under 0.1 M sucrose but slowly dies away (Fensom et al., 1968). Finally, one characteristic feature of sieve-tubes, for which there is no conceivable role in the pressure flow theory, is the existence of the slime bodies.

The second theory was suggested by De Vries and later championed by Curtis (1935). According to this concept, movement is caused by a combination of diffusion and cytoplasmic streaming. Diffusion would occur from sieve-tube to sieve-tube across the intervening cell wall. Within the sieve-tube the substances that have diffused in from the cell above would be carried downward by the cytoplasmic stream; those that diffused in from the cell below would be carried upward by the same cyclic stream. Thus this theory accounts for the known movement of substances along concentration gradients and for the apparent fact that two substances can be simultaneously translocated in opposite directions (Mason and Phillis, 1937; Palmquist, 1939). By applying two differently labeled substances to opposite ends of the phloem, many investigators have apparently demonstrated this bidirectional flow. Some of the evidence (Biddulph and Cory, 1965) indicates that this occurs in separate bundles; other evidence (Eschrich, 1967), using the aphid method, seems to demonstrate bidirectional flow within the same sieve-tube. Unfortunately there are a number of complicating factors, such as lateral movement between adjacent sieve-tubes (Ho and Peel, 1969) or even between the wood and bark (Hardy and Possingham, 1969). Interpretation of these results is therefore difficult. Even a bidirectional flow in different bundles, which appears to be accepted by proponents of the pressure flow theory (Peterson and Currier, 1969), is opposed to Münch's postulated unidirectional flow in the phloem along a turgor pressure gradient from the leaves to the roots.

In opposition to such results there is evidence from other sources that some organic substances can be translocated only in the same direction as sugars, for example, the weed killer 2,4-D and viruses. Several other objections have been raised against the cytoplasmic streaming theory, although some at least are no longer valid. Thus it was stated that no streaming occurs in sieve-tubes. But careful technique has now succeeded in demonstrating that it does occur (Thaine, 1964). Perhaps the most difficult point to reconcile with this theory is the fact that the phloem exudate is apparently cell sap and not cytoplasm, yet it contains high concentrations of the translocated substances (Table 10-3). This would seem to indicate that translocation is by way of the vacuole rather than the cytoplasm. Electron micrography, however, may have eliminated this objection because if the tonoplast really does disappear in the mature sieve-tube, cytoplasm and vacuole would become one system ("myctoplasm," Esau, 1966). Thaine, in fact, has observed in living cells what he calls "transcellular strands" through which streaming occurs. This streaming is therefore in the center of the sieve-tube and appears to traverse the sieve-tube pores, continuing from cell to cell. According to these observations, the diffusion part of the cytoplasmic streaming theory would be eliminated, and translocation would be caused by streaming not only within each cell but also from cell to cell. However, the opposing argument states that the observed rates of streaming are far too slow to account for the measured translocation rates.

A quantitative comparison, however, no longer upholds this objection. Streaming rates as high as 107 μm/sec at 27° C have been recorded (Barr and Broyer, 1964). This is 6.5 mm/min or about one eighth the maximum rate of translocation found in the phloem (5 cm/min). Since the phloem is highly specialized for translocation and possesses both ample respirable material and an active ATP–forming system, it is not unreasonable to postulate a streaming rate 8 times that in nonspecialized cells. However, Fensom et al. (1968) observed cyclosis within the sieve-tube but not across the sieve plate, and it was only at one tenth the velocity of transport known to occur in the phloem. They suggest that a fast-moving fluid passes along and over the elastic threads of protoplasm inside the sieve-tubes. They were later able to support this concept by means of a technique using Nomarski optics that permitted microscopic observation of functioning sieve-tubes in the dissected but intact phloem of *Heracleum* (Lee et al., 1971). Plastids and other organelles were visible in a rapid, bouncing motion that exceeded Brownian movement. In mature cells, neither a vacuole nor transcellular strands were detectible. The movement of organelles was not cytoplasmic streaming but appeared to be due to either the solution moving past them or to a contractile pulsatory motion. Therefore they concluded that the movement was an "activated" mass flow.

The electroosmotic theory

A third, electrokinetic (or electroosmotic) mechanism of translocation in the phloem has been proposed by Fensom (1957) and Spanner (1958). It is basically a mass flow theory, the motivating force being an electric potential gradient rather than the turgor pressure gradient of Münch. The postulated electric gradient would induce a flow of ions and therefore of current within the sap. It is known that when ions in an aqueous solution flow along an electric gradient, they drag water with them, and this movement of water is known as *electroosmosis*. The electroosmosing water would carry with it any solutes present in the phloem sap. Thus the electrokinetic or electroosmotic theory has all the disadvantages of the pressure flow theory (e.g., the large bulk of water that must be carried) but sacrifices its simplicity. Nevertheless, it has been subjected to several tests, all of which have yielded negative results. Fensom and Spanner (1969), for instance, concluded from their measurements that a fantastically high electric gradient of 28 v/cm might be necessary to account for the known flow rates in the phloem. Further tests (Tyree and Fensom, 1970) overcame some of the earlier difficulties by using the giant umbellifer *Heracleum mantegazzianum*. This plant has discrete vascular bundles of appreciable size that can be isolated from the surrounding tissue and split at the cambium into separate xylem and phloem while remaining attached to the plant at both ends. Experiments with this plant led them to conclude that an electrokinetic mechanism of sieve-tube transport is unlikely. A small electroosmotic component of at least 0.1 mv/cm was suggested by their measurements. Other experiments indicated that the large transfer of K^+ required by the theory probably does not occur (Mishra and Spanner, 1970). Despite all this completely negative evidence, Spanner (1970, 1971) still supports the theory.

Present status of theories

It is now obvious that both Münch's pressure-flow theory and De Vries' cytoplasmic streaming theory as originally stated are incorrect. The heartlike pump visualized by Münch does not exist in the mesophyll cells, since the turgor pressure gradient is from sieve-tube to mesophyll cell instead of the reverse (Arisz, 1952); furthermore, the concentration of sucrose in the sieve-tubes is 20 times that in the mesophyll cells of the leaf (Wanner, 1953). This reverse concentration gradient also eliminates Münch's concept of a closed cycle of flow that continues up the xylem into the mesophyll cells and from the mesophyll cells down the phloem.

On the other hand, many lines of evidence indicate that the flow through the sieve-tubes is dependent on metabolism. Hexoses from the mesophyll cells do move into the sieve-tubes, and evidence has been produced for the existence in sieve-tubes of hexokinase. This hexokinase is believed to participate in the transport of the hexoses into the cell (Kursanov et al., 1970). But once in the sieve-tubes, the hexoses are completely converted to sucrose by what must be a metabolic process. As a result, the osmotic concentration of the sieve-tube soon exceeds that of the mesophyll cells supplying it. Labeling has proved that some of the sucrose that moves down the sieve-tubes is respired to carbon dioxide (Canny, 1962). It is even possible to inhibit sieve-tube translocation reversibly by a localized application of cyanide, a metabolic inhibitor (Willenbrink, 1966).

Further evidence of active metabolism in the phloem is the conversion of inorganic phosphorus to nucleotide triphosphate in the phloem within 10 min (Bieleski, 1969). Both in dicotyledons and monocotyledons the phloem exudate contains a high level of ATP (an average of 615 μg/ml, Kluge et al., 1970). A correlation has been obtained between ATP level and translocation (Geiger and Christy, 1971). Of thirty enzymes investigated, twenty-four were found to occur in the sieve-tube sap of trees (Kennecke et al., 1971). Early investigators recognized the role of metabolism by calling the movement "activated diffusion."

In short, the translocation of solutes in the phloem cannot be a simple cycling as visualized by the two major theories, but it must include many processes associated with the life of the cells, for example: (1) active absorption of sugars and other substances by the sieve-tubes and (2) enzymatically controlled syntheses of larger moleculed sugars from smaller ones. Respiratory energy is required for both of these and perhaps for other associated processes. It is this participation of numerous physiological processes that makes it so difficult to determine the exact mechanism.

In the light of modern evidence, those who cling to the Münch hypothesis must modify it as follows: The initiating turgor pressure in the mesophyll cells, the passive flow across plasmodesmata between mesophyll cells and sieve-tubes, and the complete cycle through the mesophyll cells and root parenchyma would all have to be eliminated. The initiating force would now have to be the turgor pressure in the sieve-tubes themselves, which would follow a gradient with the maximum pressure at the top of the sieve-tube adjacent to the source of supply in the mesophyll cells. But the development of this pressure would depend not only on the photosynthetic production of sugars in the mesophyll cells but also on a presumably respiratory release of energy that would enable the active absorption of hexoses from the mesophyll cells and their immediate enzymatic conversion to sucrose and higher sugars. Even if this profound modification of the Münch hypothesis should prove correct, it would be only a partial explanation of the complex mechanism of translocation in the phloem, for it seems entirely inadequate to explain the demonstrated phloem transport over large distances (Zimmerman, 1969). Conclusions that a mass flow can be explained by simple fluid mechanics (Tammes et al., 1971) are based on pressure gradients 10 to 20 times those usually found (Zimmerman, 1969).

On the other hand, those who wish to champion the cytoplasmic streaming theory would also have to make modifications in the original concept, for example, elimination of diffusion as a component if transcellular streaming occurs, and the discovery of some mechanism to speed up the flow in the intact plant.

The obvious conclusion from the information explosion is that none of the proposed mechanisms by itself is an adequate explanation of translocation in the phloem. Weatherley and Johnson (1968) have come to this conclusion on the basis of a quantitative evaluation in relation to measured translocation rates. Several investigators have therefore suggested combining two concepts. Zimmerman (1969) suggests some kind of unification of transcellular protoplasmic streaming theory and the pressure flow hypothesis. Fensom and Davidson (1970) propose two mechanisms operating simultaneously: a slow mass flow of potassium and a much more rapid flow of sucrose involving waves or pulses. The two classical theories have, indeed, become so modified in recent years that they appear to be approaching each other, leading to a theory of a single mechanism that includes the remaining essential features of each. The nature of this postulated single mechanism presumably depends on the slime bodies, to which more and more investigators are turning for the missing link between the two theories. Their extreme thinness has confined explanations of their role to mechanisms involving an acceleration of the rate of flow of the bulk of sieve-tube sap between them (Wooding, 1967). Unfortunately, the flow within the living and uncut phloem cannot be observed at a high enough magnification to resolve these slime bodies. In the natural state, perhaps the tubular slime bodies may dilate to several times the dehydrated diameter observed with the electron microscope. The pressure flow could then be initiated osmotically at the top of each tubule, due to an active accumulation of sucrose formed from the hexoses of the adjacent mesophyll cells. This would be followed by an osmotic uptake of water by the tubules from the surrounding phloem sap. The hydrostatic pressure generated in this way at the top of the tubules could then produce a pulselike mass flow down the sieve-tube. It could also be interpreted as a cytoplasmic streaming mechanism, since the tubules are a kind of cytoplasmic strand or fibril. As the solution is forced through the narrow pores in the sieve plates, the water might be at least partially "wrung out" of the tubules, remaining behind in the sieve-tube sap. This would explain the frequently reported lack of movement of the phloem water with the solutes. Unfortunately, it is at present impossible to test such hypothetical mechanisms. Furthermore, the slime bodies are apparently absent from the sieve-tube elements of corn (Singh and Srivastava, 1972). More information is needed before the mechanism of translocation in the phloem can be understood.

QUESTIONS

1. What are the two translocatory tissues?
2. Which cells in each are primarily involved?
3. Are these cells dead or alive?
4. Are they normally distended or partially collapsed?
5. Are they semipermeable or freely permeable?
6. Are they thin walled or thick walled?
7. Do they have high or low sap concentrations?
8. Do they have high or low turgor pressures?
9. Do they exude or absorb on being cut?
10. How can one obtain vessel sap for analysis?
11. How can one obtain sieve-tube sap for analysis?
12. What is the major constituent of each sap?
13. Does vessel sap ever contain organic substances?
14. How can one interrupt movement in one system without affecting the other?
15. In what direction are inorganic substances mainly translocated?
16. In what direction are organic substances mainly translocated?
17. What evidence is there for this (or these) direction(s) of movement?
18. What are isotopes?
19. What do results with isotopes indicate?
20. What do the symptoms of potassium deficiency indicate about the direction of potassium translocation?
21. Is the translocation of carbohydrates in the direction expected from the laws of diffusion?
22. Would this rate be expected from the laws of diffusion?
23. What theory explains the translocation of solutes in the xylem?
24. What theory explains the translocation of solutes in the phloem?
25. What is the nature of the force according to the pressure-flow theory?
26. What would be the direction of flow according to this theory?
27. Would different substances move in different directions or the same direction?
28. What other theory is there?
29. In which theory are plasmodesmata required?
30. In which theory is diffusion involved?
31. What is the sugar that is translocated?
32. What is the evidence for metabolism?
33. What is the evidence against pressure flow?
34. What is the evidence against cytoplasmic streaming flow?

SPECIFIC REFERENCES

Anderson, R., and J. Cronshaw. 1969. The effects of pressure release on the sieve plate pores of *Nicotiana*. J. Ultrastruct. Res. **29**:50-59.

Anderssen, F. G. 1929. Some seasonal changes in the tracheal sap of pear and apricot trees. Plant Physiol. **4**:459-476.

Arisz, W. H. 1956. Significance of the symplasm theory for transport across the root. Protoplasma **46**:1-62.

Arnold, W. N. 1968. The selection of sucrose as the translocate of higher plants. J. Theor. Biol. **21**:13-20.

Baker, D. A., and J. Moorby. 1969. The transport of sugar, water, and ions into developing potato tubers. Ann. Bot. (London) **33**:729-741.

Barnes, R. L. 1963. Organic nitrogen compounds in tree xylem sap. Forest Sci. **9**:98-102.

Barr, C. E., and T. C. Broyer. 1964. Effect of light on sodium influx, membrane potential and protoplasmic streaming in *Nitella*. Plant Physiol. **39**:48-52.

Beevers, H. 1969. Metabolic sinks. p. 169-180. *In* Eastin, J. D., F. A. Haskins, C. Y. Sullivan, and C. H. M. van Bavel (eds.). Physiological aspects of crop yield. Amer. Soc. Agron., Crop Sci. Soc. Amer., Madison, Wisc.

Behnke, H.-D. 1969*a*. Aspects of sieve-tube differentiation in monocotyledons. Protoplasma **68**: 289-314.

Behnke, H.-D. 1969*b*. Contributions to fine structure and dispersal of plasmatic filaments in sieve tubes and to development and structure of sieve pores in some monocotyledons and in Nuphar. Protoplasma **68**:377-402.

Biddulph, O. 1969. Mechanisms of translocation of plant metabolites. p. 143-164. *In* Eastin, J. D., F. A. Haskins, C. Y. Sullivan, and S. H. M. van Bavel (Eds.). Physiological aspects of crop yield. Amer. Soc. Agron., Crop Sci. Soc. Amer., Madison, Wisc.

Biddulph, O., and R. Cory. 1965. Translocation of C^{14} metabolites in the phloem of the bean plant. Plant Physiol. **40**:119-129.

Biddulph, O., and J. Markle. 1944. Translocation of radiophosphorus in the phloem of the cotton plant. Amer. J. Bot. **31**:65-70.

Bieleski, R. L. 1969. Accumulation and translocation of sorbitol in apple phloem. Aust. J. Biol. Sci. **22**:611-620.

Bollard, E. G. 1957. Translocation of organic nitrogen in the xylem. Aust. J. Biol. Sci. **10**:292-301.

Bolli, H. K. 1967. Absorption und Transport markierter Aminosäuren durch Wurzeln von *Zea mays*. Ber. Schweiz. Bot. Ges. **77**:61-102.

Bonnemain, J.-L. 1969. Transport of ^{14}C assimilated by the growing leaves of the tomato in relationship to growth. Compt. Rend. Acad. Sci. (Paris) **269**:1660-1663.

Buvat, R. 1963. Infrastructure et différenciation des cellules criblée de *Cucurbita pepo*. Compt. Rend. Acad. Sci. (Paris) **256**:5193-5195.

Canny, M. J. 1962. The translocation profile; sucrose and carbon dioxide. Ann. Bot. **26**:181-196.

Canny, M. J., and O. M. Phillips. 1963. Quantitative aspects of a theory of translocation. Ann. Bot. **27**:379-402.

Crafts, A. S., and H. B. Currier. 1963. On sieve tube function. Protoplasma **57**:188-202.

Cronshaw, J., and K. Esau. 1968. P protein in the phloem of Cucurbita II. The P protein of mature sieve elements. J. Cell Biol. **38**:292-303.

Davis, J. D., and R. F. Evert. 1970. Seasonal cycle of phloem development in woody vines. Bot. Gaz. **131**:128-138.

Dunlop, J., and D. J. F. Bowling. 1971. The movement of ions to the xylem exudate of maize roots. III. The location of the electrical and electrochemical potential differences between the exudate and the medium. J. Exp. Bot. **22**:453-464.

Eschrich. W. 1967. Bidirektionelle Translokation in Siebröhren. Planta **73**:37-49.

Evert, R. F., J. D. Davis, C. M. Tucker, and F. J. Alfieri. 1970. On the occurrence of nuclei in mature sieve elements. Planta **95**:281-296.

Evert, R. F., and B. P. Deshpande. 1969. Electron microscope investigation of sieve-element ontogeny and structure in *Ulmus americana*. Protoplasma **68**:403-432.

Evert, R. F., and B. P. Deshpande. 1971. Plastids in sieve elements and companion cells of *Tilia americana*. Planta **96**:97-100.

Fensom, D. S. 1957. The bio-electric potentials of plants and their functional significance. I. An electrokinetic theory of transport. Canad. J. Bot. **35**:573-582.

Fensom, D. S., R. Clattenburg, T. Chung, D. R. Lee, and D. C. Arnold. 1968. Moving particles in intact sieve-tubes of *Heracleum mantegazzianum*. Nature **219**:531-532.

Fensom, D. S., and H. R. Davidson. 1970. Microinjection of ^{14}C-sucrose into single living sieve tubes of *Heracleum*. Nature **227**:857-858.

Fensom, D. S., and D. C. Spanner. 1969. Electroosmotic and bio-potential measurements on phloem strands of *Nymphoides*. Planta **88**:321-331.

Geiger, D. R., and A. L. Christy. 1971. Effect of sink region anoxia on translocation rate. Plant Physiol. **47**:172-174.

Geiger, D. R. J. Malone, and D. A. Catalo. 1971. Structural evidence for a theory of vein loading of translocate. Amer. J. Bot. **58**:672-675.

Geiger, D. R., and S. A. Sovonick. 1970. Temporary inhibition of translocation velocity and mass transfer rate by petiole cooling. Plant Physiol. **46**:847-849.

Hall, S. M., and D. A. Baker. 1972. The chemical composition of *Ricinus* phloem exudate. Planta **106**:131-140.

Hammel, H. T. 1968. Measurement of turgor pressure and its gradient in the phloem of oak. Plant Physiol. **43**:1042-1048.

Hansen, Poul. 1970. ^{14}C-studies on apple trees. V. Translocation of labeled compounds from leaves to fruit and their conversion within the fruit. Physiol. Plant. **23**:564-573.

Hardy, P. J., and J. V. Possingham. 1969. Studies on translocation of metabolites in the xylem of grapevine shoots. J. Exp. Bot. **20**:325-335.

Hartt, C. E., H. P. Kortschak, A. J. Forbes, and G. O. Burr. 1963. Translocation of C^{14} in sugarcane. Plant Physiol. **38**:305-318.

Hatch, M. D., and K. T. Glasziou. 1964. Direct evidence for translocation of sucrose in sugarcane leaves and stems. Plant Physiol. **39**:180-184.

Ho, L. C., and A. J. Peel. 1969. Investigation of bidirectional movement of tracers in sieve tubes of *Salix viminalis* L. Ann. Bot. **33**:833-844.

Hoefner, W. 1970. Iron- and manganese-contain-

ing compounds in the bleeding sap of *Helianthus annuus*. Physiol. Plant. **23**:673-677.
Hofstra, G., and C. D. Nelson. 1969. The translocation of photosynthetically assimilated ^{14}C in corn. Canad. J. Bot. **47**:1435-1442.
Jones, O. P., and R. W. Rowe. 1968. Sampling the transpiration stream in woody plants. Nature **219**:403.
Joy, K. W., and A. J. Antcliff. 1966. Translocation of amino acids in sugar beet. Nature (London) **211**:210-211.
Kennecke, M., H. Ziegler, and M. A. Rongine Defekete. 1971. Enzyme activities in the sieve tube sap of *Robinia pseudoacacia* L. and of other tree species. Planta **98**:330-356.
Kleinig, H., I. Dörr, C. Weber, and R. Kolemann. 1971. Filamentous proteins from plant sieve tubes. Nature **229**:152-153.
Klimovitskaya, Z. M. 1963. The role of the root system in the process of transferring manganese in plants. [Transl. title] Ref. Zh. Biol. No. 8G36. Moscow.
Kluge, M., D. Becker, and H. Ziegler. 1970. Investigations on ATP and other organic phosphorus compounds in the sieve tube sap of *Yucca flaccida* and *Salix triandra*. Planta **91**:68-79.
Kollmann, R., I. Doerr, and H. Kleinig. 1970. Protein filaments: structural components of the phloem exudate. I. Observations with *Cucurbita* and *Nicotiana*. Planta **95**:86-94.
Kriedermann, P. E. 1969. ^{14}C translocation in orange plants. Aust. J. Agr. Res. **20**:291-300.
Kursanov, A. L., S. V. Sokolova, and M. V. Turkina. 1970. Hexokinase in conducting tissues of sugar-beet and its possible connection with transport of sugars through cell membranes. J. Exp. Bot. **21**:30-39.
Läuchli, A. 1968. Untersuchungen mit der Röntgen-Mikrosonde über Verteilung und Transport von Ionen in Pflanzengeweben. II. Ionentransport nach des Fruchten von *Pisum sativum*. Planta **83**:137-149.
Läuchli, A., and E. Epstein. 1971*a*. Lateral transport of ions into the xylem of corn roots. I. Kinetics and energetics. Plant Physiol. **48**:111-117.
Läuchli, A., and E. Epstein. 1971*b*. Lateral transport of ions into the xylem of corn roots. II. Evaluation of a stelar pump. Plant Physiol. **48**:118-124.
Lee, D. R., D. C. Arnold, and D. S. Fensom. 1971. Some microscopical observations of functioning sieve tubes of *Heracleum* using Nomarski optics. J. Exp. Bot. **22**:25-38.
Martin, P. 1971. Pathways of upward translocation on nitrogen in kidney bean plants after uptake by the root. Z. Pflanzenphysiol. **64**:206-222.
Maxwell, Margaret A. B., and R. G. S. Bidwell. 1970. Synthesis of asparagine from $^{14}CO_2$ in wheat seedlings. Canad. J. Bot. **48**:923-927.
McNairn, R. B. 1970. The influence of darkness on heat-induced sieve plate callose and on axial phloem translocation in *Gossypium hirsutum*. L. Phyton. Rev. Int. Bot. Exp. **27**:69-74.
Milburn, J. A. 1970. Phloem exudation from castor bean: induction by massage. Planta **95**:272-276.
Mishra, V., and D. C. Spanner. 1970. The fine structure of the sieve tubes of *Salix caprea* (L) and its relation to the electroosmotic theory. Planta **90**:43-56.
Murmanis, L., and R. F. Evert. 1966. Some aspects of sieve cell ultrastructure in *Pinus strobus*. Amer. J. Bot. **53**:1065-1078.
Palmquist, E. M. 1939. The path of fluorescein movement in the kidney bean, *Phaseolus vulgaris*. Amer. J. Bot. **26**:665-667.
Parthasarathy, M. V., and K. Mühlethaler. 1969. Ultrastructure of protein tubules in differentiating sieve elements. Cytobiologie **1**:17-36.
Peel, A. J. 1970. Further evidence for the relative immobility of water in sieve tubes of willow. Physiol. Plant. **23**:667-672.
Peel, A. J., R. J. Field, C. L. Coulson, and D. C. J. Gardner. 1969. Movement of water and solutes in sieve tubes of willow in response to puncture by aphid stylets: evidence against a mass flow of solution. Physiol. Plant. **22**:768-775.
Peterson, C. A., and H. B. Currier. 1969. An investigation of bidirectional translocation in the phloem. Physiol. Plant. **22**:1238-1250.
Schmitz, K., and J. Willenbrink. 1968. Zum Nachweis tritierter Assimilate in den Siebrohren von *Cucurbita*. Planta **83**:111-114.
Singh, A. P., and L. M. Srivastava. 1972. The fine structure of corn phloem. Canad. J. Bot. **50**:839-846.
Spanner, D. C. 1958. The translocation of sugar in sieve tubes. J. Exp. Bot. **9**:332-342.
Spanner, D. C. 1970. The electro-osmotic theory of phloem transport in the light of recent measurements on *Heracleum* phloem. J. Exp. Bot. **21**:325-334.
Spanner, D. C. 1971. Transport in the phloem. Nature **232**:157-160.
Spanner, D. C., and R. L. Jones. 1970. The sieve tube wall and its relation to translocation. Planta **92**:64-72.
Steer, M. W. and E. H. Newcomb. 1969. Development and dispersal of P-protein in the phloem of *Coleus blumei* Penth. J. Cell. Sci. **4**:155-169.
Steucek, G. L., and H. V. Koontz. 1970. Phloem mobility of magnesium. Plant Physiol. **46**:50-52.

Stout, P. R., and D. R. Hoagland. 1939. Upward and lateral movement of salt in certain plants as indicated by radioactive isotopes of potassium, sodium, and phosphorus absorbed by roots. Amer. J. Bot. **26**:320-324.

Tammes, P. M. L., J. Van Die, and T. S. Ie. 1971. Studies on phloem exudation from *Yucca flaccida* Haw. VIII. Fluid mechanics and exudation. Acta Bot. Néer. **20**:245-252.

Tammes, P. M. L., C. R. Vonk, and J. van Die. 1967. Studies on phloem exudation from *Yucca flaccida* Haw. VI. The formation of exudate-sucrose from supplied hexoses in excised inflorescence parts. Acta Bot. Néerl. **16**:244-246.

Thaine, R. 1964. Protoplast structure in sieve-tube elements. New Phytol. **63**:236-243.

Trip, P. 1969. Sugar transport in conducting elements of sugar beet leaves. Plant Physiol. **44**: 717-725.

Trip, P., and P. R. Gorham. 1968. Translocation of radioactive sugars in vascular tissues of soybean plants. Canad. J. Bot. **46**:1129-1133.

Tyree, M. T. 1969. The thermodynamics of short-distance translocation in plants. J. Exp. Bot. **20**:341-349.

Tyree, M. T., and D. S. Fensom. 1970. Some experimental and theoretical observations concerning mass flow in the vascular bundles of *Heracleum*. J. Exp. Bot. **21**:304-324.

Van Die, J., and C. R. Vonk. 1967. Selective and stereo-specific absorption of various amino acids during xylem translocation in tomato leaves. Acta Bot. Néerl. **16**:147-152.

Walker, T. S. 1971. The isolation and some properties of a factor causing gelling in the phloem sieve-tube exudate of the squash plant. Biochem. J. **123**:6P.

Wanner, H. 1953. Die Zusammensetzung des Siebröhrensaftes; Kohlenhydrate. Ber. Schweiz. Bot. Ges. **63**:162-168.

Weatherly, P. E., and R. P. C. Johnson. 1968. The form and function of the sieve tube: a problem in reconciliation. Internat. Rev. Cytol. **24**:149-192.

Weatherley, P. E., and B. T. Watson. 1969. Some low-temperature effects on sieve tube translocation in *Salix viminalis*. Ann. Bot. (London) **33**: 845-853.

Webb, J. A. 1970. The translocation of sugars in *Cucurbita melopepo*. V. The effect of leaf blade temperature on assimilation and transport. Canad. J. Bot. **48**:935-942.

Webb, J. A. 1971. Translocation of sugars in *Cucurbita melopepo*. VI. The reversible low temperature inhibition of ^{14}C movement and cold acclimation of phloem tissue. Canad. J. Bot. **49**:717-733.

Wiersum, L. K., C. A. Vonk, and P. M. L. Tammes. 1971. Movement of ^{45}Ca in the phloem of Yucca. Naturwiss. **58**:99.

Willenbrink, J. 1966. Zur lokalen Hemmung des Assimilattransports durch Blausäure. Z. Pflanzenphysiol. **55**:119-130.

Wooding, F. B. P. 1967. Fine structure and development of phloem sieve tube content. Protoplasma **64**:315-324.

Yamamoto, T., S. Sekiguchi, and M. Noguchi. 1970. The translocation of photosynthetic products from mesophyll into midrib in tobacco plant. III. The transformation of translocated ^{14}C-sugars in the veins. Plant Cell Physiol. **11**: 367-375.

Zee, S. Y. 1969. Fine structure of the differentiating sieve elements of *Vicia faba*. Aust. J. Bot. **17**:441-456.

Ziegler, H. 1961. Wasserumsatz und Stoffbewegungen. Fortschr. d. Botanik **23**:191-205.

Zimmerman, M. 1969. Translocation of nutrients, p. 383-417. *In* M. B. Wilkins (Ed.) The physiology of growth and development. McGraw-Hill Book Co., New York.

GENERAL REFERENCES

Arisz, W. H. 1952. Transport of organic compounds. Ann. Rev. Plant Physiol. **3**:109-130.

Biddulph, O. 1959. Translocation of inorganic solutes. p. 553-603. *In* F. C. Steward. Plant physiology. Vol. II. Academic Press, Inc., New York.

Bollard, E. G. 1960. Transport in the xylem. Ann. Rev. Plant Physiol. **11**:141-166.

Crafts, A. S. 1951. Movement of assimilates, viruses, growth regulators, and chemical indicators in plants. Bot. Rev. **17**:203-284.

Crafts, A. S. 1961. Translocation in plants. Holt, Rinehart & Winston, Inc., New York.

Crafts, A. S., and C. E. Crisp. 1971. Phloem transport in plants. W. H. Freeman & Co., San Francisco.

Curtis, O. F. 1935. The translocation of solutes in plants. McGraw-Hill Book Co., New York.

Curtis, O. F., and D. G. Clark. 1950. An introduction to plant physiology. McGraw-Hill Book Co., New York.

Esau, K. 1966. Explorations of the food conducting system in plants. Amer. Sci. **54**:141-157.

Kursanov, A. L. 1961. The transport of organic substances in plants. Endeavour **20**:19-25.

Kursanov, A. L. 1963. Metabolism and the transport of organic substances in the phloem. Adv. Bot. Res. **1**:209-278.

Mason, T. G., and E. Phillis, 1937. The migration of solutes. Bot. Rev. **3**:47-71.

Münch, E. 1931. Die Stoffbewegungen in der Pflanze. Gustav Fischer Verlag, Stuttgart.

Swanson, C. A. 1959. Translocation of organic solutes. p. 481-551. *In* F. C. Steward. Plant physiology. Vol. II. Academic Press, Inc., New York.

Ziegler, H. 1963. Der Ferntransport organischer Stoffe in den Pflanzen. Naturwissenschaften **50**: 177-186.

Zimmerman, M. H. 1960. Transport in the phloem. Ann. Rev. Plant Physiol. **11**:167-190.

11

Exchange of gases

The discussion of the transfer of substances (absorption, translocation, and elimination) has so far been confined to water in the liquid state, together with the solid substances dissolved in it. The absorption, translocation, and elimination of gases remain to be dealt with. Since the gases may move freely throughout the plant, either in the gaseous phase within the intercellular space system or in solution, translocation of gases is usually not a problem. Therefore the main transfer problems are the absorption and emission of gases, the so-called *exchange of gases* between the plant and its environment.

PLANT PARTS INVOLVED

Since protoplasm is freely permeable to gases, any gas in the external atmosphere may be expected to diffuse into the plant. This is why plants have been injured or killed by leaks from gas mains, by gaseous sulfur compounds from industrial plants, or by other gases (Papetti and Gilmore, 1971). Similarly, a continuous exchange of carbon dioxide, oxygen, and water vapor takes place between all parts of the plant and its environment. Thus, the fine absorbing roots obtain oxygen from the soil and use it in respiration, which supplies the roots with all the energy for metabolism, growth, and active absorption of solutes. The carbon dioxide produced in respiration rapidly diffuses out of the cells into the surrounding soil. There are some plants, such as rice (Armstrong, 1971), that grow normally, although the root medium is deficient in oxygen, because the oxygen is transferred to the roots from the aerial part through the intercellular spaces. This is particularly true of some aquatic plants with intercellular spaces that make up as much as 70% of the volume of their tissues, as compared with much lower values (e.g., 20%) in most land plants. Oxygen can readily diffuse from the leaves that produce it down these large air-filled spaces to the roots. Conversely, there is evidence that the roots of aquatic plants may serve as a source of carbon dioxide, which diffuses up through this same intercellular system to the leaves, where it is used in photosynthesis (Wium-Andersen, 1971).

The older roots and the stem are protected against excessive gas exchange by a more or less impermeable layer of suberin or cork. Nevertheless, there is also some exchange of carbon dioxide, oxygen, and water vapor between the stem and its environment. In woody plants this occurs primarily through openings in the bark known as *lenticels,* since the corky covering of the stems is highly impermeable to these gases. As a result of this low permeability, the gas in the trunk may be of much different composition from that of the air (Table 11-1). The oxygen of the intercellular spaces becomes greatly depleted because of respiratory utilization, and the carbon dioxide content rises correspondingly. This is evidence that the stem is not an efficient gas exchanger because the exchange lags far behind the production and utilization. Other bulky organs are also inadequate gas exchangers. This may lead to metabolic anomalies, such as the case of apple fruit with a low volume of intercellular space (Henze, 1969), which may suffer from an excess accumulation of carbon dioxide or a deficiency of oxygen.

The leaf is a much more efficient gas exchanger. It is the organ responsible for the major gas exchange of the plant; and since most leaves are covered by a layer of relatively impermeable cuticle, the main exchange is usually through the *stomata.* Each stoma consists of two guard cells that impede gas exchange when tightly pressed together (stomata closed) but permit free exchange when separated by a pore (stomata open).

Table 11-1. Carbon dioxide and oxygen content of gas in trunks of paloverde *(Parkinsonia microphylla)**

Date	Carbon dioxide (%)	Oxygen (%)
Jan. 15-16, 1931	3.4	15.3
17-18	4.0	14.6
19-20	3.8	14.9
Mar. 3-4	8.5	13.8
5-6	8.8	13.7
7-8	9.4	13.1
9-10	10.7	12.9
11-12	12.0	12.6
Apr. 20-21	7.3	14.5
22-23	7.8	15.2
23-24	9.1	14.3
25-26	9.0	14.6

*Adapted from MacDougal, D. T., and E. B. Working. 1933. Carnegie Institution of Washington, Pub. No. 441. Washington, D. C.

STOMATAL MOVEMENT

Measurement

The degree of stomatal opening can be determined by direct observation under the microscope (e.g., in the case of the large stomata of *Zebrina* or *Rhoeo* species) or by observation of plastic peels deposited on the surface and removed from the leaf. Many indirect methods have also been developed. Molisch's infiltration method (Stålfelt, 1956) measures the degree of opening by the relative penetration rates of a series of liquids into the leaf. The more polar the liquid that enters the stomata and infiltrates the intercellular spaces, the wider open the stomata must be. This method has been made more quantitative by Oppenheimer and can even be used to determine the time for irrigation of plants (Ofir et al., 1968). The relative degree of opening can be measured more quantitatively by the rate of flow of a gas through the leaf, either under slight pressure or under a diffusion gradient. The instrument that does this is called a *porometer.* Many models have been developed over the years, and recent developments (e.g., Byrne et al., 1970, and Stiles, 1970) eliminate some of the problems experienced with the older models. The best instruments are complex (Monteith and Bull, 1970) and will give erroneous results unless the many sources of error are understood and eliminated (Meidner, 1970; Morrow and Slatyer, 1971*a*, *b*; Domes, 1971).

Periodicity

In general, the stomata tend to show a diurnal periodicity, closing at night and opening during the day (Fig. 11-1). That this periodicity is related to light is easily shown, since darkening the plant leads to

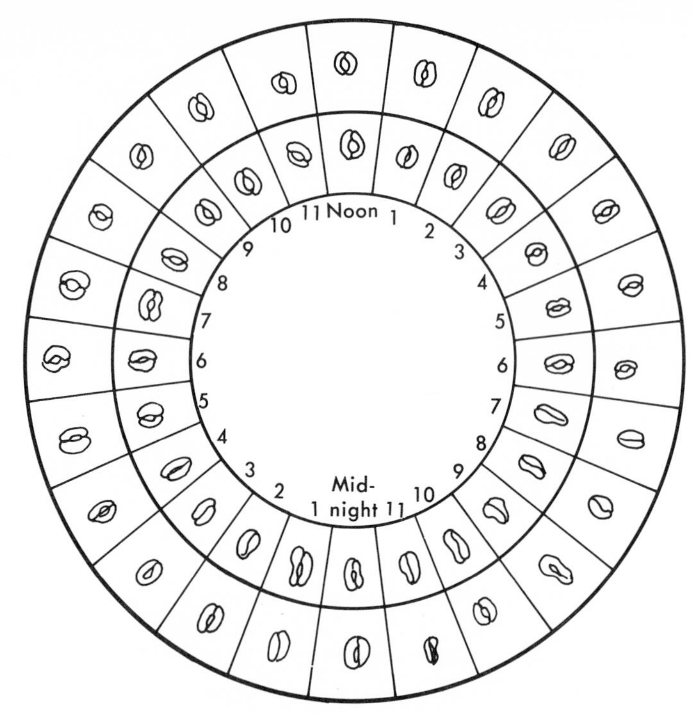

Fig. 11-1. Stomatal opening in alfalfa at different times of the day. Inner circle represents the lower leaf surface, and the outer circle represents the upper leaf surface. (Redrawn from Loftfield, J. V. G. 1921. Carnegie Institution of Washington, Pub. No. 314. Washington, D. C.)

closure; yet there are many exceptions (Loftfield, 1921). Some stomata open normally at night and others close frequently at noon when the light intensity is at a maximum. Wilting may cause closure regardless of the light factor. There is also evidence of an internal rhythm in plants that favors opening at certain times of the day and closure at others (Meidner and Mansfield, 1968).

Even when not subjected to severe soil moisture deficiency, plants may show stomatal closure in late afternoon (Balasubramaniam, 1967). The effect of light on stomatal opening is apparently indirect through its control of some other factor. To explain the action of such indirect factors, the mechanism of stomatal movement must first be understood.

Physical forces

The walls of the two guard cells of a stoma are unevenly thickened. Increased turgor leads to opening by producing a bulge in the thinner outer walls and a consequent pulling apart of the thicker inner walls bordering the aperture. Decreased turgor leads to closure by the springlike recoil of the thicker inner walls (Fig. 11-2). Experiments with models (Aylor et al., 1973) demonstrate that asymmetrically thickened guard cell walls are not necessary, provided that there are two physical constraints causing the guard cells to bend and thus open the stomatal pore. Since turgor is caused by hydrostatic pressure, the stomatal movement is controlled by water exchange, and any factor that alters the water content of the guard cells will affect their turgor and therefore the stomatal opening. The maximum possible turgor pressure of a cell is almost equal to the osmotic pressure of the cell sap. Any increase in the osmotic pressure resulting from an increase in solutes will therefore favor opening, and any decrease will favor

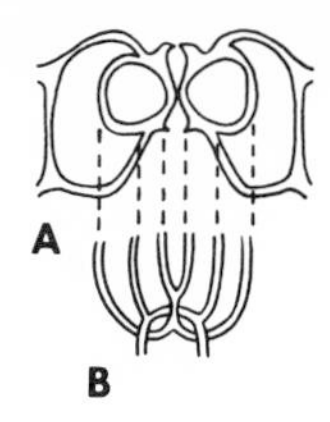

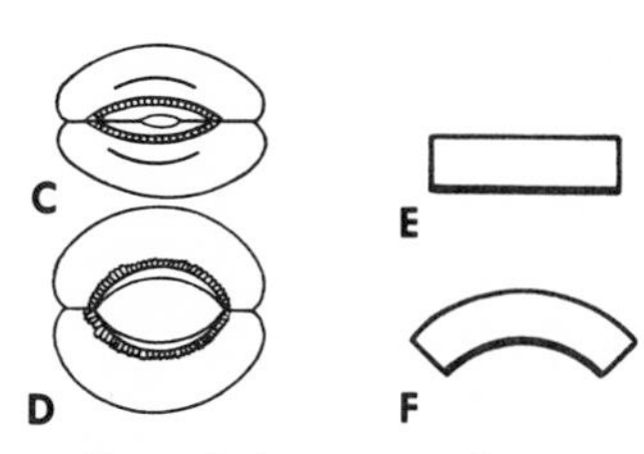

Fig. 11-2. Effect of changes in the turgor pressure of guard cells on stomatal opening. **A** and **B,** Cross section and half surface view of the stoma of *Amaryllis formosissima.* **C** and **D,** Surface views of closed and open stoma. **E** and **F,** Diagrammatic scheme of turgor curvature of a guard cell when the inner wall is thicker than the outer wall. (Redrawn from Benecke, W., and L. Jost. 1924. Pflanzenphysiologie. Gustav Fischer Verlag, Stuttgart.)

closure. However, it must be remembered that a change in solutes by itself affects only the potential turgor pressure, and the actual change in turgor pressure can occur only as a result of water movement into or out of the cell. Any real change in turgor pressure, on the other hand, must be accompanied by a change in degree of opening, provided that the cell is capable of a further stretch or shrinkage. But there is another factor: Stomatal opening is possible only if the guard cells can push back the epidermal cells adjacent to them. These other epidermal cells, as long as they are turgid, exert a back pressure on the guard cells; and if all the epidermal cells were to increase equally in osmotic potential and turgor pressure, no stomatal opening would result. Therefore we can generalize that in all cases of stomatal opening

$$P_{tg} > P_{te}$$

where P_{tg} = turgor pressure of guard cells
P_{te} = turgor pressure of surrounding epidermal cells

and the net opening pressure is

$$P_{tg} - P_{te}$$

This relationship can be demonstrated by immersion of leaf discs in solutions of graded concentrations. Maximum stomatal aperture is then obtained when the adjacent epidermal cells are incipiently plasmolyzed (Glinka, 1971) and are therefore at zero turgor pressure.

Since stomatal opening normally occurs when the leaves are at maximum turgor (e.g., in early morning) and since the maximum possible turgor pressure is given by the osmotic pressure, it follows that:

$$P_{og} > P_{oe}$$

where P_{og} = osmotic pressure of guard cells
P_{oe} = osmotic pressure of surrounding epidermal cells

Since a cell's osmotic pressure is proportional to its cell sap concentration, this means that in a turgid leaf with open stomata, the cell sap concentration of the guard cells must be higher than that of the surrounding epidermal cells. This relationship is favored by the existence of chloroplasts in the guard cells and by their absence from the other epidermal cells. The guard cells can therefore synthesize organic solutes; the other nonphotosynthesizing epidermal cells must receive them from photosynthesizing cells.

Similarly, if all the epidermal cells showed corresponding decreases in turgor pressure, no closure could occur. A simultaneous water loss from all the epidermal cells may result in an increased stomatal opening if the guard cells lose less than the other epidermal cells. For this reason, *incipient* wilting may cause increased stomatal opening, although more pronounced wilting will cause closure. This is because the guard cells are the last of the epidermal cells to reach zero turgor pressure. Finally, excessive loss of water may cause stomatal opening because of drying and consequent collapse of the guard cells.

Control mechanism

Control of water flux. Although the physical aspects of stomatal movement are clear, the chemical changes that control the physical movement are still incompletely understood. Whatever their nature, they must operate by controlling the water flux into or out of the guard cells. Thus the opening and closing of stomata may be controlled solely by water movement into or out of the guard cells without an accompanying change in solutes. Stålfelt (1956) calls this *hydroactive* (opening or closure) as opposed to *photoactive* (opening or closure), in which these water movements are themselves controlled by light through its action on cell solute content. On the same basis, opening or closure in the dark, because of its control of cell solute content, may be called *scotoactive.* There is no question as to the nature of the force involved, since the only one available is that caused by turgor pressure. Also, since there is no such thing as active water absorption (Chapter 8), the turgor changes must be controlled by changes in solutes. There is ample direct evidence of this, since the osmotic pressure of the guard cells is high (10 to 30 atm or higher) when the stomata are open and low (5 to 10 atm) when they are closed. Three lines of evidence indicate that this photoactive opening is controlled by photosynthesis: (1) the relation of the two to light is similar (Fig. 11-3); (2) the relative effectiveness of different wavelengths of light is similar, although blue seems to be more effective for stomatal opening; and (3) no opening occurs in the absence of chlorophyll (e.g., in albino plants, only guard cells with chlorophyll are functional). In fact, treatments that decrease the chlorophyll content by creating nitrogen or iron deficiency markedly reduce stomatal aperture (Shimshi, 1967). This fact has been corroborated in the case of oats (Frommhold, 1971). Fully developed stomata without chlorophyll were unable to open. The overall effect of light and

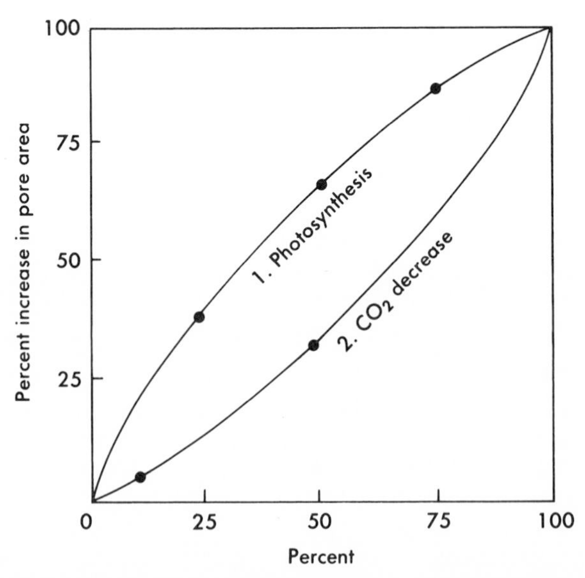

Fig. 11-3. Increase in stomatal aperture with (1) increase in photosynthesis as light is increased and (2) artificial decrease in CO_2 concentration of air at constant light of 50 footcandles. (From Scarth, G. W., and M. Shaw. 1951. Plant Physiol. 26:207-225.)

dark may be represented as in Fig. 11-4, *A*.

The major problem is the explanation of step 2. Photosynthesis may conceivably increase the solute concentration directly because of the synthesis of soluble carbohydrates. But the large increases in concentration (2 to 3 times) that occur in the short time (5 to 30 min) usually required for stomatal opening can hardly be caused by photosynthesis, particularly in view of the small amount of chlorophyll in the guard cells.

Scarth's theory. In many (but not all) plants, the guard cell chloroplasts are full of starch when the stomata are closed and free of starch when they are open. Until recently it was assumed that the starch loss was due to a conversion to sugar. Unfortunately, it is difficult to determine the sugar content of guard cells; therefore this assumption was not tested experimentally. It was also shown that artificial increases in guard cell pH, induced by ammonia or ammonium hydroxide, could lead to stomatal opening. It was therefore suggested (Scarth, 1932) that a similar rise in pH occurs on exposure to light, due to photosynthetic fixation of carbon dioxide

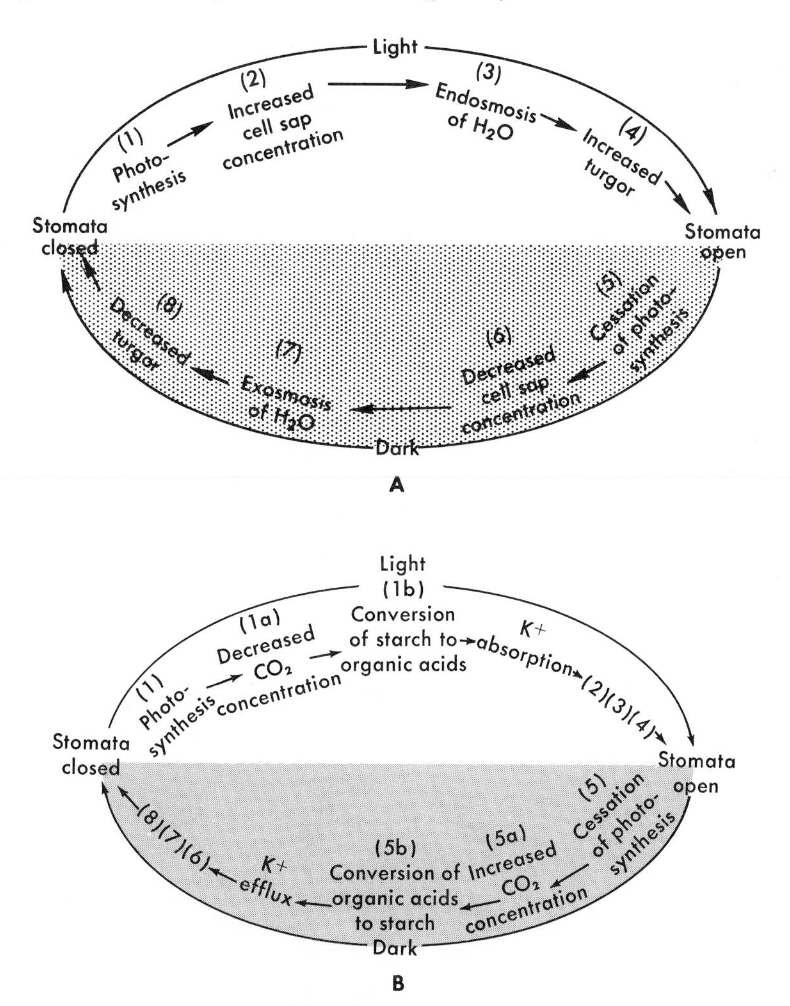

Fig. 11-4. A, Changes in the guard cells leading to photoactive opening (light) and scotoactive closing (dark) of the stomata. **B,** Hypothesis of the mechanism of stomatal opening and closing, based on a combination of Scarth's and the K^+ ion absorption hypotheses.

and a consequent decrease in carbonic acid, and that this rise in pH would lead to a hydrolysis of starch to sugar. This theory was supported by the effect of carbon dioxide concentration. Stomata can be induced to open or can be prevented from closing in the dark by replacing the intercellular atmosphere of the leaf with air containing 0.01% or less carbon dioxide instead of the normal 0.03%. Under normal conditions, stomatal opening is associated with a similar drop in carbon dioxide concentration (Table 11-2). On the basis of these facts, Scarth's theory of the control of stomatal movement was the best explanation for many years.

Scarth's hypothesis, however, was proposed to explain only photoactive opening. It was unable to explain *scotoactive* opening. The scotoactive opening of succulent plants is apparently caused by the accumulation of organic acids synthesized from carbon dioxide in the dark (Nishida, 1963). They accumulate to a sufficient degree to produce osmotic opening of the stomata. In the daytime the carbon dioxide in the leaf is photosynthesized, leading to decarboxylation of the accumulated organic acids. When their concentration drops to a sufficient degree, the stomata close. Therefore these succulents are able to continue photosynthesis during daylight, although their stomata remain closed for all or most of the daylight hours. This is, of course, an adaptive character that prevents excessive loss of water in dry climates.

Table 11-2. Relation of stomatal opening to carbon dioxide concentration outside and inside the leaf*

External CO_2 concentration (ppm)	Internal CO_2 concentration (ppm)	Stomatal openings (% > 1 μm)
8		100
100	40 (app)	70
200	100	26
400	240	12

*From Moss, D. N., and S. L. Rawlins. 1963. Nature (London) **197**:1320-1321.

Potassium accumulation theory. Since the late 1960s, Scarth's theory has been shown to be inadequate, even for photoactive stomatal opening because of two fundamental factors: (1) K^+ ions and (2) organic acids.

It has long been known that K^+ ions stimulate stomatal opening and that potassium-deficient plants do not seem able to open their stomata normally in the light, although some investigators have failed to obtain this result (Mingeau, 1969). On the basis of new experiments, Fischer (1968) and his co-workers (Fischer and Hsiao, 1968) propose that light stimulates the active absorption of K^+ ions by the guard cells and that the consequent increase in cell sap concentration leads to osmotic opening of the stomata. In favor of this concept, ATP was shown to stimulate stomatal opening (Thomas, 1971), and an inhibitor of adenosine triphosphatase was shown to inhibit opening (Thomas, 1970*b*). At the same time, an elegant quantitative analysis, using the electron probe microanalyzer, permitted quantitative measurement of the K^+ ion content of a single guard cell of *Vicia faba* (Humble and Raschke, 1971). When the stomata opened, the solute content of the guard cells increased by 4.8×10^{-12} osmoles per stoma, and 4.0×10^{-12} gram equivalents of K^+ were transported into each pair of guard cells. In opposition to Fischer's concept, however, their measurements of chlorine, phosphorus, and sulfur revealed that the K^+ ion uptake was not balanced by a concurrent uptake of inorganic anions. Therefore they concluded that the K^+ ion uptake must have been accompanied by a sufficient increase in a bivalent organic anion (e.g., malate) to balance it electrically and that the increase in content of this organic salt of potassium was sufficient to account for the measured increase in osmotic concentration of the guard cells. They also suggest that the starch breakdown which normally takes place when stomata open may supply the anions.

This conclusion has now been supported experimentally by Pallas and Wright (1973). Their analyses revealed a doubling of the organic acid content of epidermal strips from *Vicia faba* as a result of stomatal opening induced by exposure to low concentrations of carbon dioxide. That these organic acids are derived from guard cell starch and vice versa was further indicated by Fischer and Hsiao (1968) and by Mansfield and Jones (1971). The former showed that stomatal opening induced by K^+ ions was well correlated with the decrease in starch content of the guard cells; the latter demonstrated histochemically a loss of potassium from guard cells during stomatal closure. Paralleling this loss, there was a marked increase in starch content.

Present status of theories. These investigations have clearly established that the true explanation of stomatal movement involves a combination of both the old and the new information. The starch (or other carbohydrate) that disappears in the guard cells as a result of illumination, low carbon dioxide concentration, or any other stoma-opening factor is replaced not by sugars but by organic acids. At the same time that the organic acids accumulate, K^+ ions are absorbed, neutralizing the acids and accumulating with them in the guard cells. The increase in pH associated with stomatal opening would therefore be due mainly to this accumulation of basic potassium salts of organic acids.

The K^+ ion accumulation, however, does not appear to be as fundamental as the accumulation of organic acids. In some cases, K^+ ions are apparently not required for stomatal opening, such as when stomata open in the dark in the presence of sodium chloride (Thomas, 1970*a*). In the case of *Kalanchoe marmorata,* a succulent that opens its stomata in the dark and closes them in the light, the stomata were insensitive to K^+ ions, but opening seemed to be caused by an influx of Na^+ ions in the dark and an efflux in the light. In the *scotoactive* or night opening of another succulent *(Bryophyllum calycinum),* the organic acids were apparently sufficient by themselves to account for the night opening (Levitt, 1967). In the case of apple leaves, trees with many fruits show a higher degree of stomatal opening, yet the concentration of potassium in the leaves is lower because of accumulation in the fruits (Hansen, 1971). Finally, air free of carbon dioxide causes stomatal opening in the dark, comparable with that in bright light and ordinary air (Heath and Mansfield, 1969), yet there is no light-stimulated K^+ uptake.

Therefore it is assumed that the main function of K^+ absorption, which is even more important than its osmotic effect, is to enable the continued breakdown of starch to organic acids. This is not necessary in succulents because they are "acid plants," which simply means that they have some mechanism for continuing the accumulation of organic acids despite the large accumulation of H^+.

What, then, is the role of low carbon dioxide concentration in stomatal movement? According to the original Scarth concept, it is the trigger that initiates the conversion of starch to organic acids by raising the pH. This initial formation of organic acids is followed by K^+ absorption, which continues to raise the pH and therefore continues the conversion of starch to organic acids. The two processes then work simultaneously until the increase in cell sap concentration and the consequent water absorption result in stomatal opening. Instead of acting by controlling the pH, the K^+ ions may be thought of as an enzyme activator, but these two alternatives may simply be different ways of saying the same thing.

This concept of stomatal movement not only explains the roles of both the starch and the K^+ in stomatal opening, but it also leads to a single process for stomatal opening in both succulents and nonsucculents, for both scotoactive and photoactive opening. In both cases, starch is converted to organic acids. The proposed concept of stomatal movement may be represented by Fig. 11-4, *B*.

This is still an unproved and incomplete explanation of the control of stomatal movement. Only when the specific chemical reactions and the controlling enzymes are determined will it be possible to explain all the steps leading to stomatal opening and closure.

The metabolism of carbohydrates is a complex series of chemical reactions (Chapters 14 and 15). It is not surprising that several metabolic inhibitors greatly affect stomatal opening. Metabolic inhibitors such as alpha-hydroxysulfonates and sodium azide prevent opening. Others (phenylmercuric acetate, alkenylsuccinic acids) also prevent opening but apparently by a different mechanism.

There is now considerable evidence that stomatal aperture may be controlled by growth regulators (Chapter 19). This is also in accord with the above hypothesis, for GA (gibberellic acid) appears to stimulate opening (Tal et al., 1970; Horton, 1971), and it also stimulates a hydrolysis of starch (Chapter 19). Cytokinins also induce opening, but auxins and abscisic acid cause closure (Tal and Imber, 1971; Horton, 1971). Scopoletin and chlorogenic acid also cause stomatal closure in tobacco within 1 day of treatment (Einhellig and Kuan, 1971).

Most investigators consider only the role of stomatal opening in permitting photosynthesis by allowing carbon dioxide to diffuse into the leaf. They completely overlook the equally important role of permitting respiration by allowing oxygen to diffuse into the leaf. Even when night opening is recognized, no mention is made of the fact that oxygen deficiency in the dark can lead to wide opening. It is therefore necessary to explain the mechanism of stomatal opening resulting from oxygen deficiency in the dark as well as that resulting from carbon dioxide deficiency in the light. Unfortunately, little work has been done on this problem, and except for the case of succulents just mentioned, little or nothing is known of the mechanism of night opening, which may or may not be caused by oxygen deficiency.

DIFFUSION OF GASES THROUGH SMALL OPENINGS

The rapid rate of diffusion of gases into and out of the leaf is understandable on the basis of Fick's law of diffusion: The diffusion constant of gases is high (diffusion for water vapor is about 10,000 times diffusion for liquid water), the area across which diffusion is occurring is large, the concentration difference is large, and the distance is small. But the area is large only if the whole leaf area is considered. When, however, one remembers that most of the gas movement occurs through the open stomata, the high rate of diffusion may be questioned because the area of the stomata is commonly only about 1% of the leaf area. Yet a leaf may lose almost as much water as a free water surface of the same total area. Similarly, evaporation from a stand of well-watered vegetation is nearly as great as that from open water (Parlange and Waggoner, 1970). This means that the rate of water loss per unit stomatal area is almost 100 times that from a free water surface. It has also been shown that a photosynthesizing leaf absorbs carbon dioxide about as rapidly as potassium hydroxide solution of the same total surface. One must therefore conclude that either the laws of diffusion must be modified or else some other process speeds up the diffusion from the stomata. Brown and Escombe (1900) settled this problem by showing that the loss of water from a free water surface is almost unaltered when it is covered by a perforated sheet, even though the area across which diffusion occurs is reduced to a small fraction. This can be explained by the overlapping of the diffusion fronts (Fig. 11-5). The molecules passing through the pores at the perimeter can diffuse more freely in all directions because of the absence of other interfering molecules. They therefore spread out laterally and soon meet with similar molecules from adjacent pores, forming a solid front a short distance above the pores. Expressed quantitatively, this means that the diffusion across small openings is proportional to the perimeter rather than the area of the opening.

The reason for this is that the area factor in Fick's law is only an approximation which really consists of two components: (1) the molecules that move in a direction normal to the surface (or aperture) and (2) those that can "spill over" the perimeter. Since the former varies with the area and the latter varies with the perimeter, Fick's law should be changed from

$$\frac{s}{t} \propto a$$

to

$$\frac{s}{t} \propto (a + p)$$

where a = area
p = perimeter
s = amount of substance diffusing (moles)
t = time (sec)

In the case of circular apertures, since $a = \pi r^2$ and $p = 2\pi r$:

$$a \propto r^2$$
$$p \propto r$$

If the proportions of the two components are about equal in the case of an aperture 1 mm in diameter, so that half of the molecules diffuse normal to the surface and half "spill over" the perimeter then increasing the aperture diameter to 100 mm would increase the area component 100^2 or 10^4 times and the perimeter component only 100 or 10^2 times; and the area component will now account for 99% of the diffusion. Conversely, if the aperture is decreased to 0.01 mm, the area component is decreased to $(0.01)^2$, or 10^{-4} times, the perimeter component to 0.01, or 10^{-2} times, and the perimeter component will now account for 99% of the diffusion. The conclusion, therefore, is that strictly speaking:

$$\frac{s}{t} \propto (a + p)$$

But for large apertures p is negligible and

$$\frac{s}{t} \propto a$$

For small apertures a is negligible and

$$\frac{s}{t} \propto p$$

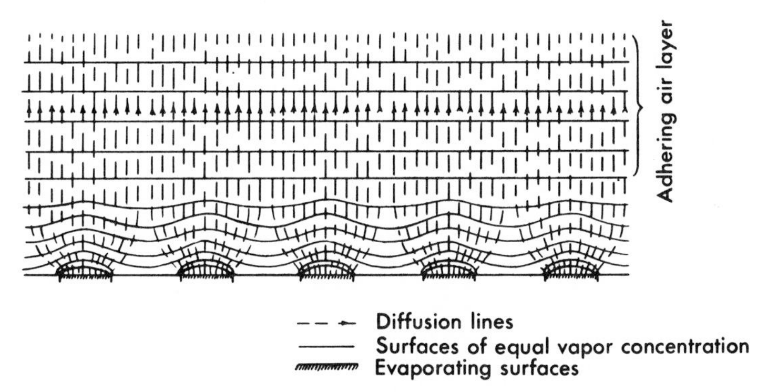

Fig. 11-5. Overlapping of diffusion fronts above pores in a septum. (From Bange, G. G. J. 1953. Acta Bot. Neerl. **2**:255-297.)

However, it must be understood that as soon as overlapping occurs, the diffusion limit is reached, and any further increase in the perimeter of the apertures cannot lead to an increase in diffusion rate. Thus if the stomata are as close together as 10 to 20 times their diameters, maximum diffusion occurs. Doubling the diameter of their openings will have no effect on the diffusion rate. This fact must be taken into consideration when explaining the control of gas exchange by stomatal movement.

It must be emphasized that the preceding perimeter (or diameter) law (derived from Stefan's law) applies strictly only when the thickness of the membrane (or the length of the pore) is approximately zero or when the ratio of the diameter to the pore length is constant (Lee and Gates, 1964). Stomata do not meet these requirements because the pore length is really the distance from the stomatal aperture to the evaporating cell surface at the bottom of the substomatal cavity. This distance is many times the widest stomatal aperture. However, if the vapor-pressure drop in this substomatal cavity is small enough, the stomata may obey the perimeter law reasonably well. Although no reliable measurements of this drop have yet been made, it seems reasonable to accept the assumption that it is small in a turgid plant under normal conditions. This follows from the fact that the boundary layer of moist air adjacent to the leaf surface (Fig. 11-5) in still air is about 1500 times the distance from the stomatal pore to the evaporating cell surface. Therefore the vapor-pressure drop within the substomatal cavity must be of the order of 1% of the total vapor-pressure drop between the leaf cells and the external atmosphere.

TRANSPIRATION

Since the quantitative aspects of oxygen and carbon dioxide exchange have been investigated primarily from the point of view of respiration and photosynthesis and will be considered under these headings, water vapor is the only gas that will be discussed at this point. The loss of water vapor from the plant has long attracted the attention of physiologists and is called *transpiration.* This process can be defined as gaseous water diffusion driven by the net radiation absorbed by the leaf (Gates, 1966). Other parts of the plant besides the leaf can transpire, and transpiration can occur without absorption of radiation. However, the major transpiration is from the leaves, and over any considerable period of time the energy utilized is balanced by absorption of radiant energy. The diffusion may occur through the leaf cuticle or through the stomata. The former is called *cuticular transpiration* and the latter, *stomatal transpiration.* As a general rule, transpiration is higher when the stomata are open, lower when they are closed. In other words, the stomatal transpiration is usually higher than the cuticular transpiration—commonly about 4 times as high on a sunny day. But there are all gradations from a maximum of several hundred to a minimum of nearly one. In the case of succulents, cuticular transpiration is essentially zero. A joint of a prickly pear can be kept in a dry room for a year without losing all of its water. Stomatal control of transpiration in such succulents is so effective that the transpirational curve at midday may be the reverse of the evaporation curve from a free water surface (Fig. 11-6). Pineapple shows this same reversal (Ekern, 1965). On the other hand, a shade plant (such as wood sorrel) may lose as much water through the epidermal cells as through the open stomata because the cuticle layer is almost nonexistent.

Many early workers believed that stomatal movement completely controlled transpiration. But this was soon found to be untrue. The transpiration rate may actually increase when stomata begin to close and decrease when they open wider. This is sometimes due to the fact that when the

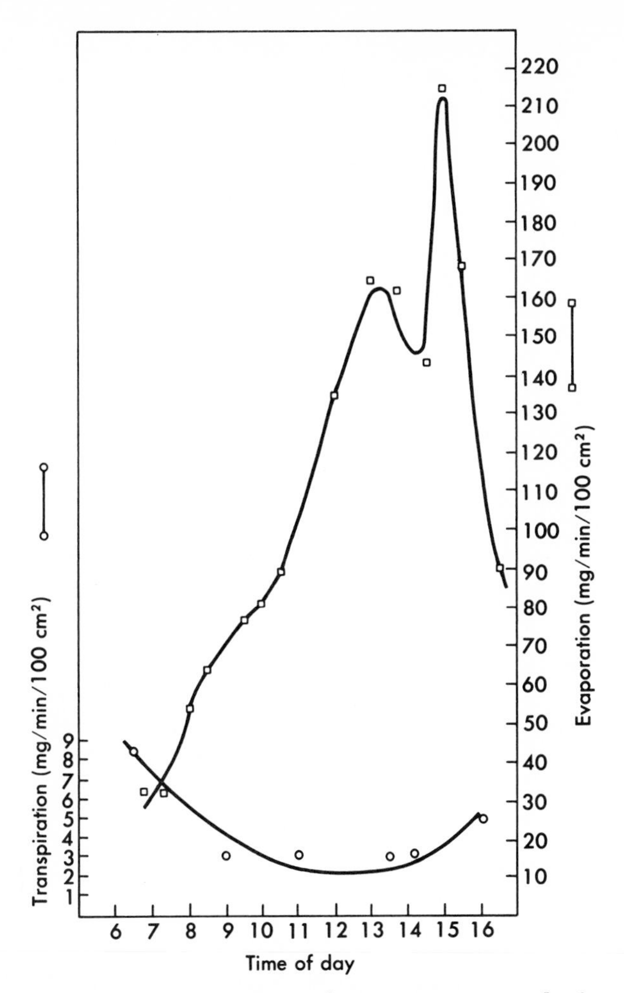

Fig. 11-6. Diurnal course of evaporation (Piche evaporimeter) and of transpiration from *Jatropha phyllacantha.* (From Ferri, M. G. 1955. Faculdad de Filosofia Ciencias e Letras, Universidade de São Paulo, São Paulo, Brazil. Bol. 195. Bot. 12.)

diameters of the stomatal openings are one tenth to one twentieth their distance apart, no further opening can affect the rate of diffusion through them, e.g., in sugar cane (Irvine, 1971). But it may also result from another variable factor that always controls the rate of diffusion—the concentration difference ($C_1 - C_2$ in Fick's law), or in the case of transpiration, the vapor-pressure difference. The larger the difference between the vapor pressure of the leaf and that of the surrounding air the greater is the transpiration rate. Both the relative humidity and the temperature may affect the vapor pressure of the air and the leaf. Consequently, they may both alter the transpiration rate.

Wind may also affect the transpiration rate. This is because of the boundary layer of moist air adjacent to the leaf surface. In still air this layer will be 1.57 cm thick above a circular leaf 10 cm in diameter (Lee and Gates, 1964). In turbulent air it is much thinner. Since the distance x in Fick's law must include this boundary layer and since diffusion is inversely proportional to x, transpiration should be (and usually is) much higher in turbulent than in still air. However, the reverse effect may be obtained if the turbulence results in stomatal closure. This effect varies with the species. Wind has little effect on the stomata of *Pinus cembra*, which grows on windswept ridges, but causes rapid closure of all the stomata of *Rhododendron ferrugineum*, which grows in protected ravines (Caldwell, 1970). Furthermore, even in turbulent air there remains a stationary boundary layer of moist air (Lee and Gates, 1964), and once a minimum layer is reached, any further increase in wind velocity is without effect. Wind will also affect the relationship between stomatal aperture and transpiration. In still air, stomatal transpiration may reach its maximum at about one fourth of full stomatal aperture because of the overlapping vapor fronts. In turbulent air, transpiration will continue to increase up to full stomatal aperture (Fig. 11-7).

There has been some controversy as to whether transpiration is a necessary evil or whether it is in some way essential for the welfare of the plant. From the former point of view, the function of the stomata is to permit exchange of carbon dioxide and oxygen in the processes of photosynthesis and respiration. At the same time, water is lost through the open stomata because there is no way of stopping the passage of water vapor while permitting the carbon dioxide and oxygen exchange. It may be objected that transpiration is useful to the plant by inducing the rise of water and indirectly its absorption from the soil. But it must be realized that although by far most of the water absorption is caused by transpirational pull, this part of the water is all wasted because transpiration can induce water absorption only by reducing cell turgor. The initial cell turgor, which is required for growth and other processes, results from root pressure. Transpiration must cause a loss of some of this water before it can lead to its replacement. The net gain in water caused by transpiration therefore can never be greater than zero. Thus transpiration cannot be thought of as essential for water uptake.

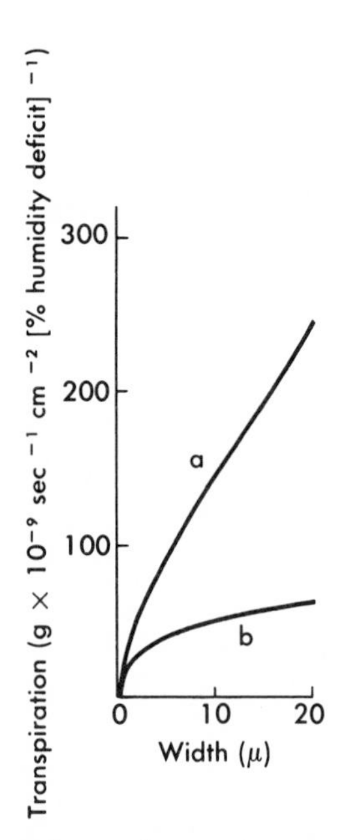

Fig. 11-7. Relation between transpiration and stomatal width in *Zebrina* in **a**, wind, and **b**, still air. (From Bange, G. G. J. 1953. Acta Bot. Neerl. 2:255-297.)

From the other point of view, transpiration fulfills three main functions: it helps maintain optimum turgidity, reduces leaf temperature, and promotes translocation and uptake of solutes.

Transpiration helps maintain optimum turgidity

When a land plant is grown in a saturated atmosphere, it develops a softer, more watery type of growth. This presumably results from the larger turgor pressure, causing excessive cell enlargement. Potato slices kept in distilled water demonstrate this effect. If kept aseptically, their cells will enlarge about 50% within a week. Not only will this increase the water content of the cells and dilute their contents, it will also lead to thinner, weaker cell walls. Aquatic plants avoid this danger by having low osmotic pressures (3 to 5 atm in many cases of freshwater plants). Terrestrial plants avoid it because they are seldom at or near full turgor, and even though their cell sap concentrations may be high (osmotic pressures up to 200 atm in the case of some halophytes), transpirational loss keeps the turgor pressure well below the osmotic pressure.

Transpiration reduces leaf temperature

In full sunlight on a hot day, a leaf's temperature may be in danger of rising to the killing point. The evaporation of water is a cooling process resulting in a loss of about 580 cal/g evaporated at the normal leaf temperature, and it may help to prevent such heat injury. It has been shown that the actual cooling effect of transpiration in most climates is not so great as some workers have thought. At the high temperatures most commonly found (30° to 40° C) this will not amount to more than 3° to 5° C (Curtis, 1938), but even this small quantity may save a plant from heat killing. It is particularly important, since the surface temperature of a sunlit leaf may rise to as much as 20° C above the air temperature (Gates, 1963); and the lower surface may have a mean temperature only 1.5° C lower than that of the upper surface. The temperature of shaded leaves may be as much as 3° to 4° C below the air temperature. In exceptionally hot climates, however, where air temperatures may reach 50° C, some specially adapted plants are able to cool their leaves 10° to 15° C below the air temperature by having high transpiration (and absorption) rates (Lange, 1959). In the case of these few exceptional plants, transpiration is unquestionably essential for survival, since their killing points are below the maximum atmospheric temperatures.

Transpiration promotes translocation and uptake of solutes

The promotion of translocation and uptake of solutes by transpiration has been the subject of the most controversy. As already seen, the mineral elements absorbed from the soil are translocated mainly, if not solely, in the vessels of the xylem, rising in the transpiration stream. The question is sometimes raised concerning whether the speed of translocation of these solutes depends on the rate of transpiration. By an analogy (first proposed by van den Honert), with an endless belt delivering coal, the rate of delivery normally depends not on the speed of rotation of the endless belt but on the rate of transfer to it (Fig. 11-8). If the substances enter the plant by active absorption, the amount reaching the leaves per unit time will depend not on the speed of the transpiration stream (or of the endless belt) but on the rate of absorption (or of deposit on the belt). On the other hand, if individual branches of a plant coming off the main transpiration stream are considered, the one with the most rapid tributary will tap off the largest amount of solutes from the main stream. Consequently, a more rapidly transpiring branch of a plant might be expected to receive a larger fraction of the absorbed nutrients. However, if the transpiration (or endless belt) is so slowed down that the solutes accumulate in roots (or the coal piles up at the beginning of the belt), the absorption rate will be stopped or at least

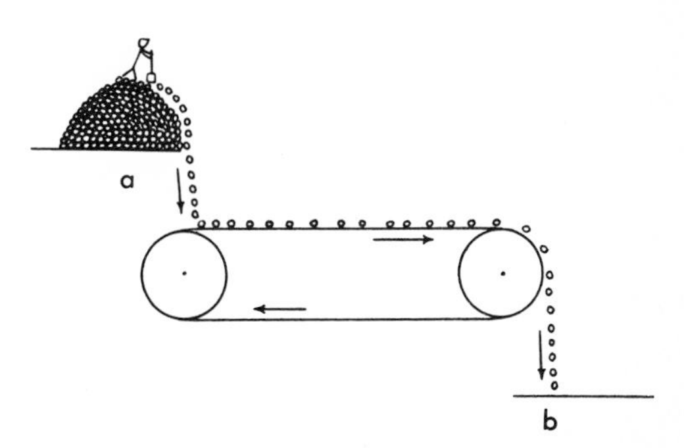

Fig. 11-8. Effect of transpiration on the rate of translocation of solutes in xylem: the analogy with an endless belt delivering coal. Rate of delivery to **b** (translocation) depends on the rate of drop from **a** (absorption), regardless of the rate of movement of the endless belt (transpirational stream).

greatly retarded because of a backflow of accumulated solutes into the soil. Under these conditions, transpiration rate may be crucial to absorption by preventing this excessive accumulation.

Many investigators have reported that increased transpiration rate does increase the uptake of substances from the root medium, but others have failed to find any such effect. The problem appears to have been resolved as a result of the many investigations in recent years. The following generalizations now appear valid:

1. At low external ion concentrations (1 mM or less) the effect of transpiration is negligible; at high external concentrations it may be significant.
2. In the case of ions absorbed actively, especially those which are quickly metabolized, the effect of transpiration is slight even at high concentrations (80 mM).
3. In the case of some plants (e.g., maize, beans) the effect of transpiration seems to be slight, even at relatively high concentrations; in other plants it appears greater.

When significant effects of transpiration on ion uptake have been found, this has led to the conclusion that active absorption occurs only in excised roots and that passive absorption is characteristic of the whole plant as a result of transpiration. Actually, the effect of transpiration is probably mainly, if not solely, a result of its ability to reduce the amount of energy needed to absorb the solutes actively. It does this by quickly transporting the actively absorbed solutes away from the absorbing organs (the roots) to the leaves. This lowers the gradient against which the absorption would have to take place and therefore decreases the energy expended per ion absorbed. It has been suggested (Hylmö, 1953) that the increased tension produced by higher transpiration rates pulls the external solution through the intercellular spaces along the cell walls of the epidermal and cortical cells, in this way bypassing the more resistant semipermeable membranes of the protoplasm in all these cells. It would then be necessary for the solution to pass through the protoplasm of only one layer of cells, the endodermis, before being pulled into the xylem. However, the protoplasm would not be completely bypassed. On the contrary, the nutrient solution would be dragged through the cell walls by the transpirational pull and would be brought into contact with the protoplasm of many more cells (Fig. 11-9).

On the basis of Arisz's symplasm concept (Chapter 10), as soon as an ion enters the protoplasm of a single cortical cell, it can then move by way of the plasmodesmata from cell to cell until it is excreted into the vessels. In support of this interpretation, Ginsburg and Ginzburg (1970) have concluded from an analysis of the diffusion coefficients of ions that radial ion movement in corn root cortex must be through, rather than between, the cells. On their way through the symplasm, some of the ions can be secreted into the vacuole of each cell, removing them from the translocation stream. These two processes (movement through the symplasm and secretion into the vacuole) are essentially independent of each other, according to Arisz, and can be separately controlled. In agreement with Arisz, Bowling and Weatherly (1965) have divided the uptake of potassium into two components: (1) an accumulation by the cells of the root and (2) a passage through the root to the shoot by way of the vessels. They were found to be entirely independent of each other. Accumulation of potassium in the root was unaffected by water flux, but passage of potassium through the root was linearly related to it. In these plants, however, there was no continuous mass flow pathway (e.g., through the cell walls) between the medium and the xylem.

These interactions between transpiration and absorption point to an important principle: Numerous physiological processes occur in the plant simultaneously. Therefore no matter how much we learn about each process by studying it independently, it must be ultimately investigated when under the influence of the others in order to understand it fully. This principle recurs in plant physiology again and again.

The rate of transpiration may be measured in several ways. The water given off may be collected in drying towers (e.g., by a desiccant such as calcium chloride), and the increase in weight of the towers will be equal to the weight of water lost. This method has been replaced by the infrared absorption method, which permits a continuous record of the transpiration rate. The water content of the air is mea-

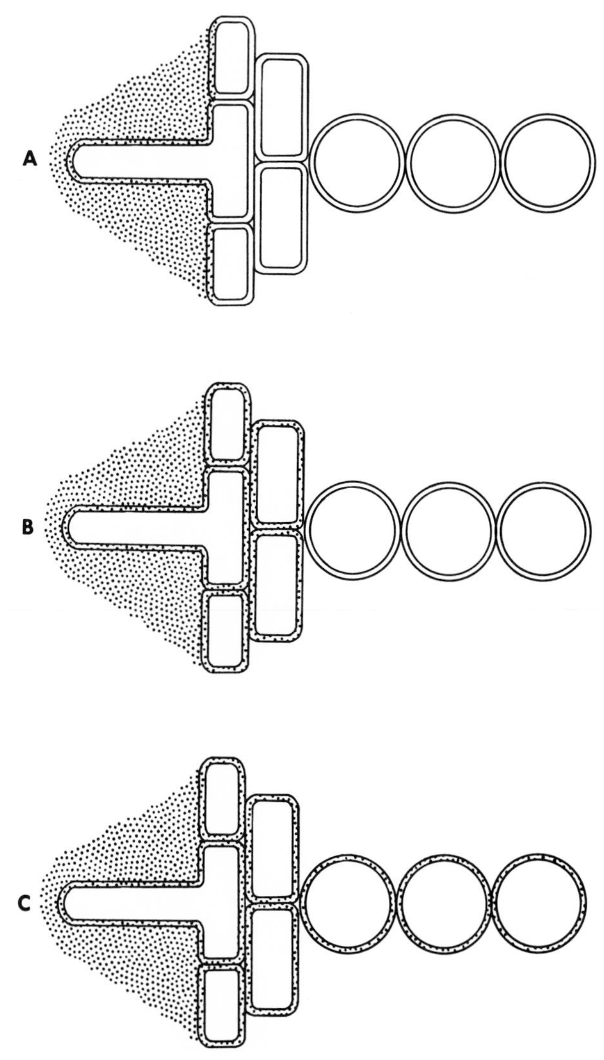

Fig. 11-9. Effect of transpiration on active absorption of solutes by roots. **A,** No transpiration. Only outer surface of epidermal cells come in contact with soil solutes. **B,** Slow transpiration. Outermost two layers in contact with soil solutes. **C,** Rapid transpiration. Several layers of cells in contact with soil solutes. Stippled cell walls contain soil solution, but the others do not.

sured before and after it passes over the plant by the amount of infrared radiation absorbed. But these methods require enclosing the plants, which is difficult to do without altering the transpiration rate. Other methods are also available, but the simplest is to weigh a potted plant at intervals. The loss of weight is essentially equal to the loss of water from the plant, since other factors (e.g., carbon assimilation) usually have a negligible effect on the weight compared to the effect of water loss. In the field, similar weight measurements have been made on shoots immediately after excision from the plant. The validity of this method has been the subject of much controversy. Transpiration has been measured in the field without disturbing the trees by (1) use of tritiated water (Kline et al., 1970) or (2) use of a "hygrosensor transpirometer," which measures the water content of the air and requires enclosure of a branch for not more than 1 min (Stark, 1969).

Attempts to control transpiration have led to the development of antitranspirants–substances that reduce the rate of transpiration. These are of three main kinds: (1) apolar substances (e.g., cetyl alcohol, oxyethylene docosanol) as well as the semi-apolar silicone (Parkinson, 1970); (2) plastics (e.g., polyethylenes); and (3) substances that lead to stomatal closure (e.g., phenylmercuric acetate [Zelitch, 1963] and SH inhibitors [Mouravieff, 1971]). All of these have led to some success, although their usefulness is still a matter of controversy, and they may also retard closure in the dark (Davenport et al., 1971). Another group of the third type of substances—alkenylsuccinic acids (e.g., decenyl succinic acid)—has been found to increase the permeability of the root cells (Zelitch, 1964; Kuiper, 1964). However, the exact mechanism of their effect on stomata is unknown. Nearly all are toxic and must be used in exactly the correct concentration to benefit the plant (Cocking and Tukey, 1970; Majernik, 1970). Another possible source of injury is the rise in leaf temperature (Brooks and Thorud, 1971) due to the decreased transpiration. As mentioned earlier, the natural plant constituent abscisic acid (Chapter 19) also causes marked stomatal closure, and its use as an antitranspirant has been suggested in place of the above toxic substances (Jones and Mansfield, 1970). Of course, carbon dioxide uptake and thus photosynthesis will also be prevented by stomatal closure. The plant has apparently learned how to overcome this problem to some extent. In the case of Sitka spruce, epicuticular wax in the stomatal antechamber reduces the rate of transpiration by about two thirds, but it reduces photosynthesis only by about one third (Jeffree et al., 1971).

RESISTANCE TO FLOW OF WATER

Water enters the roots from the soil, flows up the plant, and evaporates from the shoot surface into the atmosphere. The question has long been asked: What portion of this continuous stream offers the major resistance to flow? To answer this question, an analogue of Ohm's law has been applied (van den Honert, 1948):

$$\frac{dm}{dt} = \frac{P}{R}$$

where $\frac{dm}{dt}$ = amount of electricity or water flowing per unit time

P = electrical potential or water pressure drop

R = resistance to flow

In other words, at any point in the "stream," the amount of water flowing must be directly proportional to the pressure and inversely proportional to the resistance to flow. Since the stream is continuous, if the plant is in the steady state (loss equals absorption), the amount of water flowing past any cross-sectional area per unit time $\left(\frac{dm}{dt}\right)$ must be the same at all cross sections of the plant normal to the stream. Therefore $\frac{P}{R}$ must be constant throughout the plant, that is:

$$\frac{P_1 - P_0}{R_r} = \frac{P_2 - P_1}{R_x} = \frac{P_3 - P_2}{R_l} = \frac{P_4 - P_3}{R_g}$$

where $P_1 - P_0$, $P_2 - P_1$, $P_3 - P_2$, and $P_4 - P_3$ are the pressure drops across the various resistances

R_r = resistance across the root
R_x = resistance across the xylem
R_l = resistance across the leaf
R_g = resistance across the gaseous phase above the leaf

In the case of water flowing through the plant, the various pressure drops were taken to represent the decreases in water potential. Since the greatest decrease in water potential occurs in the gaseous phase, it was concluded that the major resistance to flow must reside just above the leaf surface, where the largest drop in vapor pressure occurs. This conclusion is opposed to all physiological experience (Kramer, 1949), which indicates that the major resistance to the liquid flow is at the plasma membranes of those living cells across which the water must flow, that is, in the root and the leaf. In many plants, in fact, it is definitely in the roots (Barrs and Klepper, 1968; Stoker and Weatherley, 1971; Molz, 1971), external to the vascular tissue (Boyer, 1971).

Direct measurements have revealed that the external, boundary layer resistance is generally less than 10% of the internal leaf resistance in corn (Impens et al., 1967). The clearest evidence is produced by stripping the epidermis from *Senecio* plants (Rawlins*). This causes rapid wilting (because of increased vapor flow), showing that the greatest resistance to flow is provided by the epidermal layer and not by the gaseous phase above it, which remains at the leaf surface regardless of whether the epidermis is present. It also proves that once the epidermal resistance is removed (e.g., by wide-open stomata), the major resistance to flow must occur in the liquid phase of water flow behind the surface of the evaporating leaf cells. Since the only resistance to flow in the vessels is the viscosity of the water, the major resistance must be at the plasma membranes in the living cells of the root and leaf, which offer a resistance to flow above that resulting from the viscosity of the water.

Since the experimental evidence categorically and consistently opposes the conclusion arrived at from the analogy with an electric current, there must be an error or fallacy in this concept. What is this fallacy? The analogy is a reversal of the usual one, for physics texts commonly explain the flow of electricity through a number of resistances by analogy with the flow of water through a number of pipes of different diameter. But it is always liquid water, not water in the vapor phase diffusing through air. Furthermore, it is liquid flow downhill due to gravity. If an analogy is to be made with upward flow in the plant, the simpler one would be with flow due to root pressure. It is obvious that resistance in the vapor phase plays no role whatsoever and that a decapitated stump will exude water at the same rate regardless of the vapor pressure above it. The only factors affecting the rate of flow in the plant will be the pressure gradient and the resistance to *liquid* flow *within* the plant. The same must be true of flow due to transpirational pull. The vapor flow within and above the plant is a diffusion flow and is therefore controlled by the vapor pressure gradient. Since the maximum possible vapor pressures (0.04 atm) are much smaller than the hydrostatic pressures (several atmospheres), the resistance to the vapor flow must also be much smaller than the resistance to the liquid flow. The real reason for the effect of the vapor phase on flow rate in the

*Rawlins, S. L. (U. S. Salinity Laboratory, Riverside, Calif.): Personal communication.

plant is that it controls the capillary force in the microcapillaries of the cell walls and therefore the transpirational pull, which controls the flow within the plant. If the major resistance to flow were in the vapor phase, there would be no such pull.

QUESTIONS

1. What gases are exchanged between the plant and its environment?
2. What plant parts are involved in this exchange?
3. Does the plasma membrane interfere with this exchange?
4. What barriers do interfere with this exchange?
5. Do the cuticular or suberized layers completely stop exchange?
6. What evidence is there that exchange is impeded?
7. What plant organ is able to control the exchange?
8. How does the plant organ do this?
9. What force is responsible for the opening of the stomata?
10. What force is responsible for the closing?
11. What must occur in order for the opening force to change?
12. What controls the movement of water into or out of the guard cells?
13. In general, what is the relation of stomatal opening to light?
14. What are the exceptions to this relationship?
15. Is the degree of opening in any way affected by cells other than the guard cells?
16. What effect does wilting have on stomatal opening?
17. What chemical change in the guard cells may lead to stomatal opening?
18. What change in gas content may initiate this chemical change?
19. Trace the series of steps leading to stomatal opening when a leaf is exposed to light.
20. How does the rate of transpiration (per unit of leaf area) compare with the rate of evaporation from a free water surface?
21. What fraction of the leaf area is covered with stomata?
22. How does the rate of water loss per unit stomatal area compare with that from a free water surface?
23. Does this mean that transpiration is too rapid to be a diffusion process?
24. How can this rate of water loss per unit of stomatal area be explained?
25. Suppose two beakers have diameters of 3 inches and 6 inches, respectively. If both are filled with water, how much weight would the 6-inch one lose during the time when the 3-inch one lost 2 g water?
26. Suppose a single stoma of 1 μm opening loses 2 g water, how much would it lose in the same time when open to 2 μm?
27. Does this mean that transpiration is always proportional to the stomatal aperture?
28. When is transpiration not proportional to the stomatal aperture?
29. What is meant by transpiration (as opposed to guttation)?
30. What is the difference between stomatal and cuticular transpiration?
31. Under what conditions could transpiration increase without the occurrence of a change in stomatal opening?
32. What effect does wind have on transpiration?
33. What is the primary role of stomatal opening in the physiology of the plant?
34. Can transpiration benefit the plant?
35. In what ways can transpiration benefit the plant?

SPECIFIC REFERENCES

Arisz, W. H. 1956. Significance of the symplasm theory for transport in the root. Protoplasma 46:5-62.

Aylor, D. E., J.-Y. Parlange, and A. D. Krikorian. 1973. Stomatal mechanics, Amer. J. Bot. **60:** 163-171.

Armstrong, W. 1971. Oxygen diffusion from the roots of rice grown under non-waterlogged conditions. Physiol. Plant. **24**:242-247.

Balasubramaniam, S. 1967. Diurnal stomatal movements of *Stachytarpheta indica* (L.). Vahl. Ceylon J. Sci. Biol. Sci. **7**:116-123.

Bange, G. G. J. 1953. On the quantitative explanation of stomatal transpiration. Acta Bot. Neerl. **2**:255-297.

Barrs, H. D., and B. Klepper. 1968. Cyclic variations in plant properties under constant environmental conditions. Physiol. Plant. **21**:711-730.

Benecke, W., and L. Jost. 1924. Pflanzenphysiologie. I. Gustav Fischer Verlag, Stuttgart.

Bowling, D. J. F., and P. E. Weatherley. 1965. The relationship between transpiration and potassium uptake in *Ricinus communis*. J. Exp. Bot. **16**:732-741.

Boyer, J. S. 1971. Resistance to water transport in soybean, bean, and sunflower. Crop Sci. **11**:403-407.

Brooks, K. N., and D. B. Thorud. 1971. Antitranspirant effects on the transpiration and physiology of tamarisk. Water Resour. Res. 499-510.

Brown, H. T., and F. Escombe. 1900. Static diffusion of gases and liquids in relation to the assimilation of carbon and translocation in plants. Phil. Trans. Roy. Soc. [B] **193**:223-291.

Byrne, G. F., C. W. Rose, and R. O. Slatyer. 1970. An aspirated diffusion porometer. Agr. Meteorol. **7**:39-44.

Caldwell, M. M. 1970. The effect of wind on stomatal aperture, photosynthesis, and transpiration of *Rhododendron ferrugineum* L. and *Pinus cembra* L. Centralbl. Gesamte Forstw. **87**:193-201.

Cocking, W. D., and H. B. Tukey, Jr. 1970. Reduction of water loss from *Chrysanthemum morifolium* following foliar application of phenylmercuric acetate. J. Amer. Soc. Hort. Sci. **95**:382-384.

Curtis, O. F. 1938. Wallace and Clum, "Leaf temperatures": a critical analysis with additional data. Amer. J. Bot. **25**:761-771.

Davenport, D. C., M. A. Fisher, and R. M. Hagan. 1971. Retarded stomatal closure by phenylmercuric acetate. Physiol. Plant. **24**:330-336.

Decker, J. P., and B. F. Wetzel. 1957. A method for measuring transpiration of intact plants under controlled light, humidity, and temperature. Forest Sci. **3**:350-354.

Domes, W. 1971. Different CO_2-sensitivities in the gas exchange of both leaf surfaces of *Zea mays*. Planta **98**:186-189.

Einhellig, F. A., and L. Y. Kuan. 1971. Effect of scopoletin and chlorogenic acid on stomatal aperture in tobacco and sunflower. Bull. Torrey Bot. Club **98**:155-162.

Ekern, P. G. 1965. Evapotranspiration of pineapple in Hawaii. Plant Physiol. **40**:736-739.

Ferri, M. G. 1955. Contribuição ao conhecimento da ecologia do cerrado e da caatinga. Faculdad de Filosofia Ciencias e Letras, Universidade de São Paulo, São Paulo, Brazil. Bol. 195. Bot. 12.

Fischer, R. A. 1968. Stomatal opening in isolated epidermal strips of *Vicia faba*. I. Response to light and to CO_2-free air. Plant Physiol. **43**:1947-1952.

Fischer, R. A. 1971. Role of potassium in stomatal opening in the leaf of *Vicia faba*. Plant Physiol. **47**:555-558.

Fischer, R. A., and T. C. Hsiao. 1968. Stomatal opening in isolated epidermal strips of *Vicia faba*. II. Responses to KCl concentration and the role of potassium absorption. Plant Physiol. **43**:1953-1958.

Frommhold, I. 1971. Ontogenetic and functional development of the stomata of oats (*Avena sativa* L.), Biochem. Physiol. Pflanz. **162**:410-416.

Gale, J., and R. M. Hagan. 1966. Plant antitranspirants. Ann. Rev. Plant Physiol. **17**:269-282.

Gates, D. M.: 1963. Leaf temperature and energy exchange. Arch. Meteorol. Geophys. Bioklim. **12**:321-336.

Gates, D. M. 1966. Transpiration and energy exchanges. Quart. Rev. Biol. **41**:353-364.

Ginsburg, H., and B. Z. Ginzburg. 1970. Radial water and solute flows in roots of *Zea mays*. II. Ion fluxes across root complex. J. Exp. Bot. **21**:593-604.

Glinka, Z. 1971. The effects of epidermal cell water potential on stomatal response to illumination of leaf discs of *Vicia faba*. Physiol. Plant. **24**:476-479.

Hansen, P. 1971. The effect of fruiting upon transpiration rate and stomatal opening in apple leaves. Physiol. Plant. **25**:181-183.

Heath, O. V. S., and F. L. Milthorpe. 1950. Studies in stomatal behavior. V. The role of carbon dioxide in the light response of stomata. J. Exp. Bot. **1**:227-243.

Heath, O. V. S. and T. A. Mansfield. 1969. The movements of stomata, p. 303-332. *In* M. B. Wilkins (Ed.). The physiology of plant growth and development. McGraw-Hill Book Co., New York.

Henze, J. 1969. Relations between respiration and internal atmosphere of CO_2-sensitive pome fruits. Qual. Plant. Mat. Veg. **19**:229-242.

Honert, T. H. van den. 1948. Water transport in

plants as a catenary process. Discuss. Faraday Soc. **3**:146-153.

Horton, R. F. 1971. Stomatal opening: the role of abscisic acid. Canad. J. Bot. **49**:583-585.

Humble, G. D., and K. Raschke. 1971. Stomatal opening quantitatively related to potassium transport: evidence from electron probe analysis. Plant Physiol. **48**:447-453.

Hylmö, B. 1953. Transpiration and ion absorption. Physiol. Plant. **6**:333-405.

Impens, I. I., D. W. Stewart, L. H. Allen, Jr., and E. R. Lemon. 1967. Diffusive resistance at, and transpiration rates from leaves in situ within the vegetative canopy of a corn crop. Plant Physiol. **42**:99-104.

Irvine, J. E. 1971. Photosynthesis and stomatal behavior in sugarcane leaves as affected by light intensity and low air flow rates. Physiol. Plant. **24**:436-440.

Jeffree, C. E., R. P. C. Johnson and P. G. Jarvis. 1971. Epicuticular wax in the stomatal antechamber of Sitka spruce and its effects on the diffusion of water vapor and carbon dioxide. Planta **98**:1-10.

Jones, R. J., and T. A. Mansfield. 1970. Suppression of stomatal openings in leaves treated with abscisic acid. J. Exp. Bot. **21**:714-719.

Kline, J. R., J. R. Martin, C. F. Jordan, and J. J. Koranda. 1970. Measurement of transpiration in tropical trees with tritiated water. Ecology **51**:1068-1073.

Kuiper, P. J. C. 1964. Water transport across root cell membranes; effect of alkenyl-succinic acids. Science **143**:690-691.

Lange, O. L. 1959. Untersuchungen über Wärmehaushalt und Hitzeresistenz mauretanischer Wüsten und Savannenpflanzen. Flora **147**:595-651.

Lee, R., and D. M. Gates. 1964. Diffusion resistance in leaves as related to their stomatal anatomy and microstructure. Amer. J. Bot. **51**:963-975.

Loftfield, J. V. G. 1921. The behavior of stomata. Carnegie Institution of Washington, Pub. No. 314. Washington, D. C.

MacDougal, D. T., and E. B. Working. 1933. The pneumatic system of plants, especially trees. Carnegie Institution of Washington, Pub. No. 441. Washington, D. C.

Majernik, O. 1970. Responses of stomata of barley and maize to phenylmercuric acetate. Biol. Plant. (Praha) **12**:419-423.

Mansfield, T. A., and R. J. Jones. 1971. Effects of abscisic acid on potassium uptake and starch content of stomatal guard cells. Planta **101**:147-158.

Meidner, H. 1970. A critical study of sensor element diffusion porometers. J. Exp. Bot. **21**:1060-1066.

Mingeau, M. 1969. Effect of mineral nutrition on water behavior in the plant. I. Absorption-transpiration balance. Ann. Agron. (Paris) **20**:263-276.

Molz, F. J. 1971. Interaction of water uptake and root distribution. Agron. J. **63**:608-610.

Monteith, J. L., and T. A. Bull. 1970. A diffusive resistance porometer for field use. II. Theory, calibration and performance. J. Appl. Ecol. **7**:623-638.

Morrow, P. A., and R. O. Slatyer. 1971*a*. Leaf resistance measurements with diffusion porometers: precautions in calibration and use. Agric. Meteorol. **8**:223-233.

Morrow, P. A., and R. O. Slatyer. 1971*b*. Leaf temperature effects on measurements of diffusive resistance to water vapor transfer. Plant Physiol. **47**:559-561.

Moss, D. N., and S. L. Rawlins. 1963. Concentration of carbon dioxide inside leaves. Nature (London) **197**:1320-1321.

Mouravieff, I. 1971. Les inhibiteurs des groupes thiols empêchent l'ouverture des ostioles stomatiques: importance probable des glyceraldehyde-phosphate deshydrogenases. Physiol. Vég. **9**:109-118.

Nishida, K. 1963. Studies on stomatal movement of crassulacean plants in relation to the acid metabolism. Physiol. Plant. **16**:281-298.

Ofir, M., E. Shmueli, and S. Moreshet. 1968. Stomatal infiltration measurement as an indicator of the water requirement and timing of irrigation for cotton. Expl. Agric. **4**:325-333.

Pallas, J. E., Jr., and B. G. Wright. 1973. Organic acid changes in the epidermis of *Vicia faba* and their implication in stomatal movement. Plant Physiol. **51**:588-590.

Papetti, R. A. and F. R. Gilmore, 1971. Air pollution. Endeavour **30**:107-114.

Parkinson, K. J. 1970. The effects of silicone coatings on leaves. J. Exp. Bot. **21**:566-579.

Parlange, J.-Y., and P. E. Waggoner. 1970. Stomatal dimensions and resistance to diffusion. Plant Physiol. **46**:337-342.

Raschke, K. 1970. Temperature dependence of CO_2 assimilation and stomatal aperture in leaf sections of *Zea mays*. Planta **91**:336-363.

Scarth, G. W., and M. Shaw. 1951. Stomatal movement and photosynthesis in Pelargonium. Plant Physiol. **26**:207-225.

Shimshi, D. 1967. Leaf chlorosis and stomatal aperture. New Phyt. **66**:455-461.

Slatyer, R. O., and J. F. Bierhuizen. 1964. The influence of several transpiration suppressants on transpiration, photosynthesis, and water use efficiency of cotton leaves. Aust. J. Biol. Sci. **17:** 131-146.

Stålfelt, M. G. 1966. The role of the epidermal cells in the stomatal movement. Physiol. Plant. **19**:241-256.

Stark, N. 1969. Transpiration of *Abies concolor*. Bot. Gaz. **130**:143-149.

Stiles, W. 1970. A diffusive resistance porometer for field use. I. Construction. J. Appl. Ecol. **7:** 617-622.

Stoker, R. and P. E. Weatherley. 1971. The influence of the root system on the relationship between the rate of transpiration and depression of leaf water potential. New Phytol. **70**:547-554.

Tal, M., D. Imber, and C. Itai. 1970. Abnormal stomatal behavior and hormone imbalance in *flacca,* a wilty mutant of tomato. Plant Physiol. **46**:367-372.

Tal, M. and D. Imber. 1971. Abnormal stomatal behavior and hormonal imbalance in *Flacca,* a wilty mutant of tomato. III. Hormonal effects on the water status in the plant. Plant Physiol. **47**:849-850.

Thomas, D. A. 1970*a*. The regulation of stomatal aperture in tobacco leaf epidermal strips. I. The effect of ions. Aust. J. Biol. Sci. **23**:961-979.

Thomas, D. A. 1970*b*. The regulation of stomatal aperture in tobacco leaf epidermal strips. II. The effect of ouabain. Aust. J. Biol. Sci. **23:** 981-989.

Thomas, D. A. 1971. The regulation of stomatal aperture in tobacco leaf epidermal strips. III. The effect of ATP. Aust. J. Biol. Sci. **24**:689-707.

Wium-Andersen, S. 1971. Photosynthetic uptake of free CO_2 by the roots of *Lobelia dortmanna.* Physiol. Plant. **25**:245-248.

Zelitch, I. 1964. Reduction of transpiration of leaves through stomatal closure induced by alkenylsuccinic acids. Science **143**:692-693.

GENERAL REFERENCES

Kramer, P. J. 1949. Plant and soil water relationships. McGraw-Hill Book Co., New York.

Levitt, J. 1967. The mechanism of stomatal action. Planta **74**:101-118.

Meidner, H., and T. A. Mansfield. 1968. Physiology of stomata. McGraw-Hill Book Co., New York.

Scarth, G. W. 1932. Mechanism of the action of light and other factors on stomatal movement. Plant Physiol. **7**:481-504.

Stålfelt, M. G. 1956. Die stomatäre Transpiration und die Physiologie der Spaltöffnungen. p. 351-426. *In* W. Ruhland. Handbuch der Pflanzenphysiologie. Vol. 3. Springer-Verlag, Berlin.

Zelitch, I. (Ed.). 1963. Stomata and water relations in plants. Conn. Agric. Exp. Sta. Bull. No. 664.

PART THREE

Biochemistry

12

Nutrition

AUTOTROPHY AS OPPOSED TO HETEROTROPHY

No organism is completely self-sufficient. Each must obtain materials ("foods") from outside of itself to live and grow. The nutrition of any organism is the supply from its environment of all those substances required for the normal completion of its life cycle. Those organisms that require both organic and inorganic substances are called *heterotrophic;* those that need be supplied only with the inorganic substances are called *autotrophic* ("self-feeding"), since they make their own organic substances. Animals and nongreen plants are heterotrophic and so are the albino mutants of green plants, which can be grown to maturity if supplied with sugar, for example, by immersing the cut ends of their leaves in sugar solutions and allowing the sugar to be taken up in the transpiration stream. Of course, they must also be supplied with the same mineral nutrients as the normal green plant. When nongreen parts (e.g., roots, stem tips, embryos) are excised (severed) from the green parts of the plant, they can be made to grow only when supplied with at least a sugar and, in many cases, several additional organic substances (Table 12-1). Therefore these parts of the green plant are heterotrophic. This fact has become more important to the plant physiologist in recent years because of the usefulness of such organ, tissue, or cell cultures as tools for investigating the growth and development of the plant. With favorable material and suitable techniques, a single culture of maize endosperm has yielded more than 5000 g tissue in 5 months (Graebe and Novelli, 1966). In many cases it has not yet been found possible to grow such cultures in a completely defined medium, and organic nutrients of incompletely known chemical constitution must be added (e.g., coconut milk). The heterotrophic nutrition of such cultures, therefore, is not completely understood. In some cases the cultures may be maintained and subcultured indefinitely. In others (e.g., wheat roots) the cultures eventually die for unknown reasons. For rapid growth of some tissue cultures a rather complex nutrient medium is sometimes needed (Table 12-2). On the other hand, no substance tested was capable of supporting the growth of a diatom in the dark, and it was therefore concluded that the organism is obligately *phototrophic* (Hayward, 1968). This term implies that light itself is an essential nutrient that cannot be replaced by the supply of any known chemical substance. (The suffix

Table 12-1. White's nutrient medium for growing excised roots*

Substance	Milligrams per liter
$MgSO_4$	360.0
$Ca(NO_3)_2$	200.0
Na_2SO_4	200.0
KNO_3	80.0
KCl	65.0
$NaH_2PO_4 \cdot H_2O$	16.5
$Fe_2(SO_4)_3$	2.5
$MnSO_4$	4.5
$ZnSO_4$	1.5
H_3BO_3	1.5
KI	0.75
Sucrose	2000.0
Glycine	3.0
Nicotinic acid	0.5
Pyridoxine	0.1
Thiamine	0.1

*From White, P. R. 1943. A handbook of plant tissue culture. The Ronald Press Co., New York.

-trophic, meaning feeding, must not be confused with the suffix *-tropic*, meaning bending; see Chapter 18.)

The normal green plant, on the other hand, is autotrophic; that is, it can synthesize all its essential organic substances, provided it is supplied with all the essential inorganic elements and grown under normal conditions (e.g., in the light). Nevertheless, the line between autotrophism and heterotrophism is not a sharp one. The difference may sometimes be quantitative rather than qualitative because a plant may be able to synthesize a substance but in insufficient quantity or too slowly to support normal growth and development. For this reason, even the

Table 12-2. Composition of revised medium (RM-1965) for tissue cultures*

Mineral salts†						
Major elements				Minor elements		
Salts	mg/l	mg atoms/l		Salts	mg/l	μg atoms/l
NH_4NO_3	1650	N	41.2	H_3BO_3	6.2	100
KNO_3	1900	N, K	18.8	$MnSO_4 \cdot 4H_2O$	22.3	100
$CaCl_2 \cdot 2H_2O$	440	Ca	3.0	$ZnSO_4 \cdot 4H_2O$	8.6	30
$MgSO_4 \cdot 7H_2O$	370	Mg, S	1.5	KI	0.83	5.0
KH_2PO_4	170	P, K	1.25	$Na_2MoO_4 \cdot 2H_2O$	0.25	1.0
Na_2EDTA	37.3‡	Na	0.20	$CuSO_4 \cdot 5H_2O$	0.025	0.1
$FeSO_4 \cdot 7H_2O$	27.8‡	Fe	0.10	$CoCl_2 \cdot 6H_2O$	0.025	0.1

Organic constituents			
Sucrose	30 g/l	3-Indoleacetic acid	2.0 (1-30) mg/l§
Agar	10 g/l	Kinetin	0.03 or 0.2 (0.001-10) mg/l§
		Thiamine · HCl	0.400 mg/l
		myo-Inositol	100 mg/l

Optional constituents			
p-Aminobenzoic acid	0.1 mg/l	Braun's supplement (Braun, A. C.):	
Gibberellic acid	1 mg/l	Cytidylic acid	200 mg/l
Adenine (adenine sulfate)	40 mg/l	Guanylic acid	200 mg/l
Tyrosine	100 mg/l	L-Asparagine	500 mg/l
Amino acid mixture (see LaMotte, 1961)		L-Glutamine	500 mg/l
Casein hydrolysate (Edamin)	1-3 g/l	Tobacco leaf extract (Murashige and Skoog, 1962)	

*From Linsmaier, E. M., and F. Skoog. 1965. Physiol. Plant. 18:100-127.
†pH adjusted to 5.6 with 1 N NaOH.
‡5 ml/l of a stock solution containing 5.57 g $FeSO_4 \cdot 7H_2O$ and 7.45 g Na_2EDTA per liter of dd H_2O.
§IAA and kinetin levels vary with the type of growth desired.

normally autotrophic higher green plant may become partially heterotrophic under unfavorable growing conditions. When, for instance, such a plant is grown autotrophically at too high a temperature (but below a directly lethal temperature), its growth may be impaired or even stopped completely. In some cases, at least, this is caused by its inability to synthesize certain organic compounds (e.g., vitamins or amino acids) in adequate quantities at high temperatures (Langridge, 1963), and it must be supplied with these substances to grow normally. Injuries resulting from these metabolic deficiences have been called "biochemical lesions."

Under normal growing conditions, however, all higher green plants are autotrophic, and their nutrition is therefore solely inorganic. On the basis of this definition, the inorganic nutrition of higher plants should include carbon dioxide, oxygen, and water, since these substances are obtained from their environment. The absorption of these substances, however, is so closely related to their metabolism and to the transfer of materials that they are best considered in connection with these processes. Therefore the nutrition of higher plants usually refers to mineral nutrition, and the elements absorbed by the roots (except carbon, hydrogen, and oxygen) are called mineral elements, since they are obtained either directly or indirectly from the minerals in the soil. The mineral nutrition of plants therefore embraces all the elements essential to the plant's existence, with the exception of carbon, hydrogen, and oxygen. Nitrogen, strictly speaking, is not a mineral element, but it is included with these, since it is obtained by the plant from the soil solids.

MINERAL NUTRITION

Essentiality

The first and perhaps most fundamental problem of mineral nutrition is to determine which chemical elements are essential to the life and growth of the plant. As a start, one might analyze the plant to find out which are present (Table 12-3). When this is done, the number of elements found depends on the plant, the medium in which it was grown, and the completeness of the analysis. However, this tells us merely that the elements not found in the plant (assuming that the methods of analysis are adequate) are not essential in appreciable quantities. The mere presence of an element in a plant does not prove its essentiality. Thus if a large enough number of plants from different regions were analyzed, certain elements might sometimes be present or sometimes be absent from the plant. Actually, some sixty elements have been found in plants (Robinson and Edgington, 1945), and undoubtedly all the rest of the now 100-odd known elements could be made to occur in the plant by supplying them to the roots.

A more direct approach to the problem of essentiality is to grow plants in the complete absence of a given element. If the plant grows normally, the element is evidently not essential; if it fails to grow nor-

Table 12-3. Analysis of stems, leaves, cob, grain, and roots of corn*

Element	Percent of total dry weight
O	44.4
C	43.6
H	6.2
N	1.46
P	0.2
K	0.92
Ca	0.23
Mg	0.18
S	0.17
Fe	0.08
Si	1.17
Al	0.11
Cl	0.14
Mn	0.035
Undetermined	0.93

*From Miller, E. C. 1938. Plant physiology. McGraw-Hill Book Co., New York.

mally, the element is usually considered essential. To be certain that the element is truly essential, it is necessary to show not only that (1) a deficiency of the element makes it impossible for a plant to complete its vegetative or reproductive cycle, but also that (2) it cannot be replaced by any other element, and that (3) the element must produce its effect within the plant. This means that the effect is not simply the result of interaction with (e.g., detoxification of) other nonessential elements, organisms, etc. outside the plant. These three requirements have been called the *criteria of essentiality* (Arnon and Stout, 1939). However, it has recently been shown that some of the accepted essential elements can at least be partially replaced by others (e.g., magnesium by manganese, potassium by rubidium, etc., Hewitt, 1951).

The search for essential elements has required the development of a technique for growing the plant in the absence of impurities. This has meant the elimination of the natural root medium—the soil. In its place a pure sand, gravel, or best of all, simple solutions have been used. The latter so-called water or solution cultures offer the greatest difficulty, since they fail to supply the normal support for the roots and the normal amounts of oxygen. This requires development of adequate artificial support and aeration. A successful water or solution culture therefore requires a certain amount of engineering. This system of growing plants in the absence of soil has been used to some extent on a practical scale and has been named *hydroponics* (Gericke, 1940).

Using only the first criterion of essentiality (i.e., no normal growth in the absence of the element), the following seven elements were long ago found to be essential: nitrogen, phosphorus, potassium, calcium, magnesium, sulfur, and iron. Thus it was discovered that plants grow normally in four-salt solutions containing potassium nitrate, monopotassium phosphate, calcium chloride, and magnesium sulfate (or other salts supplying these six elements) with a small amount of ferric chloride (or other iron salt). Even three-salt solutions (monopotassium phosphate, calcium nitrate, magnesium sulfate) plus some iron were soon found adequate. Many formulas for such solutions have been adopted, some of them named after workers who developed them (e.g., Knop's, Cronin's, Tottingham's, Shive's). However, the early solutions were made with salts that were not pure. Small quantities of other salts were therefore invariably present. Since it was already shown that one element (iron) was needed only in a small quantity, it was logical to suspect that other elements might also be essential in trace amounts. It was soon found that the better the purification of the basic three or four salts the poorer was the growth. This was followed up by direct evidence that other elements are needed in small amounts. They are just as essential for the life and growth of the plants as the preceding seven elements, and therefore no growth is possible in their complete absence. However, in view of the small amounts needed, they are usually grouped (with iron) separately from the preceding six major, or macronutrient, elements and are called trace, minor, or micronutrient elements. By the late 1930s the following six had been conclusively proved to be trace essential elements: iron, manganese, copper, zinc, boron, and molybdenum. They are needed in the nutrient solution in amounts ranging from 0.5 to 0.01 ppm (Table 12-4). Larger quantities may cause injury or death. It has been indicated that chlorine (one of the most ubiquitous of elements) is a seventh trace essential element, although usually present in the plant in as large quantities as the major essential elements (Broyer et al., 1954). There is even evidence of a chloride deficiency disease (see following discussion) in the case of coconut palms (Ollagnier and Ochs, 1971). There are reasons for

Table 12-4. Quantities of mineral elements used in complete nutrient solutions*

Major ion	Quantities (moles)	Trace element	Quantities (ppm)
NO_3^-	0.005 -0.010	Fe	0.5
$PO_4^{\equiv}$	0.00025-0.002	B	0.25 (or more)
$SO_4^=$	0.001 -0.010	Mn	0.25
K^+	0.002 -0.010	Zn	0.25
Ca^{++}	0.002 -0.005	Cu	0.02
Mg^{++}	0.001 -0.010	Mo	0.01

*Adapted from Robbins, W. R. 1946. Soil Sci. **62**:3-22.

believing that cobalt is also essential, although the classical methods have failed to prove this because of the extremely small traces in which it seems to be needed. It is certainly essential for leguminous plants when these are not supplied with nitrogen compounds. But this is really a requirement for the symbiotic bacteria, since no cobalt need be knowingly added when nitrogen compounds are supplied (Evans and Kliewer, 1964). Cobalt is also known to be essential for at least some algae, including even the large marine brown algae (Pedersen, 1969). In the case of higher plants it is possible that enough cobalt is present as impurities to satisfy the needs of the plant, and therefore the amount deliberately added has no effect. Indirect evidence favors this conclusion for the following reasons:

1. The amounts known to be present as impurities even in the most carefully purified media are of the order of the amount needed to supply the plant with the cobalt-containing vitamin B_{12}.
2. The addition of cobalamin (vitamin B_{12}) under certain conditions furthers the synthesis of an amino acid (serine) in the primary root of *Zea mays* (Graser, 1965).

Direct evidence has also been produced in one case. Experiments that carefully excluded contamination by traces of cobalt in the air have succeeded in demonstrating a significant decrease in growth when cobalt was not knowingly added (Wilson and Nicholas, 1967).

Some investigators have produced evidence for the essentiality of a few other elements (e.g., aluminum, silicon, selenium) at least for certain plants (Table 12-5), but other workers have failed to obtain positive results. This is partly explainable by the frequently observed stimulating effect of nonessential elements on the growth of plants, for instance by counteracting the toxicity of other elements present in the nutrient medium. On the other hand, two elements that are unequivocally essential for all higher plants do not seem to be essential for at least some fungi (calcium) and algae (boron). Elements such as aluminum and silicon have been called *ballast elements* (Frey-Wyssling, 1945) because they are normally present in large amounts, although the plant can be grown perfectly normally without them (as far as can be determined). Some plants are *accumulators;* that is, they concentrate large quantities of certain elements (e.g., selenium, aluminum) in their tissues.

Complete absence of any one of the essential elements (major or trace) will completely stop growth, but under normal conditions the elements are never completely absent. If, however, they are present in less than optimum quantities, growth will occur but abnormalities known as *deficiency diseases* will develop. In animals this term is used for diseases caused by vitamin deficiencies. Since plants are autotrophs,

Table 12-5. Essentiality of elements

Essential for all higher plants		Essential for all higher plants but not for all algae or fungi	Apparently essential for some species		
			Element	Species	Reference
1. Major (or macro)					
Cationic	Anionic				
K	N	Ca	V	*Scenedesmus*	Arnon and Wessel (1953)
Mg	P	B	Al	Ferns	Tauböck (1942)
Ca	S	Cl(?)	Si	Diatoms	Lewin (1955)
2. Trace (or micro)			Na	Blue-green alga	Allen and Arnon (1955)
Cationic	Anionic			*Atriplex*	Brownell (1965)
Fe	B		Se	*Astragalus* sp.	Trelease and Trelease (1938)
Mn	Mo		I	*Polysiphonia*	Fries (1966)
Cu	Cl*				
Zn					
Co*					

*Chlorine and cobalt have not been tested on enough higher plants yet, but available results indicate that chlorine is essential for all plants (Johnson et al., 1957) and that cobalt may be essential Wilson and Nicholas, 1967).

they show disease symptoms only for deficiencies of mineral elements. The symptoms of such diseases are more or less specific for each element (Bear and Coleman, 1941), although it is sometimes difficult to distinguish the differences. Furthermore, different plants will show somewhat different deficiency symptoms.

The commonest type of deficiency symptom is *chlorosis,* a reduction in the amount of green pigment in the leaf. But chlorosis can occur as a result of any one of several deficiencies (nitrogen, magnesium, iron, manganese, etc.). However, the type of chlorosis is sometimes different for different elements. Thus it is a uniform loss of color in the case of nitrogen deficiency but more pronounced between the veins in the case of iron deficiency. Deficiencies of the trace elements have given rise to a particularly large number of diseases, for example, celery crack, heart rot and dry rot of sugar beets, internal browning of cauliflower (boron deficiencies), little leaf or rosette of fruit trees (zinc deficiency), marsh spot of pea seeds (manganese deficiency), etc.

The deficiencies found in the plant may not correspond with those in the soil. Thus a lime-induced chlorosis may actually be an iron deficiency because the high pH has made the iron in the soil unavailable to the plant. Sometimes the ratios of the elements in the root medium may be more important than the absolute quantities (e.g., the iron-manganese and the calcium-boron ratios). A method that overcomes the unavailability of iron because of high pH is to apply chelated iron, that is, iron combined with an organic compound in such a way that it cannot ionize and therefore cannot be precipitated (Stewart and Leonard, 1952). One of the most commonly used of such substances is ethylenediaminetetraacetate (EDTA).

The immediate cause of cell injury, when a specific mineral element is deficient, is usually not known. In the case of potassium deficiency, there seems to be a relation to a toxic metabolic product. Putrescine and cadaverine are two diamines produced from proteins when decomposed by putrefying bacteria. In some higher plants (e.g., Solanaceae), they are normal products of metabolism. Putrescine also accumulates in wheat and barley, when they are suffering from potassium deficiency. The necrotic spots formed on the potassium-deficient leaves are due to the toxic effects of putrescine, and can be produced on leaves adequately supplied with potassium if putrescine is introduced into the plant (Coleman and Richards, 1956). Other elements, such as sulfur, phosphorus, and iron, may also affect the accumulation of putrescine when they are deficient.

Some deficiency symptoms occur first on the oldest leaves (e.g., nitrogen, potassium, phosphorus, magnesium); others occur first on the youngest leaves (sulfur, calcium, iron, manganese, boron, copper, zinc). Elements of the first group are mobile and are transferred from the older inactive leaves to the younger growing leaves. The elements of the second group are immobile and remain in the old leaves, causing the newly developing leaves to show the deficiency although the older ones do not (Müller, 1949). However, the preceding separation into clear-cut groups of mobile and immobile elements is not to be taken too rigidly, since some elements are partially mobile (e.g., sulfur) and others are completely mobile (e.g., potassium). The differences in mobility, therefore, are quantitative rather than qualitative, and if a grouping is desirable, the following three groups would perhaps serve as a rough approximation: nearly completely mobile—K, Cl; largely mobile—N, P, Mg, S; largely immobile—Ca, Fe, Mn, B, Cu, Zn, Mo.

The following quantitative evaluation of mobility in the phloem of *Yucca flaccida* has been calculated from the ratio of $\frac{\text{concentration in exudate}}{\text{concentration in leaves}}$ (Tammes and van Die, 1966):

1. K	0.89		7. Na	0.12
2. P	0.81		8. Cu	0.09
3. N	0.61		9. Mg	0.03
4. B	0.29		10. Ca	0.01
5. Zn	0.27		11. Mn	0.01
6. Fe	0.12			

But mobility to regions of exudation may be different from mobility to growing regions of the plant. This seems to be particularly true of magnesium, which is largely mobile to growing regions and immobile in the preceding test. Iron is one of the most immobile ions in the plant according to Burleson and Cunningham (1963). Manganese, on the other hand, may be immobile in old leaves but is mobile in young leaves (Klimovitskaya and Vizir, 1963).

Even some of the most immobile elements can be retranslocated, although only in extremely small amounts (e.g., calcium, Biddulph et al., 1959). Calcium is translocated almost solely in the xylem and therefore accumulates markedly in the transpiring organs (Ziegler, 1961). Potassium is translocated readily in the phloem; therefore its translocation in the plant is much less dependent on transpiration. Consequently, plant parts that are supplied mainly by the phloem (nectaries, pollen, fruit, seed) normally contain ample potassium but are deficient in calcium. This is particularly pronounced in seeds (Dickmann and Kozlowski, 1969), which therefore may not be able to germinate normally unless supplied with calcium. The mobility of an element may or may not be altered by its presence in excess. In the case of some plants, at least, the excess of calcium accumulated is not readily mobilized to the developing tissues, and the plant becomes calcium deficient as soon as the rate of calcium absorption falls below the functional requirement of the plant (Loneragan and Snowball, 1969). On the other hand, sulfur may be mobile when present in excess but not when present in deficient quantities. Manganese, one of the "immobile" elements, apparently can be redistributed, although to a small degree compared to potassium and nitrogen (Vose, 1963).

Roles of elements

Difficult as it sometimes has been to prove the essentiality of a nutrient element, it is often much more difficult to find out why it is essential, that is, its role in the physiology of the plant. If the element is a constituent of a substance whose function in the plant is known, this, of course, explains why that element is essential. However, it must always be remembered that a single element may play many different roles in the physiology of the plant. Nitrogen, for instance, is one of the elements in proteins and nucleic acids; life is therefore impossible without nitrogen. But it is also a constituent of phospholipids, some vitamins, and chlorophyll, all of which play definite roles in the physiology of the plant. Nitrogen also occurs in many substances with unknown roles in the plant (e.g., plant bases). Phosphorus, like nitrogen, is a necessary constituent of many vitally important substances: nucleoproteins, phospholipids, enzyme components, etc. Sulfur occurs in proteins and some vitamins as well as in other substances such as glutathione. The importance of sulfolipids has been discovered by Davies and co-workers (1965). Magnesium is a constituent of chlorophyll, but this cannot be its only function, since it is also essential for nongreen plants, primarily because it is an activator of many enzymes. Calcium occurs in the pectate that cements the walls of adjacent cells together, but unlike magnesium, it is not essential for some nongreen lower plants. However, its role is obviously much more fundamental than as a mere cementer of cells. There is reason to believe that protoplasm cannot exist in the absence of certain cations. In most organisms calcium is undoubtedly one of these ions and potassium is another. Since protoplasm probably functions only at a pH of about 6.5 to 8, the protoplasmic proteins with their relatively low isoelectric points (mostly 4.5 to 5) must be in the acidic or anionic form. Therefore they must form salts with the cations. Apparently, K^+ and Ca^{++} and perhaps Mg^{++} are the cations primarily involved, and they must be present in a certain ratio to maintain the normal, native, and active state of the proteins. Unlike the other major essential elements, potassium is not a constituent of any known essential organic substance but occurs almost solely in the ionic form. The K^+ ion is, however, essential as an activator in protein synthesis (Evans and Sorger, 1966). The fact that it is needed in so much larger quantity than ions such as Ca^{++} and Mg^{++} may be related to the fact that a balanced solution for living cells consists of about 10 parts of potassium to 1 part of calcium or magnesium. It is also needed in unusually large quantities (0.01 to 0.1 N, Evans and Sorger, 1966) as an enzyme activator (Chapter 13) because so small a percentage remains attached to the enzyme on account of the high dissociation of its salts. It has been proposed that its function is to maintain the enzyme in a specific conformation necessary for catalytic activity (Wildes et al., 1970). The Mg^{++} ion is also needed in large quantities as an enzyme activator, not because of the needs of a single enzyme, but because it is required for a far larger number of enzymes than the other major, cationic, essential elements. The Ca^{++} ion is required in the smallest amount as an enzyme activator. According to earlier concepts, a "balanced ion atmosphere" is needed to maintain the properties of protoplasm. This view was based primarily on the success of balanced solutions in keeping cells alive and on colloidal principles. However, modern evidence points to the organelles of protoplasm as requiring protection by cations; for example, K^+ protects ribosomes (Ts'o, 1962). This may result from neutralization of the negative charge of the proteins, enabling them to attach to the strongly negative RNA. Mg^{++} stabilizes the configuration of the ribosomal RNA (Bayley, 1966; Sager and Hamilton, 1967). The Ca^{++} ion appears to play some essential role in the membrane (Chapter 5), for which it cannot be replaced by Mg^{++} (Dainty et al., 1970).

The major cationic elements are therefore needed in much larger quantities than the trace cationic elements, probably because of their role in neutralizing the acid proteins and other organic acids produced continuously in the metabolism of the cell (Chapter 14). In the case of magnesium a second role is as a constituent of chlorophyll.

The role of the trace elements is largely understood: Iron, copper, and zinc are constituents of many enzymes (e.g., peroxidase, catalase, and cytochromes in the case of iron; tyrosinase and ascorbic acid oxidase in the case of copper; carbonic anhydrase in the case of zinc), and manganese is an activator of several others. It is now known that molybdenum is a cofactor of the enzymes of nitrate assimilation, and this may be its sole role (Vega et al., 1971). However, in spite of the great amount of work on boron, its functions are completely unknown. According to one theory, boron forms a complex with sugars that penetrates into living cells more readily than do free sugars and is therefore more readily translocated to the growing cells (Gauch and Dugger, 1953); but this theory is opposed by most experimental evidence, and many other explanations have been proposed. According to Lee and Aronoff (1967), borate inhibits an enzyme system, which in the absence of boron, leads to excess formation of phenolic acids and to necrosis and eventual death of the plant. According to another proposal, it is required for flavonoid synthesis (Rajaratnam et al., 1971). The exact role of chlorine is not understood, partly because it occurs almost solely in the inorganic form. However, four chlorine-containing organic compounds have now been found in higher plants (Marumo et al., 1968), which may conceivably lead to an understanding of its role.

The main substances required for the life of the plant and the essential nutrient elements they contain are summarized in Table 12-6.

Toxic effects

All the trace essential elements are toxic when supplied to the plant in quantities considerably above the optimum. Borates have actually been used as weed killers, and copper has long been known to be a cause of the toxicity of standard distilled water to living cells. Some plants are much

Table 12-6. Organic substances required by living plant cells and tissues and the essential mineral elements contained by them

Element	Cell wall	Protoplasm				
	Pectins	Proteins (enzymes)	Nucleic acids	Lipids*	Chlorophyll	Enzyme cofactors
1. Major						
N		X	X	X	X	X
P			X	X		X
S		X		X		X
Ca	X	X (protein salt?)				X
Mg	(X)	X (protein salt?)			X	X
K		X (protein salt?)				X
2. Trace						
Fe, Cu, Mn, Zn, Mo,						X
B						Inhibitor?

*The listed elements are needed only for specific lipids.

more sensitive to such effects than others. Algae are readily killed by low concentrations of copper (e.g., 1 ppm). Higher plants suffer injury in acid soils frequently because of excess absorption of manganese. Even the major essential elements may be injurious in high concentrations, sometimes because of the osmotic effect leading to dehydration and perhaps even plasmolysis of the cells. The toxicity of single salt solutions (e.g., potassium chloride) has already been discussed (Chapter 7) and is overcome by antagonism (e.g., by calcium chloride). Even nonessential elements may counteract (or antagonize) such toxicity. This is one reason why some nonessential elements can sometimes stimulate the growth of plants. Thus aluminum stimulates growth when copper is present in toxic quantities. On the other hand, some nonessential (or possibly essential) elements are highly toxic (e.g., mercury, selenium, iodine, cadmium, cobalt, nickel, thallium).

It must therefore be concluded that all the mineral elements, whether essential or not, are capable of being toxic to plants if supplied in sufficient concentration and in the absence of a sufficient quantity of an antagonizing element. For optimum nutrition a mineral nutrient solution must therefore consist of both an optimum concentration of each salt and an optimum balance of the ions composing the salts.

QUESTIONS

1. What is an autotroph?
2. What is a heterotroph?
3. To which groups do higher plants belong?
4. Are there any exceptions?
5. Are any parts of the higher plant heterotrophic?
6. What must be supplied to excised roots beside inorganic substances for them to grow?
7. What is usually meant by the nutrition of higher plants?
8. How many elements are found in a normal plant?
9. How many salts must the higher plant be supplied with to provide all the major essential elements?
10. How many essential "mineral" elements are known?
11. Is there any difference in essentiality between major and trace elements?
12. What are the criteria of essentiality?
13. What elements are likely to be added to the essential list?
14. Will these be major or trace essential elements?
15. What is a deficiency disease?
16. What are some of the symptoms of a deficiency disease?
17. Why, in general, are elements essential?
18. Give an example of why elements are essential.
19. Why are the trace elements essential?
20. Why are trace elements essential in smaller quantities than the major elements?
21. Are there any elements whose essentiality is not understood?
22. Are any essential elements toxic?
23. Which are toxic in lower concentrations, the major or the trace essential elements?
24. Are any nonessential elements toxic?
25. Can nonessential elements counteract toxicity of essential elements?
26. Of what importance is toxicity in a search for new essential elements?

SPECIFIC REFERENCES

Allen, M. B., and D. I. Arnon. 1955. Studies on nitrogen-fixing blue-green algae. I. The sodium requirement of *Anabaena cylindrica*. Physiol. Plant. **8**:653-660.

Arnon, D. I., and P. R. Stout. 1939. The essentiality of certain elements in minute quantity for plants with special reference to copper. Plant Physiol. **14**:371-375.

Arnon, D. I., and G. Wessel. 1953. Vanadium as an essential element for green plants. Nature (London) **172**:1039-1040.

Bayley, S. T. 1966. Composition of ribosomes of an extremely halophilic bacterium. J. Molec. Biol. **15**:420-427.

Biddulph, O., R. Cory, and S. B. Biddulph. 1959. Translocation of calcium in the bean plant. Plant Physiol. **34**:512-519.

Brownell, P. F. 1965. Sodium as an essential micronutrient element for a higher plant *(Atriplex vesicaria).* Plant Physiol. **40**:460-468.

Broyer, T. C., A. B. Carlton, C. M. Johnson, and P. R. Stout. 1954. Chlorine—a micronutrient element for higher plants. Plant Physiol. **24**:526-532.

Burleson, C. A., and H. G. Cunningham. 1963. Iron status and needs of the southern region. Plant Food Rev. **9**:9-13.

Coleman, R. C., and F. Y. Richards. 1956. Physiological studies in plant nutrition. XVIII. Some aspects of nitrogen metabolism in barley and other plants in relation to potassium deficiency. Ann. Bot. **20**:393-409.

Dainty, J., R. J. Lannoye, and S. E. Tarr. 1970. Voltage-current characteristics of *Chara australis* during changes of pH and exchange of Ca-Mg in the external medium. J. Exp. Bot. **21**:558-565.

Davies, W. H., E. I. Mercer, and T. W. Goodwin. 1965. The occurrence and intracellular distribution of the plant sulfolipid in maize, runner beans, plant tissue cultures and *Euglena gracilis.* Phytochemistry **4**:741-749.

Dickmann, D. I., and T. T. Kozlowski. 1969. Seasonal changes in the macro- and micro-nutrient composition of ovulate strobili and seeds of *Pinus resinosa.* Canad. J. Bot. **47**:1547-1554.

Evans, H. S., and M. Kliewer. 1964. Vitamin B_{12} compounds in relation to the requirements of cobalt for higher plants and nitrogen-fixing organisms. Ann. N. Y. Acad. Sci. **112**:735-755.

Fries, L. 1966. Influence of iodine and bromine on growth of some red algae in axenic culture. Physiol. Plant. **19**:800-808.

Gauch, H. G., and W. M. Dugger, Jr. 1953. The role of boron in the translocation of sucrose. Plant Physiol. **28**:457-466.

Graebe, J. E., and G. D. Novelli. 1966. A practical method for large scale plant tissue culture. Exp. Cell Res. **41**:509-520.

Graser, H. 1965. Förderung der Serin-synthese durch Vitamin B_{12} (Cobalamin). Naturwissenschaften **52**:108-109.

Hayward, J. 1968. Studies on the growth of *Phaeodactylum tricornutum.* II. The effect of organic substances on growth. Physiol. Plant. **21**:100-108.

Johnson, C. M., P. R. Stout, T. C. Broyer, and A. B. Carlton. 1957. Comparative chlorine requirements of different plant species. Plant & Soil **8**:337-353.

Klimovitskaya, Z. M., and L. K. Vizir. 1963. The content and distribution of manganese in plants. [Transl. title.] Ref. Zh. Biol. No. 16872. Moscow.

Langridge, J. 1963. Biochemical aspects of temperature response. Ann. Rev. Plant Physiol. **14**: 441-462.

Lee, S., and S. Aronoff. 1967. Boron in plants; a biochemical role. Science **158**:798-799.

Lewin, J. C. 1955. Silicon metabolism in diatoms. II. Sources of silicon for growth of *Navicula pelliculosa.* Plant Physiol. **30**:129-134.

Linsmaier, E. M., and F. Skoog. 1965. Organic growth factor requirements of tobacco tissue cultures. Physiol. Plant. **18**:100-127.

Loneragan, J. F., and K. Snowball. 1969. Rate of calcium absorption by plant roots and its relation to growth. Aust. J. Agr. Res. **20**:479-490.

Marumo, S., H. Haltori, H. Abe, and K. Muriakata. 1968. Isolation of 4-choroindolyl-3-acetic acid from immature seeds of *Pisum sativum.* Nature **219**:959-960.

Müller, D. 1949. Die physiologische Grundlage für die Mangelsymptome der Pflanzen. Physiol. Plant. **2**:11-23.

Ollagnier, M., and R. Ochs. 1971. Le chlore: nouvel élément essentiel dans la nutrition minérale du palmier à huile et du cocotier. Compt. Rend. Acad. Sci. (Paris) **272**:2177-2180.

Pedersen, M. 1969. Marine brown algae requiring vitamin B_{12}. Physiol. Plant. **22**:977-983.

Rajaratnam, J. A., J. B. Lowry, P. N. Avadhani, and R. H. V. Corley. 1971. Boron: possible role in plant metabolism. Science **172**:1142-1143.

Robbins, W. R. 1946. Growing plants in sand cultures for experimental work. Soil Sci. **62**:3-22.

Robinson, W. O., and G. Edgington. 1945. Minor elements in plants and some accumulator plants. Soil Sci. **60**:15-28.

Sager, R., and M. G. Hamilton. 1967. Cytoplasmic and chloroplast ribosomes of *Chlamydomonas;* ultracentrifugal characterization. Science **157**: 709-711.

Stewart, I., and C. D. Leonard. 1952. Iron chlorosis —its possible causes and control. Citrus Mag. **14**(10):22-25.

Tammes, P. M. L., and J. van Die. 1966. Studies on phloem exudation from *Yucca flaccida* Haw. IV. Translocation of macro- and micronutrients by the phloem sap stream. Proc. Kon. Nederl. Akad. Wet. [Biol. Med.] **69**:655-659.

Tauböck, K. 1942. Über die Lebensnotwendigkeit des Aluminiums für Pteridophyten. Bot. Arch. **43**:291-304.

Trelease, S. F., and H. M. Trelease. 1938. Selenium as a stimulating and possibly essential element for indicator plants. Amer. J. Bot. **25**:372-380.

Ts'o, P. O. P. 1962. The ribosomes—ribonucleoprotein particles. Ann. Rev. Plant Physiol. **13**: 45-80.

Vega, J. M., J. Herrera, P. J. Apacio, A. Paneque, and M. Losada. 1971. Role of molybdenum in nitrate reduction by *Chlorella*. Plant Physiol. **48**:294-299.

Vose, P. B. 1963. The translocation and redistribution of manganese in *Avena*. J. Exp. Bot. **14**: 448-457.

White, P. R. 1943. A handbook of plant tissue culture. The Ronald Press Co., New York.

Wildes, R. A., H. J. Evans, and R. R. Becker. 1970. The effect of univalent cations on the circular dichroism of pyruvate kinase. Plant Physiol. **46**(Supp.):125.

Wilson, S. B., and D. J. D. Nicholas. 1967. A cobalt requirement for non-nodulated legumes and for wheat. Phytochemistry **6**:1057-1066.

Ziegler, H. 1961. Wasserumsatz und Stoffbewegungen. Fortschr. d. Bot. **23**:191-205.

GENERAL REFERENCES

Bear, F. E., and R. Coleman (Ed.). 1941. Hunger signs in crops. American Society Agronomy and National Fertilizer Association, Washington, D. C.

Evans, H. S., and G. J. Sorger. 1966. Role of mineral elements with emphasis on the univalent cations. Ann. Rev. Plant Physiol. **17**:47-76.

Frey-Wyssling, A. 1945. Ernährung und Stoffwechsel der Pflanzen. Buchergilde Gutenberg, Zurich.

Gericke, W. F. 1940. Soilless gardening. Prentice-Hall, Inc., Englewood Cliffs, N. J.

Hewitt, E. J. 1951. The role of the mineral elements in plant nutrition. Ann. Rev. Plant Physiol. **2**:25-52.

Hewitt, E. J. 1952. Sand and water culture methods used in the study of plant nutrition. Tech. Commun. Commonw. Bur. of Hort. Plantn. Crops No. 22, East Malling.

Hoagland, D. R. 1944. Inorganic plant nutrition. Chronica Botanica Co., Waltham, Mass.

Miller, E. C. 1938. Plant physiology. McGraw-Hill Book Co., New York.

Stiles, W. 1946. Trace elements in plants and animals. Cambridge University Press, London.

Wallace, T. 1953. The diagnosis of mineral deficiencies in plants by visual symptoms. Chemical Publishing Co., Inc., New York.

13

Metabolism

ANABOLISM AND CATABOLISM

The nutrition of the plant should include all the raw materials that must be supplied to support its normal growth and development. Beside the mineral elements already discussed, the autotrophic plant requires an adequate supply of carbon dioxide, oxygen, and water. However, the plant must first convert all these raw materials into complex organic substances before they can support its growth and development. Conversely, many of these complex organic substances must be subsequently broken down to simpler substances in order to support certain phases of the plant's growth and development. All these chemical changes that take place in the plant are referred to as its *metabolism*. The metabolism of the plant may be conveniently divided in two parts (Fig. 13-1). All the chemical changes leading to syntheses (building of more complex from less complex substances) comprise its *anabolism*. All those leading to breakdown of more complex to less complex substances are included in its *catabolism*. The many intermediate substances produced in the process are the intermediate metabolites. The many chemical reactions in which these are involved constitute the intermediate metabolism of the plant. The tremendous number of chemical changes that constantly must be taking place becomes evident from the vast number of chemical substances found in the plant. A total of 1385 plant substances of known chemical structure are listed in a monograph by Mentzer and co-workers (1968); but this includes only a small part of the total, and there are as many as 4000 different substances in one chemical group that have been found in plants (Chapter 16). All except the few that have not yet changed on entering from the environment have arisen from synthesis or breakdown in the plant—primarily in the protoplasm. Many of the substances are so complex that their synthesis may have required dozens of chemical reactions. Furthermore, although many of the substances in the plant are stable (e.g., substances in the cell wall, such as cellulose), others, and in particular the fundamental structural and functional units of the protoplasm (some of the nucleic acids and proteins), *are continuously being broken down and resynthesized.* This is proved by feeding the plant with radioactive carbon or heavy nitrogen because the substances formerly present in the protoplasm soon become labeled by these isotopes and therefore must have been broken

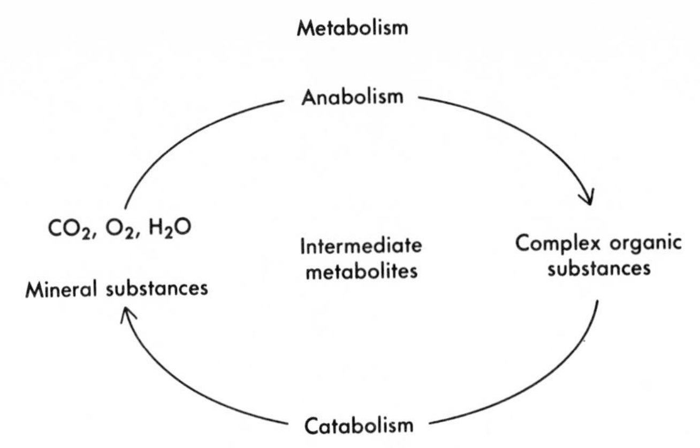

Fig. 13-1. The interdependence of anabolism and catabolism in the overall metabolism of the plant.

down and resynthesized. By these methods, protein turnover rates of 0.5% to 1.5% per hour have been found in rapidly growing leaves (Hellebust and Bidwell, 1963).

An actively growing plant enlarges its body by rapidly building new cells and tissues. It must therefore rapidly synthesize from the raw materials that it absorbs from its environment the building blocks needed to produce these new cells and tissues. Hence, an actively growing plant must metabolize actively. But what if it is not producing new cells and tissues and therefore is not growing? Does metabolism then cease? According to the laws of thermodynamics, free energy tends toward a minimum. The complex chemical substances that make up the body of even a nongrowing plant are at a higher free energy level than the raw materials in its environment from which they were synthesized. Therefore catabolic (breakdown) reactions must take place, leading to a decrease in free energy and ultimately to death of the plant. Even in the dead (aseptic) plant, the breakdown reactions can be expected to continue until equilibrium is reached with its environment. At the normal temperatures of a living plant these changes would lead to relatively rapid death. Consequently, in order for the nongrowing plant to survive, it must prevent or at least slow down this approach to equilibrium. Therefore even in a nongrowing plant, anabolic changes must be constantly taking place, counteracting the destructive catabolic changes. When the two are occuring at the same rates, the plant is in the steady state and is neither growing nor decreasing in mass. When anabolic changes exceed the catabolic, it is either storing complex reserves or growing. When catabolic changes exceed the anabolic, it is losing mass; that is, it is either using up its reserves more rapidly than it is growing or it is dying. An exact steady state is, of course, difficult to attain. It is approached most closely in dormant (nongrowing) plants. But even then, catabolism may slightly exceed anabolism (e.g., in seeds and tubers), and death will eventually occur; or anabolism may slightly exceed catabolism (e.g., some evergreens in winter on bright sunny days), and the reserves will increase.

It must be concluded that regardless of whether a plant is growing, as long as it is alive it must be metabolizing; and this must include both catabolism and anabolism. Even when it dies, chemical changes will continue; but these will involve only breakdown reactions, and even they will not be in the ordered, controlled manner characteristic of the living plant. The term *metabolism* should therefore be restricted to the ordered series of chemical changes under the control of the living cell. It is usually assumed that no life is possible without metabolism, and no metabolism is possible without life.

Both of these assumptions may be questioned, however. It is now possible to follow, with cell extracts and therefore in nonliving material, many of the individual metabolic reactions that normally occur in living cells. This means that portions of the plant's metabolism are possible in the absence of life. Conversely, some biologists have asked the philosophical question: If metabolism comes to a complete stop, is the organism dead? A possible test case is provided by dormant, air-dry seeds, which metabolize extremely slowly even at normal temperatures. It is now known that such seeds can be cooled to the lowest temperatures so far attained (a small fraction of a degree above absolute zero) without injury. When rewarmed and supplied with water, they germinate and grow normally. Presumably, if absolute zero were attained, they would also survive; yet all chemical reactions within them would cease and, by definition, so would metabolism. It is a philosophical question whether the seeds would be dead or dormant at absolute zero and whether they would be revived from death or merely activated from the dormant state when rewarmed and moistened. From the practical point of view, however, the preservation of life at such low temperatures is of prime importance. Seeds kept in the frozen condition (in the permafrost) for 10,000 years are still alive and are now growing normally (Chapter 22).

Despite the preceding apparent exception, it is certainly true that under normal conditions there is no life without metabolism; therefore life cannot be understood without a thorough knowledge of metabolism. Two aspects of metabolism are intimately interrelated but are more conveniently discussed separately: the changes in chemical structure and the energy changes.

CHANGES IN CHEMICAL STRUCTURE

The vast number of chemical substances found in plants might lead to the expectation of a correspondingly large number of kinds of chemical reactions. Fortunately, there are only six main types of reactions.

Hydrolysis-condensation reactions

Hydrolysis is a breakdown reaction in which water participates and a large molecule is broken down into two or more smaller ones. Condensation is the reverse synthesis of a large molecule from two smaller ones, with the release of water.

$$R^1OR^2 + H_2O \underset{\text{Condensation}}{\overset{\text{Hydrolysis}}{\rightleftharpoons}} R^1OH + R^2OH$$

R^1 and R^2 = different chemical radicals (e.g., a methyl group, a benzene ring).

At equilibrium, when the rate of the reaction in the forward direction ($\rightharpoonup$) is equal to the rate in the backward direction ($\leftharpoondown$), reactants and products exist in proportions characteristic of each specific reaction. Because of the high concentration of water in normally metabolizing cells, the hydrolytic products on the right will be at much higher concentration than the condensed reactants on the left *when the reaction is at equilibrium.* Therefore the hydrolysis tends to go to completion. But this may not hold true in the plant, since the metabolic process of which this reaction is a component may not be at equilibrium. Many different kinds of substances may be hydrolyzed, for example, esters, glycosides, peptides. Consequently, hydrolyses are responsible for the cleavage of many kinds of macromolecules. When the macromolecules are insoluble food reserves (e.g., fats, carbohydrates, proteins) and the hydrolysis converts them into soluble substances, the process is called *diges-*

tion. Many carbohydrates are broken down in this way, for example:

$$\text{Starch} + H_2O \rightarrow \text{Maltose}$$

It was formerly thought that the reverse reaction (condensation) was responsible for the syntheses of macromolecules; but it is now known that these syntheses are generally caused by the group transfer, addition, or synthesizing reactions described next. There are, however, some cases in which reversal of hydrolysis is the normal mode of synthesis (e.g., ATP synthesis, Chapter 14).

Group transfer reactions

A group transfer reaction, in its simplest form, is basically similar to a hydrolysis, but instead of water, a more complex chemical hydroxide is involved.

$$R^1OR^2 + R^3OH \rightleftharpoons R^1OH + R^3OR^2$$

Many different kinds of chemical groups may be transferred: C-groups (methyl, hexose, etc.), N-groups (amino, etc.), P-groups, S-groups. One of the most important of these is the transfer of P-groups, a vast number of which occur in the plant; for example:

$$\text{ATP} + \text{Ribose} \rightleftharpoons \text{ADP} + \text{Ribose phosphate}$$

where ATP = adenosine triphosphate
ADP = adenosine diphosphate

Group removal or addition reactions

In this case a single large molecule is broken down to one relatively large and one relatively small molecule. The small molecule is said to be removed or added:

$$R^1R^2H \rightleftharpoons R^1H + R^2$$

One of the most important of these involves the formation and cleavage of C-C bonds:

$$\text{RCOOH} \underset{\text{Carboxylation}}{\overset{\text{Decarboxylation}}{\rightleftharpoons}} \text{RH} + CO_2$$

These reactions are basic to many metabolic processes. The synthesis of carbohydrates in photosynthesis is initiated by a carboxylation. Besides carbon dioxide, other simple molecules (e.g., water, ammonia) may also be removed from or added to the more complex organic molecules in the plant.

Electron transfer or oxidation-reduction reactions

The previous reactions involve the splitting off, transfer, or removal of whole chemical groups. In opposition to them, oxidation-reduction reactions occur as a result of a transfer of electrons:

$$A + B \rightleftharpoons A^+ + B^-$$

These reactions are among the commonest in all metabolism. A series of electron transfers occurs in each of the most important metabolic processes—respiration, photosynthesis, N-metabolism, S-metabolism. This sequence of electron transfers between adjacent molecules has led to the term *electron transport,* indicating successive transfers between a series of molecules rather than a single transfer from one molecule to a second neighboring one.

The substance (A) that loses the electron is the electron donor and is said to be oxidized (or is the reducing agent or reductant). The substance (B) that gains the electron is the electron acceptor and is said to be reduced (or is the oxidizing agent or oxidant). For such a transfer to occur between atoms, the electron acceptor must be more electronegative than the electron donor (Table 4-3).

Ions are not necessarily formed as a result of the electron transfer because the electron may be accompanied by a proton in the form of a hydrogen atom, resulting in hydrogenation of the electron acceptor and dehydrogenation of the electron donor:

$$AH_2 + B \rightleftharpoons A + BH_2$$

The substance AH_2 is oxidized to A, whereas B is reduced to BH_2. This kind of oxidation-reduction reaction depends on the lower electronegativity (Table 4-3) of the hydrogen atom than that of the other atoms (carbon, nitrogen, oxygen, sulfur) with which it is normally combined in plant substances. The removal of a hydrogen atom is therefore equivalent to removal of the electron that it had donated to the carbon (or other) atom to which it was bonded, and the addition of a hydrogen atom to a carbon (or nitrogen, oxygen, sulfur) atom is equivalent to a transfer of an electron to the carbon or other atom. The apparent hydrogen transfer may actually be a true electron transfer, since in some cases only the electron may be passed from substance to substance, the proton remaining temporarily in solution as an H^+:

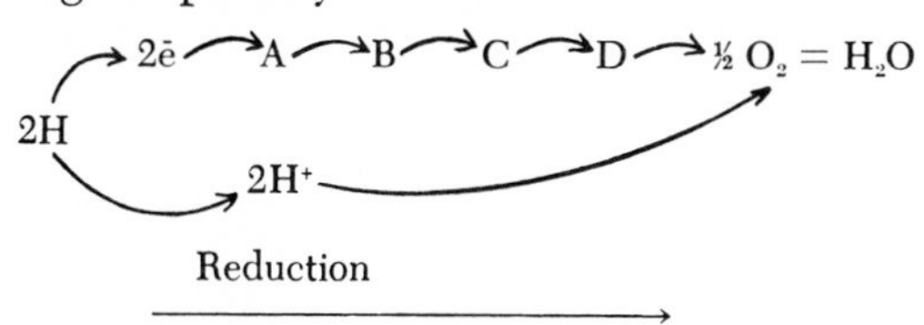

The equivalence of the different terminologies is shown in Fig. 13-2.

Isomerization

In some cases there is no change in the empirical formula of a substance, but there is a change in the relative positions of some of the atoms or groups within the molecule:

$$\text{Isomer 1} \rightleftharpoons \text{Isomer 2}$$

Isomerization may simply involve the conversion of a D-isomer to an L-isomer, or it may result from the shift of a chemical group, for example, in the conversion from a 2-phospho-compound to a 3-phospho-compound or the conversion of one pentose to another.

Synthesizing reactions

Synthesis of complex substances may involve transfer and addition reactions, but they are also commonly caused by the following synthesizing reaction:

$$A + B + NPPP \rightleftharpoons AB + NP + PP$$

where N = nucleoside (a purine or pyrimidine combined with a pentose)
P = phosphate
Nucleoside phosphate is called a nucleotide

Thus (1) the making and breaking of the electron pairs of covalent bonds and (2) the electron transfer between atoms are the fundamental characteristics of metabolism and therefore of life. The six elements (hydrogen, carbon, nitrogen, oxygen, sulfur, phosphorus) have mobile electrons and therefore form multiple bonds more easily than all the other elements in the periodic table (Pullman and Pullman, 1963). It is therefore not surprising that the elements hydrogen, carbon, nitrogen and oxygen make up 90% of living systems and that

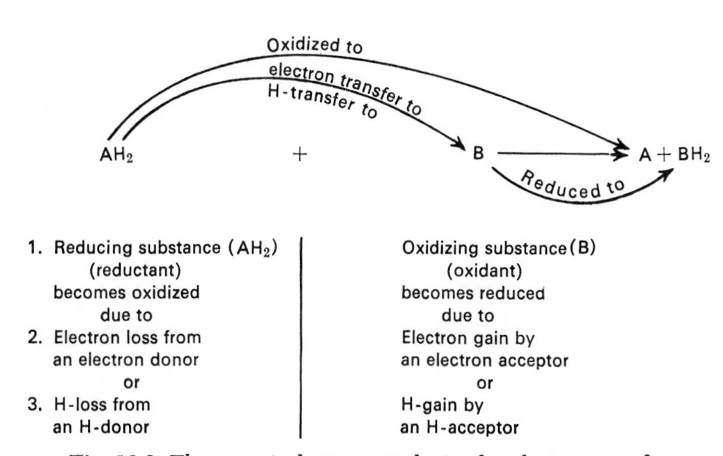

Fig. 13-2. Three equivalent terminologies for electron transfer.

sulfur and phosphorus are next in quantity. Since double bonds are caused by mobile electrons, this also explains the common occurrence of conjugated substances (i.e., substances with alternating single and double bonds) in living cells. Conjugated substances are typically colored.

ENERGY CHANGES

As stated earlier, no life has been found without metabolism. In the case of a growing plant this follows logically from energy considerations. There are four kinds of work associated with growth: (1) mechanical work such as the growth upward of the plant body against gravity and the creation of new surfaces, (2) "osmotic" work, or the absorption of substances against a diffusion gradient followed by osmosis; (3) electrical work, when the substances are in the ionic form and are absorbed along an electrical potential gradient; and (4) chemical work involved in the synthesis of the many substances in its cells. Probably 99% of the plant's energy requirements are for the last of these—chemical work. According to the first law of thermodynamics, all these four kinds of work must involve a transfer of energy from some other source. The only source directly available to the plant is the energy released in its metabolism. According to the second law of thermodynamics, free energy (energy available to do work) tends toward a minimum. Yet the free energy of the living plant increases during growth because of its increased content of many substances at a high level of free energy. This is especially true of protoplasm, the vital properties of which depend on its content of many substances in a highly unstable state. The second law dictates that these substances must be continuously changing to the more stable, lower free energy states. This would lead to death of the plant if unopposed. But the plant is a thermodynamically open system; it can absorb both mass and free energy from its environment. Therefore the continuously formed substances of low free energy are just as continuously reconverted to the high energy state by the expenditure of metabolic energy (the only source of extra free energy constantly available to the plant). This metabolic energy, in turn, is derived by the green plant from the radiant energy of sunlight, which is absorbed daily and stored as metabolizable, chemical energy. A specific example is provided by the proteins of the living protoplasm. In the native state (in which they function normally) they exist as large molecules folded in a highly ordered manner (Chapter 16). From the second law of thermodynamics we would expect them to change to the unfolded (less ordered) denatured state. These denatured proteins must then be renatured, commonly by hydrolysis to amino acids, followed by recombination to form the native proteins. A steady state of breakdown and resynthesis is thus maintained *at the expense of metabolic energy.*

This cycle of changes is illustrated in Fig. 13-3.

In all the preceding six classes of chemical reactions, energy changes accompany the changes in chemical composition. Thus a complete representation of the hydrolysis-condensation reaction is as follows:

$$R^1OR^2 + H_2O \underset{\text{Condensation}}{\overset{\text{Hydrolysis}}{\rightleftharpoons}} R^1OH + R^2OH + \text{Heat energy}$$

The reversibility of the reaction depends on these energy changes. Hydrolyses usually involve a large release of energy and therefore go nearly to completion. This is why they are not readily reversible. Group removal reactions (e.g., decarboxylations) usually involve a small release of energy and are readily reversible. Thus these reactions can go in either direction and may be controlled by the end product. A decrease in carbon dioxide content of the air may be expected to lead to decarbox-

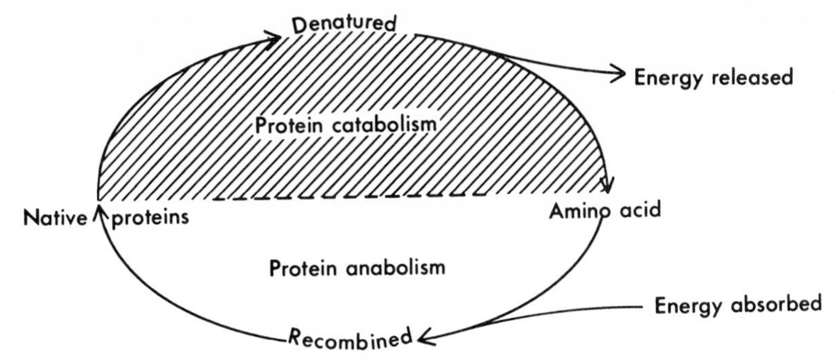

Fig. 13-3. Protein metabolism in living cells. Native proteins unfold to the denatured form followed by hydrolysis to amino acids accompanied by a loss of free energy. This is counteracted by a continuous synthesis of native protein at the expense of amino acids and free energy.

ylation and an increase to lead to carboxylation. These energy changes are readily explained because changes in chemical structure involve electron displacements and therefore the performance of work. Thus an addition of energy to an atom may be expected to result in an electron displacement followed by a change in structure. When at its lowest free energy level, the molecule is said to be in the "ground state," and all the electrons are distributed in pairs with antiparallel spins. If an electron is displaced from its orbit ("excited"), an "empty" molecular orbit results and the free energy of the molecule increases, since it can do work by allowing an electron to fall into this empty orbit. It can therefore undergo chemical reaction.

In any one chemical reaction there is normally both a displacement of electron orbits *from* one atomic nucleus and a displacement *toward* another atomic nucleus. If the former involves more work than the latter, the net result will be a gain in free energy by the products formed as compared with the reactants, and the reaction is *endergonic,* or *energy is absorbed.* Endergonic reactions cannot take place unless free energy is supplied from some external source. Conversely, if the displacement toward the second nucleus involves more work than displacement from the first nucleus, there will be a net loss of free energy by the products formed, and the reaction is *exergonic,* or *energy is released.* Exergonic reactions are *spontaneous.* This means that they can occur without a *net* supply of free energy. The syntheses of more complex from less complex molecules in the plant are endergonic; the breakdown reactions are exergonic. We can now define anabolism and catabolism energetically. The part of the metabolism that results in net syntheses and therefore consists mainly of endergonic reactions is called *anabolism;* the part consisting of net breakdown (exergonic) reactions is called *catabolism.* However, the two are intimately related because (1) the endergonic reactions cannot take place unless free energy is supplied from some other source, and (2) many reactions are taking place in the plant at any one time. Therefore it is at least conceivable that the endergonic reactions may take place at the expense of the free energy released by the exergonic reactions. When this happens, endergonic reactions are called *driven reactions;* and since the two kinds are energetically linked, they are also called *coupled reactions.* Thus the fixation of nitrogen by microorganisms

$$N_2 + 3H_2O \rightleftharpoons 2NH_3 + 3/2O_2$$

is an endergonic reaction that cannot go of itself because there is a free energy *increase* of 162 kcals (Wohl and James, 1942). But if the oxygen produced can be used to oxidize glucose, the free energy *released* by this reaction is 172 kcals. The total (coupled) reaction is

$$N_2 + 3/2H_2O + 1/4C_6H_{12}O_6 \rightleftharpoons 2NH_3 + 3/2CO_2$$

where $C_6H_{12}O_6$ = glucose. This combination of reactions is accompanied by a net free energy decrease of 10 kcal (172 — 162 kcal) and is therefore thermodynamically spontaneous.

Thus, in general, the anabolism of plants is possible only if catabolism is simultaneously taking place. This must always be remembered, although the two are studied separately for purposes of classification. It is, of course, true that the free energy needed to drive some endergonic reactions may be obtained from sources other than exergonic reactions. In the plant this is true of the photochemical reactions in the photosynthetic process, the free energy for which is obtained from sunlight.

Plant metabolism, and therefore plant life, is thus driven by the electronic energy of exergonic reactions

Linked or coupled reactions
1) A → B + C + energy (exergonic)
2) Energy + D + E → F (endergonic or driven)

where A, D, E are reactants and B, C, F are products. But the energy is not transferred directly. When an exergonic reaction occurs by itself, the electronic energy is converted to heat and is therefore lost. In the plant much of the energy of exergonic reactions can be trapped in the unstable, or "high energy," bonds of high energy phosphates, for instance, by linking two inorganic orthophosphates (molecules of low free energy) in an anhydride link of high free energy with the formation of pyrophosphate (P-O-P). The major form of this pyrophosphate is in adenosine triphosphate (ATP). The high energy pyrophosphate can then combine with other substances in such a way as to induce them to undergo endergonic reactions at the expense of the high energy pyrophosphate, which is then converted back to low energy inorganic phosphate. The preceding reactions would therefore have to be linked to another pair of reactions

$$ADP + P_i \rightleftharpoons ATP$$

where P_i = inorganic phosphate. The energetic relationship between these reactions is indicated in Fig. 13-4.

The ATP contains the exergonic energy that would otherwise be given off as heat and transfers it to D + E, driving the reaction to product F and regenerating the low free energy inorganic phosphate. The cycle can be repeated over and over, and the high energy phosphates can receive the energy from many different substances and

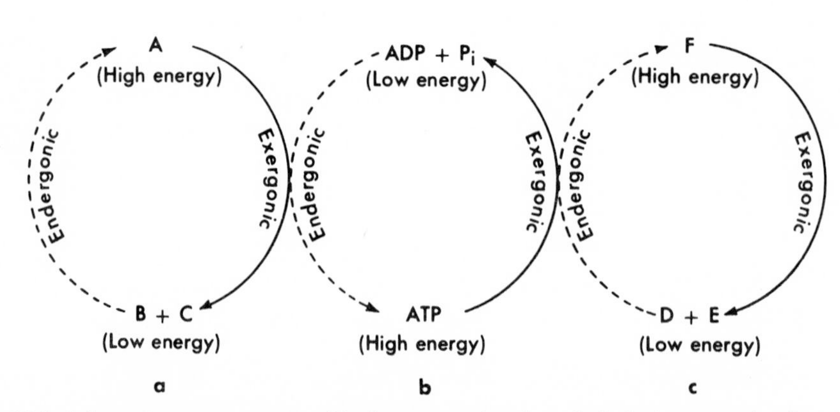

Fig. 13-4. Schematic representation of high energy phosphate link, **b**, between the driving, **a**, and the driven, **c**, reaction. The exergonic reaction in **a** drives the endergonic reaction in **b**. The exergonic reaction in **b** drives the endergonic reaction in **c**.

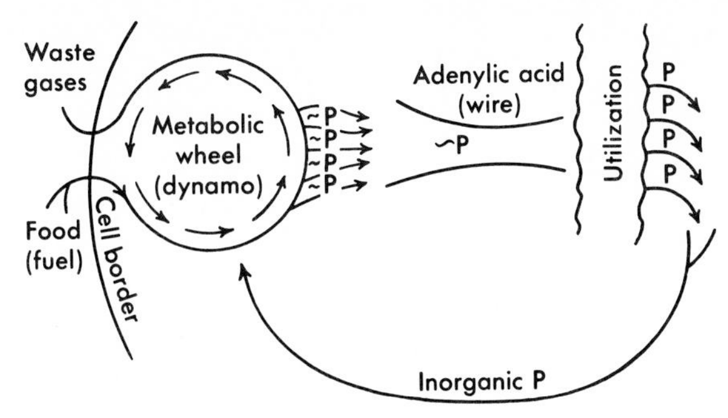

Fig. 13-5. Analogy between utilization of metabolic energy by means of high energy phosphates (~ **P**) and conversion of mechanical energy of dynamo into electrical energy. (From Lipmann, F. 1941. Adv. Enzymol. **1**:99-162.)

can transfer it to many others. By analogy with a dynamo (Fig. 13-5), the energy released in metabolism can be thought of as being "brushed off" as high energy phosphate groups and conducted along the adenylic acid (AMP) or adenosine diphosphate (ADP) "wire" to drive the "motors" of the endergonic reactions. The waste, low energy inorganic phosphate must then be reconverted to high energy phosphate by the metabolic wheel ("the dynamo"). This explains why the syntheses of macromolecules are not produced in the plant by a simple reversal of hydrolysis, that is, by condensation. The condensation reactions are endergonic, and since they are not linked with reactions involving high energy phosphates, they are not receiving the free energy needed to drive them.

The transfer of energy by means of high energy phosphates is required for many energy-consuming processes in the plant—the movement of substances against a gradient (active absorption), the streaming of cytoplasm, the creation of new surfaces, etc. But because of their instability, the high energy phosphates cannot be stored in large enough quantities for future needs of the plant. Therefore the excess high energy phosphate is used to synthesize the more stable carbohydrates, fats, etc., which can retain the chemical energy until it is needed, at which time it must be reconverted to high energy phosphates before being used. The transfer of metabolic (chemical) energy over long distances is by means of these stable storage products, for example sucrose (Chapter 10).

But what of these stable substances? How can they retain large quantities of chemical energy and how can they liberate this energy when it is needed? It must be realized that exergonic reactions are *spontaneous* only from the thermodynamic point of view; that is, they can take place without a *net* supply of energy. This does not mean that the reactions will take place unaided but simply that they are downhill reactions as far as the net energy change is concerned. However, the potentially reactive molecules are made up of both positively and negatively charged components and therefore may either attract or repel each other (Linford, 1966). When the molecules of two substances capable of reacting approach each other, the first effect is an attraction, which results in a closer approach (Fig. 13-6, from *c* to *b*). Even if the reaction between the two is exergonic and therefore spontaneous, the forces of repulsion now prevent a closer approach and the reaction does not occur. For them to approach close enough (Fig. 13-6, from *b* to *a*) to react chemically, work must be done on them by some outside force. The energy just sufficient to do this work is called the *activation energy* (Fig. 13-7, ΔE_1^*). When the energy level of the molecules is raised by this amount, the reaction takes place and the

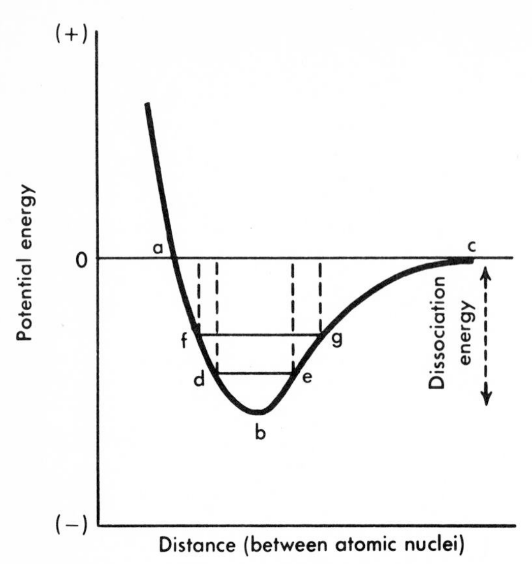

Fig. 13-6. Attraction and repulsion between molecules (or other chemical particles) as they approach each other. On approaching from *c* to *b* there is a decrease in energy; that is, they attract each other. From *b* to *a* there is an increase in energy; therefore no further approach is possible without adding energy to the system, that is, doing work on the molecules. (From Linford, J. H. 1966. An introduction to energetics with applications to biology. Butterworth & Co. [Publishers], Ltd., London.)

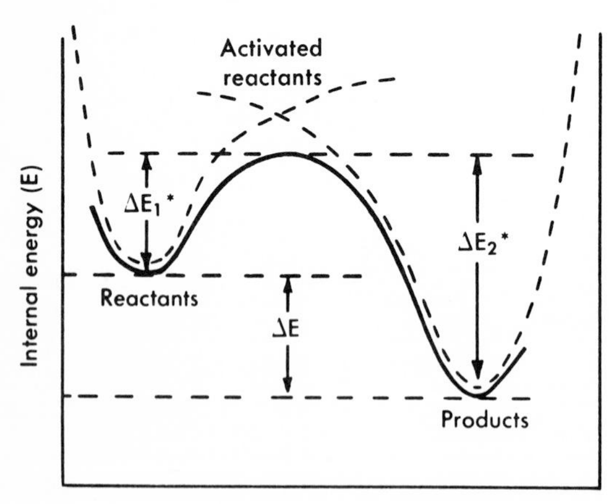

Fig. 13-7. Energy changes in a spontaneous reaction. The internal energy of the molecules must be increased by ΔE_1^* (the activation energy) before the reaction can begin. (From Linford, J. H. 1966. An introduction to energetics with applications to biology. Butterworth & Co. [Publishers], Ltd., London.)

products are formed with a net *release* of energy—the *energy of reaction* (Fig. 13-7, ΔE). The total energy released is the sum (Fig. 13-7, ΔE_2^*) of the activation energy and the energy of reaction. The reverse (endergonic) reaction will occur only if this total sum of the two amounts of energy is supplied to the products. Therefore the activation energy constitutes the barrier to the progress of spontaneous (exergonic) reaction, and the more stable the reactants the larger is the energy barrier. The activation energy may be supplied by heating the reactants.

ENZYMES

Catalytic nature

Stable substances can be made to undergo exergonic reactions without heating in the presence of a highly reactive substance acting in extremely small quantities and known as a *catalyst.* If the reaction involves two substances, the catalyst may speed it up by simply adsorbing these substances, bringing them close enough together to reduce the energy barrier. Thus when a stove is turned on, the gas does not burn until the activation energy is supplied by a match. Instead of a match, platinum black may be used. It acts as a catalyst by adsorbing the gases from the stove and the air, in this way starting the reaction without a supply of heat because it lowers the activation energy.

The catalyst may form an intermediate compound with the substance, converting it into a substance that is unstable. The energy barrier is in this way reduced to a small enough value for the energy of the molecules to overcome and they can now form products spontaneously by dropping to a lower energy level (Fig. 13-7). When two substances react, the steps will then be as follows:

$$A + \text{catalyst} \rightleftharpoons \text{cat } A$$
$$\text{cat } A + B \rightleftharpoons C + D + \text{catalyst} + \text{energy}$$
$$\text{Net reaction: } A + B \xrightleftharpoons{\text{catalyst}} C + D + \text{energy}$$

The net result is that the catalyst has permitted the reaction to take place by removing or lowering the energy barrier (energy of activation). It is active in small quantities because it is regenerated after each reaction and able to combine with another molecule.

In the presence of such a catalyst, the exergonic reaction proceeds more or less rapidly with the release of free energy. The chemical substance or substances resulting from the first reaction may in turn be capable of reacting exergonically with another substance in the presence of a second catalyst, which is different from the first. Thus a whole chain of reactions may occur in a definite order, each link in the chain controlled by a specific catalyst. As a result, a complex substance may be broken down, step by step, to simpler and simpler substances, and at each step a packet of energy will be released, which may be used to drive an endergonic reaction

$$A \overset{cat_a}{\rightleftharpoons} 2B + \text{energy}$$

$$B \overset{cat_b}{\rightleftharpoons} C + D + \text{energy}$$

$$C \overset{cat_c}{\rightleftharpoons} E + F + \text{energy}$$

$$D \overset{cat_d}{\rightleftharpoons} G + H + \text{energy}$$

Net result: $$A \overset{cat_{a-d}}{\rightarrow} E + F + H + 4 \text{ energy packets}$$

where cat_a, cat_b, etc. are each specific catalysts for a specific reacting substance.

By this gradual, orderly release of energy at the right time and in the right place, the plant is able to perform all its energy-consuming processes: synthesis of substances, active absorption of substances, cell growth, maintenance of gradients, protoplasmic streaming, etc. But none of these energy-consuming processes would be possible in the absence of the catalysts necessary for the energy-releasing reactions; and since there are many such reactions, there must also be many catalysts because they are highly specific. The organic catalysts that control the metabolism of living organisms are known as *enzymes.* They can induce reaction velocities that are as much as 10 times as rapid as those produced by the best of the known nonenzymatic catalysts. The rapid action of enzymes may be illustrated by catalase, which controls the following reaction:

$$2H_2O_2 \overset{\text{Catalase}}{\rightleftharpoons} 2H_2O + O_2$$

where H_2O_2 = hydrogen peroxide

One molecule of catalase can decompose 1000 molecules of hydrogen peroxide per second! This is neither a maximum nor a minimum rate, since the rates for different enzymes vary from 10^2 to 10^6 moles of product per minute per mole enzyme (Mahler and Cordes, 1968).

The concentration of the enzyme affects the speed of the reaction but not the position of equilibrium, and the final quantity of product is unchanged, except when the enzyme concentration is high. In some cases, however, despite its catalytic role, the enzyme concentration may be unexpectedly high. In acetate-adapted cells of the alga *Chlorella pyrenoidosa,* the enzyme isocitrate lyase may constitute 7% of the total soluble protein in the cell, or 1% of the cell dry weight (John and Syrett, 1968). The enzyme ribulose diphosphate carboxylase may account for as much as 70% of the total leaf protein (Chapter 15). One possible function for such high concentrations is the elimination of substrate inhibition (Wuntch et al., 1970).

Because of their high activity, their presence can be easily revealed. In many cases it is necessary only to show that substance A will not break down of itself but will when mixed with a plant juice. This indicates that the juice contains an enzyme that can catalyze the breakdown of A. For more rigorous proof this preliminary test is followed by other more exacting ones. Some reactions may conceivably occur in

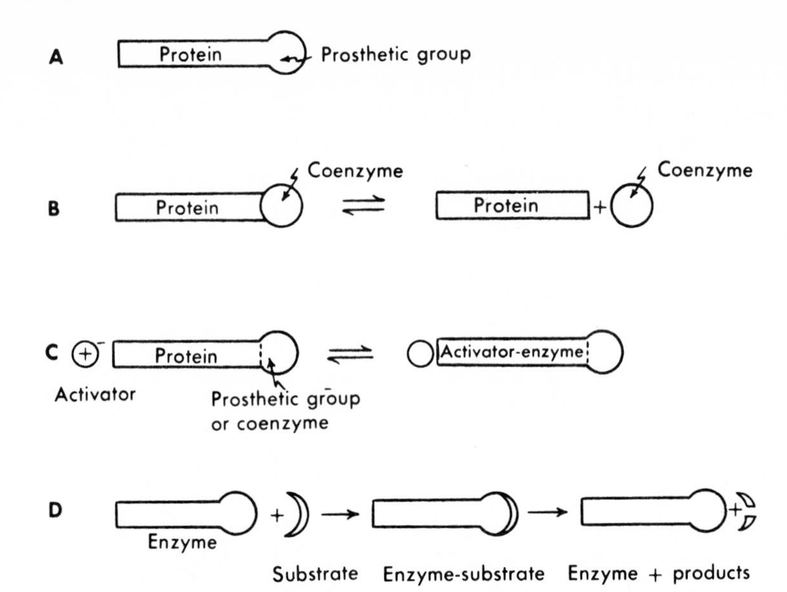

Fig. 13-8. Diagrammatic representation of enzyme systems. **A,** Protein plus nonremovable component. **B,** Protein plus removable component. **C,** Protein plus one or two removable components. **D,** Enzyme-substrate reaction.

Table 13-1. Some plant enzymes*

EC no. (Enzyme Commission)	Group	Systematic name	Trivial name	Active groups and cofactors	Reaction
	Hydrolases				
3.1.1.3		Glycerol ester hydrolase	Lipase	Ca^{++}	A triglyceride + H_2O = diglyceride + fatty acid
	Transferases				
2.4.1.13		UDP glucose: D-fructose 2-glycosyltransferase	UDP glucose-fructose glucosyltransferase, sucrose-UDP glycosyl-transferase		UDP glucose + D-fructose = UDP + sucrose
	Oxidoreductases				
1.10.3.1		o-Diphenol : O_2 oxidoreductase	Catechol oxidase	Cu	2 o-diphenol + O_2 = 2 o-quinone + $2H_2O$
	Lyases				
4.1.1.1		2-Oxo-acid carboxy-lyase	Pyruvate decarboxylase (formerly known as carboxylase)	$T.M^{++}$	A 2-oxo-acid = aldehyde + CO_2

*From Dixon, M., and E. C. Webb. 1964. Enzymes. Academic Press, Inc., New York.
T. = thiamine pyrophosphate; M^{++} = divalent cation.

the plant without catalysis by enzymes; but new enzymes are still being discovered, and some may later be found to control reactions now believed to be nonenzymatic.

In recent years new enzymes have been discovered at the rate of thirty to forty a year. About 900 are known and more than 100 of these have been sufficiently purified to be crystallized. These have all proved to be globular proteins with molecular weights from about 10,000 to millions.

Components and terminology

Some enzymes are active without any nonprotein component, but most consist of two components: a protein and a non-protein, or cofactor (Fig. 13-8). The non-protein component may be firmly attached to the protein, in which case it is called a *prosthetic group;* or it may be readily removed (e.g., by dialysis), in which case it is called a *coenzyme.* If the nonprotein component is a metal ion, it is called an *activator,* since the protein is inactive in the absence of the metal ion. The hydrolases (referred to later) generally do not have coenzymes, although they frequently require metal activators. The chemical substance(s) that reacts in the presence of the enzyme to form some other substance(s), called the *product(s),* is known as the *substrate.*

The terminology of enzymes was arbitrary at first. Later, the tendency grew to name an enzyme after either the substrate or the kind of reaction, adding the suffix -ase. Now all the known enzymes have been given "systematic names" by the 1961 Commission of the International Union of Biochemistry (Dixon and Webb, 1964), consisting of the substrate followed by the kind of reaction plus the ending *-ase* (Table 13-1). Since many of these names are unwieldy, the old "trivial names" may still be used. The commission established six main classes of enzymes based on the main kinds

Table 13-2. Some examples of enzyme cofactors

Kind	Name	Abbreviation	Kind of enzyme associated with cofactor
1. Coenzyme	Nicotinamide-adenine dinucleotide (coenzyme I)	NAD^+ (DPN)	O-R
	Nicotinamide-adenine dinucleotide phosphate (coenzyme II)	$NADP^+$ (TPN)	O-R
	Ascorbic acid	AA	O-R
	Glutathione	GSH	O-R
	Coenzyme A	CoA	Transferase
2. Prosthetic groups	Flavin-adenine mononucleotide	FMN	O-R
	Flavin-adenine dinucleotide	FAD	O-R
	Thiamine pyrophosphate	TPP	Lyase
	Pyridoxine phosphate	PP	Transferase, hydrolase
3. Activators		K^+	Hydrolase, transferase
		Mg^{++}	Transferase, hydrolase, lyase, ligase
		Ca^{++}	Hydrolase
		Zn^{++}	Hydrolase
		Cu^{++}	O-R
		Mn^{++}	O-R, transferase, hydrolase, lyase
		Fe^{++}	O-R, hydrolase
		Co^{++}	O-R
		Mo^{++}	O-R

of reactions they control. Ninety percent of these enzymes control the first four main types of reactions mentioned earlier:

1. Hydrolytic reactions are controlled by hydrolases (3)
2. Group transfer reactions by transferases (2)
3. Oxidation-reduction reactions by oxidoreductases (1)
4. Group removal and addition reactions by lyases (4)
5. Isomerization reactions by isomerases (5)
6. Synthesizing reactions by ligases (previously called synthetases) (6)

The numeral in parentheses after each name is the group number agreed on by the Committee on Enzyme Nomenclature.

In contrast to the proteins of enzymes that are numbered in the many hundreds, there are only twenty-four to thirty-six cofactors (Table 13-2). Most of the coenzymes are nucleotides (purine or pyrimidine—sugar phosphates, see Chapter 16) or related substances, and each may be a cofactor for a large number of different enzymes. The amounts present may therefore be considerable, for example, as high as 1 mg NADPH (nicotinamide-adenine dinucleotide phosphate) per gram fresh weight in maize seedlings (Fritz et al., 1963). Other cofactors, however, are present in much lower quantities. In the case of heterotrophic plants or plant parts (as well as other organisms) some of these coenzymes must be supplied in the form of vitamins in the nutrient medium. The following example illustrates some of the preceding terminology:

The distribution of the known cofactors among the six groups of enzymes is summarized in Table 13-3.

Activity, specificity, isozymes, inhibition

Because of the combination of the protein with the substrate, the latter is activated. Since most enzyme molecules have surface areas much larger than that of the substrate molecules, only a part of the enzyme, known as the *active site,* is in direct contact with the substrate and reacts with it (Fig. 13-9). NMR studies have shown that the substrates are confined at the active site, where they have a relatively long residence time and the two tumble together in solution as the enzyme-substrate (ES) complex (Reuben, 1971). In many cases, the enzyme-substrate intermediate is covalently bonded (Bell and Koshland, 1971). The active site consists of amino acid residues that are not necessarily adjacent to each other on the polypeptide chain of the enzyme but that may be brought together from different parts of the chain because of the folds in the three-dimensional structure of the enzyme (Chapter 16). Certain polar amino acids such as histidine, lysine, serine, tyrosine, cysteine, and asparagine play especially important roles in the active sites of a variety of enzymes (Sizer, 1966). It appears that the thiol (SH) group of the cysteine residue and the imidazole (3C + 2N ring) group of the histidine residues (which are close to each other in the protein molecule) form the active site of proteolytic enzymes such as papain, ficin, and bromelin (Husain and Lowe, 1968).

Malate (substrate) + Malate dehydrogenase (enzyme) + $NADP^+$ (oxidized coenzyme) + Mn^{++} (activator) =

Pyruvate + CO_2 (products) + NADPH (reduced coenzyme) + Malate dehydrogenase (enzyme) + Mn^{++} (activator)

where $NADP^+$ = oxidized nicotinamide-adenine dinucleotide phosphate
NADPH = reduced nicotinamide-adenine dinucleotide phosphate
Mn = manganese

Table 13-3. Relative distribution of enzymes and prosthetic groups*

Enzyme group	Number of enzymes known	Pyri- doxal	Flavin	Metal	Haem	Thiamine pyrophos- phate	Biotin	Cobal- amin
I. Oxido- reductases	222	1	34	24	9	2	0	0
II. Trans- ferases	238	18	0	0	0	1	1	0
III. Hydrolases	213	1	0	4	0	0	0	0
IV. Lyases	117	29	1	1	0	4	0	1
V. Isomerases	47	4	0	0	0	0	0	1
VI. Ligases	47	0	0	0	0	0	4	0
Total	884	53	35	29	9	7	5	2

*From Smith, M. H. 1967. Nature (London) **213**:627-628.

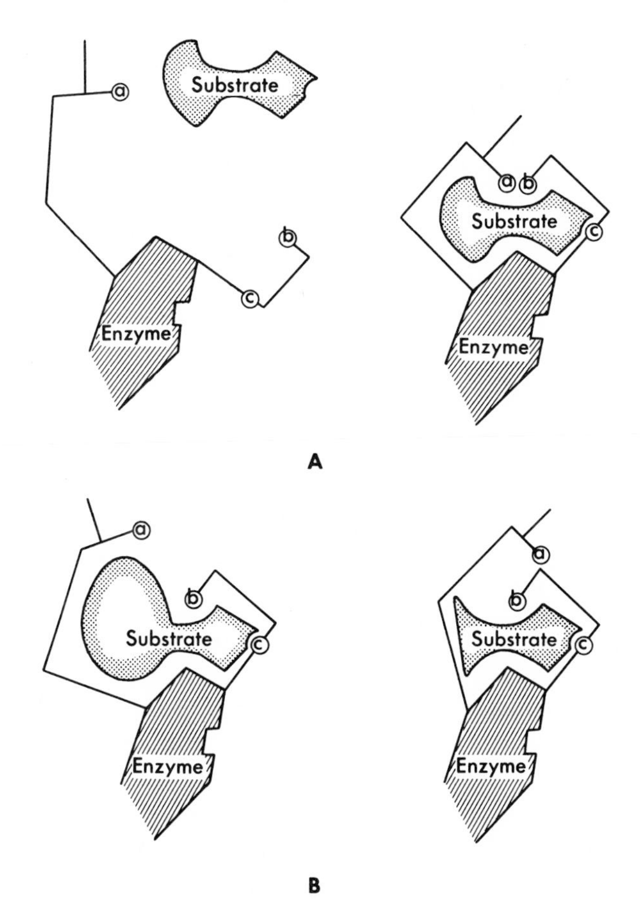

Fig. 13-9. Specificity of enzyme for substrate. **A**, Good fit between the substrate and reactive groups of the enzyme; reaction proceeds. **B**, Poor fit between substrate and reactive groups of the enzyme; no reaction. *a*, *b*, and *c* represent either cofactors or reactive groups of the protein molecule ("active sites"). (From Koshland, D. E., Jr. 1963. Sympos. Quant. Biol. **28**:473-482.)

The exact mechanism of enzyme catalysis is still not completely understood. Lysozyme is the first enzyme for which the relation between structure and function has become clear (Chipman and Sharon, 1969). The enzyme binds its substrate in a cleft running the whole length of its molecule. The complex is stabilized by a large number of H-bonds and of nonpolar interactions. This distorts the substrate from its most stable configuration, raising its internal energy and therefore lowering the activation energy necessary to initiate the reaction. The role of the cofactor, when there is one, is twofold. It may participate directly in the catalytic process by combining with the substrate, it may induce a structural change in the enzyme, or it may do both.

It is the precise arrangement of these active sites that is responsible for the *specificity* of the enzyme, that is, its ability to catalyze the reaction of a single substance or up to a few dozen related substances (Fig. 13-9). Thus several enzymes may contain the same prosthetic group (Table 13-2), but because of the difference in the protein component, each will act specifically on a different substance. For instance, tyrosinase and ascorbic acid oxidase both have copper as a prosthetic group, yet the former acts on tyrosine (and several similar substances) but not on ascorbic acid; the latter acts on ascorbic acid but not on tyrosine.

Despite the specificity of enzymes, several distinct enzymes (multiple molecular forms) may control the same reaction. They are then called *isoenzymes* or *isozymes*. In this case the protein component is different, permitting the separation of the several isozymes by protein separation techniques. The different isozymes are commonly located in different organs or in different parts of the cell. There are, for instance, three isozymes of malic dehydrogenase in the green stem tissue of *Opuntia* (Mukerji and Ting, 1969). One is in the chloroplasts, another is in the mitochondria, and the third is soluble in the cytoplasm. Although their physical properties are different, their kinetic properties are similar. The number of isozymes found in different species or varieties may be great—twenty-four peroxidase isozymes were found among 250 varieties of maize (Hamill and Brewbaker, 1969). One suggested role for isozymes is to cover a range of conditions. One isozyme may be active under conditions that inactivate another; for instance, in the case of two peroxidases of the tomato, one operates under acidic conditions (Evans, 1970).

Because of the protein portion, enzymes are *heat labile;* that is, they are irreversibly inactivated by 60° to 100° C. The coenzyme component is heat stable, being unaffected by boiling. The enzyme is frequently inactive at temperatures near freezing or in an acid pH in the case of some (e.g., trypsin). Low temperature inactivation is reversible, and so is pH inactivation in some cases if the pH change is not too extreme. Heat inactivation may be reversible in the case of some enzymes, but only if denaturation is not followed by aggregation. Poisons may inactivate many enzymes when present in small quantities; for example, 0.0001 M cyanide inactivates many respiratory enzymes (containing heavy metals). These substances are called *inhibitors*. Other inhibitors are sulfides, azides, fluorides, iodoacetates, etc. Some inhibitors are somewhat specific; that is, they inhibit only one or a few enzymes. Frequently, they inhibit by combining with the prosthetic group or with the sulfhydryl group of the protein, thus preventing it from combining with the substrate. Some of the newer inhibitors actually prevent synthesis of the enzyme protein (Chapter 16). Besides the above artificial enzyme inhibitors, natural inhibitors are found in many plants, for instance, proteinase inhibitors in seeds (Ogawa et al., 1971) and others in roots (Chen and

Boll, 1971*a*). Similarly, many phytotoxic substances owe their toxicity to enzyme inhibition (Koch, 1970).

An important group of inhibitors are the structural analogues—substances so similar chemically to the substrate that the enzyme cannot distinguish between them. To be an inhibitor, the structural analogue must not be made to react by the enzyme but must remain attached to it. If the normal reaction is represented as

$$1.\ E + S \rightleftharpoons ES; \quad ES \rightarrow E + P$$

then inhibition may occur in the following two ways:

$$2.\ E + I \rightleftharpoons EI$$
$$3.\ E + I \rightarrow EI$$

where E = enzyme
S = substrate
P = products
I = inhibitor

In 2 the inhibition is reversible (that is, it can be overcome by increasing the amount of substrate); in 3 it is irreversible regardless of substrate concentration.

The structural analogue is also called an antimetabolite if it causes signs of deficiency of the substrate (or essential metabolite, Hochster and Quastel, 1963).

Location in the cell

One of the most amazing facts about enzymes is the tremendous number of different kinds that must coexist in the extremely small space occupied by the protoplasm of a single cell. The number is so large that it has even been suggested that all the cytoplasmic proteins are enzymes. Each must be able to control its own particular reaction without interference from—indeed, with the aid of—other enzymatically controlled reactions. This orderly arrangement would not be possible if all the enzymes were mixed haphazardly in solution. It is not surprising, then, that so many of them have been found attached to organelles in the cytoplasm, for example, the mitochondria. If each one occupies its own place on such an organelle, an orderly series of reactions can take place there.

Many of these enzymes must therefore be active in the insoluble state, as components of the membrane system of the organelle. Even normally soluble enzymes may be insolubilized without loss of their activity (Line et al., 1971; Epton and Thomas, 1971). Other protoplasmic organelles, besides the mitochondria, possess their own specific enzymes (Table 13-4). Some unidentified protoplasmic particulates also possess enzyme activity; there is one from castor beans that synthesizes oleic acid from acetyl coenzyme A (acetyl CoA) or malonyl CoA (Drennan and Canvin, 1969). There are even enzyme aggregates below the particulate size, such as one with a molecular weight of 120,000 that contains five enzymatic activities (Berlyn et al., 1970). Such groups have been called multienzyme systems. Even a single enzyme may consist of two or more components or subunits, such as tryptophan synthase (Chen and Boll, 1971*b*), which may or may not be active individually.

The enzyme content of a cell is not fixed but varies with the conditions. Many enzymes are *adaptive;* that is, they accumulate in the presence of substrate and disappear in its absence. Thus the enzyme nitrate reductase increases in quantity if nitrate is supplied to the plant and decreases in quantity as the nitrate is metabolized and unreplaced. Other enzymes are *constitutive*—they occur regardless of the presence or absence of substrate.

Enzymes have frequently been reported outside the protoplasm of the higher plant cell and even outside the plant organ; but it is difficult to prove that this occurs in the case of the normal, uninjured cell. The frequently reported occurrence of cytochrome oxidase on the walls of plant cells is believed to be an artifact, although in its stead, peroxidase has been identified on the walls of lignifying cells (Lipetz, 1965).

Plant metabolism includes such a vast

Table 13-4. Specific location of some enzymes within the cell

Cell structure or organelle	Enzymes	Reference
Cell wall	α-glucosidase	Klis, 1971
	β-glycerophosphatase	Hall, 1969
	β-glucosidase	Nevins, 1970; Chkanikov et al., 1969
	Peroxidase	Barnett and Curtis, 1970 (Chapter 1)
	Acid phosphatase	Poux, 1970
Lysosomes	Hydrolytic enzymes	Matile et al., 1965 (Chapter 1) Semadeni, 1967 (Chapter 1)
Mitochondria	Enzymes of aerobc respiration (Chapter 14)	
Chloroplasts	53 enzymes of photosynthesis	Smillie and Scott, 1969
Plasma membrane	Enzymes for synthesis of pectic substances	Kauss et al., 1969
	Adenosine triphosphatases and other phosphorylases	Arif and Koningsberger, 1969
Golgi membranes (or dictyosomes)	β-glucan synthetase	Ray et al., 1969
	Acid phosphatase	Poux, 1970
Peroxisomes	Glycolate oxidase	Tolbert et al., 1969
	Glycolate reductase	
	Catalase	
	Malic dehydrogenase	
Microbodies	Catalase	Vigil, 1969
Cytoplasm (free)	Fatty acid synthetase	Pirson and Lynen, 1971
Starch particles	3-phosphoglycerate Phosphatase	Randall and Tolbert, 1971

number of chemical reactions that it is neither possible nor desirable to consider each reaction individually. But it is possible to deal with them in groups according to their relation to specific metabolic processes. An understanding of metabolism requires the study of these individual subsidiary processes, into which it can be divided. The most important of these can be classified as shown.

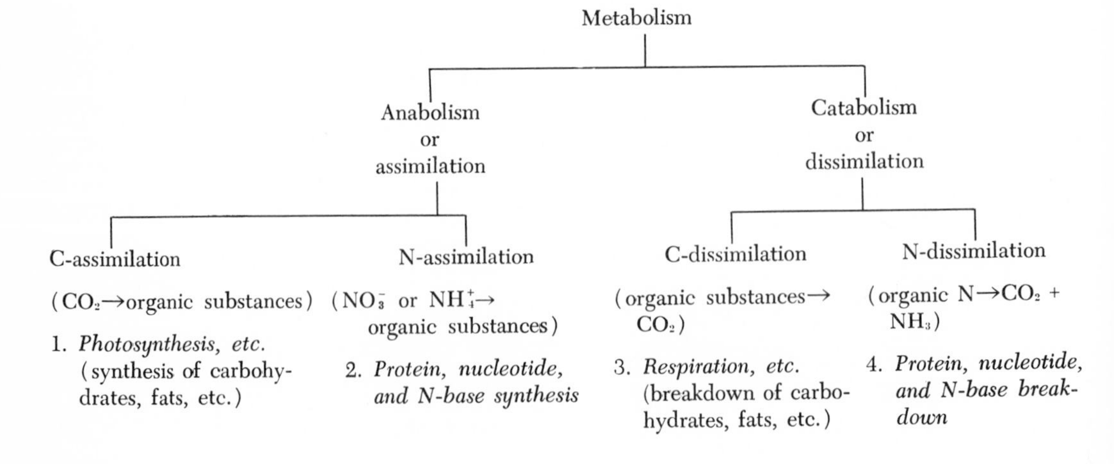

A complete understanding of plant metabolism would also include S-metabolism and P-metabolism, as well as that of all the other essential elements. Although some of these metabolic branches are distinct from each other, others are intimately related. The preceding classification must therefore be considered as one of convenience rather than representing completely separate categories in the plant.

QUESTIONS

1. What is meant by metabolism?
2. What is meant by anabolism and catabolism?
3. How many different kinds of reactions occur in the plant?
4. Name one kind of reaction.
5. Give an example of this kind of reaction.
6. Name a second reaction and give an example.
7. Which kinds of reactions are not readily reversible? Why?
8. What is an oxidation?
9. What is a reduction?
10. Is there any necessary relation between oxidation and reduction?
11. What chemical element is commonly transferred in an oxidation?
12. What are exergonic and endergonic reactions?
13. How are they related to anabolism and catabolism?
14. How are they related to each other?
15. What are spontaneous reactions?
16. Do they take place unaided?
17. What is activation energy?
18. What role do catalysts play in reactions?
19. What are the catalysts in the plant?
20. What is the chemical nature of pure enzymes?
21. What other component does it usually require for activity?
22. What are these components called?
23. Which part is primarily responsible for the chemical activity?
24. What is enzyme specificity?
25. Which component is responsible for it?
26. What is the effect of high temperature on enzymes?
27. What is the effect of low temperature on enzymes?
28. How do these effects differ?
29. Are there any other factors that affect enzyme activity?
30. Which component is heat labile?
31. Can two different enzymes have the same cofactor?
32. What is the substrate?
33. How are enzymes proved to be present in the plant?
34. How are enzymes named?

SPECIFIC REFERENCES

Arif, H., and V. V. Koningsberger. 1969. Membrane-bound ATP-ase and other phosphohydrolase activities in *Saccharomyces carlsbergensis*. I. Introductory experiments with a crude membrane preparation. Proc. Kon. Ned. Akad. Wetensch. Ser. B. Phys. Sci. **72**:230-245.

Barron, E. S. 1951. Thiol groups of biological importance. Adv. Enzymol. **11**:206-266.

Bell, R. M., and D. E. Koshland, Jr. 1971. Covalent enzyme-substrate intermediates. Science **172**:1253-1256.

Berlyn, M. B., S. I. Ahmed, and N. H. Giles. 1970. Organization of polyaromatic biosynthetic enzymes in a variety of photosynthetic organisms. J. Bacteriol. **104**:768-774.

Bernfeld, P. 1951. Enzymes of starch degradation and synthesis. Adv. Enzymol. **12**:379-428.

Chen, J., and W. G. Boll. 1971*a*. Tryptophan synthase: purification, and some properties of an inhibitor from pea roots. Canad. J. Bot. **49**: 821-832.

Chen, J., and W. G. Boll. 1971*b*. Tryptophan synthase: a two-component enzyme from pea plants (*Pisum sativum* cv. Alaska). Canad. J. Bot. **49**:1155-1163.

Chipman, D. M., and N. Sharon. 1969. Mechanism of lysozyme action. Science **165**:454-465.

Chkanikov, D. I., G. A. Tarabrin, A. M. Shabanova, and P. F. Konstantinov. 1969. Localization of β-glucosidase in the cells of higher plants. Fiziol. Rast. **16**:322-325.

Drennan, C. H., and D. T. Canvin. 1969. Oleic acid synthesis by a particulate preparation from developing castor oil seeds. Biochim. Biophys. Acta **187**:193-200.

Epton, R., and T. H. Thomas. 1971. Improving nature's catalysts. Aldrichimica Acta **4**:61-65.

Evans, J. J. 1970. Spectral similarities and kinetic differences of two tomato plant peroxidase isoenzymes. Plant Physiol. **45**:66-69.

Fritz, G. J., E. R. Stout, and D. E. Leister. 1963. Estimation of nicotinamide nucleotide coenzymes in etiolated maize seedlings. Plant Physiol. **38**: 642-648.

Hall, J. L. 1969. Histochemical localization of β-glycerophosphatase activity in young root tips. Ann. Bot. **33**:399-406.

Hamill, D. E., and J. L. Brewbaker. 1969. Isoenzyme polymorphism in flowering plants. IV. The peroxidase isoenzymes of maize *(Zea mays)*. Physiol. Plant. **22**:945-958.

Hellebust, J. A., and R. G. S. Bidwell. 1963. Protein turnover in wheat and snapdragon leaves. Canad. J. Bot. **41**:969-983.

Husain, S. S., and G. Lowe. 1968. Evidence for histidine in the active sites of ficin and stembromelin. Biochem. J. **110**:53-57.

John, P. C. L., and P. J. Syrett. 1968. The estimation of the quantity of isocitrate lyase protein in acetate-adapted cells of *Chlorella pyrenoidosa*. J. Exp. Bot. **19**:733-741.

Kauss, H., A. L. Swanson, R. Arnold, and W. Odzuck. 1969. Biosynthesis of pectic substances: localization of enzymes and products in a lipid-membrane complex. Biochim. Biophys. Acta **192**:55-61.

Klis, F. M. 1971. α-Glucosidase activity located at the cell surface in callus of *Convolvulus arvensis*. Physiol. Plant. **25**:253-257.

Koch, H. 1970. Phytotoxic substances and their possible action mechanism. Sci. Pharm. **38**: 79-89.

Koshland, D. E., Jr. 1963. The role of flexibility in enzyme action. Sympos. Quant. Biol. **28**:473-482.

Line, W. F., A. Kwong, and H. H. Weetall. 1971. Pepsin insolubilized by covalent attachment to glass: preparation and characterization. Biochim. Biophys. Acta **242**:194-202.

Lipetz, J. 1965. Peroxidase interference with cytochrome oxidase localization in plant cells. J. Histochem. Cytochem. **13**:300-301.

Lipmann, F. 1941. Metabolic generation and utilization of phosphate bond energy. Adv. Enzymol. **1**:99-162.

Mahler, H. R., and E. H. Cordes. 1968. Basic biological chemistry. Harper & Row, New York.

Mentzer, C., O. Fationoff, and C. Deschamps-vallet. 1968. Actualités de phytochimie fondamentale. Vol. 3. Masson et Cie, Paris.

Mukerji, S. K., and I. P. Ting. 1969. Malic dehydrogenase isoenzymes in green stem tissue of *Opuntia*: isolation and characterization. Arch. Biochem. Biophys. **131**:336-351.

Ogawa, T., T. Higasa, and T. Hata. 1971. Proteinase inhibitors in plant seeds. IV. Trypsin-inhibitor complex formation. Agr. Biol. Chem. **35**:717-723.

Nevins, D. J. 1970. Relation of glycosidases to bean hypocotyl growth. Plant Physiol. **46**:458-462.

Pirson, W., and F. Lynen. 1971. The location of fatty acid synthetase in the yeast cell. Hoppe-Seylers Z. Physiol. Chem. **352**:797-804.

Poux, N. 1970. Localization of enzyme activities in the root meristem of *Cucumis sativus* L. III. Acid phosphatase activity. J. Microsc. (Paris) **9**:407-434.

Randall, D. D., and N. E. Tolbert. 1971. 3-Phosphoglycerate phosphatase in plants. III. Activity associated with starch particles. Plant Physiol. **48**:488-492.

Ray, P. M., T. L. Shininger, and M. M. Ray. 1969. Isolation of β-glucan synthetase particles from plant cells and identification with Golgi membranes. Proc. Nat. Acad. Sci. USA **64**:605-612.

Reuben, J. 1971. Substrate anchoring and the catalytic power of enzymes. Proc. Nat. Acad. Sci. USA **68**:563-565.

Sizer, I. W. 1966. The mechanism of enzyme catalysis. Chem. Eng. Prog. Symp. Ser. **62**(69): 1-10.

Smillie, R. M., and N. S. Scott. 1969. Organelle biosynthesis: the chloroplast. *In* F. E. Hahn (Ed.). Progress in molecular and submolecular biology. **1**:136-202. Springer-Verlag New York Inc.

Smith, M. H. 1967. Relative distribution of enzymes and prosthetic groups. Nature (London) **213**:627-628.

Theorell, H. 1956. Nature and mode of action of oxidation enzymes. Science **124**:467-472.

Tolbert, N. E., A. Oeser, R. K. Yamazaki, R. H. Hageman, and T. Kisaki. 1969. A survey of plants for leaf peroxisomes. Plant. Physiol. **44**: 135-147.

Vigil, E. L. 1969. Intracellular localization of catalase (peroxidatic) activity in plant microbodies. J. Histochem. and Cytochem. **17**:425-428.

Wohl, K., and W. O. James. 1942. The energy changes associated with plant respiration. New Phytol. **41**:230-256.

Wuntch, T., R. F. Chen, and E. S. Vesell. 1970. Lactate dehydrogenase isozymes: kinetic properties at high enzyme concentrations. Science **167**: 63-65.

GENERAL REFERENCES

Dixon, M., and E. C. Webb. 1964. Enzymes. Academic Press, Inc., New York.

Enzyme nomenclature; recommendations (1964) of the International Union of Biochemists. 1965. Elsevier Publishing Co., Amsterdam.

Florkin, M., and E. M. Stotz. 1962. Comprehensive biochemistry. Elsevier Publishing Co., Amsterdam, New York, 15 vol (+12 in preparation).

Green, D. E. 1940. Mechanisms of biological oxidations. Cambridge University Press, New York.

Hochster, R. M., and J. H. Quastel, 1963. Metabolic inhibitors. I. Academic Press, Inc., New York.

Lehninger, A. L. 1965. Bioenergetics. W. A. Benjamin, Inc., New York.

Linford, J. H. 1966. An introduction to energetics with applications to biology. Butterworth & Co. (Publishers), Ltd., London.

Northrop, J. H., M. Kunitz, and R. M. Herriott. 1948. Crystalline enzymes. Columbia University Press, New York.

Pullman, B., and A. Pullman. 1963. Quantum biochemistry. Interscience Publishers, New York.

Schwimmer, S., and A. B. Pardee. 1953. Principles and procedures in the isolation of enzyme. Adv. Enzymol. **14**:375-409.

Sumner, J. B., and K. Myrback. 1950-1952. The enzymes. Academic Press, Inc., New York. 4 vol.

Walter, C. 1965. Steady-state applications in enzyme kinetics. The Ronald Press Co., New York.

14 Respiration

Overall process
Glycolysis
Krebs cycle
Electron transport
Oxidative phosphorylation
Respiratory enzymes
Respiration rates

OVERALL PROCESS

The most important catabolic, energy-releasing process in the plant is the *oxidative breakdown of organic substances,* which is called respiration. Sugars are most commonly the starting point, and carbon dioxide and water are usually the end substances formed, although there are many exceptions. Respiration consists of a large number of chemical reactions, all of which are enzymatically controlled. For some purposes, however, it is possible to ignore the enzymes and even the individual chemical reactions and to consider the overall process. Thus if sugar is respired, respiration can be represented by the following:

$$C_6H_{12}O_6 + 6O_2 \longrightarrow 6CO_2 + 6H_2O + 678 \text{ kcal}$$

where $C_6H_{12}O_6$ = glucose

This means that 180 g (1 mole) glucose (or some other hexose) is oxidized by 192 g (6 moles) of oxygen, releasing 264 g (6 moles) of carbon dioxide, 108 g (6 moles) of water, and 678 kcal of heat. Thus each gram of glucose respired releases 3.74 kcal of heat. It must be recognized that this equation represents only a *balance sheet,* showing the quantitative relations between the raw materials respired and the products formed. That it does *not* represent the true chemical changes is obvious from the now known fact that all the oxygen absorbed in the molecular form is reduced to water, although the preceding general equation does not include enough water to account for this relation. The value of 678 kcal means that 113 kcal are released per mole of oxygen absorbed. This, however, is a theoretical value, and measured values may be significantly different, for example, 126, 141, and 97 kcal for germination of seeds of *Prosopis juliflora, Kochia scoparia,* and *Oxyria digyna,* respectively (Lacher et al., 1966). The difference from the theoretical value undoubtedly results from secondary reactions such as hydration.

If this equation is correct as a balance sheet, it should be possible to measure respiration by determining any one of the preceding five quantities. In practice it is usually most convenient to determine either the carbon dioxide released or the oxygen absorbed. If the correctness of the preceding equation is in doubt, it is necessary to measure both quantities. The *molar* ratio of carbon dioxide evolved per mole oxygen absorbed is then called the *respiratory quotient* (R.Q.):

$$\text{R.Q.} = \frac{\text{Moles } CO_2 \text{ evolved}}{\text{Moles } O_2 \text{ absorbed}}$$

If this ratio is 1, the preceding equation would appear to be correct; that is, glucose or some other carbohydrate is apparently being respired to carbon dioxide and water.

If substances other than glucose are respired, a different respiratory quotient usually results. Thus a fat would give:

$$2C_{51}H_{98}O_6 + 145O_2 \longrightarrow 102CO_2 + 98H_2O$$

where $C_{51}H_{98}O_6$ = glycerol tripalmitate

The respiratory quotient is therefore 0.7. Similarly, it can readily be calculated for organic acids, for example, 4 for oxalic acid ($C_2H_2O_4$) if it were respired. The value for proteins varies between 0.5 and 1.0, depending on the particular protein and the substances produced. The quotient can vary from zero (no carbon dioxide evolved) to infinity (no oxygen absorbed). Some values that have been found experimentally are listed in Table 14-1.

A respiratory quotient of less than 1 may result from one of several causes:

1. A substrate of lower oxygen content than carbohydrate is being used.
2. A carbohydrate is incompletely oxidized, for example, with the formation of organic acids of higher oxygen content than that of the carbohydrates. This is characteristic of succulents such as *Opuntia* (Table 14-1).
3. Oxygen uptake occurs in processes other than respiration.
4. Carbon dioxide is absorbed in the process of dark carbon dioxide assimilation (Chapter 15). This is also characteristic of succulents.

Table 14-1. Respiratory quotients (CO_2/O_2) in different plants*

	Respiratory quotient
Leaves rich in carbohydrates	1
Darkened shoots of *Opuntia* (prickly pear)	0.03
Germinating starchy seeds	1
Germinating linseed (high fat)	0.64
Germinating buckwheat (high protein)	0.50
Germinating peas	1.5-2.4

*Adapted from Thomas, M., et al. 1960. Plant physiology. J. & A. Churchill, Ltd., London.

A respiratory quotient greater than 1 may mean that some high oxygen substances, such as organic acids, are being respired or that respiration is partially incomplete and therefore does not utilize molecular oxygen.

Despite these many exceptions, the respiratory quotient of higher plants is commonly 1. This means that the equation given on p. 192 is probably correct for most higher plants under normal conditions, and carbohydrates are the main substrate for respiration. Many other lines of evidence confirm this conclusion. Even when fats are the main reserve material, they are first converted to sugars, which are then respired. Consequently, an understanding of respiration requires some knowledge of the carbohydrates.

Many different carbohydrates occur in the plant (Fig. 14-1)—some soluble (sugars), others insoluble (most polysaccharides). Each has an empirical formula consisting of carbon, hydrogen, and oxygen, with the hydrogen and oxygen in the proportions in which they occur in water. But the properties of carbohydrates depend on their structural formulas. Carbohydrates consist of polyhydroxy aldehydes or ketones, which means that they have many OH groups and a CHO or CO group. In the case of the more complex carbohydrates, which are not themselves aldehydes or ketones, they give rise to polyhydroxy aldehydes and ketones (and to nothing else) when hydrolyzed. Complete hydrolysis of starch, cellulose, and maltose yields only the monosaccharide glucose. Hydrolysis of sucrose yields both glucose and fructose.

Because of their hydroxyl groups, carbohydrates form esters with acids. The most important of these in metabolism are the phosphate esters that are formed by the following group transfer reaction:

$$RCHOH + ATP \rightleftharpoons RCHOH_2PO_3 + ADP$$

where ATP = adenosine triphosphate
ADP = adenosine diphosphate
$RCHOH_2PO_3$ = sugar phosphate

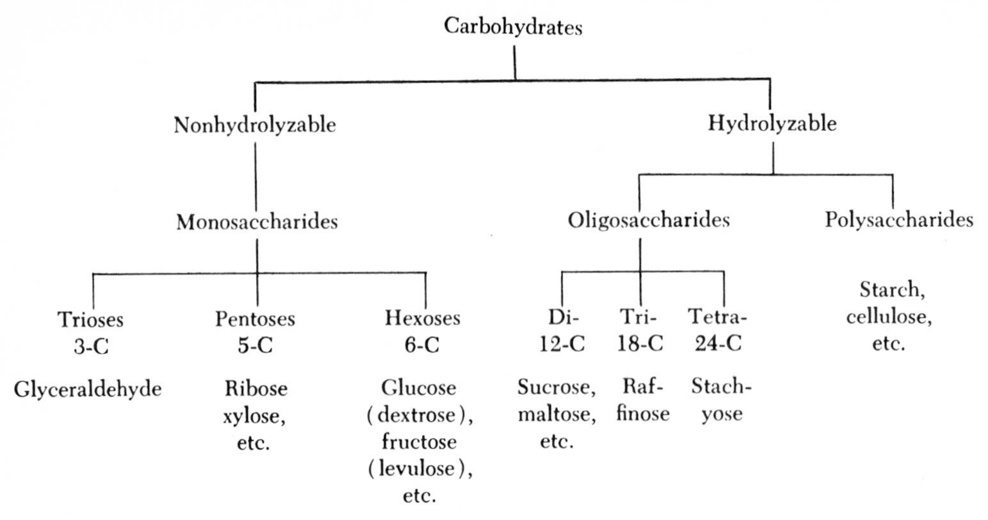

Fig. 14-1. Classification of carbohydrates and examples of those most commonly found in the largest quantities in higher plants.

This formation of a phosphate ester of the sugar is the first step in respiration. It is called *phosphorylation.*

GLYCOLYSIS

Many organisms are able to break down (or lyse) carbohydrates to carbon dioxide and some other substance (e.g., an alcohol or organic acid) without utilizing molecular oxygen but with the release of energy. This process has been called by several names: *anaerobic respiration* because it may occur in the absence of air; *fermentation* because it is so often controlled by microorganisms or their enzymes (previously known as ferments); or *glycolysis* (*glyco-* means sugar) because it is a breakdown of sugars. It is now known that all higher plants are capable of respiring anaerobically for varying lengths of time; but most die within 1 to 3 days in the complete absence of molecular oxygen. One exception is the latex of *Hevea brasiliensis,* in which glycolysis was the only metabolic process functioning even when exposed to air (Jacob, 1970). Since anaerobic respiration may occur even in the presence of oxygen, it must be defined as respiration without the *participation of* molecular oxygen.

When higher plants respire in the absence of oxygen, their respiration is usually partly or wholly alcoholic fermentation:

$$C_6H_{12}O_6 \longrightarrow 2C_2H_5OH + 2CO_2 + 28\text{kcal}$$

where C_2H_5OH = ethyl alcohol

Much less energy is released in this way than in complete aerobic oxidation to carbon dioxide and water. Substances other than ethyl alcohol may accumulate, for example, glycerol, lactic acid, and acetaldehyde. In the case of some plants (e.g., potato tubers), acids accumulate but little or no alcohol. Just as in animal cells and some microorganisms, the acid has been identified as lactate, for instance in the roots of *Zea mays* (Kohl and Matthaei, 1971). If only alcohol is produced, the weight ratio of alcohol to carbon dioxide is 1.05. The actual ratios found vary from this value all the way down to zero (Table 14-2).

Since so many plants accumulate ethyl alcohol under anaerobic conditions, early

Table 14-2. Ratios of carbon dioxide to alcohol produced in plants respiring anaerobically*

	Ratio
Carrot roots	100:102
Oranges	100:70
Apples	100:42
Potato tubers	100:7
Alcoholic fermentation (theoretical)	100:105

*Adapted from Kostytchev, S. R. 1926. Lehrbuch der Pflanzenphysiologie. I. Springer-Verlag, Berlin.

workers suggested that the normal aerobic respiration involves, first, the anaerobic breakdown to alcohol and carbon dioxide, and, second, a further breakdown of the alcohol to carbon dioxide and water. That the argument does not hold for some plants (e.g., the potato) follows from the near absence of alcohol under anaerobic conditions. Furthermore, it was soon shown that higher plants do not oxidize alcohol readily.

It is now known that alcohol is not produced when the higher plant is respiring aerobically, although the first stage of aerobic respiration is the same as in alcoholic fermentation, up to a point preceding alcohol formation. Alcohol is thus an end product that occurs on a branch line instead of on the main line of reactions. This side branch operates only when respiration is anaerobic. The sequence in alcoholic fermentation consists of some twelve chemical reactions, including the following steps:

Hexose →(1) Hexose phosphates →(2) Triose phosphates →(3) Phosphoglyceric acid →(4) Phosphopyruvic acid →(5) Pyruvic acid →(6) Acetaldehyde →(7) Ethyl alcohol

Each step in the reaction is controlled by an enzyme. The variety of kinds of enzymes involved is indicated by the number of each step or group of steps, each number corresponding to a specific enzyme group as follows:

(1) Transferase (transfer of phosphate from ATP to hexose)
(2) Lyase (aldehyde-lyase: removal of glyceraldehyde)
(3) Oxidoreductase (oxidation of aldehyde to carboxyl)
(4) Lyase (hydro-lyase; removal of molecule of water)
(5) Transferase (phosphotransferase: transfer of phosphate to ADP)
(6) Lyase (carboxyl-lyase: removal of carbon dioxide)
(7) Oxidoreductase (reduction of aldehyde to alcohol)

The preceding series of reactions is called the Embden-Meyerhof-Parnas Pathway (EMP).

KREBS CYCLE

When the respiration is aerobic, glycolysis goes only as far as pyruvic acid. Acetaldehyde and alcohol are on the anaerobic branch line. Pyruvic acid is therefore the pivot. When aerobic respiration occurs, this substance is converted to another organic acid, which in its turn is changed to a third one, etc. (Fig. 14-2). Each molecule of pyruvic acid is thus broken down to carbon dioxide and water with the release of energy, by passing through a series of some twelve chemical reactions, converting it successively into 6-, 5-, and 4-carbon acids. This cyclic series of reactions is known as the Krebs (or tricarboxylic acid, TCA, or citric acid) cycle. It was first established in microorganisms and animal cells. The general operation of the Krebs cycle in higher plants is indicated by the identification of all the organic acids of the cycle in the leaves of all the ten species tested in one investigation (Clark, 1969).

As in glycolysis, each step in the Krebs cycle is controlled by an enzyme; but not as many kinds of enzymes are required for the Krebs cycle proper as for glycolysis.

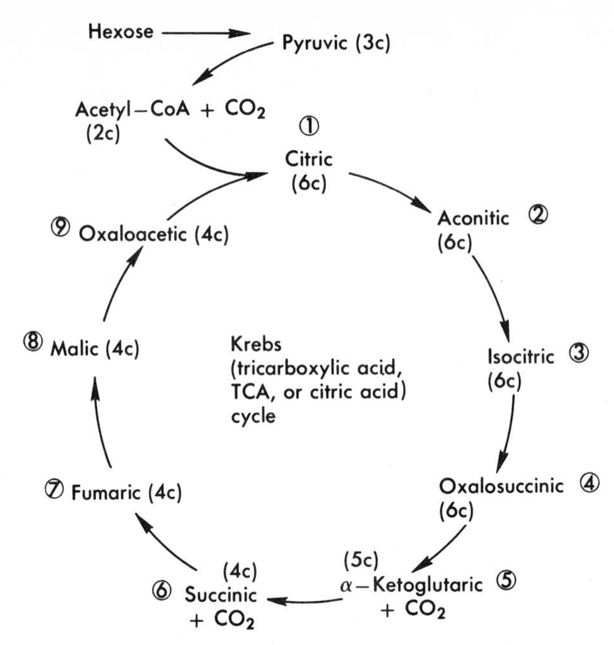

Fig. 14-2. Simplified version of Krebs cycle. The number of C-atoms in each organic acid is indicated in parentheses. The water molecules removed and the enzymes involved are not shown.

Since the pyruvic acid is essentially broken down to carbon dioxide and water, the main enzymes are oxidoreductases and lyases (both carboxy-lyases and hydro-lyases).

The first part of respiration sometimes goes through a pathway called the pentose phosphate pathway (PPP), or hexose monophosphate pathway (or shunt-HMP or HMS). It differs from glycolysis by involving pentose phosphates as intermediates between the formation of hexose phosphate and triose phosphate. It also includes an oxidation of the hexose phosphate (instead of the triose phosphate as in glycolysis), and this oxidation depends, indirectly, on aerobic respiration. The whole process of respiration may be represented as shown in Fig. 14-3.

Glycolysis seems to be the sole or nearly the sole pathway for the first stage of respiration in young tissues and undifferentiated cells. As leaves age, their synthetic ability decreases, and the first stage of respiration changes from the glycolytic to the pentose phosphate pathway (Faust et al., 1968). Similarly, in the root tips of peas, respiration is apparently mainly by means of glycolysis, but differentiation seems to involve an increase in activity of the PPP (Fowler and Ap Rees, 1970). The PPP is apparently favored when the energy requirement is low, although it also appears to become activated in cut regions of etiolated plants (Yukio et al., 1968).

The oxidative breakdown of carbohydrate can therefore be described more correctly as follows:

Anaerobic phase:

1. $C_6H_{12}O_6 \longrightarrow 2CH_3{\cdot}CO{\cdot}COOH + 4H$

Aerobic phase:

2. $2CH_3{\cdot}CO{\cdot}COOH + 5O_2 \longrightarrow 6CO_2 + 4H_2O$
3. $4H + O_2 \longrightarrow 2H_2O$

Total: $C_6H_{12}O_6 + 6O_2 \longrightarrow 6CO_2 + 6H_2O$

where $CH_3{\cdot}CO{\cdot}COOH$ = pyruvic acid

ELECTRON TRANSPORT

The net result of the Krebs cycle is thus the breakdown of pyruvic acid to carbon dioxide and water. The carbon dioxide re-

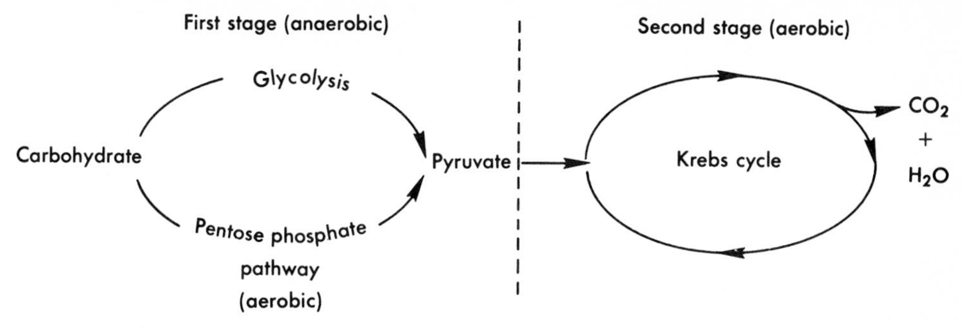

Fig. 14-3. The two main stages in the respiratory breakdown of carbohydrate to CO_2 + H_2O.

lease is a group removal reaction, in which carboxylic acids are decarboxylated in the presence of carboxy-lyases. The water is formed by oxidation-reduction reactions (or dehydrogenations) involving the removal of hydrogen atoms from the organic acids in the presence of oxidoreductase enzymes. This occurs, for instance, in the case of succinate:

$$\text{Succinate} \underset{}{\overset{\text{O-Rase}}{\rightleftharpoons}} \text{Fumarate} + 2\text{H}$$

These hydrogen atoms are ultimately combined with oxygen to form water. The dehydrogenation is also a removal of electrons, since each hydrogen atom possesses an electron. It is therefore called an electron transport. Since neither the hydrogen atom nor its electron can exist free, each hydrogen or electron donor must be linked to a hydrogen or electron acceptor in order to become oxidized by the electron loss. Such a link cannot occur directly between the electron-donating organic acids and the electron-accepting oxygen because the oxidation-reduction potentials of the two systems are too far apart; it is only when the potentials of two systems overlap, that one can oxidize or reduce the other. Therefore a series of oxidation-reduction systems must mediate the electron transport from the organic acids to molecular oxygen. Each of these systems accepts electrons from another system at a higher reduction potential (lower oxidation potential) and transports them to a third system at a lower reduction potential (higher oxidation potential). In this way the electrons are transported step by step from a region of higher free energy to one of lower free energy, and the net result is a transfer of hydrogen atoms from the organic acids to molecular oxygen, forming water as follows:

$$\text{Organic acid} \xrightarrow{\text{H}} \text{System 1} \xrightarrow{\text{H}} \text{System 2} \xrightarrow{\text{H}} O_2$$

Each of these steps is controlled by a specific oxidoreductase enzyme, and in each case a cofactor is successively reduced by the system above it and oxidized by the system below it. There are many such systems, for example:

$$\text{NADP}^+ + 2\text{H} \rightleftharpoons \text{NADPH}$$
$$\text{Cytochrome}_{ox} + 2\text{H} \rightleftharpoons \text{Cytochrome}_{red}$$
$$\text{Quinone} + 2\text{H} \rightleftharpoons \text{Hydroquinone, etc.}$$

where $NADP^+$ = nicotinamide-adenine dinucleotide phosphate
NADPH = reduced nicotinamide-adenine dinucleotide phosphate

OXIDATIVE PHOSPHORYLATION

There are therefore many side reactions associated with the Krebs cycle proper, without which the cycle would soon come to a stop. Most of these are electron transfer reactions. There are also some electron transfers, although fewer in number, associated with glycolysis. In both cases each individual electron-transfer reaction releases a small amount of free energy; and because of the greater number associated with the Krebs cycle, much more total free energy is released in the aerobic than in the anaerobic phase. However, the free energy of any chemical reaction is given off as heat, unless it is coupled to an energy-absorbing reaction. Therefore the plant must have some mechanism for trapping this free energy before it becomes converted to unavailable heat energy and for using it to drive its energy-requiring processes. The trapping reaction is the conversion of low energy (e.g., inorganic) phosphates to high energy organic phosphates. Since the trapped energy is derived from the oxidative reactions of respiration, the process is called *oxidative phosphorylation.* This is the process illustrated earlier by analogy with a dynamo (the metabolic wheel), from which energy is brushed off and transported to the region of utilization by way of an organic phosphate known as adenylic acid (Fig. 13-2). When the phosphate gives up its energy, low energy inorganic phosphate is regenerated, which may be again converted into the high energy form by the metabolic wheel. The phosphates therefore take part in the Krebs cycle as well as in glycolysis, although this is not shown in Fig. 14-2. The importance of the high energy phosphates is dramatically illustrated by the use of certain substances (e.g., 2,4-dinitrophenol) that act as *uncouplers.* They prevent the conversion of respiratory energy to high energy phosphates by inhibiting the enzymes controlling these reactions. But they do not prevent the respiratory breakdown of the carbohydrates, and in fact the rate of this breakdown may increase, although the energy released is wasted, and therefore growth and other energy utilizing processes are inhibited.

The minimum overall conversion of respiratory energy to high phosphate energy may be represented as follows (Lehninger, 1964):

$$C_6H_{12}O_6 + 6O_2 + 38ADP + 38H_3PO_4 \rightarrow 6CO_2 + 38ATP + 44H_2O + 382 \text{ kcal}$$

where H_3PO_4 = phosphoric acid (actually in the form of a phosphate)
ADP = adenosine diphosphate
ATP = adenosine triphosphate

The release of heat energy (382 kcal) is much less than either the heat of reaction (678 kcal, p. 192) or the slightly larger free energy change. This means that 300 kcal of energy that would have been given off as heat is trapped in the ATP and can be used to drive endergonic processes. The efficiency of energy conversion according to the preceding equation is therefore about 44%, but the exact relation in plants is unknown. It has even been suggested that the plant may be 90% efficient in the conversion of respiratory energy to high energy phosphate (Beevers, 1961). Recent measurements have led to values between the preceding two figures. Lacher and associates (1966) made use of the fact that dinitrophenol (DNP) is an uncoupler of oxidative phosphorylation. It prevents the phosphates from trapping any of the energy released by respiratory electron-transport reactions; and all the free energy is released as heat. By comparing the energy values in normal and uncoupled germinating seeds, they showed that 62% of the respiratory energy is trapped by oxidative phosphorylation.

More commonly, however, the efficiency of oxidative phosphorylation is measured directly by the P:O ratio—the ratio of the number of moles of ADP fixed per mole of oxygen absorbed. Mitochondria from young wheat seedlings have a high efficiency, yielding ratios of P:O of 6, 4, and 5 with the three Krebs cycle acids, α-ketoglutarate, succinate, and malate, respectively (Sarkissian and Srivastava, 1970).

Because of its high free energy, ATP is unstable and must be utilized shortly after it is formed. Therefore the plant cannot store its energy in this form and must depend on the more stable carbohydrates as its main store of energy. When needed, these can be readily converted to the available (ATP) form by the process of respiration.

It must be emphasized, therefore, that the main function of respiration is the production of high energy phosphates (mainly ATP), which may then be used to drive the endergonic processes of the plant. For some time, scientists have sought the mechanism by means of which the energy released by the exergonic (electron transport or oxidative) reactions of respiration is transferred to the phosphate, yielding ATP. In the case of glycolysis the mechanism is simple, since the phosphates are directly involved in the electron transport in steps 3 and 7 (p. 195). The high energy phosphates (e.g., glycerophosphate) are therefore intermediate products. In the case of the Krebs cycle, it is the free organic acids (or their salts) that transfer the electrons to a series of cofactors (p. 196). There is no direct connection between the electron transport chain and the phosphates. Therefore either there must be some unknown phosphate intermediates or another mechanism must be involved. One such mechanism has been proposed by Mitchell (1967). He calls it the "chemiosmotic hypothesis." It can be simply described as follows:

1. The electron transport leads to an accumulation of the oppositely charged protons (H^+) within the membrane, which is impermeable to it.
2. This leads to a higher $[H^+]$ inside than outside the membrane and a higher $[OH^-]$ outside than inside it.
3. This activity gradient "pulls" the H^+ and OH^- out of the ADP and inorganic phosphate, leading to a condensation reaction (i.e., removal of water) and the formation of ATP.

In plants, at least part of the mechanism has been shown to operate in the case of chloroplast thylakoids, in which the membrane acts as a proton pump (Kreutz, 1969). In the mitochondrion, the separation is in the opposite direction.

This hypothesis has the added advantage that it can also explain the mechanism by means of which respiratory energy is utilized to produce ion transport (i.e., active absorption) which would essentially be a reversal of this process, the released H^+ ion, for instance, exchanging for a K^+ ion from the outside.

RESPIRATORY ENZYMES

Since glycolysis and the Krebs cycle proper each involves some ten to twelve reactions, and this does not include the many side reactions of electron transport and phosphorylation, there must be several dozen enzymes that participate in the respiration of a single sugar molecule to carbon dioxide and water. The detailed treatment of all these enzymes is the province of a more advanced course in biochemistry.

Plant physiologists, however, have long been preoccupied with the problem of the *terminal enzymes* (Beevers, 1961). These are the enzymes that control the final electron (or hydrogen) transfer to molecular oxygen. There are four main groups of terminal enzymes:

1. When oxygen is the hydrogen acceptor, the enzymes are called oxidases.
2. If the oxygen molecule is incorporated into the substrate, the enzymes are oxygenases.
3. If hydrogen peroxide is the acceptor, they are peroxidases.
4. A final exceptional group (catalases) breaks down any hydrogen peroxide formed.

The second group is not important in plants. The other reactions may be represented as follows:

$$R(OH)_2 + O_2 \xrightarrow{\text{Oxidase}} R(O)_2 + H_2O_2$$

$$AH_2 + H_2O_2 \xrightarrow{\text{Peroxidase}} A + 2H_2O$$

$$H_2O_2 \xrightarrow{\text{Catalase}} H_2O + \tfrac{1}{2}O_2$$

Several oxidases are known, each of which controls the oxidation of a specific substrate:

$$\text{Ascorbic acid} \xrightarrow{\text{Ascorbic acid oxidase}} \text{Dehydroascorbic acid}$$

$$\text{Phenol} \xrightarrow{\text{Phenolase}} \text{Quinone}$$

$$\text{Reduced cytochrome} \xrightarrow{\text{Cytochrome oxidase}} \text{Oxidized cytochrome}$$

All the oxidases are heavy metal enzymes because of the presence of iron or copper as cofactors. They are therefore inhibited by cyanides, since these combine with heavy metals. This inhibition becomes progressively lower in aging organs. Cytochrome oxidase is the one involved in the terminal oxidation reaction, leading to the formation of high energy phosphates. The exact role of the other oxidases in plant metabolism is still not well understood. Inada (1967) suggests that ascorbic acid oxidase or peroxidase (which is activated by root aging) may participate in the defense mechanism as an alternative oxidase for cytochrome oxidase. Similarly, phenolase is thought to function mainly as a defense against wounding and infection, by the oxidation of chlorogenic acid (MacCrae and Duggleby, 1968).

The enzymes controlling glycolysis and the pentose phosphate pathway appear to occur free in the cytoplasm. Some of them, however, appear to be associated with a particulate preparation (Goh, 1971). On the other hand, those responsible for the Krebs cycle reactions have been found to occur only in the mitochondria. When these organelles are isolated from the plant, the oxidation by them of the intermediates in the Krebs cycle requires the addition of oxygen, ADP, and inorganic phosphate. This is because the previously described phosphorylation is linked to the electron transport of the oxidation reactions.

The mitochondrion is a complex structure in spite of its small size (Fig. 14-4). Some of the enzymes are loosely associated with it, others are tightly enclosed in it, and a third group are structurally bound to it (Klingenberg and Pfaff, 1966). It is presumably because of its specialized structure and the specific location of its many enzymes that the mitochondrion is capable of producing all the steps of the Krebs cycle and associated reactions in the correct order and at the necessary rates.

On the basis of the overall equation, the respiration of higher plants in its strictest sense consists only of the energy-releasing breakdown of sugars to carbon dioxide and water. Therefore this process cannot include all of catabolism. There must frequently also be a hydrolytic breakdown (or *digestion*) of more complex, insoluble carbohydrates to sugars, of lipids to fatty acids, glycerol, etc., and of proteins to amino acids. The products of digestion may then be respired, or they may be resynthesized to other complex substances, or they may be translocated to other parts of the plant. Just as in the case of respiration, digestive processes are controlled by enzymes, most of which are hydrolases. In some cases it is impossible to distinguish between digestion and respiration, for example, when starch is phosphorylated to hexose phosphate, which is then broken down in glycolysis. Similarly, although respiration in its narrower sense has been described previously as beginning with carbohydrates, in its broader sense this is not always the case, as shown earlier by the variation in respiratory quotients. In many seeds, for instance, the reserves are in the form of fats or proteins, and during germination these are respired. The reactions involved are less well understood than in the case of the more common respiratory breakdown of carbohydrates. Proteins, however, if respired to completion, may give rise to ammonia as well as to carbon dioxide, and this may lead to injury.

RESPIRATION RATES

Plant respiration releases large quantities of carbon dioxide into the atmosphere. The forest has been estimated to release about

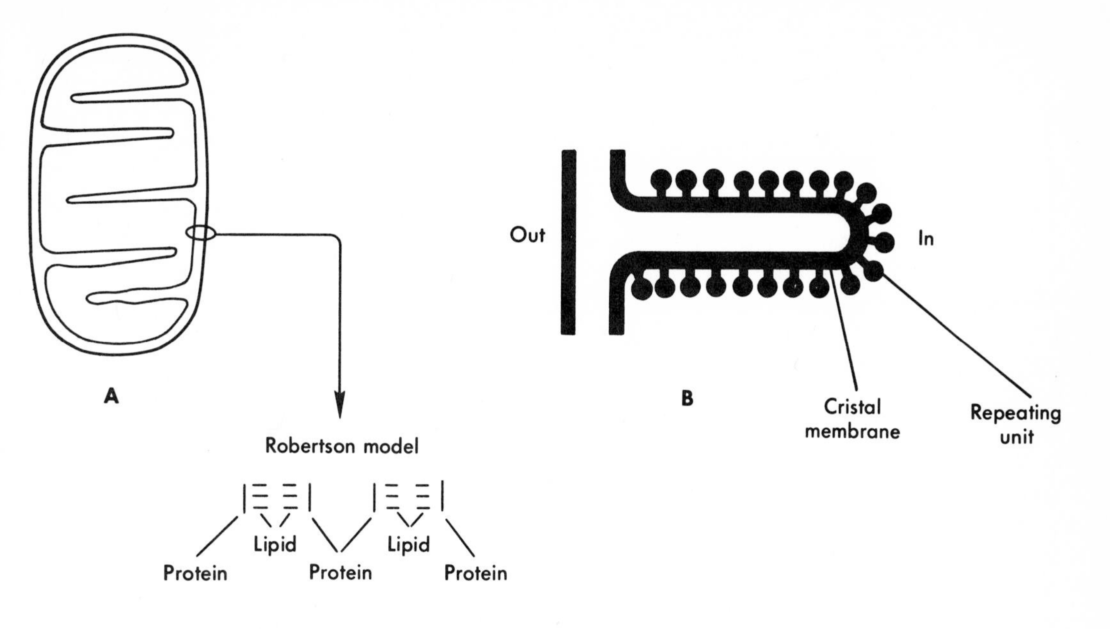

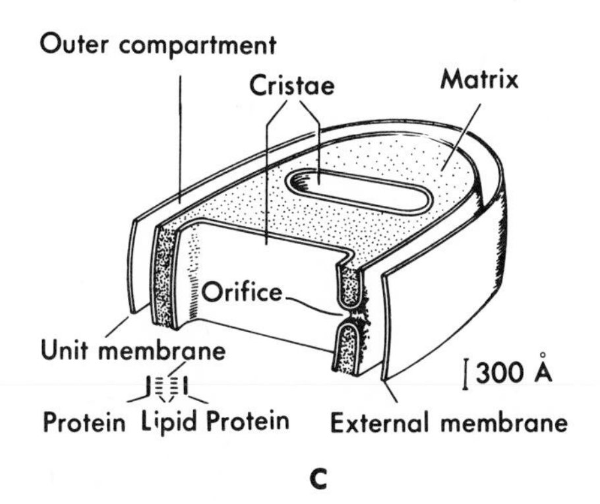

Fig. 14-4. Mitochondrion structure. **A** and **B,** Two-dimensional structure. **C,** Three-dimensional structure (**A** and **B** from Lee, C. P., and L. Ernster and **C** from V. P. Whittaker. 1966. *In* J. M. Tager, et al. Regulation of metabolic processes in mitochondria. Elsevier Publishing Co., Amsterdam.)

3400 g carbon dioxide per square meter per year and therefore to respire some 2100 g carbohydrate per square meter per year (Woodwell and Dykeman, 1966). Yet at certain times of the year the rate drops to a low value, and at other times it rises to a correspondingly high value because the rate of respiration is affected by many factors, both internal and external. From the equation for the overall process (p. 192) it is obvious that an ample supply of hexoses or other respirable material is essential for a rapid rate of carbon dioxide evolution. The same is true of the oxygen supply to the respiring cells, at least if respiration is aerobic. Yet it has been shown that, in many cases, oxygen actually decreases the rate of breakdown of sugars as compared with anaerobic respiration. This is called the *Pasteur effect.* An accumulation of the end products also reduces the rate. This fact is made use of in the storage of fruit. Air containing 10% carbon dioxide retards respiratory breakdown and therefore reduces sugar loss and prolongs the life of the fruit; but the oxygen content of the air must be maintained as high as normal (at the expense of the nitrogen) to prevent anaerobic respiration, which leads to injury.

The rate of respiration is, of course, greatly affected by temperature. A rise of 10° C usually increases the rate 2 to 3 times. This is the main basis for the low-temperature preservation of plant parts. Dropping the temperature from 30° to 0° C may reduce the rate of respiration to 1/27 (i.e. $[1/3]^3$), in this way preserving the reserves 27 times as long. Injury may greatly increase the respiration rate.

A striking variation in respiration rate may occur with age. Thus a developing fruit may show a significant drop during the early stages, followed by a rise, then another drop (Fig. 14-5). The first drop is perhaps more apparent than real, since the cell enlargement results in about as much of a decrease in percent protoplasm (which is the only part of the cell capable of respiring) as the decrease in respiration per gram of total dry matter (Kidd and West, 1945). The subsequent rise and fall, however, are real changes, since they occur when the fruit is already full size. The rise that accompanies ripening is called the *climacteric* and is associated with the production of ethylene. Other volatile products responsible for the flavor and aroma also reach a maximum at this time (e.g., methyl, ethyl, and amyl esters of formic, acetic, caproic, and caprylic acids). Artificial ripening may be induced by treating the unripe fruit with ethylene gas. Unripe bananas when treated for 5 days with air containing 1 part ethylene in 10,000 at 20° to 22° C may ripen 20 days sooner than the untreated fruit. Even when artificially produced in this way, the ripening is always accompanied by an increased rate of respiration. This climacteric period is followed by the final period of *senescence*, during which the respiration process seems to be thrown out of gear. Not only does the total rate drop rapidly but ethyl alcohol and acetaldehyde steadily accumulate, the latter giving the off-taste. This seems to indicate that reactions involving molecular oxygen are hindered and that anaerobic respiration becomes more pronounced. Other metabolic changes also occur; for example, insoluble pectins that normally cement cells together are converted to soluble pectin by the enzyme protopectinase. As a result, the cells separate and the flesh becomes mealy. In the case of fruit such as tomato and grapes, even the cell walls dissolve and the free protoplasts are released. These later break down. This whole self-digestive process is known as *autolysis*.

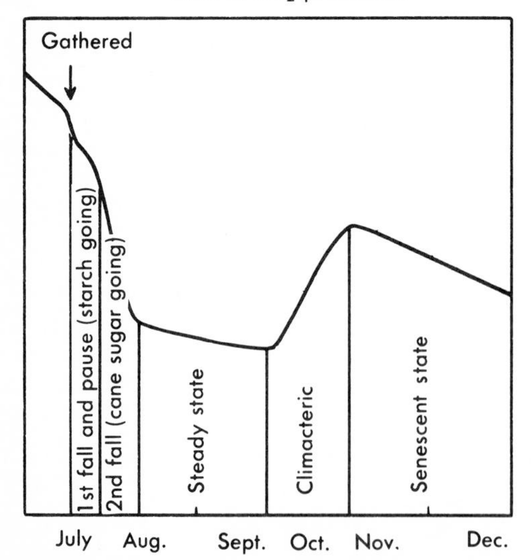

Fig. 14-5. Changes in respiration rate of apple fruit during growth and development. (From Kidd, F. R. 1935. Nature [London] **135**:327-330.)

The effect of light on respiration rate has been a subject of much controversy for some time. It was once suggested that green cells cease respiring when they begin photosynthesizing in the light. Experiments with radioisotopes soon disproved this concept. Since the light effects on respiration are intimately related to photosynthesis, they will be considered in the next chapter.

QUESTIONS

1. If a higher plant in the dark gives off 12 moles carbon dioxide in a certain time, how much oxygen would you expect it to have absorbed?
2. What substrate would have been used up?
3. How much substrate would have been used up?
4. How much heat would have been liberated?
5. What respiratory quotient is assumed?
6. What does this mean?
7. Between what limits may the respiratory quotient vary?
8. What is the respiratory quotient if fats are respired?
9. What is the respiratory quotient if proteins are respired?
10. What is the respiratory quotient if organic acids are respired?
11. What does a respiratory quotient of zero mean?
12. What does a respiratory quotient of infinity mean?
13. What is the major substrate for the respiration of higher plants?
14. What is a carbohydrate?
15. What kinds of carbohydrates are there?
16. Which carbohydrates are respired?
17. Which carbohydrates are not respired?
18. What is the first chemical step in respiration?
19. What is anaerobic respiration?
20. What is alcoholic fermentation?
21. What is glycolysis?
22. Is more, less, or the same amount of energy released in glycolysis compared to aerobic respiration?
23. What are the usual end products in higher plants?
24. How is respiration measured?
25. What relation is there between glycolysis and aerobic respiration?
26. How many chemical steps are there in glycolysis?
27. What mineral element is essential for glycolysis?
28. What are some of the intermediate substances formed?
29. What is the "pivot" substance?
30. What is the name for the second or aerobic half of (aerobic) respiration?
31. Is there any other name for it?
32. What kinds of chemical substances are involved?
33. Approximately how many chemical reactions are included?
34. What is the net result of the Krebs cycle?
35. What is the pentose phosphate pathway?
36. What role does electron transport play in respiration?
37. What is oxidative phosphorylation?
38. What role does it play in respiration?
39. How many enzymes are probably involved in respiration?
40. Where are the Krebs cycle enzymes located?
41. What effect does temperature have on respiration rate?
42. What is meant by the climacteric?
43. By senescence?
44. By autolysis?

SPECIFIC REFERENCES

Clark, R. B. 1969. Organic acids from leaves of several crop plants by gas chromatography. Crop Sci. **9**:341-343.

Faust, M., B. C. Smale, and H. J. Brooks. 1968. Metabolic pathways and synthetic ability of leaves of several *Pyrus* species. Phytochem. **7**:1519-1522.

Fowler, M. W., and T. Ap Rees. 1970. Carbohydrate oxidation during differentiation in roots of *Pisum sativum*. Biochim. Biophys. Acta **201**: 33-44.

Goh, C.-J. 1971. Respiratory enzyme systems in cultured callus tissues. New Phytol. **70**:389-395.

Inada, K. 1967. Physiological characteristics of rice roots, especially with the viewpoint of plant growth stage and root age. Bull. Nat. Inst. Agr. Sci. (Japan) Ser. D. **16**:19-151.

Jacob, J.-L. 1970. Characteristics of glycolysis and its regulation in the latex of *Hevea brasiliensis*. Physiol. Veg. **8**:395-411.

Kidd, F. R. 1935. Respiration of fruits. Nature (London) **135**:327-330.

Kidd, F. R., and C. West. 1945. Respiratory activity and duration of life of apples gathered at different stages of development and subsequently maintained at a constant temperature. Plant Physiol. **20**:467-504.

Klingenberg, M., and E. Pfaff. 1966. Structural and functional compartmentation in mitochondria, p. 180-201. *In* J. M. Tager, S. Papa, E. Quagliariello, and E. C. Slater. Regulation of metabolic processes in mitochondria. Elsevier Publishing Co., Amsterdam.

Kohl, J.-G., and U. Matthaei. 1971. On the role of lactic acid in the metabolism of the seedling roots of *Zea mays* L. Biochem. Physiol. Pflanz. **162**:119-126.

Kreutz, W. 1969. The molecular architecture of thylakoids: first indications of the existence of a proton pump. Ber. Deut. Bot. Ges. **82**:459-474.

Lacher, J. R., A. Amador, and K. Snow. 1966. Effect of dinitrophenol on the heats of respiration of germinating seeds of *Prosopis juliflora, Kochia scoparia,* and *Oxyria digyna.* Plant Physiol. **41**:1435-1438.

Lee, C. P., and L. Ernster. 1966. The energy-linked nicotinamide nucleotide transhydrogenase reaction, p. 218-234. *In* J. M. Tager, S. Papa, E. Quagliariello, and E. C. Slater. Regulation of metabolic processes in mitochondria. Elsevier Publishing Co., Amsterdam.

MacRae, A. R., and R. G. Duggleby. 1968. Substrates and inhibitors of potato tuber phenolase. Phytochem. **7**:855-861.

Mitchell, P. 1967. Proton-translocation phosphorylation in mitochondria, chloroplasts, and bacteria; natural fuel cells and solar cells. Fed. Proc. **26**:1370-1379.

Sarkissian, I. V., and H. K. Srivastava. 1970. High efficiency of oxidative phosphorylation in mitochondria of wheat. Canad. J. Biochem. **48**:692-698.

Whittaker, V. P. 1966. The ultrastructure of mitochondria, p. 1-27. *In* J. M. Tager, S. Papa, E. Quagliariello, and E. C. Slater. Regulation of metabolic processes in mitochondria. Elsevier Publishing Co., Amsterdam.

Woodwell, G. M., and W. R. Dykeman. 1966. Respiration of a forest measured by carbon dioxide accumulation during temperature inversions. Science **154**:1031-1034.

Yukio, M., A. Komamine, and M. Shimokoriyama. 1968. Studies on the changes in respiratory metabolism in the cut region of etiolated *Vicia faba* epicotyl. Bot. Mag. (Tokyo) **81**:434-444.

GENERAL REFERENCES

Beevers, H. 1961. Respiratory metabolism in plants. Harper & Row, Publishers, New York.

James, W. O. 1946. The respiration of plants. Ann. Rev. Biochem. **15**:417-434.

Kostytchev, S. R. 1926. Lehrbuch der Pflanzenphysiologie. I. Chemische Physiologie. Springer-Verlag, Berlin.

Lehninger, A. L. 1964. The mitochondrion; molecular basis of structure and function. W. A. Benjamin, Inc., New York.

Thomas, M., S. L. Ranson, and J. A. Richardson. 1960. Plant physiology. J. & A. Churchill, Ltd., London.

15

Photosynthesis

ENERGY RELATIONS

Photosynthesis is a unique process in two distinct ways. Unlike respiration, which occurs in essentially all living cells of all living organisms, it is unique to the green cells of plants (including some bacteria). It is also unique in being the single energy-trapping process on which all life (except that of a few chemosynthetic bacteria) depends, since it is the only biological process that converts large quantities of light energy into chemical energy. In the case of nonliving systems most of the photochemical reactions known to chemists are, in fact, exothermic and are therefore valueless for storing energy (Daniels, 1967). In these cases the light simply catalyzes the reaction. Similarly, the many other photophenomena that occur in all organisms do not convert light energy into chemical energy. Most of them are primarily processes of a cybernetic character—the photoproducts control chain reactions that affect or determine the response of an organism to the environment (Thomas, 1965).

The central role of photosynthesis in the energy cycle of life is illustrated in Fig. 15-1. Photosynthesis can be defined as the *light-energy-trapping process on which all life depends.* Ever since its importance to all life was first realized, it has attracted the attention of many of the world's top scientists. The story of photosynthetic research is one of the most fascinating in the whole realm of biology. In an elementary text it is impossible to do justice to the romance of the history of research into the problem, to the many controversies that have raged and that still exist, or to the many details of the process that are now known.

Historically, photosynthetic research can be divided into two eras: B.I. (before isotopes) and A.I. (after isotopes). The first era extends to the mid-1930s. The information available in this era was so meager that it could be adequately covered in an elementary text. During the second era, however, and especially in the past two decades, research has been so intense and varied that even active investigators of photosynthesis have difficulty in adequately explaining all of what is known about the subject. Furthermore, interpretations

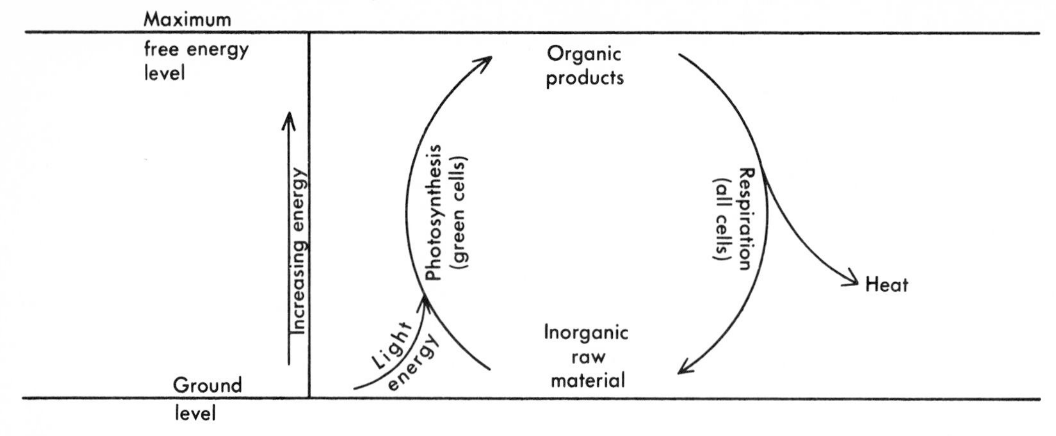

Fig. 15-1. Role of photosynthesis in the energy cycle of life.

vary with the research school. In order to explain their interpretations, investigators of photosynthesis are prone to propose "schemes." There is a dangerous tendency for textbook writers to forget that they *are* just schemes (i.e., hypotheses) and to accept them as scientific laws. It must, therefore, be emphasized at this point that those schemes that seem best supported by the evidence at this writing are being adopted. They must not be accepted as the final word on the subject.

The chemical energy that the plant produces photosynthetically must be stable enough to store for future use. There are two main stages in the process: (1) the energy-trapping stage and (2) the energy-stabilizing stage. These two stages differ both with respect to the nature of their chemical changes and to their location in the cell (Fig. 15-2).

Historically, only the second stage was investigated during the first era. It was common to find a chapter entitled "Carbon Assimilation" but none labelled "Photosynthesis," especially in the European texts. For this historical reason and because of the relation of this second stage to respiration, which is covered in the previous chapter, it will be discussed first.

C-ASSIMILATION

"Dark" C-assimilation

The synthesis stage of photosynthesis is the anabolic counterpart of respiration because it is the synthesis of organic substances from carbon dioxide. It is therefore called photosynthetic C (carbon)-assimilation. This process is possible only in green plants; but all organisms—colorless as well as green plants and even animals—are able to assimilate carbon nonphotosynthetically in the dark. This is called "dark" C-assimilation, since it can occur in the continuous absence of light, whereas photosynthetic C-assimilation can occur in the dark only for about 10 min (Miyachi and Hogetsu, 1970) after a period of light.

In both cases, carbon dioxide is the source of carbon, but the end products are normally different in higher plants. Photosynthetic C-assimilation results in the formation of carbohydrates, dark C-assimilation in organic acids. Whether or not it occurs as a component of photosynthesis, the synthesis of complex substances from carbon dioxide involves endergonic reactions that must be driven by a continuous supply of free energy. In the case of dark C-assimilation, the only energy available is respiratory energy, and this drains the reserves of the organism. Dark C-assimilation is therefore a wasteful process. Photosynthetic C-assimilation, on the other hand,

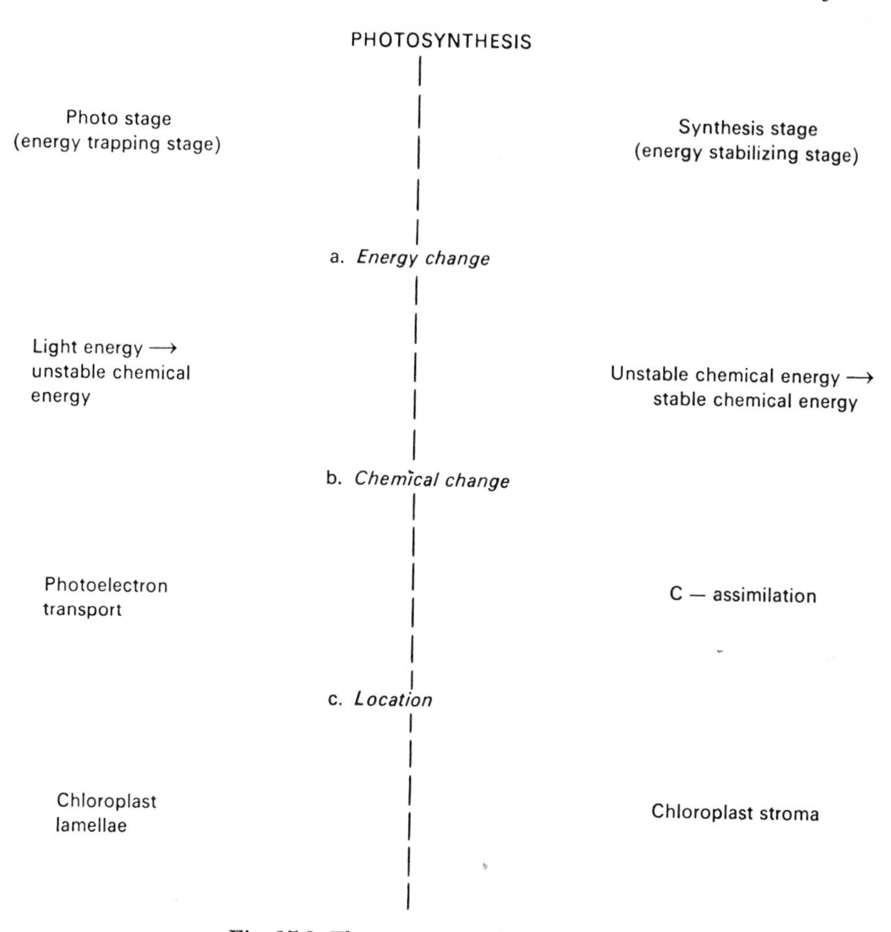

Fig. 15-2. The two stages of photosynthesis.

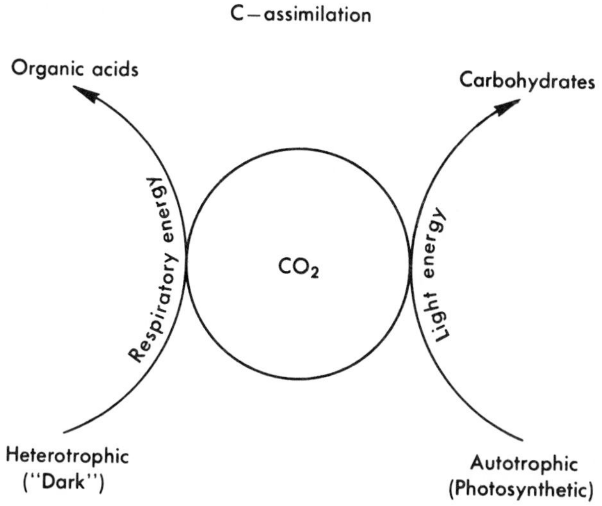

C—assimilation

Organic acids

Carbohydrates

Respiratory energy

CO_2

Light energy

Heterotrophic
("Dark")

Autotrophic
(Photosynthetic)

Fig. 15-3. "Dark" as opposed to photosynthetic C-assimilation.

increases the reserves by utilizing light energy. The two processes are also called heterotrophic and autotrophic C-assimilation, respectively (Fig. 15-3). Dark, or heterotrophic, C-assimilation does, however, play a specific role in the metabolism of the plant in spite of its wastefulness. Succulents, for instance, which grow under conditions of water deficiency, retain their high water content by keeping their stomata closed during the daytime. This reduces transpiration to a minimum. However, it also prevents the uptake from the atmosphere of the carbon dioxide needed for photosynthesis. At night, when there is little danger of water loss, they open their stomata and accumulate carbon dioxide by dark C-assimilation, forming a carboxylic acid (RCOOH). This accounts for the accumulation of organic acids in succulents at night, leading to an unusually low pH (Chapter 3). But a small amount of carbon dioxide remains in equilibrium with the carboxylic acid as follows:

$$RCOOH \rightleftharpoons RH + CO_2$$

In the light the small amount of carbon dioxide is photosynthesized to sugars. This destroys the equilibrium, and carbon dioxide is formed from RCOOH. Photosynthesis can in this way continue in the light without an external supply of carbon dioxide (and therefore with the stomata closed) until all the RCOOH accumulated at night is used up in the synthesis of carbohydrates. Since the photosynthetic gain in energy exceeds the respiratory loss of the dark C-assimilation, the net result is a gain:

This is, of course, only making the best of a poor situation, and these succulents, as a rule, cannot assimilate as much carbon per 24 hr period as can a normally photosynthesizing plant when supplied with adequate water. Pineapple seems to be an exception, since it is as high yielding a plant as any nonsucculent.

The dark C-assimilation of succulents was first studied intensively in members of the family Crassulaceae. It has, therefore, been called Crassulacean acid metabolism, or CAM. It also occurs in succulents that are not members of this family. However, not all succulents are equally effective. Of twenty-one species tested, eleven took up carbon dioxide in the dark for periods varying from a few hours to the whole night and longer (Holdsworth, 1971). Furthermore, CAM is not strictly related to succulence but appears to be characteristic of xeromorphic (dry land) plants (McWilliams, 1970). The substance that fixes (i.e., combines with) carbon dioxide is phosphoenolpyruvate (PEP), and the enzyme controlling the reaction is phosphoenolpyruvate carboxylase (PEP carboxylase). This enzyme has now been identified in the chloroplasts of the cactus, indicating that they are the site of the dark C-assimilation in succulents (Mukerji and Ting, 1968). The central reaction in dark C-assimilation is therefore:

$$\underset{(3C)}{PEP} + CO_2 \xrightarrow[\text{Dark}]{\text{PEP carboxylase}} \underset{(4C)}{\text{Oxaloacetic acid}}$$

Most commonly, the oxaloacetic acid is then converted to malic acid, which is ac-

$$\xleftarrow[\text{(dark)}]{\text{Night (N)}} \qquad \xrightarrow[\text{(light)}]{\text{Day (D)}}$$

$$\text{Vacuole} \xleftarrow[\text{energy}]{\text{Respiratory}} \boxed{RCOOH \rightleftharpoons RH + CO_2} + H_2O \xrightarrow[\text{energy}]{\text{Light}} (CH_2O)$$

Small energy loss | Large energy gain

24 hr sum: small energy loss (N) + large energy gain (D) = moderate energy gain where (CH_2O) = carbohydrate

cumulated. It is becoming apparent that dark C-assimilation may play an important role even in nonsucculents and nonxeromorphic plants. Epiphytes and terrestrial herbs of the rain forest (Coutinho, 1965), for instance, show the typical Crassulacean acid metabolism. *Avena* coleoptiles show greater growth in the dark in the presence of 0.5% to 10% carbon dioxide than in its absence (Harrison, 1965). Excised tomato roots growing in liquid culture show 50% more growth when aerated with air than when aerated with carbon dioxide–free air (Splittstoesser, 1966). Similarly, dark C-assimilation occurs in cotton roots, leading to synthesis of the amino acid asparagine (Ting and Zsroche, 1970), and in potato tubers (Clegg and Whittingham, 1970). It is now believed that dark C-assimilation occurs in all plants (Walker, 1966). Furthermore, it may occur in the light as well as in the dark, although in the light it accounts for only 3% to 12% of the total assimilated. It is becoming apparent that stomatal opening in all plants, whether it occurs in the light or the dark, is due to this dark C-assimilation, resulting in the accumulation of organic acids in the vacuole, in sufficient quantity to increase turgor pressure to the point that the guard cells are pushed apart (Chapter 11).

Photosynthetic C-assimilation

In many respects photosynthetic C-assimilation is the direct opposite of respiration and can be crudely represented by the same equation turned around:

$$6CO_2 + 6H_2O \xrightarrow[\text{Chlorophyll}]{\text{Light}} C_6H_{12}O_6 + 6O_2$$

where $C_6H_{12}O_6$ = glucose

This equation is also frequently written in the simplified form:

$$CO_2 + H_2O \xrightarrow[\text{Chlorophyll}]{\text{Light}} (CH_2O) + O_2$$

Instead of carbohydrate being broken down into carbon dioxide and water (as happens in respiration), the latter are built up to carbohydrate; instead of oxygen being absorbed, it is released; instead of energy being released, it is absorbed in the form of light energy and converted into the chemical energy of the carbohydrate molecule; instead of oxidation of a carbon-containing compound and reduction of oxygen to water, carbon dioxide is reduced to hexose, and water is oxidized to oxygen.

Just as in the case of respiration, the rate of photosynthesis can be theoretically measured by determining any one of the quantities in the preceding equation; but the most easily determined are the carbon dioxide absorbed and oxygen released. The photosynthetic quotient is O_2/CO_2, and, as in the case of the converse respiratory quotient (CO_2/O_2), is usually equal to 1. However, the photosynthetic quotient has not proved to be as valuable a ratio as the respiratory quotient partly because photosynthesis, unlike respiration, normally goes all the way or not at all (as far as gross measurements of oxygen and carbon dioxide are concerned) and partly because carbohydrates are nearly always the end point.

Just as in the case of respiration, the preceding equation for photosynthesis is simply a statement of the relative quantities of raw materials and final products. It reveals nothing about the individual chemical reactions that follow each other, step by step, in an orderly and integrated fashion. During the two decades from the 1930s to the 1950s, intensive investigations largely revealed what these chemical reactions are. This great progress in understanding of the photosynthetic intermediates resulted from the use of isotopes combined with the newly developed chromatographic techniques. The important contributions of these "tagged" elements to the understanding of permeability and translocation have already been mentioned. In the case of photosynthesis, they have been responsible for opening up a whole new field of investigations.

The normal isotope of carbon is ^{12}C, the heavy carbon (the stable isotope) is ^{13}C, and the two radioactive isotopes are ^{11}C and ^{14}C. Only radioactive isotopes have been used in studies of photosynthesis. The first used (in the 1930s) was ^{11}C, the only one available at this early date, since only the cyclotron was producing isotopes. By use of this isotope, Ruben and associates were able to develop the techniques and concepts that served as the bases for the later, more successful investigations using ^{14}C. Their lack of success resulted solely from the extremely short half-life of ^{11}C—22 min. This resulted in a useful life of about 3 hr, during which the whole investigation had to be completed. On the other hand, ^{14}C has a half-life of over 5000 years, and therefore continues to label substances for an unlimited time as far as experiments are concerned. This isotope did not become available until the first atomic pile was built in the mid-1940s. The general procedure is as follows.

The plant is supplied with $^{14}CO_2$ in the light for a few seconds, or minutes, then quickly killed and analyzed. Those substances that contain ^{14}C must have been synthesized by the plant from the $^{14}CO_2$. If the plant is allowed to use the $^{14}CO_2$ for different lengths of time, the substance formed in the shortest period is obviously the first in the chain of reactions. In this way it was proved that a phosphoric ester of a 3-carbon compound, *phosphoglycerate* (PGA), is the first detectable substance formed in photosynthesis, appearing within 2 to 5 sec. The carbon dioxide actually is fixed in the first reaction before being converted to phosphoglycerate. This substance must then be reduced to the triose, and two molecules of triose are finally combined to form hexose (Fig.15-4). But again, as in respiration, many intermediate chemical reactions are involved.

Enough information is now available to construct a hypothetical series of reactions known as the Benson-Calvin (or simply, Calvin), or reductive pentose phosphate, cycle (Fig. 15-5). As the latter name implies, it is to a large degree a reversal of the respiratory oxidative pentose phosphate path. Unfortunately, it has been worked out under nonphysiological conditions, that

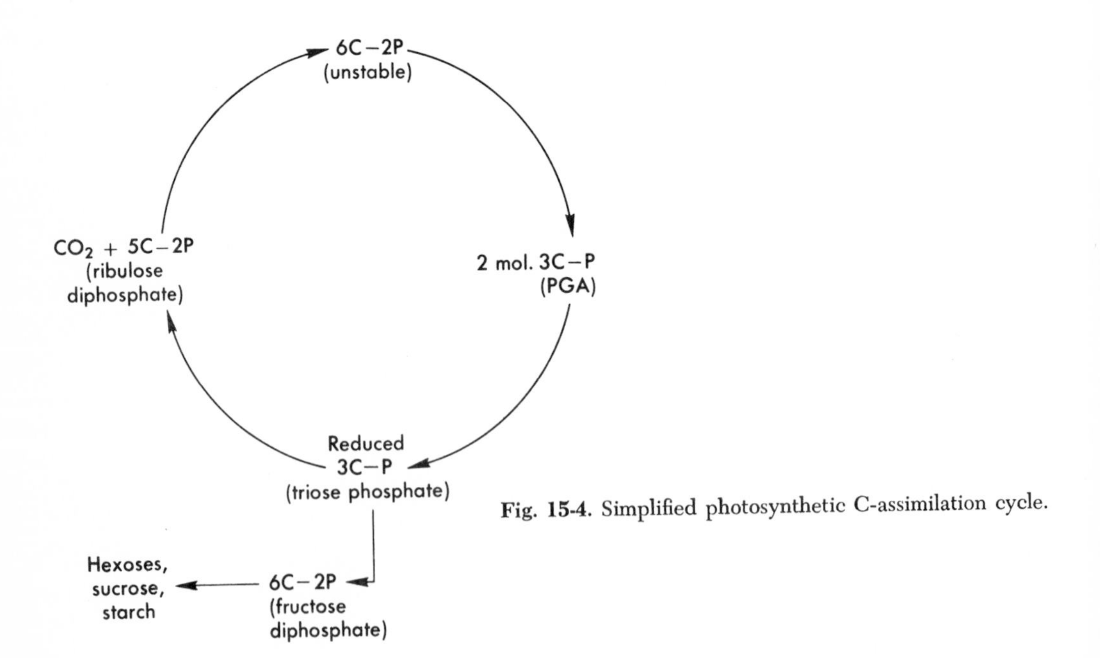

Fig. 15-4. Simplified photosynthetic C-assimilation cycle.

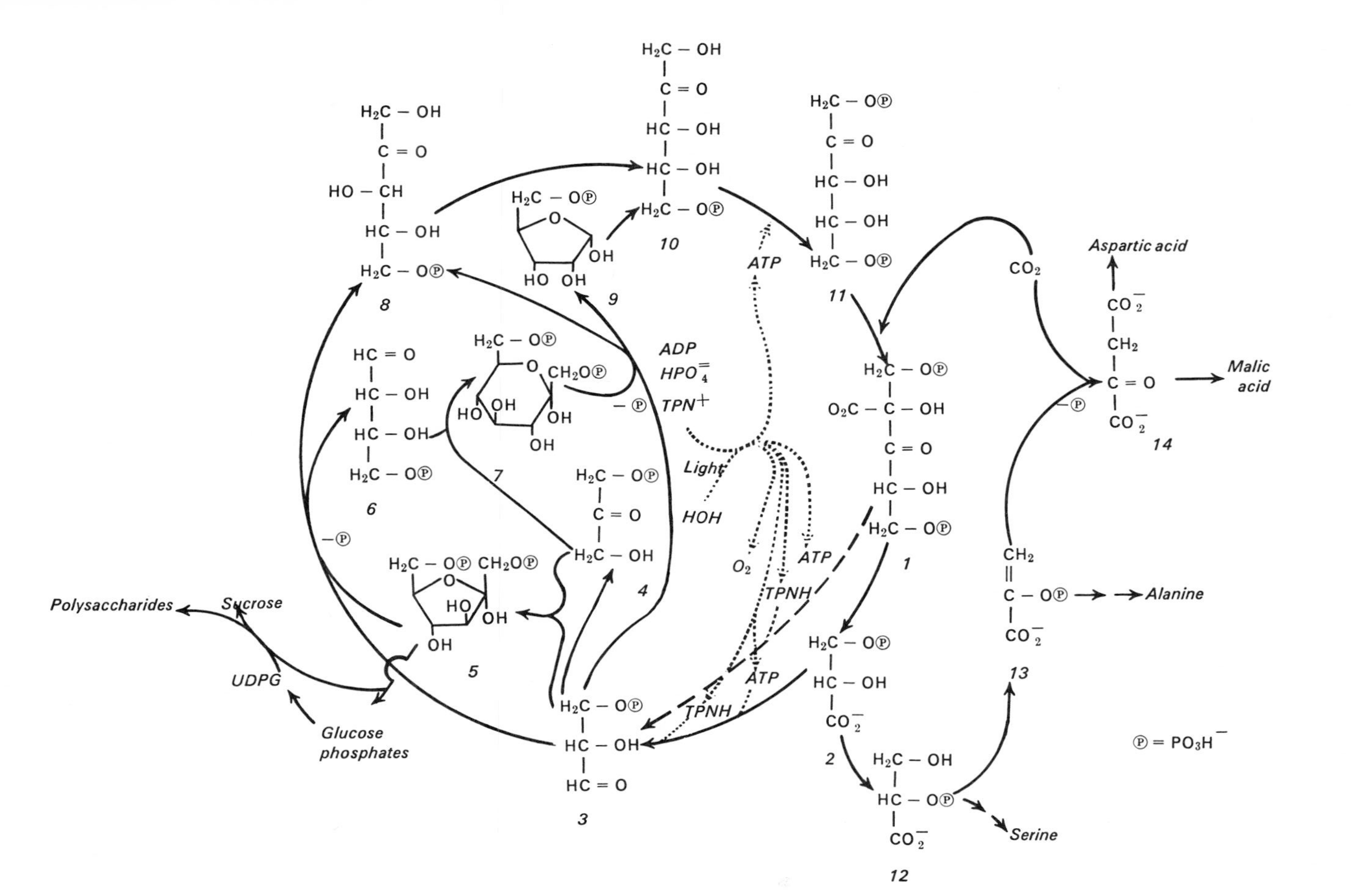

Fig. 15-5. Benson-Calvin, or reductive pentose phosphate, cycle. The order of the reactions is from reaction 1 (right side of circle) to the successively higher numbers. TPN and $TPNH^+$ are used here for NADP and $NADPH^+$, respectively. The RuDP of reaction 1 must be regenerated from five molecules of photosynthetically fixed CO_2, by way of a series of reactions involving tetrose (4-C), pentose (5-C), and even sedoheptulose (7-C) sugar phosphates. (From Calvin, M., and J. A. Bassham. 1962. The photosynthesis of carbon compounds. W. A. Benjamin, Inc., New York.)

is, with much higher carbon dioxide concentrations than are normally available. At physiological carbon dioxide concentrations a glyoxylate cycle has also been found to occur in leaves (Zelitch, 1965). Other work, however, reinforces the theory of the Benson-Calvin cycle and leads to the conclusion that the role of glycolic acid in photosynthesis is of limited importance (Plamondon and Bassham, 1966). The formation of this 2C acid, in fact, leads to a loss in photosynthetic efficiency, since it is the substrate for photorespiration (see later discussion).

C_4 plants and the Hatch and Slack pathway

The Benson-Calvin cycle has been accepted as the universal path of photosynthetic C-assimilation. In sugar cane, however, the first stable compounds are malic and aspartic acid, which are converted to sucrose by way of 3-phosphoglycerate and hexose phosphate (Kortschak et al., 1965). The evidence therefore indicates that, in this plant, the carboxylation reaction gives rise only to 4C-dicarboxylic acids (Hatch and Slack, 1966), as in the case of dark C-fixation. Instead of combining with RuDP, as in the Calvin cycle, carbon dioxide combines with PEP by β-carboxylation, forming oxaloacetate, which is then reduced to malate or aminated to aspartate according to the following equation:

$$\underset{\leftarrow(3C)}{\text{PEP} + CO_2} \xrightarrow[\text{Light}]{\text{PEP carboxylase}} \underset{(4C)}{\text{Oxaloacetate}}$$

On the basis of the above equation, the only difference between this C_4-photosynthetic C-assimilation and so-called "dark" C-assimilation is that the former occurs normally in the light, the latter in the dark. What this really means is that the C_4-photosynthetic C-assimilation involves the same central reaction as the dark C-assimilation, but the former is dependent on photosynthetically trapped light energy and the latter is dependent on respiratory energy. PEP is presumably formed during glycolysis.

The enzyme PEP carboxylase has been isolated from leaves in which the above reaction occurs (Waygood et al., 1969) and is bound to the chloroplast membrane (Baldry et al., 1969). This newly discovered kind of C-assimilation has been called the *Hatch and Slack pathway.* It has been found in several species of *Gramineae* but not in others and in only one species of *Cyperaceae,* among sixteen species from other families (Hatch et al., 1967). It has also been found in four dicotyledonous families: Amaranthaceae, Chenopodiaceae, Portulacaceae, and Euphorbiaceae (Crookston and Moss, 1970), and in at least one species of the Compositae (Schoch, 1971).

The C_4 pathway is apparently related to the environment, occurring only in tropical and arid zones, and only in herbaceous plants. Both high light intensity and high temperature seem necessary for these plants; therefore they occur mainly at lower rather than high altitudes (Hofstra et al., 1972). Because the first substances labeled in plants photosynthesizing $^{14}CO_2$ are 4C organic acids, they are called *C_4 plants* to distinguish them from all other plants that show the label first in the 3C organic acid PGA and are called *C_3 plants.* In at least four genera, one species of the genus may be a C_4 plant, although another species is a C_3 plant (Black and Mollenhauer, 1971). For example, *Atriplex hastata* is a typical C_3 plant with high RuDP carboxylase activity, whereas *A. spongiosa* (from arid climates) is a C_4 plant with high PEP carboxylase activity (Osmond et al., 1969).

It has now become clear that the Hatch and Slack pathway is not a substitute for the reductive pentose phosphate (Calvin) cycle, but is simply a C-fixation process that is tacked onto the Calvin cycle (Fig. 15-6). The C_4 plants therefore have two consecutive carbon dioxide fixation reactions, which must follow each other in a specific order (Berry, 1971). It is now apparent that this ordered arrangement is dependent on a specific anatomical feature of these plants. They invariably contain two chlorophyllous tissues: a bundle sheath of thick-walled cells containing abundant chloroplasts and a layer of thin-walled palisade or spongy mesophyll cells (Fig. 15-7). The two tissues are arranged concentrically around the vascular bundles of the leaf. Although the mechanism linking the two cycles is not definitely established, it is now known that the PEP-carboxylase is primarily (probably solely) in the external mesophyll cells, and the C_4 cycle must therefore take place in these cells. The RuDP carboxylase is primarily in the internal, bundle-sheath cells, and the Calvin cycle must take place in these cells. It appears probable that malate formed in the mesophyll cells is translocated by way of plasmodesmata to the bundle-sheath cells, where it is decarboxylated in the presence of malic enzyme or another enzyme (Edwards et al., 1971); and the released carbon dioxide is immediately carboxylated to 3-PGA and carried through the Calvin cycle (Berry et al., 1970). In some species the transfer is from aspartic acid (Downton, 1971). In spite of the apparent added complexity, the C_4 plants are more efficient photosynthesizers than the C_3 plants.

Just as in the case of respiration, photosynthesis includes not only changes in chemical composition but energy changes as well. The endergonic changes of the photostage are light induced and cannot be

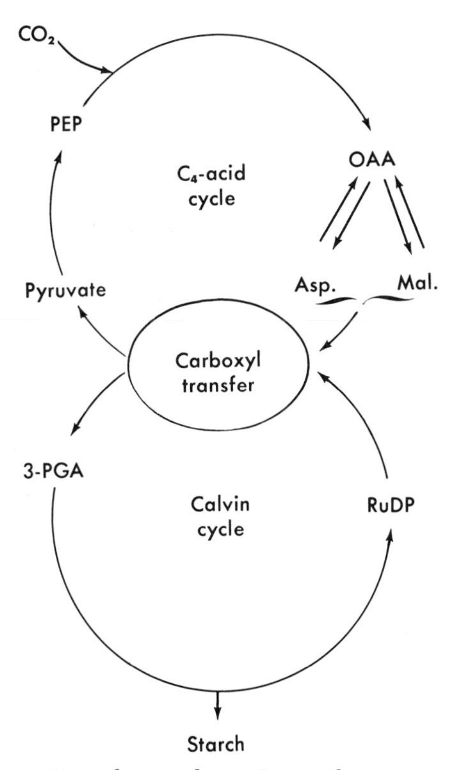

Fig. 15-6. Photosynthetic C-assimilation in a C_4 plant. **OAA,** Oxaloacetate; **Asp.,** aspartate; **Mal.,** malate; **PEP,** phosphoenolpyruvate; **3-PGA,** 3-phosphoglyceric acid; **RuDP,** ribulose-1,5-diphosphate. (From Berry, J. A. 1971. Carnegie Institution Yearbook **69:**649-655.)

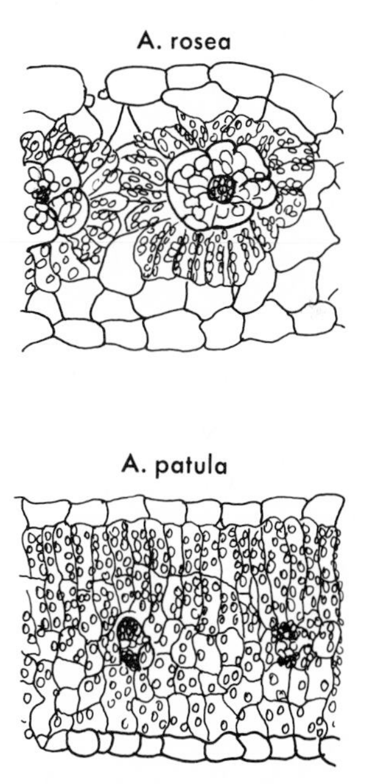

Fig. 15-7. Camera lucida drawings of living leaf sections of *A. rosea* (a C_4 plant) and *A. patula* ssp. *hastata* (a C_3 plant). (From Boynton, J. E., M. A. Nobs, O. Björkman, and R. W. Pearcy. 1971. Carnegie Institution Yearbook **69:**629-632.)

studied by use of ^{14}C, as was done in the case of C-assimilation, which per se does not include a photochemical process and does not require light.

PHOTOELECTRON TRANSPORT

As in the case of any electron transport, the photoelectron transport of photosynthesis results from the presence of a series of substances capable of acting alternately as electron donors and acceptors (p. 175). At one end of the chain of donors, oxygen is liberated.

Oxygen evolution

Although chlorophyll-containing bacteria can photosynthesize anaerobically without releasing oxygen, all the green, higher plants evolve oxygen during photosynthesis. From the general, overall equation given on p. 209, it is not clear whether this oxygen comes from the carbon dioxide, the water, or both. This question was answered by the first experiment on photosynthesis (by Ruben's group) in which isotopes were used. When heavy oxygen (^{18}O) was supplied to the plant as part of the water molecule, the oxygen evolved was found to be enriched in ^{18}O. When it was supplied as part of the carbon dioxide molecule, the oxygen evolved was not enriched in ^{18}O. Although this evidence is not as conclusive as it at first appeared (Brown and Frenkel, 1953), it agrees with other evidence indicating that the source of the oxygen evolved in photosynthesis is the water and not the carbon dioxide. Consequently, the equation as given on p. 209 is not correct and must be modified as follows:

$$6CO_2 + 12H_2O^* + \text{Light energy} \longrightarrow C_6H_{12}O_6 + 6H_2O + 6O^*_2$$

The asterisk indicates the same atoms of oxygen. To balance the equation, twice as many water molecules must be involved as in the first overall equation. Even this equation is not necessarily correct. It simply shows the *minimum* number of water molecules. As far as the preceding evidence is concerned, there may be 18 or 24 or some other multiple of 6 on the left side of the equation and 12 or 18, etc., respectively, on the right side of the equation (Rabinowitch, 1945).

The oxygen-evolving reaction is therefore as follows:

$$4H_2O \xrightarrow[\text{Chlorophyll}]{\text{Light}} 4H + O_2 + 2H_2O$$

Since this implies the splitting of water by light into hydrogen and hydroxyl radicals, it was originally called the "photolysis" of water. However, it must not be assumed that this reaction is truly photochemical, since the light is not absorbed by the water. More recently, therefore, the process has been thought of as a photo-oxidation of hydroxyl ions accompanied by a photoreduction of H^+, that is, a photoelectron transport. It has been argued, however, that there is no unequivocal direct evidence for the participation of water in the primary photochemistry of photosynthesis (Clayton, 1965).

Unlike photosynthetic C-assimilation, the oxygen evolution can occur only in the light. It can, in fact, occur in the absence of any appreciable C-assimilation, in the so-called *Hill reaction.* This is the name given to the *photosynthetic release of molecular oxygen by free, illuminated chloroplasts* accompanied by reduction of an oxidizing substance (such as ferric salt and quinones). The free chloroplasts are obtained by grinding or "homogenizing" green leaves in buffer solutions and centrifuging the homogenate successively at two speeds, the first at a very slow speed to remove sediment, the second at a slightly more rapid speed to precipitate the chloroplasts. The latter are then resuspended in solutions of various kinds and exposed to light under conditions permitting measurement of oxygen evolution.

In this way it has been found possible to obtain nearly as rapid evolution of oxygen from isolated chloroplasts as from the normal leaf, although no carbon assimilation occurs. But it must be emphasized that the free chloroplast obtained in this manner is not a complete system in itself even for oxygen evolution. It must be supplied with an oxidizing agent which it can reduce in the light. That the evolution of oxygen is controlled by enzymes has been shown by the use of inhibitors such as urethane, which also inhibit photosynthesis in the normal leaf. By supplying the necessary enzymes and substrates not present in the free chloroplasts, it is possible to obtain carbon dioxide assimilation with free chloroplasts in the light (Vishniac and Ochoa, 1952). By the use of modern, improved methods of isolating chloroplasts in the unaltered state, Bidwell and coworkers (1969) were able to obtain photosynthetic C-assimilation rates by the isolated chloroplasts of the alga *Acetabularia mediterranea* essentially equal to those of the whole cells.

The Hill reaction can be represented as follows:

$$A + H_2O \xrightarrow[\text{Chloroplasts}]{\text{Light}} AH_2 + \tfrac{1}{2}O_2$$

where A = oxidizing substance

The Hill reaction, of course, also proves that the oxygen is evolved from water and not from the carbon dioxide, since little or no carbon dioxide is absorbed in the process.

$NADP^+$ reduction

The Hill reaction results in the oxidation of water to molecular oxygen. Although many substances can serve artificially as the electron acceptor or oxidant (becoming reduced in the process), the natural substance is $NADP^+$:

$$NADP^+ + H_2O \xrightarrow[\text{Chloroplasts}]{\text{Light}} NADPH + \tfrac{1}{2}O_2 + H^+$$

where $NADP^+$ = nicotinamide–adenine dinucleotide phosphate

NADPH = reduced nicotinamide–adenine dinucleotide phosphate

One result of the photostage of photosynthesis is therefore the formation of reduced NADP, which can then bring about the reduction of carbon dioxide to carbohydrate. But several other substances are also involved as intermediates in the electron transport from water to NADP (e.g., quinones, cytochromes, ferredoxin).

Photophosphorylation—the energy trap

It is now clear that although the C-assimilation reactions of the Calvin cycle can all occur in the dark, they must all be preceded in the plant by photooxidation-reduction (or photoelectron transport) reactions, known as "light reactions." These culminate in the synthesis of high energy phosphate, which can then be used in the dark to drive the reactions of the Calvin cycle. Since this production of high energy phosphate differs from its respiratory formation by a dependence on light energy, it is called *photo*phosphorylation and may be represented as follows:

$$ADP + P_i \xrightarrow[\text{Chloroplasts}]{\text{Light}} ATP$$

where ADP = adenosine diphosphate
ATP = adenosine triphosphate
P_i = inorganic phosphate

Two kinds of photophosphorylation have been identified in experiments with free chloroplasts. Cyclic photophosphorylation occurs independently of the photooxidation of water and is therefore not accompanied by oxygen evolution. Noncyclic photophosphorylation is linked to both the formation of reduced pyridine nucleotide and the photooxidation of water as follows:

$$NADP^+ + ADP + P_i + H_2O \xrightarrow[\text{Chlorophyll}]{\text{Light}} NADPH + ATP + \tfrac{1}{2}O_2 + H^+$$

There has been some controversy as to whether or not the cyclic process occurs in the normal photosynthetic process (van

Niel, 1962). Simonis (1967), however, has produced evidence of the occurrence of both cyclic and noncyclic photophosphorylation in intact cells just as in isolated chloroplasts. According to Ramirez and co-workers (1968), cyclic photophosphorylation may serve as the source of ATP for biosynthetic processes occurring in chloroplasts that are not on the main photosynthetic path of carbohydrate synthesis but that branch off from this path, for example, protein, DNA, and RNA synthesis. In agreement with this interpretation, the cyclic process can be blocked without affecting photosynthesis (Tanner et al., 1969; Raven, 1970). Raven has suggested that some of the ATP produced by cyclic photophosphorylation can be used to support active K^+ influx.

The hydrogen from the reduced pyridine nucleotide (NADPH) may then initiate the reduction of carbon dioxide, and the high energy phosphate (ATP) may supply the energy to drive the endergonic reactions of the Calvin cycle (see later discussion). The two substances formed (NADPH and ATP) cannot be appreciably stored and are present only in catalytic amounts (Arnon, 1961).

PHOTOCHEMICAL REACTIONS

The visible radiations known as "light" occur from the violet at about 380 nm (nanometers, formerly called millimicrons [mμ]) to the far red at about 770 nm. Photosynthesis is controlled by light because it can occur only when exposed to radiation within this range of wavelengths, except in photosynthetic bacteria, which can photosynthesize in the short infrared up to about 900 nm. Infrared radiation has no effect on the photosynthesis of all other plants, other than the indirect temperature effect. Ultraviolet irradiation inhibits photosynthesis because of the destruction of essential substances such as plastoquinone (Mantai and Bishop, 1967). The true "light reactions" are, of course, the photochemical reactions. These are the direct result of the absorption of a photon of light by a molecule. The added energy from this photon lifts an electron from the "ground" (i.e., low energy) state to the "excited" (i.e., high energy) state, leaving behind a positive "hole." While in this excited state, the molecule can undergo a photochemical reaction. The molecules in the plant directly involved in the photochemical reactions are the photosynthetic pigments that absorb the light. They then transfer the light energy, via electron transfers, and set in motion all the above-mentioned reactions—pyridine nucleotide reduction, photophosphorylation, photooxidation of water, and C-assimilation. The light absorption leads to activation of chlorophyll in an extremely rapid process, probably occurring in about 10^{-15} sec (Kamen, 1963):

$$\text{Chl} + h\nu \longrightarrow \text{Chl}^*$$

where Chl = a chlorophyll molecule
hv = a photon of light
Chl* = an excited chlorophyll molecule

The excitation of the chlorophyll molecule may result in an electron transfer, ultimately to NADP:

$$2\text{Chl}^* + \text{NADP}^+ + \text{H}^+ \longrightarrow 2\ \text{Chl}^+ + \text{NADPH}$$

The chlorophyll molecule may then be returned to the unexcited state by electron transport, ultimately from water (or its OH^-):

$$\text{Chl}^+ + \text{OH}^- \longrightarrow \text{Chl} + \tfrac{1}{2}\text{H}_2\text{O} + \tfrac{1}{4}\text{O}_2$$

In this way the chlorophyll molecule would be regenerated in the low energy or nonexcited state after absorbing a quantum of light energy and using it to reduce a molecule of $NADP^+$ and to oxidize a molecule of water (Fig. 15-8). *The primary photochemical act is therefore the separation of oxidizing and reducing entities by light within the chlorophyll molecule.*

It has long been suspected that there is more than one photochemical reaction in photosynthesis. Direct evidence for this has now been produced by the "Emerson enhancement (or second Emerson) effect," discovered by Emerson and his co-workers in 1957 (see Rabinovitch and Govinjee, 1969). This name refers to the greater photosynthetic yield in the simultaneous (or quickly alternating) presence of light of two wavelengths than the summation of the separate yields in each wavelength. The exact way in which these two photochemical reactions fit into the scheme is still being investigated. The most widely accepted current hypothesis explains the two primary photoacts as depending on two photosystems (Fig. 15-9). Photosystem I produces a strong reductant (leading to NADP reduction) and a weak oxidant. Photosystem II produces a weak reductant and a strong oxidant (leading to oxygen evolution). The two operate in series, the weak oxidant from photosystem I oxidizing the weak reductant from photosystem II by way of an electron transport system involving intermediate carriers (Schwartz, 1967). Photosystem I is most efficiently sensitized by red light of longer wavelengths (above 690 nm but not above 720

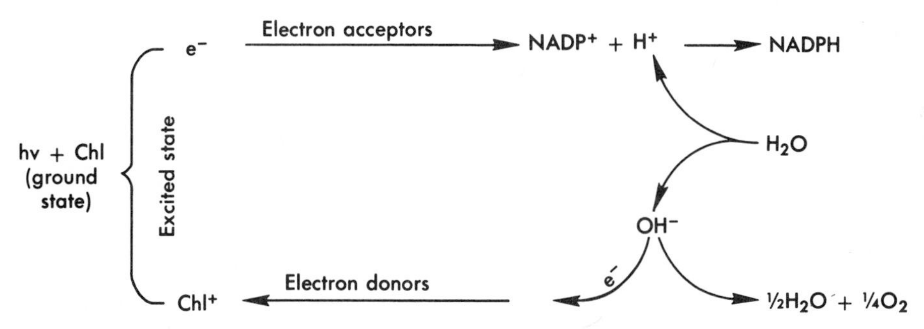

Fig. 15-8. Primary photochemical act. Arrows represent the direction of electron transport. (Redrawn from Clayton, R. K. 1965. Molecular physics in photosynthesis. Blaisdell Publishing Co., Division of Ginn & Co., Waltham, Mass.)

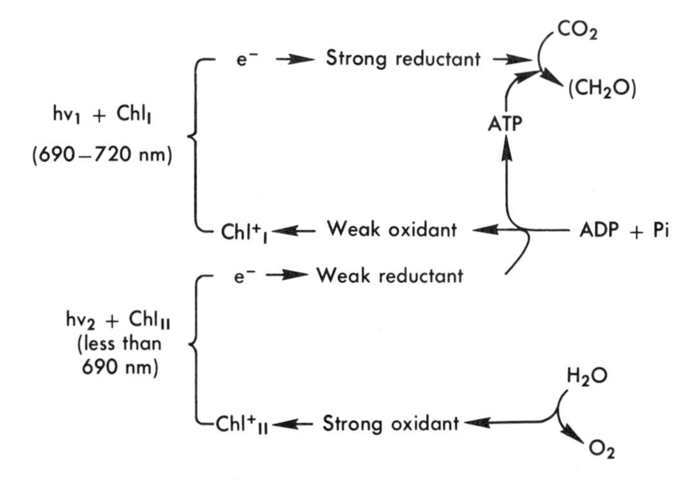

Fig. 15-9. The two photochemical reactions of photosynthesis. Arrows represent the direction of electron transport. (Redrawn from Clayton, R. K. 1965. Molecular physics in photosynthesis. Blaisdell Publishing Co., Division of Ginn & Co., Waltham, Mass.)

nm), photosystem II by red light of shorter wavelengths (below 690 nm). The two photosystems are presumably different chlorophyll-protein complexes. Photosystem I is supposed to be the more primitive one, since it occurs in photosynthetic bacteria, whereas photosystem II does not (Olson, 1967). Two chlorophyll-protein complexes have, in fact, been isolated from the chloroplasts of several plants, and they are believed to represent photosystems I and II (Thornber et al., 1967). The net effect of this photoelectron transport system is to raise the oxidation-reduction potential from a maximum oxidized level in water (+ 800 mv) to a maximum reduced level in NADPH (actually in ferredoxin, − 400 mv). This involves an increase in energy because of the absorption of the two photons of light energy by the two chlorophyll systems (Fig. 15-10). Although accepted by most investigators, this "Z-scheme" is inadequate according to Arnold and Azzi (1968), and they have replaced it by another. Arnon's group has suggested three light reactions instead of two, the extra one associated with photosystem II (Knaff and Arnon, 1969).

The further oxidation-reduction reactions leading to the photooxidation of water, photophosphorylation, and NADP reduction are complete within about 10^{-4} sec from the instant of light absorption. It is only after these initial light-controlled reactions that the dark chemical reactions of C-assimilation (the Calvin cycle) can occur.

A simplified version of the relations between the two stages of photosynthesis is shown in Fig. 15-11. If this representation of the photosynthetic process is correct, 3 molecules ATP and 2 molecules NADPH are required for every 1 molecule carbon dioxide assimilated.

The photooxidation of water, the reduction of pyridine nucleotide, and photophosphorylation are represented as light reactions as opposed to the reactions of C-assimilation which can all occur in the dark immediately after exposure to light. However, even these so-called light reactions have dark components (e.g., in the case of photophosphorylation, Hind and Jagendorf, 1963). Thus they are light reactions only in the sense that they are more closely dependent on light in time and therefore have not been found to occur after the light exposure. For this reason

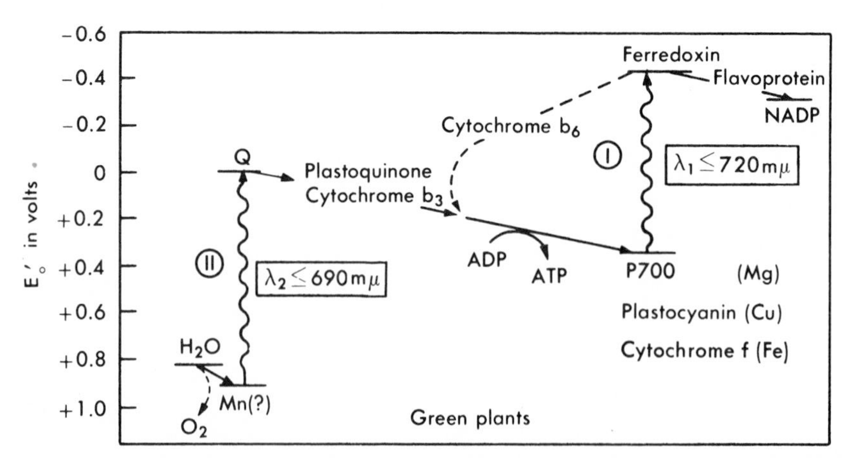

Fig. 15-10. Photosynthetic electron transfer ("Z-scheme") resulting in a rise in reduction potential from that of water to that of NADPH. P700 is a portion of chlorophyll *a* peculiar to photosystem I. (From Olson, J. M. 1967. Brookhaven Lect. Ser., Pub. No. 62, Brookhaven National Laboratory, Upton, N. Y.)

they take place in the quantasomes, together with the primary light reactions, whereas the C-assimilation occurs in the stroma of the chloroplasts (Smith and French, 1963) and is therefore separated from the photochemical reactions in space as well as in time. It has been suggested that quantasomes are the "photosynthetic unit," a name given by Emerson and Arnold for the packet of chlorophyll molecules, all of which are capable of transferring the light energy they trap to a single reacting carbon dioxide molecule. According to Arnold and Azzi (1968), the photosynthetic unit is a solar battery with three terminals: a negative terminal at the level of ferredoxin and two positive terminals at the levels of cytochrome f and $H_2O—O_2$.

This photosynthetic unit is supposed to contain about 2500 chlorophyll molecules per evolved molecule of oxygen (1560 to 2350 according to Myers and Graham, 1971), or 250 to 300 molecules for a one-quantum reaction center (Wild and Egle, 1967). It consists of four small units of pigment system I and one large unit of

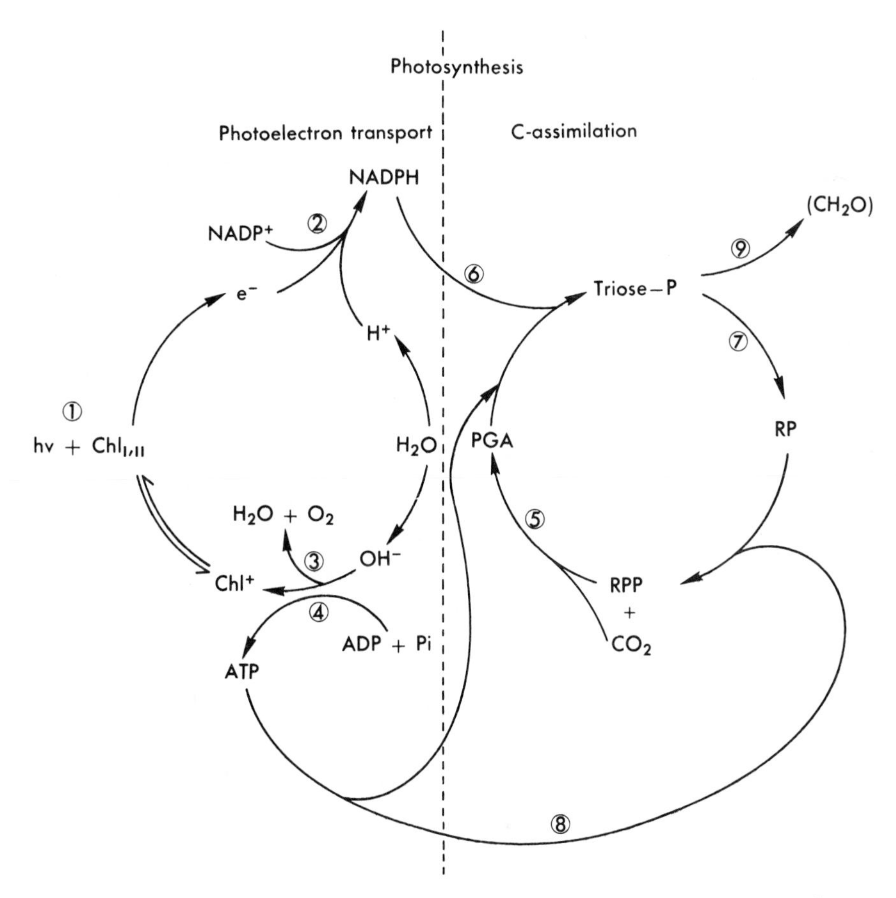

Fig. 15-11. Interrelations between photostage (photoelectron transport) and synthesis stage (C-assimilation) of photosynthesis. The numbers correspond to the following reactions: **1,** primary photochemical reactions; **2,** NADP reduction; **3,** photolysis of water (or photo-oxidation of hydroxyl ions); **4,** photophosphorylation; **5,** carboxylation; **6,** reduction (of PGA) with dephosphorylation of ATP; **7,** regeneration (of ribulose phosphate); **8,** phosphorylation (of ribulose phosphate to ribulose diphosphate); **9,** condensation of triose phosphate to hexose phosphate and release of carbohydrate.

pigment system II (Gibbs, 1967). It may not, however, be constant in size for all plants. In the normal *Chlorella* cells, 460 chlorophyll molecules are available for a one-electron transport system, consisting of photosystems I and II, or 230 molecules per quantum absorbed. In a *Chlorella* mutant, however, there were only 80 to 100 molecules, or 40 to 50 per quantum absorbed (Wild et al., 1971).

A heavy particle has been isolated that is active in oxygen evolution and enriched in chlorophyll *b* and manganese and is believed to be pigment system II. A lighter particle has been separated that is active in NADP reduction, is enriched in chlorophyll *a*, iron, and copper, and is believed to be pigment system I.

The exact function of the quantasome is, however, still in question. Its size does not agree with the suggested size of the photosynthetic unit. It is 100 Å square with a molecular weight of 2×10^6 and is a complex of about 50% protein and 17% chloroplast pigments, having only 230 chlorophyll *a* molecules (Vernon and Seeley, 1966). It is membrane-bound and possesses adenosine triphosphatase activity. It has therefore also been called a photophosphorylase (Howell and Moudrianakis, 1967). It is, however, composed of smaller units that are thought to be enzyme complexes (Mühlethaler, 1966). The membrane of the chloroplast is layered on both sides with quantasomes and is therefore thought to be the actual site of photosynthesis.

According to another concept (Junge et al., 1968) there are three photosynthetic functional units:

1. The functional unit for the photoelectron transport system. This contains 10^3 chlorophyll molecules. The light absorbed by this unit is conducted to the two photochemically active chlorophyll molecules that generate the electron flow.
2. The functional unit for the proton transporting system. This includes the photoelectron transport system that generates the proton transport but is 100 times larger (10^5 chlorophyll molecules). It corresponds to one *thylakoid*, that is, one disc-shaped, closed vesicle surrounded by a membrane.
3. The functional unit of photophosphorylation. This also consists of one thylakoid, because the proton transport promotes ATP formation (according to the chemiosmotic theory, (Chapter 14).

One thylakoid therefore contains many photoelectron transport units, presumably in the plane of its membrane.

PIGMENTS

There are two families of pigments involved in photosynthesis (Kouchkovsky, 1968). Members of the first family (chlorophylls and carotenoids) are able to absorb light and to sensitize photochemical reactions. Members of the second family (cytochrome, ferredoxin, plastocyanin, flavoproteins, plastoquinones) are electron carriers of the redox chain. Although colored, their role does not depend on their ability to absorb light. Therefore the terms *photosynthetic* or *chloroplast pigments* normally refer only to the first family.

There are two main kinds of chlorophylls in higher plants (although each may occur in more than one form)—one bluish green (chlorophyll *a*), the other yellowish green (chlorophyll *b*). The empirical formulas are $C_{55}H_{72}O_5N_4Mg$ for chlorophyll *a*, the same with a CHO group in place of a CH_3 group for chlorophyll *b*. The chlorophyll molecule cannot be synthesized by the cell in the absence of iron. The exact structure of chlorophyll is now known (Fig. 15-12) and has been confirmed by its total synthesis by Woodward (Smith and French, 1963). The portion of the molecule that contains magnesium is chemically related to the heme of the blood. Both chlorophylls occur only in the chloroplasts (in higher plants), usually in the ratio of 3*a*:1*b*. With

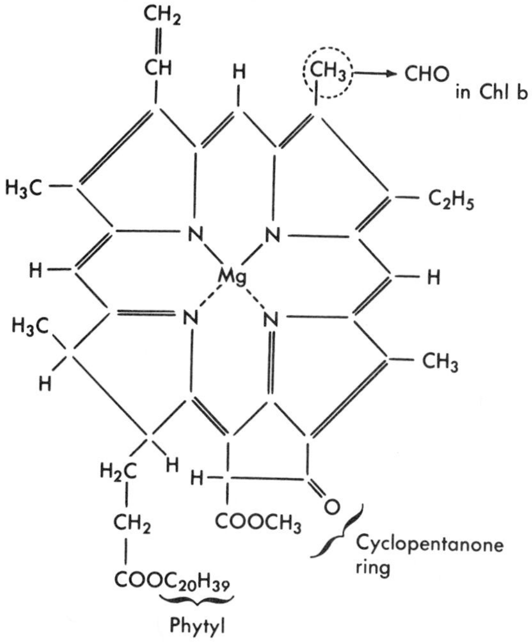

Fig. 15-12. Chlorophyll molecule. (Redrawn from Clayton, R. K. 1965. Molecular physics in photosynthesis. Blaisdell Publishing Co., Division of Ginn & Co., Waltham, Mass.)

them are two yellow carotenoid pigments, carotene and xanthophyll, in about the same molar quantity as chlorophyll *b*. There may, however, be as many as nine carotenoid pigments in the chloroplasts (Strain et al., 1968). Carotene is a polyene hydrocarbon ($C_{40}H_{52}$) and xanthophyll is an oxygen derivative of it ($C_{40}H_{52}O_2$). The carotenoids are just as efficient harvesters of light as are the chlorophylls, and they apparently play two roles in photosynthesis (Clayton, 1965).

1. The light absorbed by β-carotene (but not by xanthophyll) is transferred to chlorophyll with nearly 100% efficiency (Goedheer, 1969).
2. Carotenoid-deficient mutants suffer severely from bright illumination, due to bleaching of chlorophyll, which is presumably prevented in the normal plant by the carotenoids (Metzner, 1969). On the other hand, the low photosynthetic efficiency of Scotch pine needles in blue light has been ascribed to the screening absorption by photosynthetically inactive carotenoids (Linder, 1971).

But chlorophyll is irreplaceable as the initiator of the chemistry of photosynthesis. It is apparently chlorophyll *a* specifically that initiates these reactions, since a barley mutant lacking chlorophyll *b* showed no decrease in photosynthetic efficiency (Highkin and Frenkel, 1962). Similarly, the lack of chlorophyll *b* in mutants of *Chlorella pyrenoidosa* had no noticeable effect on their photosynthesis (Wild and Egle, 1967). Nevertheless, when present, chlorophyll *b* did participate in the primary reaction of photosynthesis.

It has long been known that chlorophyll can be extracted from leaves as a chlorophyll-protein complex, and purified complexes have been obtained in water-soluble form (Murata et al., 1971). The chlorophyll apparently must be combined with protein as a chlorophyll-protein complex to function photosynthetically (Gregory et al., 1971). It has been possible to separate several of these chlorophyll-protein complexes, some of which are the major components of photosystem I, others of photosystem II (Renny, 1971).

That chlorophyll is the energy-absorbing substance can be shown in several ways. The *absorption spectrum* (the curve showing the percent absorption of light at each wavelength) of chlorophyll (Fig. 15-13) is similar to the *action spectrum* of photosynthesis (the curve showing the relative rates of the reaction when supplied with the same amount of radiant energy at each wavelength). In other words chlorophyll absorbs most intensely in the red and blue at the wavelengths that permit the most rapid photosynthesis; it absorbs least in the green, and this light (when at the same intensity as the other wavelengths) permits only the slowest rate of photosynthesis. It is interesting that all wavelengths of light

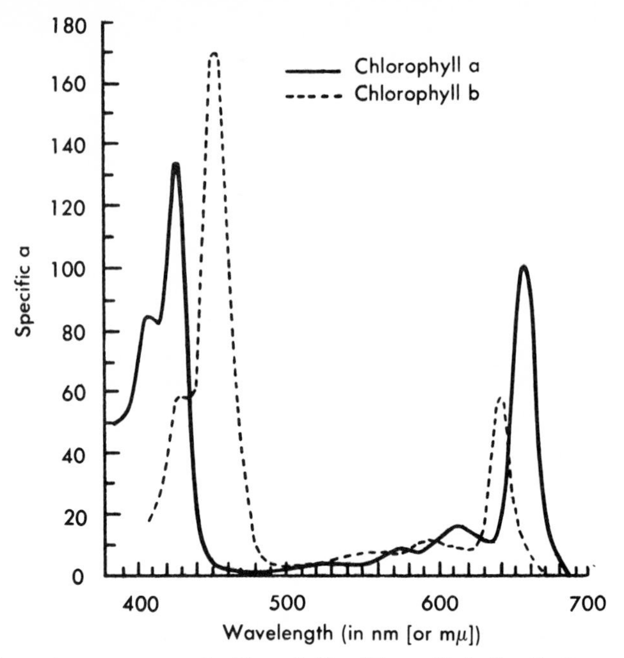

Fig. 15-13. Absorption spectrum of chlorophylls. (From Zscheile, F. P., and C. L. Comar. 1941. Bot. Gaz. **102**:463-481.)

do permit photosynthesis, although in view of the preceding absorption characteristics of chlorophyll, red is the most efficient. Chlorophyll possesses the property of *fluorescence*—it is able to absorb light of one wavelength and to emit light of a longer wavelength. Whatever wavelength it absorbs, it always fluoresces in the red (with a maximum at 684 nm). For this reason chlorophyll appears green by transmitted light, red by reflected light. This is another reason for the high efficiency of red light. Only that fraction of the quantum of absorbed light which can be reemitted as fluorescent light is available for photosynthesis. In the case of a quantum of red light of 684 nm, all the energy can be reemitted and is therefore available for photosynthesis. In the case of a quantum of blue light, only about two thirds can be reemitted; the remaining one third is given off as heat and is therefore wasted.

ENZYMES

As in the case of respiration, the dozens of reactions that comprise the process of photosynthesis are all controlled by as many enzymes, each one specific for each reaction. Many of the reactions are the same as those involved in respiration, although the direction of the reactions is reversed. This is particularly true of the Calvin cycle, which consists largely of a reversal of the pentose phosphate pathway, the alternate path for glycolysis. Some reactions, however, are unique to the Calvin cycle, for example, the first reaction in which carbon dioxide is combined with ribulose diphosphate to yield two molecules of phosphoglycerate:

$$\underset{\text{(Ribulose diphosphate)}}{\text{RuDP} + CO_2} \xrightarrow{\text{RuDP carboxylase}} \underset{\text{(3-phosphoglycerate)}}{\text{2 3-PGA}}$$

The enzyme controlling this reaction has the trivial name of carboxydismutase, or ribulose diphosphate carboxylase. It has a molecular weight of 550,000 and is indistinguishable from the "protochlorophyll protein holochrome complex," or the "Fraction I protein," which has been isolated from leaves (Gibbs, 1967). In the pure form, this enzyme has a low affinity for carbon dioxide. Therefore to photosynthesize rapidly, the plant must produce other proteins that overcome this low affinity (Latzko and Gibbs, 1968); or it produces another enzyme system with a much higher affinity for carbon dioxide (PEP carboxylase, see previous discussion). There is also a series of enzymes and cofactors involved in the electron transport from chlorophyll to NADP and from water to chlorophyll. These include Mn^{++}, several plastoquinones, at least two cytochromes (B3 or B559 and f), a copper protein (plastocyanin), the pigment P700, an iron protein (ferredoxin), NADP reductase, and NADP (Bishop, 1966). Their probable interrelations are shown in Fig. 15-10.

All the photosynthetic enzymes are present in the chloroplasts (carboxydismutase in large quantities), since this is where the whole process of photosynthesis occurs. Therefore the chloroplast (like the mitochondrion) is a powerhouse of enzymes. Because some of these enzymes are specific for photosynthetic reactions, there are also substances that are specific inhibitors of photosynthesis. This has proved to be of great practical importance, since some of these substances can be used as herbicides (weed killers) that act specifically on autotrophic photosynthesizing organisms and have no effect on heterotrophic (nonphotosynthesizing) organisms. Two examples of these are the phenylureas (CMU and DCMU) and the aminotriazines (simazine and atrazine). These herbicides have also proved to be of great value in investigating the individual photosynthetic processes that they inhibit (Krogmann and Powers, 1965). They specifically inhibit oxygen evolution (i.e., the Hill reaction). Several other herbicides also appear to inhibit photosynthetic reactions closely connected to the evolution of oxygen (Nishimura, 1967). The bipyridilium herbicides probably compete for electron flow from the primary electron acceptor of photosystem I (Dodge, 1971). Others (e.g., phlorizin) uncouple photophosphorylation (Izawa et al., 1967).

The division of labor between the two powerhouses (the chloroplast and the mitochondrion) is sharp. The mitochondrion is unable to photosynthesize, and the chloroplast is unable to support dark respiration. Photorespiration does occur in the chloroplast (see following discussion), but its role is completely unrelated to that of dark respiration. The chloroplast does not possess (1) a cytochrome c oxidase system, (2) a citric acid cycle, (3) pathways of hexose breakdown, or (4) as great an ability to reduce acetyl coenzyme A to fatty acids (Gibbs, 1967).

PHOTOSYNTHETIC RATES

Efficiency of photosynthesis

Modern man's population explosion has led to continuous attempts to increase the food supply. This, of course, cannot exceed the amount of photosynthate produced by green plants. The relation is demonstrated in corn, for hybrid vigor or heterosis leading to higher yields parallels a higher photosynthetic potential in the hybrids than in the parent lines (Muresan et al., 1967). It is therefore of practical importance to know what are the limits to the production of photosynthate because any attempt to exceed these limits is useless. In spite of the high efficiency of modern agriculture, only about 0.2% of the annual sunshine is stored by photosynthesizing plants (Daniels, 1967), 0.4% in a natural ecosystem (Egunjobi, 1971). In sunny climates, solar radiation averages about 500 cal/cm^2 per day. An idealized photochemical reaction effective at a wavelength of

500 nm, which absorbed and converted into chemical energy one tenth of the sunlight, could therefore produce 3.6 tons per day per acre (Daniels, 1967). This is as much as most crops produce in a year.

The efficiency of photosynthesis is therefore low under normal conditions. In fact, it is commonly stated that the animal is more efficient than the plant, since only 1% to 2% of the sun's energy (during the growing season) is used for photosynthesis. But this is equivalent to saying that if the animal is fed wood, its efficiency would be less than 1% because just as the chemical substances in wood are mostly indigestible, so also most of the radiant energy from the sun is in the nonvisible wavelengths and therefore not usable by the plant. Even the usable wavelengths are not normally supplied in quantities that can be used efficiently. Therefore only about 2% of the incident light energy is ordinarily used in photosynthesis, or 3% of the absorbed light energy. However, there are plants that under normal conditions have efficiencies considerably higher than this (Table 15-1). Bulrush millet may accumulate 54 g/m^2 per day, storing as dry matter 4.2% of the total radiation or 9.5% of the visible radiation (Begg, 1965). In weak light and with an ample supply of carbon dioxide the efficiency may rise to 30% of the absorbed radiation. Many other examples of high efficiency have been found (Wassink, 1964). Under optimal conditions in small cultures of algae, an energy yield of 15% to 20% can be reached. Even in the case of seedlings of higher plants, under light limitations, yields of 10% to 15% have been obtained.

External and internal factors

The rate of photosynthesis is controlled by many factors, both internal and external. Therefore published values always include the levels of the three important external factors—temperature, light, and carbon dioxide concentration (Table 15-2). A change in any one of these external factors may markedly change the rate of photosynthesis.

As in the case of respiration, the effect of temperature is pronounced. When the temperature is varied, the rate drops to zero at the so-called *minimum* temperature, rises to a maximum at the *optimum* temperature, and drops again to zero at the *maximum* temperature (Fig. 15-14). These three are the *cardinal* temperatures. When temperature is the limiting factor, a rise of 10° C may increase the rate of photosynthesis 2 or 3 times or sometimes even more. This is typical of ordinary chemical processes. Since physical processes are less affected by temperature, photochemical reactions show little change; for example, a 10° C rise in temperature usually increases the rate of a photochemical reaction by only about one tenth. This fact was long taken as evidence that the dark chemical reactions of photosynthesis are slower than the photochemical reactions; this is the reason why the former are limiting. Indeed, the significant tem-

Table 15-1. Percentages of absorbed solar radiation retained in photosynthate

Species	Percent absorbed	Percent of incident radiant energy retained as photosynthate	Percent of absorbed radiant energy retained	Reference
Potato	52	1.65		Kaminski, 1969
Rye		2.34		Kaminski, 1970
Wheat		6 (during growth phase) 1 (average during whole period)		Baldry and Jonard, 1969
Rice		0.8-3.7	4.1-6.9	Murata et al., 1968
Bulrush millet		4.2	9.5	Begg, 1965

Table 15-2. Rate of photosynthesis in some plants at temperatures of 18° to 20° C at maximum or near-maximum light intensities and with normal carbon dioxide content of the air (0.03% or 0.56 mg per liter)*

Plant	Milligrams CO_2 per 50 cm^2 leaf surface per hr	Carbohydrate ($C_6H_{12}O_6$) synthesized in g/m^2 leaf surface per hr
Potato	9.57	1.30
Tomato	8.42	1.15
Sugar beet	9.26	1.26
Spinach	9.78	1.33
Vicia faba	8.83	1.20
Phaseolus vulgaris	9.27	1.26

*Adapted from Lundegårdh, H. 1925. Gustav Fischer Verlag, Stuttgart.

perature effect was the earliest evidence that not all of the photosynthetic process requires light. In memory of the man who first observed this, the term *Blackman reaction* was used for the dark chemical reactions.

Another early proof of the existence of dark chemical reactions was the fact that intermittent light can permit just as rapid a rate of photosynthesis as continuous light if the light and dark periods are short enough, even though the dark periods between flashes reduce the total light received to about half as much per hour. However, if the length of the alternating light and dark periods is increased to about 1 min each, almost no photosynthesis takes place in some plants. This is because of the *induction period*—a lag that occurs after exposure to darkness for a minute or more. On exposure to the light it may take the plant a full minute (the induction period) to reach its normal rate of photosynthesis. When the light intensity is varied from zero to full sunlight, the rate continues to increase throughout the range, or else it

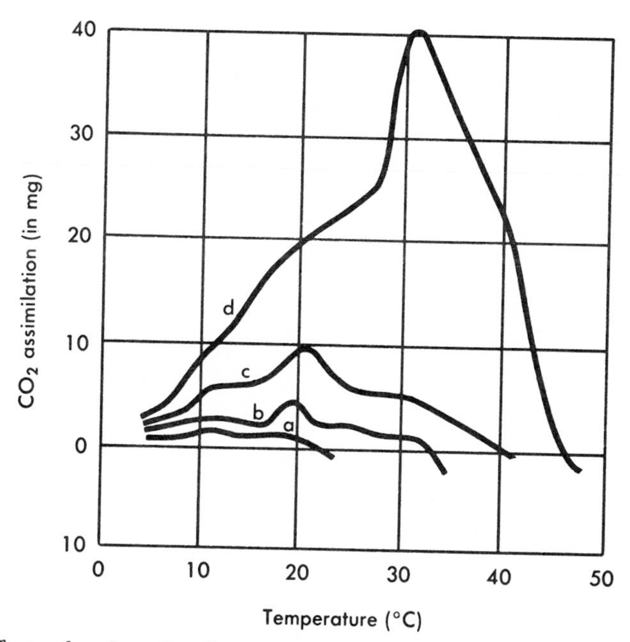

Fig. 15-14. Effects of carbon dioxide concentration, light, and temperature on photosynthesis of potato leaves. Curve **a**, a weak light and low carbon dioxide concentration. Curve **b**, 20% normal light and 0.03% carbon dioxide. Curve **c**, normal light and 0.03% carbon dioxide. Curve **d**, normal light and 1.22% carbon dioxide. (Redrawn from Lundegårdh, H. 1925. Klima und Boden. Gustav Fischer Verlag, Stuttgart.)

levels off in the upper range (Fig. 15-15). The changes with carbon dioxide concentration are more complex.

In the case of the alga *Chlorella,* the concentration of carbon dioxide in the nutrient medium that produces maximal photosynthesis is in the range of 1.5% to 2.5% by volume (Ammann and Lynch, 1967). In the case of *Scenedesmus,* a concentration of 5% produced the maximum rate (Das and Gupta, 1970). When higher plants are exposed to concentrations of carbon dioxide above that normally found in air, this may lead to stomatal closure, bringing photosynthesis to a stop. Therefore the higher plant cannot take advantage of the higher carbon dioxide concentrations as can the alga. Because of their ability to utilize higher carbon dioxide concentrations, the algae can convert a higher proportion of the incident radiant energy into chemical energy than can higher plants (see previous discussion). Nevertheless, even in the case of higher plants, carbon dioxide enrichment of the air above the normal content may increase the rate of photosynthesis (Fig. 15-14). In the case of *Callistephus chinensis,* increases up to 600 ppm markedly increased the amount of dry matter produced (Hughes, 1969). Further increases to 900 ppm produced smaller and more variable increases. Saturation occurred at 1500 to 3000 ppm (0.15% to 0.3%) for all the plants studied by Nilovskaya (1968), increasing the rate of photosynthesis by 1.25 to 2.75 times the rate in the controls. Attempts to increase yield by carbon dioxide applications in the field, however, have not been successful because of the rapid loss of carbon dioxide to the atmosphere (Allen et al., 1971).

In each of these cases just mentioned, to obtain the maximum response to a factor, the others must be present at the optimum level. If one factor is present at a low enough level, only this one limits the rate of photosynthesis; that is, this one factor when increased will greatly increase the rate of photosynthesis, whereas the others when increased have little effect on the rate of the process (Figs. 15-14 and 15-15). This relation has been called the *law of limiting factors.* Thus although light intensity, temperature, and carbon dioxide concentration are all important factors, if the

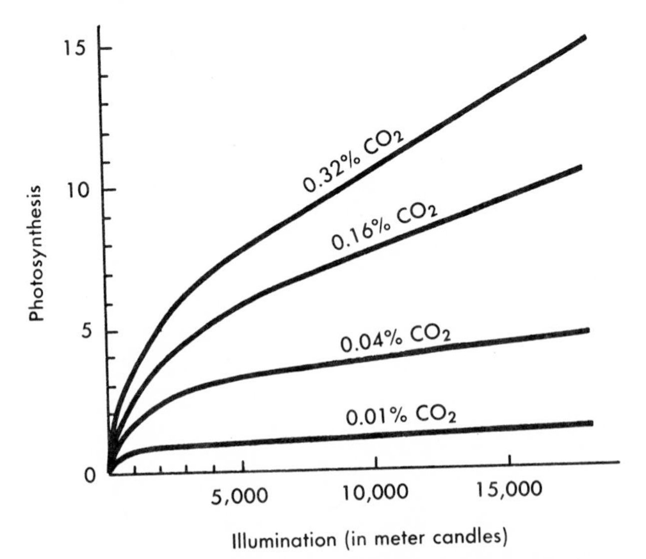

Fig. 15-15. Photosynthesis of aquatic plant *(Fontinalis)* at different light intensities and carbon dioxide concentrations. (From Harder, R. 1921. Jb. Wiss. Bot. **60**:531-571.)

temperature on a normal day is around 0° C, no increase in light intensity or carbon dioxide concentration will affect the rate of the process. Since temperature is the limiting factor, a small increase in temperature will produce a corresponding increase in the rate of photosynthesis (Fig. 15-14).

Other external factors beside light intensity, temperature, and carbon dioxide concentration may affect the rate of photosynthesis. Low moisture supply may lead to a cessation of photosynthesis, but this is really because of the low carbon dioxide concentration after stomatal closure. Higher than normal oxygen concentrations have long been known to inhibit photosynthesis. This is called the *Warburg effect.* In the case of land plants (but not in algae) even the oxygen of normal air strongly inhibits photosynthesis (Björkman, 1966). Thus when compared to normal air, absence of oxygen resulted in a 44% enhancement of photosynthesis in seven species at 30° C and an 85% enhancement in three species at 40° C (Hesketh, 1967). No enhancement occurred in true tropical grasses and in one dicotyledon. In maize there was no enhancement, even in light of low intensity.

Internal factors may also markedly affect the photosynthetic rate, although again subject to the law of limiting factors. Photosynthesis cannot occur without chlorophyll, and therefore the rate must parallel chlorophyll content up to a point. This relation is most commonly found in young, developing leaves or in mature leaves at low light intensity. In wheat, the photosynthetic rate was related to chlorophyll *a* content only at the maximum temperature for photosynthesis (Planchon, 1971). Enzyme activity may also be the limiting factor. There are key enzymes, such as RuDP carboxylase within the C-reduction cycle, that regulate the process (Bassham, 1971). The activity of PEP carboxylase (see following discussion) per unit leaf area was correlated with the photosynthetic rate in *Cenchrus ciliaris,* although other enzymes tested showed no such correlation (Treharne et al., 1971). The rate of removal of assimilate may also affect the rate of photosynthesis. In the flag leaf of wheat, the rate is regulated directly by the demand for assimilate by grain filling, by darkened leaves, and by roots (King et al., 1967). Similarly, $^{14}CO_2$ uptake/cm^2 in the leaf may be as much as 50% greater in fruit-bearing shoots than in nonbearing shoots of apple (Hansen, 1970). In both cases a more rapid translocation from source to sink may speed up photosynthesis by removing end-product inhibition (Chapter 16). Black and Mollenhauer (1971) concluded that the most reliable anatomical criterion for determining photosynthetic capacity is the number and concentration of chloroplasts, mitochondria, and peroxisomes in bundle-sheath cells (in plants that possess such cells).

Relation to respiration—photorespiration

The most important internal factor affecting the photosynthetic yield is the respiration rate. This is true for several reasons.

1. The yield of a plant depends on the total gain in carbohydrate over a 24 hr period. This, in turn, depends not only on the measured rate of photosynthesis during daylight but also on the night respiration, since this uses up some of the accumulated carbohydrate. Thus the yield of a crop may sometimes be greatly increased by lowering the night temperature, since this will increase the net assimilate per 24 hr period, for instance in rice (Bhattacharya, 1970). There is even evidence that the southern limit to the natural range of a species *(Trientalis borealis)* is due to the warm nights, resulting in a high enough respiratory rate to deplete the carbohydrate reserves (Anderson, 1970). Ponderosa pines in their natural environment lose about 12% of their photosynthate at night (Helms, 1970). The low optimum temperature (15° to 20° C) for net assimilation of

wheat at constant temperature is due to the effect of temperature on respiration (Friend, 1969).

2. The measured carbon dioxide absorption (or oxygen evolution) in the light is really a result of the excess of photosynthesis over respiration and is therefore a measurement of the *net rate of photosynthesis*. To get the true, or "actual," rate of photosynthesis it would be necessary to add to the measured carbon dioxide absorption the carbon dioxide evolution (i.e., respiration) of the same plant when kept in the dark. This correction, however, does not always give the true value for photosynthesis, since light itself may alter the rate of respiration.

By supplying the plant with atmospheric oxygen containing heavy oxygen (^{18}O) and with water with the normal isotope (^{16}O), it is possible to measure respiration directly while photosynthesis is taking place. This method of measurement may show (a) an inhibition of respiratory oxygen evolution by light; (b) an acceleration; or (c) no effect, depending on the plant, or even on the light intensity (Hoch et al., 1963). Low light intensities apparently inhibit the normal dark respiration, and higher light intensities induce a new and higher oxygen uptake. The inhibition may conceivably be caused by a competition between chloroplasts and mitochondria for phosphate and ADP (Jagendorf and Kok, 1964). In the case of the nonphotosynthesizing yeast, however, light also inhibits respiration, due apparently to a destruction of the cytochrome oxidase system (Ninnemann et al., 1970).

3. The enhancement of respiration by light is, however, due to a distinct process called *photorespiration* (Tregunna et al., 1966). It must be distinguished from dark respiration (Table 15-3), which also occurs in the light and which involves a breakdown of carbohydrates by way of glycolysis, PPP, and the Krebs cycle. Photorespiration has therefore been defined as *light-stimulated carbon dioxide evolution from (and oxygen absorption by) green leaves* (Poskuta, 1969). It also occurs, however, in green cells of lower plants. There is considerable variation among species in their rates of photorespiration. Some plants, like maize, have a negligible photorespiration and are therefore more efficient photosynthetically (Zelitch, 1966). Other species may possess a rate of photorespiration up to 5 times as high as the rate of dark respiration by the same leaves (Zelitch, 1968—see Goldsworthy, 1970). Minimum rates of 5.5 mg $CO_2/hr/dm^2$ were calculated for *Amaranthus* and the sunflower (Fock et al., 1970). Photorespiration is closely related to photosynthesis, since conditions that affect photosynthesis equally affect photorespiration (Bidwell, 1968). Unlike dark respiration that occurs in the mitochondria, photorespiration occurs in the chloroplasts, as is readily proved in the large-celled alga *Acetabularia* (Bidwell et al., 1969). The metabolic path is different from that of ordinary, dark respiration because instead of carbohydrate, the substrate is glycolic acid. In the presence of high carbon dioxide concentrations, however, this path may

Table 15-3. Comparison of dark respiration and photorespiration

Criterion	Dark respiration	Photorespiration
Organs in which it occurs	All	Green leaves
Organelles in which it occurs	Mitochondria	Plastids and peroxisomes
Substrates (for carbon dioxide evolution)	Carbohydrates	Glycolate
Response to light	Inhibited	Enhanced
Gas evolved	Carbon dioxide	Carbon dioxide
Gas absorbed	Oxygen	Oxygen
Rate in fully illuminated leaves		Up to 5 times the dark respiration

$$\underset{\text{Glycolic acid}}{CH_2OH \cdot COOH} + O_2 \xrightleftharpoons{\text{Glycolic acid oxidase}} \underset{\text{Glyoxylic acid}}{CHO \cdot COOH} + H_2O_2$$

also be followed in the dark (McNaughton, 1966). The glycolic acid is oxidized to glyoxylate at the expense of the oxygen absorbed, in a reaction involving the enzyme glycolic acid oxidase as shown above.

There is sufficient glycolic acid oxidase in leaves to account for even the highest rates of oxygen absorption in photorespiration (Bidwell, 1968). It is, in fact, possible to induce tobacco to photosynthesize as efficiently as maize by inhibiting the oxidation of glycolate. This stimulates the carbon dioxide uptake threefold at 35° C, although not at 25° C (Zelitch, 1966), presumably because the rate of photorespiration is greatly increased at higher temperatures.

Photorespiration has been found in tropical grasses in spite of the absence of detectable carbon dioxide evolution in the light. It has therefore been concluded (Rehfeld et al., 1970) that photorespiration occurs in all plants but is not manifest in plants such as tropical grasses that have high levels of the C_4 dicarboxylic acid path of carbon dioxide fixation (see previous discussion). This provides a more effective carbon dioxide trapping system that captures all the carbon dioxide released in photorespiration, preventing the detection of photorespiration in the air external to the leaf. It has also been suggested that the enzyme carbonic anhydrase may trap the carbon dioxide, since it is present in photorespiring spinach chloroplasts but not in nonphotorespiring maize (Waygood et al., 1969). In weak light it has indeed been possible to detect carbon dioxide evolution by tropical grasses, approaching the rates by temperate zone plants and equaling 4.2% of photosynthesis (Irvine, 1970). This explanation, however, appears to apply only to some C_4 plants (e.g., *Amaranthus lividus*) whereas others (e.g., corn) show as little as one tenth the level of photorespiration found in C_3 plants (soybean; Laing and Forde, 1971). Similarly, in the C_4 plant *Sorghum bicolor*, the activity of the glycolate pathway is so low that little carbon dioxide release occurs (Osmond and Harris, 1971).

These newer results also explain the long-known Warburg effect—the inhibition of photosynthesis by oxygen (see earlier discussion). This can logically be explained as an inhibition only of *net* photosynthesis resulting from an increase in photorespiration by the oxygen-induced production of glycolate. In plants with active photorespiration, glycolate becomes the major photosynthetic product at high partial pressures of oxygen or low carbon dioxide concentrations. It is this shift at high oxygen concentrations of a major portion of the total assimilated carbon into glycolate that impairs the functioning of the photosynthetic C-reduction cycle (Ellyard and Gibbs, 1969).

Experimental results, however, indicate that photorespiration may be only partially responsible for the Warburg effect. In the case of soybeans, the photosynthetic rates in air (12 to 24 mg $CO_2/dm^2/hr$) are 5 to 6 mg less than the rates in a low-oxygen atmosphere (Curtis et al., 1969). Slightly more than half of this decrease was due to photorespiration. In the case of corn, no inhibiting effect of oxygen could be detected, indicating that this occurs only because of the photorespiration (Bulley et al., 1969).

Measurements in sunflowers reveal that photorespiration accounts for only one third or two thirds of this inhibition by oxygen at 300 and 50 ppm of carbon dioxide, respectively (Ludwig and Canvin, 1971). On the other hand, the inhibition of photosynthesis in cucumber leaves by oxygen concentrations above 21% was due only to an increase in carboxylation resistance (Peisker and Apel, 1971). This is probably due to photooxidation and therefore destruction of part of the photosynthetic apparatus (Kok, 1967).

These apparently contradictory results now appear to be resolved. Interactions have been demonstrated between oxygen and RuDP carboxylase (Bowes and Berry, 1972), suggesting a direct inhibition of the enzyme at light saturation. These results are consistent with the hypothesis that glycolate is formed by way of an alternate reaction catalyzed by RuDP carboxylase, which occurs when oxygen is bound to the enzyme in place of carbon dioxide. This production of glycolate from RuDP occurs in 21% oxygen and low carbon dioxide concentration, and it provides a mechanism for the consumption of energy input without a net carbon gain. The Warburg effect therefore involves both photorespiration and an inhibition of the carboxylation of RuDP.

Compensation points

The quantitative effect of respiration on the net photosynthetic yield is interrelated with the environmental factors that control the rate of photosynthesis.

Whatever the effect of light on respiration, the fact that there is normally a net photosynthesis during daylight proves that the rate of photosynthesis exceeds the rate of respiration. But if the light intensity is decreased steadily, a point will be reached at which the photosynthetic rate just equals the respiration rate, and the net rate of photosynthesis is zero. This is called the *light compensation point.* It may occur at surprisingly high illumination, such as 1.5 to 4 klux in vegetable crops (Tatsumi and Hori, 1969). Shade plants have a much lower light compensation point. This is due to a low percentage of dry matter and a large leaf area per gram fresh weight (gfw), which lead to a low respiratory rate per unit leaf area, as low as 0.05 mg CO_2/50 cm^2 leaf blade at 20° C (Löhr and Muller, 1968).

The plant, however, is able to adapt to changes in light intensity. For instance, when white clover is exposed to different levels of constant light and other factors, the plants adapt their respiration rates so as to maintain a constant proportion (about 20%) of the photosynthetic rate (McCree and Troughton, 1966). Consequently, even when the light level drops below the compensation point, this adaptation occurs within about 24 hr, the compensation point drops, and the plants are able to gain in weight.

Besides the light compensation point, there is also a *carbon dioxide compensation point*—the carbon dioxide concentration of the air at which the net rate of photosynthesis is zero. This can be simply determined by enclosing the plant material in a carbon dioxide–impermeable Mylar bag and when equilibrium is reached in the light, analyzing the air in the bag (Goldsworthy and Day, 1970). The carbon dioxide compensation point for many grasses ranges from 10 to 40 ppm at 25° C (Krenzer and Moss, 1969), compared to a carbon dioxide content of 300 ppm in normal air. In general, plants that fix carbon dioxide by way of RuDP carboxylase have relatively high carbon dioxide compensation points. Plants that fix carbon dioxide by way of PEP carboxylase have low carbon dioxide compensation points (Downton and Tregunna, 1968), due to a greater affinity for carbon dioxide at the carboxylation site (Bull, 1969). Thus the carbon dioxide compensation points for corn (Hew et al., 1969*a*) and sugar cane (Coombs, 1968) are near zero. An increase in light intensity decreases the carbon dioxide compensation point in Scotch pine, an increase in temperature increases it (Gordon and Gatherium, 1969). This interrelationship with other factors explains the wide differences in values to be found in the literature, such as the low values found by

Moss and co-workers in 1971: In the case of fifty-four genotypes, the compensation point was less than 5 ppm CO_2 and in 114 others it was near 0 ppm.

There is also a *temperature compensation point,* since respiration rate rises more rapidly with temperature than does photosynthesis, at least above a certain temperature. All these compensation points are interdependent, and any measured value is valid only for the specific environmental conditions used.

C_4 vs. C_3 plants

Several tropical grasses (sugar cane, maize, etc.) are unusually high-yielding, much higher than grasses or other plants from temperate climates. This is due largely to the higher net photosynthetic rates of the tropical grasses which are twice the rates for grasses from temperate climates at light saturation (Hofstra and Nelson, 1969). Maximum rates of near 60 mg $CO_2/dm^2/hr$ have been obtained in maize, sugar cane, and sorghums, as compared to only 20 mg for small grains and numerous other grasses (Moss, 1968). The higher yielding species are C_4 plants and the lower yielding species are C_3 plants. The difference depends on the ability of the C_4 plants to respond favorably to higher values of environmental factors. They have higher optimum temperatures for photosynthesis and their photosynthetic rate rises with light intensity up to full sunlight, whereas the C_3 plants reach their maximum at half of full sunlight or lower (Fig. 15-16).

It is the apparent absence of photorespiration that enables these C_4 tropical grasses to accumulate photosynthate at higher temperatures than can the C_3 plants of temperate climates. The latter show a decrease in *net* photosynthesis between 20° and 30° C in normal air due primarily to an increase in the rate of photorespiration and therefore of carbon dioxide evolution in the light, with relatively little effect on *actual* photosynthesis (Hew et al., 1969*b*).

One exception has been found to the preceding generalization. *Typha latifolia* is

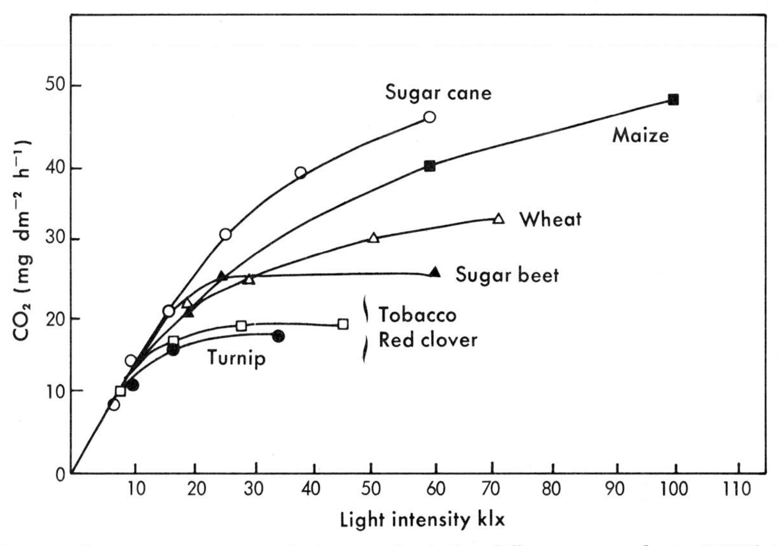

Fig. 15-16. Light response curves of photosynthesis for different crop plants. 0.03% CO_2, 20° C. (From Stoy. *In* J. D. Eastin et al. [Eds.]. Physiological aspects of crop yield. Copyright 1969 by the American Society of Agronomy and reprinted by permission of the copyright owner.)

a C_3 plant, yet its net photosynthetic rate is more than 60 mg $CO_2/dm^2/hr$ and is therefore equal to that in the C_4 plants (McNaughton and Fullem, 1970). It is the most efficient C_3 plant, and it has a much lower level of glycolate metabolism than other C_3 plants. This is further evidence that the high net photosynthetic rates of the C_4 plants are due to their ability to prevent photorespiratory loss.

Nevertheless, there are four characteristics of the C_4 plants that have been considered as possible factors in their higher photosynthetic rates (Crookston and Moss, 1970): (1) a high capacity to fix carbon dioxide, leading to a carbon dioxide compensation point near zero (Fig. 15-17); (2) the absence of any oxygen-induced inhibition of photosynthesis (the Warburg effect) (Fig. 15-18); (3) the lack of an appreciable glycolate metabolism and therefore of photorespiration; and (4) the few if any grana in their bundle-sheath chloroplasts. It is now known that the absence of photorespiration from C_4 plants may be more apparent than real, due to a high capacity of the PEP carboxylase to fix the carbon dioxide evolved by photorespiration. This factor would largely eliminate both the oxygen effect and the effect of glycolate metabolism, as long as the resulting increase in photorespiratory carbon dioxide did not exceed the high capacity of the C_4 plant to fix carbon dioxide. In the case of two C_4 plants, this was the mechanism in the dicot *Atriplex spongiosa,* in opposition to the monocot *Sorghum bicolor* in which the activity of the glycolate metabolism was so low that little photorespiratory release of carbon dioxide occurred (Osmond and Harris, 1971).

The major cause of the higher photosynthetic rates in the case of many C_4 plants is therefore their higher capacity to fix carbon dioxide. It is because of this

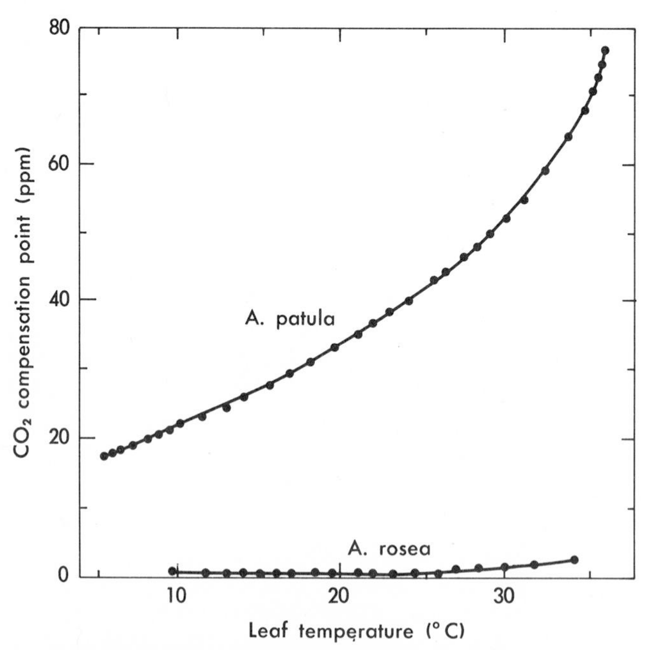

Fig. 15-17. CO_2 compensation points and their relation to temperature in a C_4 species, *A. rosea* and a C_3 species, *A. patula* of *Atriplex.* (From Björkman, O., et al. 1970. Carnegie Institution Yearbook **68**:620-632.)

characteristic that they are able to make use of higher light intensities and temperatures than the C_3 plants. The C_3 plants presumably lower the carbon dioxide concentration of the atmosphere to the limit of their capacity at the lower light intensities and temperatures, whereas the C_4 plants can continue to lower it beyond this point (even to nearly zero), at the high light and temperature levels. This difference has been explained by the PEP carboxylase of the C_4 plants, which apparently has a higher affinity for carbon dioxide than does the RuDP carboxylase of the C_3 plants (see previous discussion).

There is, however, an additional factor. The fourth characteristic mentioned is the absence of grana from the bundle-sheath chloroplasts of the C_4 plants. It has been suggested that this is accompanied by an absence of photosystem II and therefore of the Hill reaction (Woo et al., 1970). The resulting lack of oxygen would inhibit photorespiration. This mechanism appears superfluous in view of the high carbon dioxide–capturing ability of the C_4 plants. Furthermore, the presence or absence of well-defined grana is not a reliable criterion for placing a plant in the C_4 or C_3 group (Black and Mollenhauer, 1971). A more striking feature is the dense concentration of chloroplasts and other organelles in the mature bundle-sheath cells compared to the mesophyll cells.

Nevertheless, bundle-sheath cells are characteristic of C_4 plants, and they actively fix carbon dioxide through the Calvin cycle in opposition to the mesophyll cells of the C_4 plants, which fix carbon dioxide through the C_4 cycle. Since photorespiration occurs only in association with the Calvin cycle, the carbon dioxide is released only in these internal bundle-sheath cells. Before escaping from the leaf, it must then diffuse into the mesophyll cells where it is captured by the PEP carboxylase. The

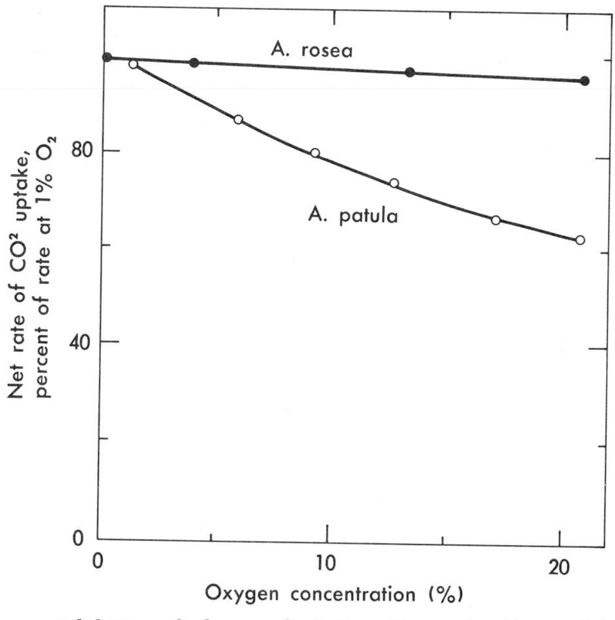

Fig. 15-18. Oxygen inhibition of photosynthesis in a C_3 species *(A. patula)* compared to lack of inhibition in a C_4 species *(A. rosea)* of the same genus, *Atriplex.* (From Björkman, O., et al. 1970. Carnegie Institution Yearbook **68**:620-632.)

bundle-sheath cells may also play another role in maintaining high photosynthetic rates in the C_4 plants. These plants show a more rapid translocation of photosynthate —70% or more during the first 6 hr after assimilation in the C_4 plants (sorghum and millet) compared to 45% to 50% in the C_3 plants (tomato, castor bean, *Nicotiana affinis,* and soybean—Hofstra and Nelson, 1969). This must be at least partly due to the shorter distance from the Calvin cycle centers in the bundle-sheath cells to the adjacent vascular tissue in the C_4 plants. In the C_3 plants all the mesophyll cells are involved in the Calvin cycle and therefore the sugars must diffuse over far larger distances before reaching the phloem.

The most important factors in the high photosynthetic rates of the C_4 plants are apparently (1) the high carbon dioxide–fixing capacity of their PEP carboxylase (2) the location of this enzyme external to the photorespiring bundle-sheath cells, and (3) the close proximity of the Calvin cycle (bundle-sheath) cells to the translocation system (the phloem).

However, not all differences in yield can be explained by differences in rates of net photosynthesis. It is possible, for instance, for a plant to compensate for a lower photosynthetic rate per unit leaf surface by development of a larger leaf area. The amount of photosynthate produced by a plant does indeed rise with increasing size of foliage per ground area (the so-called leaf area index or LAI—Stoy, 1969, see Chapter 4) up to the point of mutual shading. It must also be realized that the yield of a crop is normally not the *total* dry matter produced, but only one fraction of it, such as the fruit. Therefore a high yield may depend just as much on the fraction of photosynthate that is diverted into these organs as on total photosynthate (Stoy, 1969). Because of these complicating factors, it is not surprising that the yield differences among thirty-six varieties of soybeans could not be accounted for by the differences in their photosynthetic rates (Curtis et al., 1969).

Besides the ordinary green plants there are bacteria that photosynthesize. They also possess chlorophyll and assimilate carbon compounds at the expense of carbon dioxide in the light but do not evolve oxygen. In fact, they are able to photosynthesize only under anaerobic conditions.

Although photosynthesis is usually represented as leading to the formation of hexose sugar, it may also lead to the accumulation of many other substances, for example, starch, fats, and indirectly even proteins. This may be illustrated as shown in the following diagram.

$$CO_2 + H_2O \xrightarrow[\text{Chlorophyll}]{\text{Light}} \text{Intermediates of Calvin cycle} \rightarrow \text{Hexose phosphates} \rightarrow \text{Sucrose}$$

Intermediates of Calvin cycle ↑ Lipids; ↗ Amino acids and proteins; ↘ Pentosans, pectins
Hexose phosphates ↓ Cellulose
Sucrose ↗ Hexoses; ↘ Starch

QUESTIONS

1. What must be measured in order to determine the rate of photosynthesis?
2. What factors must be available in order for photosynthesis to occur?
3. Where does the evolved oxygen come from?
4. How do we know where it comes from?
5. What, then, would be the overall equation for photosynthesis?
6. What method (technique) has been most useful in recent years for investigating the intermediates of photosynthesis?

7. What actual isotopes have been used?
8. According to the evidence, what is the first substance formed as a result of carbon dioxide fixation?
9. What other substances are formed soon after carbon dioxide fixation, before the release of free carbohydrates?
10. Can the photosynthetic conversion (fixation) of carbon dioxide to carbohydrates occur in the dark?
11. What, then, is the difference between photosynthetic and dark carbon dioxide assimilation?
12. What is the Hill reaction?
13. What is meant by "photolysis of water"?
14. What is the difference between photophosphorylation and oxidative phosphorylation?
15. What is the difference between cyclic and noncyclic photophosphorylation?
16. What is photoelectron transport?
17. What is the Emerson enhancement effect?
18. How many photochemical reactions participate in the photosynthetic process?
19. What is a photosynthetic unit?
20. What molecule is excited during photosynthesis?
21. What is fluorescence?
22. If chlorophyll absorbs blue light, what color does it fluoresce?
23. If it absorbs green light, what color does it fluoresce?
24. What is the role of chlorophyll in photosynthesis?
25. Describe the absorption spectrum of chlorophyll.
26. What is meant by the action spectrum of photosynthesis?
27. How do the two spectra compare?
28. What are the plastid pigments?
29. Are the other plastid pigments of any importance in photosynthesis?
30. What factors affect the rate of photosynthesis?
31. Which factor is the most important?
32. Is temperature as important in photosynthesis as in respiration?
33. What is meant by net photosynthesis?
34. What is meant by actual photosynthesis?
35. How is the actual rate of photosynthesis determined?
36. Does respiration occur in the light?
37. How do we know whether or not it occurs in the light?
38. How is the yield of the plant related to photosynthesis and respiration?

SPECIFIC REFERENCES

Allen, L. H., Jr., S. E. Jensen, and E. R. Lemon. 1971. Plant response to carbon dioxide enrichment under field conditions: a simulation. Science **173**:256-258.

Ammann, E. C. B., and V. H. Lynch. 1967. Gas exchange of algae. III. Relation between the concentration of carbon dioxide in the nutrient medium and the oxygen production of *Chlorella pyrenoidosa*. Appl. Microbiol. **15**:487-491.

Anderson, R. C. 1970. The role of daylength and temperature in tuber formation and rhizome growth of *Trientalis borealis* Raf. Bot. Gaz. **131**:122-128.

Arnold, W., and J. R. Azzi. 1968. Chlorophyll energy levels and electron flow in photosynthesis. Proc. Nat. Acad. Sci. **61**:29-35.

Baldry, C. W., C. Bucke, and J. Coombs. 1969. Light/phosphoenolpyruvate dependent carbon dioxide fixation by isolated sugar cane chloroplasts. Biochem. Biophys. Res. Comm. **37**:828-832.

Baldy, C., and P. Jonard. 1969. Contributions to the photosynthetic output of wheat. Compt. Rend. Acad. Sci. (Paris) **268**:1296-1299.

Bassham, J. A. 1971. The control of photosynthetic carbon metabolism. Science **172**:526-534.

Begg, J. E. 1965. High photosynthetic efficiency in a low latitude environment. Nature (London) **205**:1025-1026.

Berry, J. A. 1971. The compartmentation of reactions in β-carboxylation photosynthesis. Carnegie Inst. Yearbook **69**:649-655.

Berry, J. A., W. J. S. Downton, and E. B. Tregunna. 1970. The photosynthetic carbon metabolism of *Zea mays* and *Gomphrena globosa:* the location of the CO_2 fixation and the carboxyl transfer reactions. Canad. J. Bot. **48**:777-786.

Bhattacharya, B. 1970. Effects of various ranges

of day and night temperatures at the ripening period on the grain production in rice plants. J. Fac. Agr. Kyushu Univ. **16**:85-140.

Bidwell, R. G. S. 1968. Photorespiration. Science **161**:79-80.

Bidwell, R. G. S., W. B. Levin, and D. C. Shephard. 1969. Photosynthesis, photorespiration, and respiration of chloroplasts from *Acetabularia mediterranea.* Plant Physiol. **44**:946-954.

Björkman, O. 1966. The effect of oxygen concentration on photosynthesis in higher plants. Physiol. Plant. **19**:618-633.

Black, C. C., Jr., and H. H. Mollenhauer. 1971. Structure and distribution of chloroplasts and other organelles in leaves with various rates of photosynthesis. Plant Physiol. **47**:15-23.

Bowes, G., and J. A. Berry. 1972. The effect of oxygen on photosynthesis and glycolate excretion in *Chlamydomonas rheinhardtii.* Carnegie Inst. Yearbook **71**:148-158.

Boynton, J. E., M. A. Nobs, O. Björkman, and R. W. Pearcy. 1971. Leaf anatomy and ultrastructure. Carnegie Inst. Yearbook **69**:629-632.

Bull, T. A. 1969. Photosynthetic efficiencies and photorespiration in Calvin cycle and C_4-dicarboxylic acid plants. Crop Sci. **9**:726-729.

Bulley, N. R., C. D. Nelson, and E. B. Tregunna. 1969. Photosynthesis: action spectra for leaves in normal and low oxygen. Plant Physiol. **44**: 673-684.

Clegg, C. J., and C. P. Whittingham. 1970. Dark CO_2 fixation by potato tuber tissue. Phytochemistry **9**:279-287.

Coombs, J. (Ed.). 1968. Photosynthesis in sugar cane. Proceedings of an international symposium. Tate & Lyle Ltd., London.

Coutinho, L. M. 1965. Algumas informacões sobre a capacidade ritmica diaria da fixaçaoe acumulaçao de CO_2 no escuro em epifitas e erbaceas terrestre da mata plurial. Bol. no. 294. Fac. Fil. Cienc. e Letras da USP Botânica 21.

Crookston, R. K. and D. N. Moss. 1970. The relation of carbon dioxide compensation and chlorenchymatous vascular bundle sheaths in leaves of dicots. Plant Physiol. **46**:564-567.

Curtis, P. E., W. L. Ogren, and R. H. Hageman. 1969. Varietal effects in soybean photosynthesis and photorespiration. Crop Sci. **9**:323-327.

Das, O. A. K., and A. B. Gupta. 1970. The effect of carbon dioxide concentration and light intensity on growth rate of *Scenedesmus obliquus.* Labdev. J. Sci. Technol. (Part B) **8**:233-236.

Dodge, A. D. 1971. The mode of action of the bipyridilium herbicides, paraquat and diquat. Endeavour **30**:130-135.

Downton, W. J. S. 1971. Further evidence for two modes of carboxyl transfer in plants with C_4-photosynthesis. Canad. J. Bot. **49**:1439-1442.

Downton, W. J. S., and E. B. Tregunna. 1968. Carbon dioxide compensation: its relation to photosynthetic carboxylation reactions, systematics of the Gramineae, and leaf anatomy. Can. J. Bot. **46**:207-215.

Edwards, G. W., R Kanai, and C. C. Black. 1971. Phosphoenolpyruvate carboxykinase in leaves of certain plants which fix CO_2 by the C_4-dicarboxylic acid cycle of photosynthesis. Biochim. Biophys. Res. Comm. **45**:278-285.

Egunjobi, J. K. 1971. Ecosystem processes in a stand of *Ulex europaeus,* L. I. Dry matter production, litter fall and efficiency of solar energy utilization. J. Ecol. **59**:31-38.

Ellyard, P. W., and M. Gibbs. 1969. Inhibition of photosynthesis by oxygen in isolated spinach chloroplasts. Plant Physiol. **44**:1115-1121.

Fock, H., J. D. Becker, and K. Egle. 1970. Use of labeled carbon dioxide for separation of CO_2 evolution from true CO_2 uptake by photosynthesizing *Amaranthus* and sunflower leaves. Canad. J. Bot. **48**:1185-1189.

Friend, D. J. C. 1969. Net assimilation rate of wheat as affected by light intensity and temperature. Canad. J. Bot. **47**:1781-1787.

Goedheer, J. C. 1969. Energy transfer from carotenoids to chlorophyll in blue-green, red, and green algae, and greening bean leaves. Biochim. Biophys. Acta **172**:252-265.

Goldsworthy, A. 1970. Photorespiration. Bot. Rev. **36**:321-340.

Goldsworthy, A., and P. R. Day. 1970. A simple technique for the rapid determination of plant CO_2 compensation points. Plant Physiol. **46**:850-851.

Gordon, J. C., and G. E. Gatherium. 1969. Effect of environmental factors and seed source on CO_2 exchange of Scotch-pine seedlings (*Pinus sylvestris*). Bot. Gaz. **130**:5-9.

Gregory, R. P. F., S. Raps, and W. Bertsch. 1971. Are specific chlorophyll-protein complexes required for photosynthesis? Biochim. Biophys. Acta **234**:330-334.

Hansen, P. 1970. ^{14}C-studies on apple trees. VI. The influence of the fruit on the photosynthesis of the leaves, and the relative photosynthetic yields of fruits and leaves. Physiol. Plant. **23**: 805-810.

Harder, R. 1921. Kritische Versuche zu Blackmans Theorie der "begrenzenden Faktoren" bei der Kohlensäure-assimilation. Jb. Wiss. Bot. **60**:531-571.

Harrison, A. 1965. Carbon dioxide effects on the extension in length of *Avena* coleoptiles. Physiol. Plant. **18**:208-218.

Hatch, M. D., and C. R. Slack. 1966. Photosynthesis by sugar-cane leaves; a new carboxylation reaction and the pathway of sugar formation. Biochem. J. **101**:103-111.

Hatch, M. D., C. R. Slack, and H. S. Johnson. 1967. Further studies on a new pathway of photosynthetic carbon dioxide fixation in sugarcane and its occurrence in other plant species. Biochem. J. **102**:417-422.

Helms, J. A. 1970. Summer net photosynthesis of ponderosa pine in its natural environment. Photosynthetica **4**:243-253.

Hesketh, J. 1967. Enhancement of photosynthetic CO_2 assimilation in the absence of oxygen, as dependent upon species and temperature. Planta **76**:371-374.

Hess, J. L., and N. E. Tolbert. 1967. Changes in chlorophyll *a/b* ratio and products of $C^{14}O_2$ fixation by algae grown in blue or red light. Plant Physiol. **42**:1123-1130.

Hew, C. S., G. Krotkov, and D. T. Canvin. 1969*a*. Determination of the rate of CO_2 evolution by green leaves in the light. Plant Physiol. **44**:662-670.

Hew, C. S., G. Krotkov, and D. T. Canvin. 1969*b*. Effects of temperature on photosynthesis and CO_2 evolution in light and darkness by green leaves. Plant Physiol. **44**:671-677.

Highkin, H. R. and A. W. Frenkel. 1962. Studies of growth and metabolism of a barley mutant lacking chlorophyll *b*. Plant Physiol. **37**:814-820.

Hind, G., and A. T. Jagendorf. 1963. Separation of light and dark stages in photophosphorylation. Proc. Nat. Acad. Sci. USA **49**:715-722.

Hoch, G., O. v. H. Owens, and B. Kok. 1963. Photosynthesis and respiration. Arch. Biochem. Biophys. **101**:171-180.

Hofstra, G., and C. D. Nelson. 1969. A comparative study of translocation of assimilated ^{14}C from leaves of different species. Planta **88**:103-112.

Hofstra, J. J., S. Aksornkoae, S. Atmowidjojo, J. F. Banaag, S. R. A. Sastrohoetomo, and L. T. N. Thu. 1972. A study of the occurrence of plants with a low CO_2 compensation point in different habitats in the tropics. Ann. Bogoriensis **5**:143-157.

Holdsworth, M. 1971. Carbon dioxide uptake by succulents. Canad. J. Bot. **49**:1520-1522.

Howell, S. H., and E. N. Moudrianakis. 1967. Function of the "quantasome" in photosynthesis; structure and properties of membrane-bound particles active in the dark reactions of photophosphorylation. Proc. Nat. Acad. Sci. USA **58**: 1261-1268.

Hughes, A. P. 1969. Effects of carbon dioxide concentration on the growth of *Callistephus chinensis* cultivar Johannistag. Ann. Bot. **33**: 351-365.

Irvine, J. E. 1970. Evidence for photorespiration in tropical grasses. Physiol. Plant. **23**:607-612.

Izawa, S., T. N. Connolly, G. D. Winget, and N. E. Good. 1967. Inhibition and uncoupling of photophosphorylation in chloroplasts, p. 169-187. *In* J. M. Olson, G. Hind, H. Lyman, and H. W. Siegelman. Energy conversion by the photosynthetic apparatus. Brookhaven Symposium in Biology, Pub. No. 19. Brookhaven National Laboratory, Upton, N. Y.

Jagendorf, A. T., and B. Kok. 1964. Photosynthesis. Science **143**:388-395.

Junge, W., E. Reinwald, B. Rumberg, U. Siggel, and H. T. Witt. 1968. Further evidence for a new function unit of photosynthesis. Naturwissenschaften **55**:36-37.

Kaminski, A. 1969. Measurement of the amount of light energy absorbed by the potato (*Solanum tuberosum* L.) Ekol. Pol. Ser. A. **17**:373-379.

Kaminski, A. 1970. Energetic balance of potato and rye plantations. Wszechswiat **2**:46.

King, R. W., I. F. Wardlaw, and L. T. Evans. 1967. Effect of assimilate utilization on photosynthetic rate in wheat. Planta **77**:261-276.

Knaff, D. B., and Daniel I. Arnon. 1969. A concept of three light reactions in photosynthesis by green plants. Proc. Nat. Acad. Sci. USA **64**:715-722.

Kok, B. 1967. Photosynthesis, p. 335-379. *In* M. B. Wilkins (Ed.). The physiology of growth and development. McGraw-Hill Book Co., New York.

Kortschak, H., C. E. Hartt, and G. O. Burr. 1965. Carbon dioxide fixation in surgar-cane leaves. Plant Physiol. **40**:209-213.

Kouchkovsky, Y. de. 1968. Nature et rôle des pigments intervenant dans le transfert d'énergie et le transport d'electrons en photosynthèse. Bull. Soc. Fr. Physiol. Veg. **14**:409-450.

Krenzer, E. G. Jr., and D. N. Moss. 1969. Carbon dioxide compensation in grasses. Crop Sci. **9**:619-621.

Krotkov, G., V. C. Runeckles, and K. V. Thimann. 1958. Effect of light on the CO_2 absorption and evolution by *Kalanchoe*, wheat, and pea leaves. Plant Physiol. **33**:289-292.

Laing, W. A., and B. J. Forde. 1971. Comparative photorespiration in *Amaranthus,* soybean and corn. Planta **98**:221-231.

Latzko, E., and M. Gibbs. 1968. Distribution and activity of the reductive pentose phosphate cycle in spinach leaves and in chloroplasts isolated by different methods. Z. Pflanzenphysiol. **59**:184-194.

Linder, S. 1971. Photosynthetic action spectra of

Scots pine needles under different nursery conditions. Physiol. Plant. **25**:58-63.

Löhr, E., and D. Müller. 1968. Blatt-atmung der hoheren Bodenpflanzen in tropischer Regenurwald. Physiol. Plant. **21**:673-675.

Ludwig, L. J. and D. T. Canvin. 1971. The rate of photorespiration during photosynthesis and the relationship of the substrate of light respiration to the products of photosynthesis in sunflower leaves. Plant Physiol. **48**:712-719.

Mantai, K. E., and N. I. Bishop. 1967. Studies on the effects of ultraviolet irradiation on photosynthesis and on 520 nm light-dark difference spectra in green algae and isolated chloroplasts. Biochim. Biophys. Acta **131**:350-356.

McCree, K. J., and J. H. Troughton. 1966. Prediction of growth rates at different light levels from measured photosynthesis and respiration rates. Plant Physiol. **41**:559-566.

McNaughton, S. J. 1966. Light-stimulated oxygen uptake and glycolic acid oxidase in *Typha latifolia* L. leaf disks. Nature (London) **211**:1197-1198.

McNaughton, S. J., and L. W. Fullem. 1970. Photosynthesis and photorespiration in *Typha latifolia*. Plant Physiol. **45**:703-707.

McWilliams, E. I. 1970. Comparative rates of dark CO_2 uptake and acidification in the Bromeliaceae, Orchidaceae, and Euphorbiaceae. Bot. Gaz. **131**:285-290.

Metzner, H. 1969. Comparative biochemistry of photosynthesis. FEBS Letters **5**:93-95.

Miyachi, S., and D. Hogetsu. 1970. Light-enhanced carbon dioxide fixation in isolated chloroplasts. Plant Cell Physiol. **11**:927-936.

Moss, D. N. 1968. Relation in grasses of high photosynthetic capacity and tolerance to atrazine. Crop Sci. **8**:774.

Mühlethaler, K. 1966. Der Feinbau des Photosynthese-Apparates. Umschau Wiss. Tech. **66**:659-662.

Mukerji, S. K., and I. P. Ting. 1968. Intracellular localization of CO_2 metabolism in *Cactus phylloclades*. Phytochem. **7**:903-911.

Murata, Y., A. Miyasaka, K. Munakata, and S. Akita. 1968. On the solar energy balance in rice population in relation to the growth stage. Proc. Crop Soc. Japan **37**:685-691.

Murata, T., F. Toda, K. Uchnio, and E. Yakushiji. 1971. Water-soluble chlorophyll protein of *Brassica oleracea* var. botrys (cauliflower). Biochim. Biophys. Acta **245**:208-213.

Muresan, T., N. Hurduc, and I. Nastasia. 1967. Correlation between photosynthesis and heterosis in corn. An. Inst. Cercet Pentru Cereale Plante Teh. Fund. Ser. C. Amelior. Genet. Fiziol. Technol. Agr. **35**:339-356.

Myers, J., and J. R. Graham. 1971. The photosynthetic unit in *Chlorella* measured by repetitive short flashes. Plant Physiol. **48**:282-286.

Nilovskaya, N. T. 1968. Photosynthesis and respiration of some vegetable plants in an atmosphere with various carbon dioxide contents. Fiziol. Rast. **15**:1015-1021.

Ninnemann, H., W. L. Butler, and B. L. Epel. 1970. Inhibition of respiration in yeast by light. Biochim. Biophys. Acta **205**:499-506.

Nishimura, M. 1967. Oxidation-reduction reactions of cytochromes in red algae. *In* J. M. Olson, G. Hind, H. Lyman, and H. W. Siegelman. Energy conversion by the photosynthetic apparatus. Brookhaven Symposium in Biology, Pub. No. 19. Brookhaven National Laboratory, Upton, N. Y.

Osmond, C. B. 1967. β-carboxylation during photosynthesis in *Atriplex*. Biochim. Biophys. Acta **141**:197-199.

Osmond, C. B., and B. Harris. 1971. Photorespiration during C_4 photosynthesis. Biochim. Biophys. Acta **234**:270-282.

Osmond, C. B., J. H. Troughton, and D. J. Goodchild. 1969. Physiological, biochemical and structural studies of photosynthesis and photorespiration in two species of *Atriplex*. Z. Pflanzenphysiol. **61**:218-237.

Peisker, M., and P. Apel. 1971. Research on the influence of oxygen on the CO_2 gas exchange of photosynthesizing leaves. Biochem. Physiol. Pflanz. **162**:165-176.

Plamondon, J. E., and J. A. Bassham. 1966. Glycolic acid labeling during photosynthesis with $C^{14}O_2$ and tritiated water. Plant Physiol. **41**: 1272-1275.

Planchon, C. 1971. Relations entre l'assimilation chlorophyllienne nette, la teneur en chlorophylle *a* et la température chez le Blé tendre (*Triticum aestivum* L.). Compt. Rend. Acad. Sci. (Paris) **242**:68-71.

Poskuta, J. 1969. Photosynthesis, respiration and postillumination fixation of CO_2 by corn leaves as influenced by light and oxygen concentration. Physiol. Plant. **22**:76-85.

Ramirez, J. M., F. F. del Campo, and D. I. Arnon. 1968. Photosynthetic phosphorylation as energy source for protein synthesis and carbon dioxide assimilated by chloroplasts. Proc. Nat. Acad. Sci. USA **59**:606-612.

Raven, J. A. 1970. The role of cyclic and pseudocyclic photophosphorylation in photosynthetic $^{14}CO_2$ fixation in *Hydrodictyon africanum*. J. Exp. Bot. **21**:1-16.

Rehfeld, D. W., D. D. Randall, and N. E. Tolbert. 1970. Enzymes of the glycolate pathway in plants without CO_2-photorespiration. Canad. J. Bot. **48**:1219-1226.

Renny, R. 1971. Resolution of chloroplast lamellar proteins by electrophoresis in polyacrylamide gels: different patterns obtained with fractions enriched in either chlorophyll *a* or chlorophyll *b*. FEBS Letters **13**:313-317.

Schoch, E. 1971. Malate and aspartate now also found in a composite as the principal products of short-term $^{14}CO_2$ fixation. Z. Pflanzenphysiol. **64**:367-368.

Schwartz, M. 1967. Quantum yield of ferrocytochrome C photo-oxidation in chloroplast particles. Nature (London) **213**:1187-1189.

Simonis, W. 1967. Zyklische und nicht zyklische Photophosphorylierung in vivo. Ber. Deut. Bot. Ges. **80**:395-402.

Splittstoesser, W. E. 1966. Dark CO_2 fixation and its role in the growth of plant tissue. Plant Physiol. **41**:755-759.

Strain, H. H., J. Sherma, and M. Grandolfo. 1968. Comparative chromatography of the chloroplast pigments. Anal. Biochem. **24**:54-69.

Tanner, W., M. Loffler, and O. Kandler. 1969. Cyclic photophosphorylation *in vivo* and its relation to photosynthetic CO_2 fixation. Plant Physiol. **44**:422-428.

Tatsumi, M., and Y. Hori. 1969. Studies on the photosynthesis of vegetable crops. I. Photosynthesis of young plant of vegetables in relation to light intensity. Bull. Hort. Res. Sta. Min. Agr. Forest Ser. A. (Hiratsuka) **8**:127-140.

Thornber, J. P., R. P. F. Gregory, C. A. Smith and J. L. Bailey. 1967. Studies on the nature of the chloroplast lamellae. I. Preparation and some properties of two chlorophyll-protein complexes. Biochemistry **6**:391-396.

Ting, I. P., and W. Zsroche. 1970. Asparagine biosynthesis by cotton roots: carbon dioxide fixation and cyanide incorporation. Plant Physiol. **45**:429-434.

Tregunna, E. B., G. Krotkov, and C. D. Nelson. 1966. Effect of oxygen on the rate of photorespiration in detached tobacco leaves. Physiol. Plant. **19**:723-733.

Treharne, K. J., A. J. Pritchard, and J. P. Cooper. 1971. Variation in photosynthesis and enzyme activity in *Cenchrus ciliaris* L. J. Exp. Bot. **22**: 227-238.

Vishniac, W., and S. Ochoa. 1952. Fixation of carbon dioxide coupled to photochemical reduction of pyridine nucleotides by chloroplast preparations. J. Biol. Chem. **195**:75-93.

Walker, D. A. 1966. Carboxylation in plants. Endeavour **25**:21-26.

Wassink, E. C. 1964. Efficiency of light energy conversion in laboratory experiments and crop growth. [Transl. title.] Meded. Landboewhogesch. Wageningen **64**:1-33.

Waygood, E. R., R. Mache, and C. K. Tan. 1969. Carbon dioxide, the substrate for phosphoenolpyruvate carboxylase from leaves of maize. Canad. J. Bot. **47**:1455-1458.

Wild, A., H.-O. Zickler, and H. Grahl. 1971. Further studies on the variability of the photosynthetic unit. Planta **97**:208-223.

Wild, A., and K. Egle. 1967. Zur Bedeutung von Chlorophyll für die Photosynthese von Chlorella. Biol. Zentralbl. **86**(supp.):495-508.

Woo, K. C., J. M. Anderson, N. K. Boardman, W. J. S. Downton, C. B. Osmond, and S. W. Thorne. 1970. Deficient photosystem II in agranal bundle sheath chloroplasts of C_4 plants. Proc. Nat. Acad. Sci. USA **67**:18-25.

Zelitch, I. 1965. The relation of glycolic acid synthesis to the primary photosynthetic carboxylation reaction in leaves. J. Biol. Chem. **240**:1869-1876.

Zelitch, I. 1966. Increased rate of net photosynthetic carbon dioxide uptake caused by the inhibition of glycolate oxidase. Plant Physiol. **41**: 1623-1631.

Zelitch, I. 1971. Photosynthesis, photorespiration, and plant productivity. Academic Press, Inc., New York.

Zscheile, F. P., and C. L. Comar. 1941. Influence of preoperative procedure on the purity of chlorophyll components as shown by absorption spectra. Bot. Gaz. **102**:463-481.

GENERAL REFERENCES

Arnon, D. I. 1961. Cell free photosynthesis and the energy conversion. *In* D. W. McElroy, and B. Glass. Light and Life. The Johns Hopkins Press, Baltimore.

Bishop, N. J. 1966. Partial reactions of photosynthesis and photoreduction. Ann. Rev. Plant Physiol. **17**:185-208.

Brown, A. H., and A. W. Frenkel. 1953. Photosynthesis. Ann. Rev. Plant Physiol. **4**:23-58.

Calvin, M., and J. A. Bassham. 1962. The photosynthesis of carbon compounds. W. A. Benjamin, Inc., New York.

Clayton, R. K. 1965. Molecular physics in photosynthesis. Blaisdell Publishing Co., Division of Ginn & Co., Waltham, Mass.

Daniels, F. 1967. Direct use of the sun's energy. Amer. Sci. **55**:15-47.

Franck, J., and W. E. Loomis. 1949. Photosynthesis in plants. Monogr. Amer. Soc. Plant Physiol. Iowa State University Press, Ames, Iowa.

Gibbs, M. 1967. Photosynthesis. Ann. Rev. Biochem. 36:757-784.
Kamen, M. D. 1963. Primary processes in photosynthesis. Academic Press, Inc., New York.
Kirk, J. T., and R. A. E. Tilney-Bassett. 1967. The plastids, W. H. Freeman & Co., Publishers, San Francisco.
Krogmann, D. W., and W. H. Powers. 1965. Biochemical dimensions of photosynthesis. Wayne State University Press, Detroit.
Lundegårdh, H. 1925. Klima und Boden. Gustav Fischer Verlag, Stuttgart.
Olson, J. M. 1967. Energy conversion in photosynthesis. Brookhaven Lect. Ser., Pub. No. 62, Brookhaven National Laboratory, Upton, N. Y.
Olson, J. M., G. Hind, H. Lyman, and H. W. Siegelman, 1967. Energy conversion by the photosynthetic apparatus. Brookhaven symposium in Biology, Pub. No. 19. Brookhaven National Laboratory, Upton, N. Y.
Rabinowitch, E. I. 1945. Photosynthesis and related processes. Interscience Publishers, New York.
Rabinowitch, E. I. 1952. Photosynthesis. Ann. Rev. Plant Physiol. 3:229-265.
Rabinowitch, E. I., and Govindjee. 1969. Photosynthesis. John Wiley & Sons, Inc., New York.
San Pietro, A. 1971. Photosynthesis: Methods in enzymology. Vol. 23. Academic Press, Inc., New York.
Smith, J. H. C., and C. S. French. 1963. The major and accessory pigments in photosynthesis. Ann. Rev. Plant Physiol. 14:181-224.
Stiller, M. 1962. The path of carbon in photosynthesis. Ann. Rev. Plant Physiol. 13:151-170.
Thomas, J. B. 1965. Primary photoprocesses in biology. Interscience Publishers, Inc., Division of John Wiley & Sons, Inc., New York.
van Niel, C. B. 1962. The present status of the comparative study of photosynthesis. Ann. Rev. Plant Physiol. 13:1-26.
Vernon, L. P., and M. Avron. 1965. Photosynthesis. Ann. Rev. Biochem. 34:269-296.
Vernon, L. P., and G. R. Seely. 1966. The chlorophylls. Academic Press, Inc., New York.

16

Other metabolic paths

DEPENDENCE ON RESPIRATION AND PHOTOSYNTHESIS

Respiration and photosynthesis involve breakdown and synthesis, respectively, of carbon-containing substances, and they are the two basic processes of all C-metabolism. But they are also the two basic processes for *all* plant metabolism (e.g., N-metabolism) for two reasons:

1. They are the only sources of high energy phosphates. All other syntheses in the plant therefore depend on them for the energy needed to drive their endergonic reactions; and since catabolism is possible only if substances have first been synthesized, catabolism is also dependent on them, at least indirectly as shown in Fig. 16-1.

2. Respiration and photosynthesis also supply the basic organic substances that are the starting points for the synthesis of all other organic substances in the plant. Thus although the beginning of respiration and the end of photosynthesis are represented as hexose, other carbohydrates may also be formed. In the case of photosynthesis, as already indicated, the hexose phosphates may lead to formation of any one of a large number of different carbohydrates. However, the metabolic path may also branch off at some earlier point in the cycle (see diagram on p. 234), leading to synthesis of fats, proteins, etc.

The metabolism of other carbon-containing substances (lipids, glycosides, terpenes, lignins, etc.) is not as well understood as that of the carbohydrates, although great

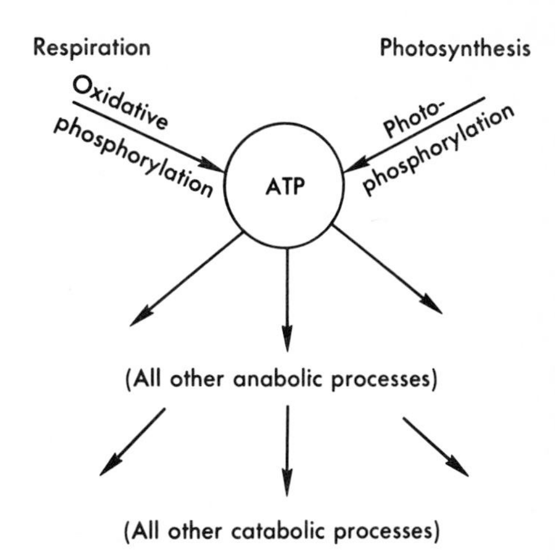

Fig. 16-1. Dependence of all other metabolic processes on respiration and photosynthesis—the only two processes that generate ATP.

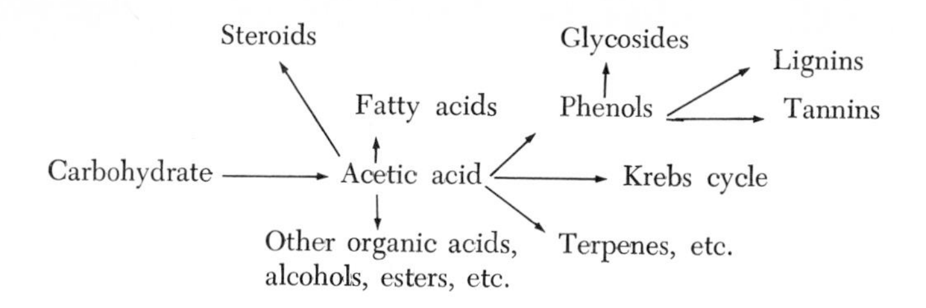

advances in our knowledge concerning it are being made daily. Each of these groups includes a large number of substances that must be synthesized step by step from simpler substances in the plant. The pivot substance in many cases seems to be acetic acid, which links the anaerobic phase of respiration with the aerobic Krebs cycle, and which may also be formed in chloroplasts from photosynthetically fixed carbon dioxide (Shah and Rogers, 1969). We may therefore visualize many metabolic pathways radiating from this point like the spokes of a wheel, each involving its own series of enzymes and each leading to a step-by-step biosynthesis of the substances at the end of the spoke. These interrelations with respiration may be represented as shown above.

There is no basic difference between the metabolism of these substances and that of the carbohydrates, so it will not be considered here. N-metabolism, however, does involve some basic differences that are fundamental to all life.

N-METABOLISM

Proteins and amino acids

The mechanism of protein synthesis is of fundamental importance to many biological disciplines—genetics, evolution, morphology, taxonomy, etc. In physiology it must be understood before the more complex mechanisms of growth and development can be explained. In fact, the true explanations of these processes are only now beginning to unfold as a result of the recent discoveries of the fundamentals of protein synthesis. The process of protein breakdown is taking place constantly in all living protoplasm, although the total protein content may remain unchanged because of a compensating synthesis. In tobacco leaves, half of the proteins may be broken down and resynthesized in about 7 days (Holmsen and Koch, 1964). Both protein catabolism and anabolism are therefore basic to all life. Although the two processes take place at the same rate in the steady state, the rates in the nonsteady state may change independently of each other. In growing cells the rate of breakdown is relatively low and the rate of synthesis is high. Nevertheless, both processes may occur rapidly in absolute terms. In the case of tobacco callus tissue growing exponentially, 18 mg protein/g protein/hr were synthesized and 11 mg were degraded, yielding a net gain of 7 mg protein/g protein/hr (Kemp and Sutton, 1971). In nongrowing cells the rate of breakdown is relatively high and the rate of synthesis may decrease to a very low value. For instance, there may be a considerable breakdown in storage cells, followed by translocation to the growing tissues, such as at the time of bud burst in the spring (Tromp and Ovaa, 1971). However, the rates may not be the same for all proteins. In tobacco leaves the so-called fraction I protein is synthesized more rapidly than other proteins during formation of the leaves and is broken down more rapidly during aging (Dorner et al., 1957). When the leaves are beginning to develop, it accounts for 17% of the protein, after 17 days for 55% of the protein, and after 45 days for only 26% of the protein.

Furthermore, its synthesis is affected by light and it is not found in albino leaves or roots.

Protein breakdown in plants was early investigated in the case of leguminous seeds (Chibnall, 1939). When these high protein seeds germinate, the proteins largely disappear from the nongrowing storage tissues and are replaced by amino acids and amides, which can then be translocated to the growing regions where proteins are resynthesized from them. The greater part of the protein loss can usually be accounted for by the two amides, *asparagine* and *glutamine*; the remainder can be accounted for by a smaller quantity of several amino acids. This would appear to be a digestion process similar to the hydrolysis of carbohydrates to sugars and of fats to fatty acids and glycerol:

Digestion of storage substances

Polysaccharides	Fats	Proteins
↓	↓	↓
Sugars	Glycerol	Amino acids
	+	↓
	Fatty acids	Amides (asparagine and glutamine)

But proteins are far more complex than these other reserves because the polysaccharides are digested to one or a few sugars and each of the fats to no more than three fatty acids, whereas even the simplest protein gives rise on hydrolysis to about twenty different amino acids. In spite of this complexity, by 1968 the amino acid sequence had been worked out for some 200 proteins, including three plant proteins (Nature **218**:1107, 1968). They all contain from fifty to 250 amino acid residues per molecule (Lienier, 1966). This is illustrated by the structure of a small protein molecule (Fig. 16-2), the first true protein to be synthesized in the laboratory.

The complexity of the molecule is indicated by the three levels of structure: the primary (the order of the amino acids in the chain), the secondary (the spiral or helical structure of the chain, not shown here), and the tertiary (the folds in the spiral molecules). Most proteins contain more than one peptide chain, and this gives rise to the quaternary structure of the molecule. The protein molecule of enzymes is usually globular but other proteins may be rod shaped or a random coil. The total of the four levels of structure gives the three-dimensional structure. Difficult as it is to determine the primary structure of proteins, it is much more difficult to determine the three-dimensional structure. By 1969 the complete three-dimensional structure had been worked out for some six proteins (Nature **221**:220, 1969).

The primary structure of a protein molecule results from the strong covalent peptide bonds. The secondary and tertiary structures may be fixed, partly by a small number of strong covalent, disulfide (SS) bonds (Fig. 16-2) but in most cases by a much larger number of weak noncovalent bonds (Fig. 16-3). It was formerly thought that the latter were primarily H-bonds, but modern evidence indicates that although the H-bonds are responsible for the secondary (helical) structure of proteins, the tertiary structure is mainly the result of the weak attraction between the hydrophobic side groups of the protein molecules. The polypeptide chains may be linked either covalently or noncovalently in the quaternary structure.

It is only when the protein possesses its normal, three-dimensional (secondary, tertiary, quaternary) folded structure that it is in the *native* state. Normally, only native proteins are enzymatically active, although there are some exceptions. When it unfolds, it is said to be in the *denatured* state. Not only is the denatured protein inactive enzymatically but many of its other properties may change as well. This denaturation may, however, be reversible, for example, when induced by low temperature (Brandts, 1967). However, if two or more denatured protein molecules combine with each other, even by weak hydrogen or hydrophobic bonds, they form *aggregates* and the denaturation is then irreversible. Denatured proteins are more readily hydrolyzed to peptides and amino acids, which can then serve as raw materials for the resynthesis of native proteins (p. 177). Denaturation may therefore be a normal component of plant metabolism because although the enzyme must be stable under a precise set of conditions, it must also be able to disintegrate when no longer required (Gutfreund, 1967).

According to modern concepts, both the secondary and tertiary structures (i.e., the folding of the molecule) are a direct consequence of the order and chemical nature of the amino acids in the chain (Schachman, 1963; Epstein et al., 1963). The template (see later discussion) is therefore needed only for the primary structure of the polypeptide. An understanding of protein breakdown and synthesis therefore depends on a knowledge of amino acid structure and synthesis. All the amino acids found in proteins are of the α-amino type (i.e., the amino group is on the so-called α-carbon atom, which is next to the carboxyl group):

Generalized structure	*Specific amino acids*
R·CH·COOH (with NH_2 on the CH)	$CH_2 \cdot NH_2 \cdot COOH$ (glycine)
	$CH_3 \cdot CHNH_2 \cdot COOH$ (alanine) etc.

Some amino acids have more than one carboxyl group and are therefore acidic; others have more than one amino group and are basic. Those with one of each group are nearly neutral. In the formation of proteins the amino acids link together

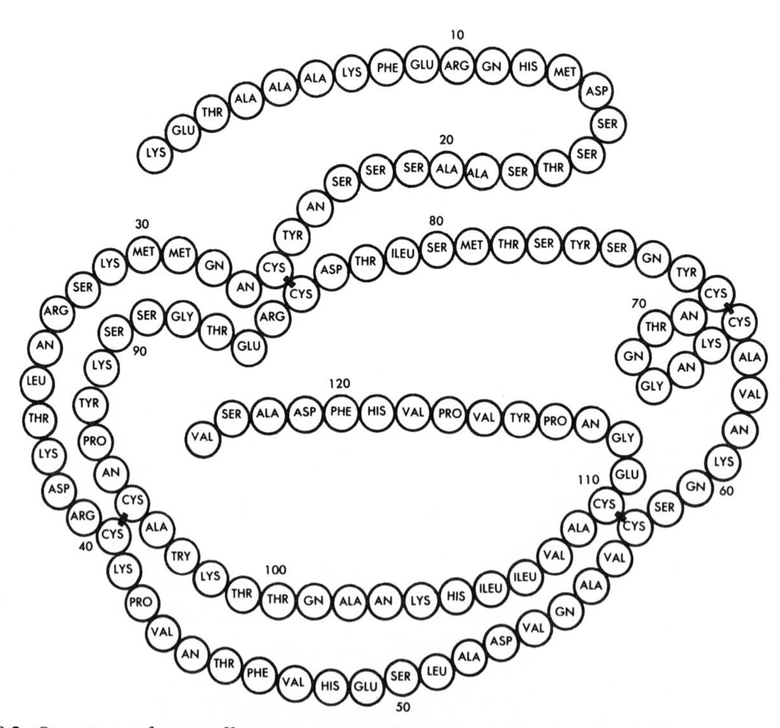

Fig. 16-2. Structure of a small protein molecule (ribonuclease). The folds are held in place by four covalent disulfide bonds joining the eight paired cysteine residues. For symbols, see p. 245. (From Anfinsen, C. R. 1967. *In* J. M. Allen. Molecular organization and biological function. Harper & Row, Publishers, New York.)

to form a peptide as follows:

$$\underset{\text{Amino acid}}{NH_2{\cdot}\overset{\overset{R_1}{|}}{C}H{\cdot}COOH} + \underset{\text{Amino acid}}{NH_2\overset{\overset{R_2}{|}}{C}H{\cdot}COOH} \rightarrow \underset{\text{Dipeptide}}{NH_2{\cdot}\overset{\overset{R_1}{|}}{C}H{\cdot}CO\text{-}NH{\cdot}\overset{\overset{R_2}{|}}{C}H{\cdot}COOH} + H_2O$$

The protein chain therefore has the following makeup:

$$\ldots\ NH \diagdown \underset{\underset{R_1}{|}}{CH} \diagup CO \diagdown NH \diagup \overset{\overset{R_2}{|}}{CH} \diagdown CO \diagup NH \diagdown \underset{\underset{R_3}{|}}{CH} \diagup CO\ \ldots$$

The average molecular weight of an individual amino acid is about 100. The smallest protein molecules contain about 100 such amino acids (less in a few cases) and therefore have molecular weights of approximately 10,000. All proteins contain essentially the same twenty amino acids, but each one differs in the proportion of each amino acid and in the order in which it occurs in the molecule (the primary structure of the molecule). If a protein consists of 200 amino acid residues consisting of the twenty different amino acids, there are 20^{200} possible combinations. The number of different proteins possible in nature staggers the imagination.

Calculations from the amino acid analyses of eighty separate proteins have led to the following numbers of molecules in an "average protein" of 233 amino acid residues and a molecular weight of 25,700 (Smith, 1966):

Asparagine*	(An)	25
Glutamine*	(Gn)	24
Leucine	(Leu)	20
Alanine	(Ala)	19
Glycine	(Gly)	17
Serine	(Ser)	17
Valine	(Val)	16
Lysine	(Lys)	15
Proline	(Pro)	13
Threonine	(Thr)	13
Isoleucine	(Ileu)	11
Arginine	(Arg)	10
Phenylalanine	(Phe)	10
Tyrosine	(Tyr)	8
Histidine	(His)	5
Methionine	(Met)	4
Cysteine	(Cys)	3
Tryptophan	(Trp)	3

*Aspartic acid (Asp) and glutamic acid (Glu) presumably included in asparagine and glutamine, respectively.

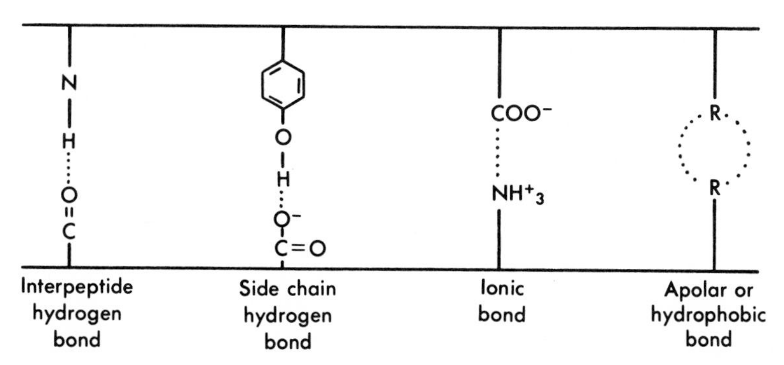

Fig. 16-3. Intramolecular noncovalent bonds (.....) responsible for the secondary and tertiary structure of proteins, and covalent bonds (———) responsible for primary structure. Noncovalent bonds are weak: 2 to 6 kcals per mole, compared to 50 to 200 kcals per mole for most covalent bonds. In some proteins, the covalent SS bond may also participate in the tertiary structure. (From Schachman, H. K. 1963. Sympos. Quant. Biol. **28**:409-430.)

N-assimilation

One characteristic of the higher plant is its ability to synthesize from inorganic nitrogen all the twenty-odd amino acids necessary for the formation of its proteins. Animals, on the other hand, must be supplied with the ten *essential amino acids,* that is, the ten that they are unable to synthesize. The others they can synthesize from these ten. They are unable to synthesize any from inorganic nitrogen. Nitrogen is absorbed by the plant in the inorganic form, either as NO_3^- or as NH_4^+. The NO_3^- must first be reduced to NH_4^+ in a series of steps, probably as follows:

$$NO_3^- \rightarrow NO_2^- \rightarrow N_2O_2^= \rightarrow NH_2OH \rightarrow NH_4^+$$

where NO_3^- = nitrate
NO_2^- = nitrite
N_2O_2 = hyponitrite
NH_2OH = hydroxylamine
NH_4^+ = ammonium

This would seem to indicate that the NH_4^+ form is a more efficient source of nitrogen, since respiratory energy must be used to reduce NO_3^- to this form. However, under suitable conditions both forms are equally satisfactory sources of nitrogen, and the NO_3^- form is suitable over a larger range of conditions (e.g., over a wider pH range).

The NH_4^+ is then converted to organic nitrogen by a reductive combination with carboxylic acids of the Krebs cycle (and pyruvic acid) to give amino acids. The reaction, in simplified form, is as follows:

$$NH_3 + \underset{\text{Pyruvic acid}}{CH_3{\cdot}CO{\cdot}COOH} + 2H \rightleftharpoons$$

$$\underset{\text{Alanine}}{CH_3{\cdot}CHNH_2{\cdot}COOH} + H_2O$$

The two hydrogen atoms are provided by reduced coenzyme. Other amino acids may then be formed from this first one by transamination. Specific enzymes are required for these reactions, for example, NO_3^- reductase for the reduction of NO_3^- and aminases and transaminases for the synthesis of amino acids from NH_4^+. The enzymes that control NO_3^- reduction require molybdenum as a cofactor.

In many cases the amino acid synthesis occurs in the roots of a plant (e.g., the apple), and the amino acids are translocated to the growing shoots, where they are built up to form proteins. In other plants (e.g., wheat), however, the inorganic nitrogen is translocated directly to the leaves and synthesized there to amino acids and proteins. The synthesis in the roots occurs, of course, in the dark and is dependent on respiration for energy and carbon chains. The synthesis in the leaves, on the other hand, may occur only in the light, at least in the case of certain plants. The tracer work on photosynthesis has shown that some amino acids are synthesized within about 5 min after photosynthesis has started. Apparently there is a close connection between photosynthesis and amino acid synthesis in green cells. Further evidence of this close connection is the existence in spinach chloroplasts of two enzymes required for reduction of NO_3^- to NH_4^+ (Losada and Paneque, 1968) and in chloroplasts of *Vicia faba* of an enzyme involved in ammonium incorporation (Leech and Kirk, 1968). Conversely, nitrate reductase is absent from the roots of *Xanthium pennsylvanicum,* and 95% of the nitrogen exported from the root is free NO_3^- (Wallace and Pate, 1967). The more advanced species (e.g., the grasses) synthesize their amino acids in the leaves and only in the light. This is a more efficient process, since the endergonic syntheses use light energy more directly than when dependent on the stored reserves of the root. This is not true, however, of all the grasses, since the amino acids are synthesized from inorganic nitrogen in the roots of maize (Ingverson and Ivanko, 1971) and rice (Oritani and Yoshida, 1970). Even wheat roots, when excised, can synthesize organic nitrogen (Björn, 1967).

Protein synthesis

The synthesis of proteins is dependent on nucleotides, which are also nitrogen-containing substances. They consist of three components:

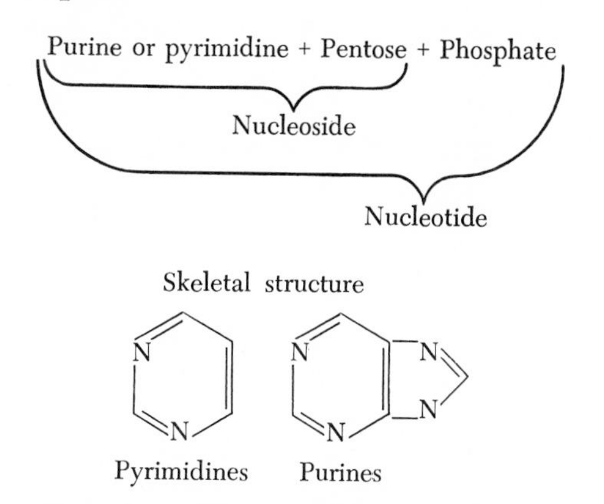

Thus the purine *adenine* combines with the pentose ribose to form the nucleoside *adenosine*, which then combines with phosphoric acid to form the nucleotides *adenosine monophosphate* (AMP), *adenosine diphosphate* (ADP), or *adenosine triphosphate* (ATP). This and other simple nucleotides such as GTP and CTP then may be combined in a long chain to form the complex nucleotides or nucleic acids, just as the amino acids are combined to form the proteins. When nucleotides contain ribose, they form the polymeric molecules of ribonucleic acid (RNA); when they contain deoxyribose, they form deoxyribonucleic acid (DNA). These nucleic acids resemble the proteins in several ways. They have large molecular weights (e.g., 10^6), they occur in the folded form, and they can be denatured (unfolded) just as in the case of proteins, for example, by heating. However, their relation to proteins is much closer than this similarity, since they are required for protein synthesis.

Unlike the carbohydrates, fats, and most other high molecular weight substances that are mainly chemically identical in different plants, proteins differ from species to species and even from variety to variety. Since the hundreds of known proteins differ from each other in the order of arrangement of the twenty amino acids, and since there are probably from 1000 to 10,000 or more different proteins in a single cell (Pridham and Swain, 1965), there must be some process other than simple enzymatic catalysis that controls this order. Indeed there could not be enough different, specific enzymes in a cell to control each of the 100 to 10,000 successive amino acid linkages required for synthesis of even a single protein. Furthermore, since enzymes are themselves proteins, the problem would multiply geometrically, and this process would be impossible under any condition. The method of synthesis must therefore be basically different from that of any other substances in the plant. This problem has been the subject of intense investigations in recent years. Most of the work has been done with microorganisms, but the mechanisms worked out for them probably also apply in general to higher plants, since this is true of metabolism as a whole. That this assumption is basically correct is indicated by the newer evidence accumulating from direct experiment with higher plants.

For decades there has been a *template* theory of protein synthesis. According to this concept, a template or pattern for each specific protein must first be made, and the amino acids are laid down on the pattern according to the order specified by it. Direct evidence for this theory has been accumulating. The more specific template concept (or "dogma") of today is as follows.

The function of nucleic acids is the storage and transcription of biological information (Spiegelman and Hayashi, 1963):

$$\text{DNA} \xrightarrow{\text{Duplication}} \text{DNA} \xrightarrow{\text{Transcription}} \text{RNA} \xrightarrow{\text{Translation}} \text{Protein}$$

The function of the protein is the expression of this information. Protein synthesis is controlled initially by a nucleic acid (DNA) in the nucleus or, more specifically, in the genes of the chromosomes. It has even been reported that a double recessive white-flowered *Petunia* was converted into a red-flowered form by treating the young seedling with DNA isolated from a red-flowered pure line (Hess, 1969). The nuclear DNA itself is a double polymer of purines (guanine and adenine) and pyrimidines (cytosine and thymine). Adenine is always paired with thymine, cytosine with guanine. Longitudinal splitting of the DNA yields reciprocal half polymers, each of which regenerates the missing half, thus multiplying the DNA. However, these half polymers can also serve as templates, on which a different nucleic acid (messenger RNA, or mRNA) is formed, with a complementary sequence of units. RNA differs from DNA by having uracil instead of thymine and ribose instead of deoxyribose. The mRNA then peels off, passes from the nucleus into the cytoplasm, and becomes associated with a polyribosome. It then acts in turn as a template on which is deposited a series of small RNA molecules—the transfer or soluble RNA (tRNA or sRNA). Beginning at one end of a messenger molecule, a series of these small tRNA molecules attach successively to it. In each case the tRNA has a structure complementary to that of the mRNA and holds a specific amino acid attached to it. In this way, as amino acids are brought in, they attach to each other by peptide linkages and finally peel off as a polypeptide chain, that is, a protein molecule (Fig. 16-4). The polypeptide chain then folds to form the native, folded protein (Fig. 16-5). More recent evidence suggests that the nascent polypeptide chain folds as it is being synthesized on the ribosome in vivo (Teipel and Koshland, 1971). Minor changes may, however, continue in vitro over much longer periods of time.

The released proteins do not necessarily remain in the soluble molecular form. They

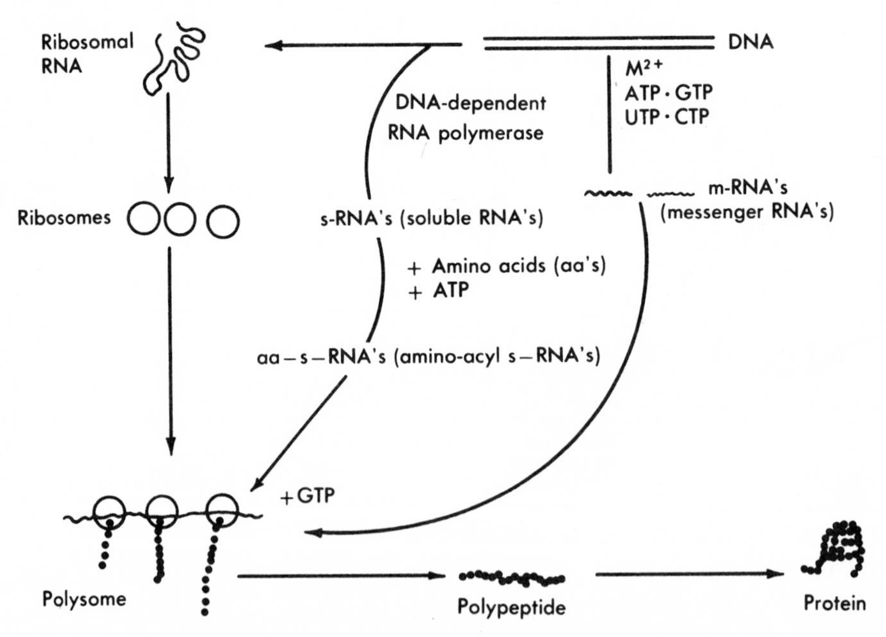

Fig. 16-4. Schematic model of protein synthesis (translation process). M^{2+} = divalent cation activator; ATP = adenosine triphosphate; CTP = cytidine triphosphate; GTP = guanosine triphosphate; UTP = uridine triphosphate; DNA = deoxyribonucleic acid; RNA = ribonucleic acid. (From Cohen, N. R. 1966. Biol. Rev. 41:503-560.)

may link up with other proteins or with other substances (lipids, carbohydrates, nucleic acids, etc.) forming *conjugated proteins*. Protein synthesis has been stated to require up to 90% of the biosynthetic energy of the cell (Lehninger, 1965). This energy is required not only for the synthesis of the amino acids but also for their incorporation into the protein molecule. It is required, for instance, for attachment of the amino acid to its specific tRNA (Novelli, 1967):

$$E + ATP + AA_1 \rightleftharpoons E - AA_1 - AMP + PP_i$$
$$E - AA_1 - AMP + tRNA^1 \rightleftharpoons AA_1 - tRNA^1 + E + AMP$$

where E = enzyme aminoacyl—tRNA synthetase specific for AA_1
AA_1 = a specific amino acid
$tRNA^1$ = its specific transfer RNA

Although most of the work on which the preceding concepts are based has been done with microorganisms and animal cells, investigations have shown that the ribosomes of higher plants have characteristics similar to those of bacteria and mammalian cells and are therefore capable of the same mechanism of protein synthesis (Marcus and Feeley, 1965). In wheat embryos, for instance, as in the case of many microorganisms, the polypeptide chains are initiated by a specific methionine-carrying tRNA molecule (Marcus et al., 1970).

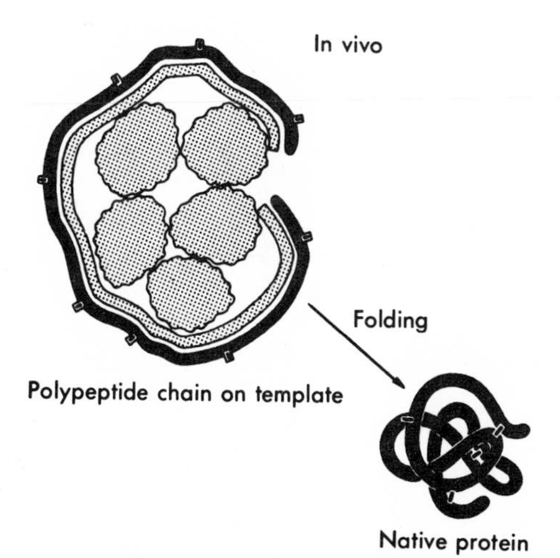

Fig. 16-5. Schematic diagram of the conversion of a newly synthesized polypeptide chain to a folded "native" protein. The eight protruding attachments on the chain represent chemical side groups which later combine to form four bonds holding the folds in place. (From Epstein, C. J., et al. 1963. Sympos. Quant. Biol. **28**:439-449.)

It is now known that protein synthesis occurs not only on the ribosomes that are free in the cytoplasm but also in the mitochondria and chloroplasts, presumably on the ribosomes that have been found within these organelles. Just as in the case of the nucleus, protein synthesis in these organelles is under the control of their own DNA (Shephard, 1965). About 9% of the total tobacco leaf DNA is in the chloroplasts (Tewari and Wildman, 1966), and 20% in bean leaves (Gyldenholm, 1968). It differs both physically and chemically from nuclear DNA. Chloroplast DNA is also capable of replication more rapidly than nuclear DNA in the case of tobacco (Green and Gordon, 1966). Chloroplast DNA is just as capable of synthesizing RNA as is nuclear DNA (Spencer and Whitfeld, 1967; Gibbs, 1967), and about 15% to 25% of the total leaf RNA is located in the chloroplasts (Wollgiehn et al., 1966).

It has been suggested that all the enzymes of photosynthesis are synthesized within the chloroplasts on chloroplast ribosomes and that chloroplast DNA is involved in the formation of this chloroplast ribosomal system (Smillie et al., 1968). In favor of this conclusion, the recovery of photosynthetic electron transport in inhibited cells of the green alga *Chlamydomonas reinhardtii* requires the transcription of chloroplast DNA (Goodenough and Levine, 1971). At least two proteins of the chloroplast lamellae are apparently synthesized by the chloroplast ribosomes (Machold, 1971), as well as RuDP carboxylase and ribosomal proteins, although other chloroplast proteins apparently are not (Ellis and Hartley, 1971). Similarly, in mitochondria, the protein of the outer membrane is apparently synthesized in the cytoplasm, but on the other hand the mitochondrial ribosomes synthesize proteins that are integrated into the inner mitochondrial membrane (Neupert and Ludwig, 1971).

Other nitrogen-containing substances

There are, of course, many nitrogen-containing substances in the plant beside proteins. Some 4000 alkaloids have been found in 3600 species of plants (Leete, 1967), and others are being discovered at a rapid rate. These plants must have a special metabolism not found in all species and even differing within those species that synthesize them (Willaman and Li, 1963).

Since all these nitrogen-containing substances in the plant (amines, betaines, alkaloids, purines, pyrimidines) are derived indirectly from the inorganic nitrogen (NO_3^- or NH_4^+) obtained from the soil, they must be related in some way to the first-formed organic substances—the amino acids. Just as acetic acid and pyruvic acid may be considered as pivot substances for C-metabolism, the amino acids must act as the pivot for N-metabolism as follows:

$$\begin{array}{c} \text{Amides Betaines Amines} \\ \nwarrow \;\; \uparrow \;\; \nearrow \end{array}$$

$$\left.\begin{array}{l} NO_3^- \\ \text{or} \\ NH_4^+ \end{array}\right. + \text{R·CO·COOH} \longrightarrow \text{Amino acids} \rightleftharpoons \text{Proteins}$$

$$\downarrow$$

$$\text{Alkaloids} \leftarrow \left\{\begin{array}{c} \text{Purines} \\ \text{and} \\ \text{pyrimidines} \end{array}\right\} \longrightarrow \begin{array}{c}\text{Nucleo-} \\ \text{tides}\end{array}$$

where R·CO·COOH = an α-keto acid

MINERAL METABOLISM

Although all the mineral elements play some role in the metabolism of the plant, it does not necessarily follow that they are all themselves metabolized. Manganese, for instance, is active in the ionic form, since it is an enzyme activator. Iron, on the other hand, becomes a component of organic substances (e.g., of the cytochrome molecule) and therefore is metabolized. The metabolism of the mineral elements is not as well worked out as the metabolism of most of the carbon- and nitrogen-containing compounds, even though a much smaller number of substances is involved. The two mineral elements with the most complex metabolism are phosphorus and sulfur. P-metabolism is an essential component of both respiration and photosynthesis and of both C-metabolism and N-metabolism. Yet its metabolism is relatively simple, since the plant is unable to change the oxidation state of phosphorus. It is absorbed in the most highly oxidized form (+5) as the inorganic phosphate, which is merely transferred to a vast number of organic substances, forming organic phosphates. Sulfur, on the other hand, is also absorbed in the most highly oxidized form (+6, inorganic sulfate), but it can be reduced in the plant to all possible levels. S-metabolism is therefore in itself complex. The most important sulfur-containing substances in the plant are the three amino acids *(cysteine, cystine,* and *methionine),* the peptide *glutathione* (glutamyl-cysteinylglycine), and certain vitamins or enzyme cofactors *(thiamine, biotin,* and *thioctic acid).* It is as the mercaptan, containing the thiol (SH, or sulfhydryl) group, and as the corresponding oxidized disulfide form that sulfur is most reactive in the plant.

The interrelations between respiration and the other phases of plant metabolism are shown in Fig. 16-6.

CONTROL OF METABOLISM

The chemical reactions in the various metabolic pathways are all controlled by enzymes and therefore by proteins. The discovery of the enzymes that catalyze the individual reactions of respiration, photosynthesis, and other metabolic paths has therefore revealed the specific proteins that enable these processes to take place. However, there is a vast number of these chemical reactions taking place in the plant, and indeed in each cell or even each organelle at any one time, and many of these are interdependent. How does the plant succeed in producing each of these enzymes in the right quantity, in the right place, and at the right time in order to ensure the continuation of each process at the speed and in the order needed for the normal metabolism of the plant?

The importance of this question goes far beyond an understanding of metabolism itself. Most physiological processes, including the growth and development of the plant (Chapters 19 and 20), are dependent on metabolic energy. The physiology of the plant is therefore largely controlled by its metabolism. The factors that control the metabolism of the plant also control all the physiological processes that are dependent on metabolic energy. Hence, no true understanding of plant physiology is possible without an understanding of the control of metabolism.

Since the metabolic reactions are all catalyzed by enzymes, any control of enzyme activity will control metabolism. There are two basic kinds of enzyme control that may, however, be interdependent:

1. All enzymes can be inhibited or con-

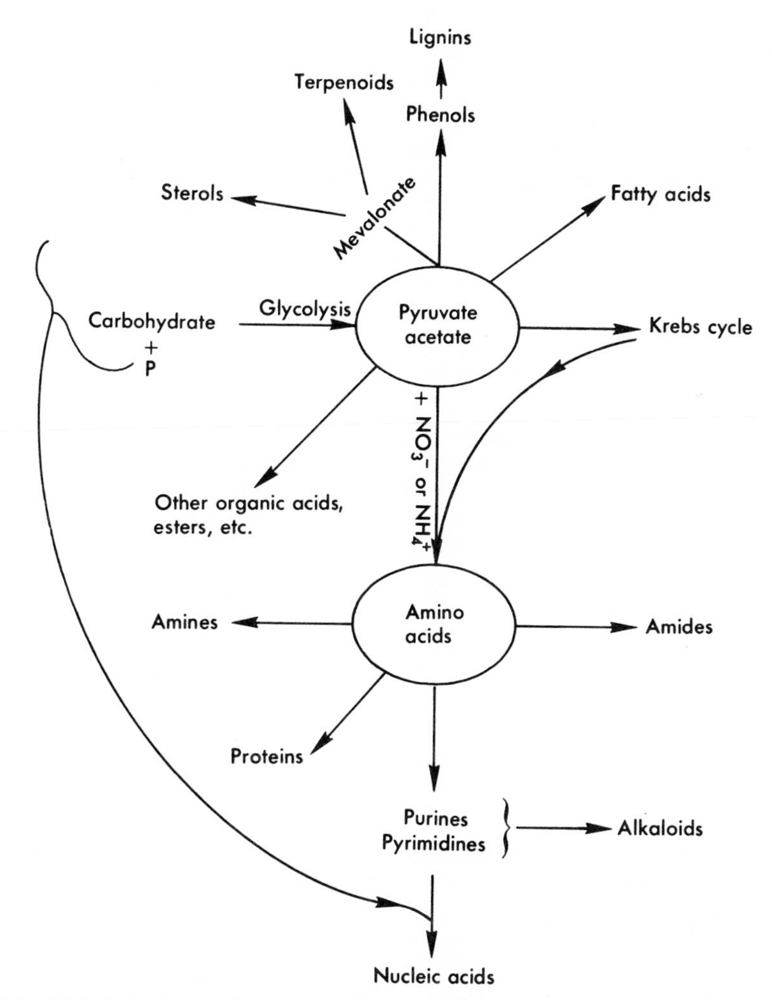

Fig. 16-6. Interrelations between respiration and other phases of plant metabolism.

versely, deinhibited or activated in various ways, and these inhibiting and activating factors can therefore control metabolism.

2. Since enzymes are proteins, any factor that affects the rates of synthesis and breakdown of proteins will also control metabolism.

From this second control mechanism, it is now obvious why the mechanism of protein synthesis and its relation to DNA and RNA are of such fundamental importance to the physiologist. As a result of the tremendous research explosion in this area in recent years, the mechanism of the control of metabolism is gradually unfolding.

Inhibition and activation of enzymes

The first mechanism, enzyme inhibition or activation, is believed to be a part of all control devices (Perutz, 1967), since the synthesis or breakdown of inactive enzymes can have no effect by itself on metabolism. The protein molecules, comprising the enzymes, are able to change their structure reversibly from an active to an inactive form in response to chemical stimuli. Thus sulfhydryl enzymes (e.g., many dehydrogenases) are inactivated reversibly by oxidation of the sulfhydryl group to disulfide. Conversely, disulfide enzymes (e.g., ribonuclease) are reversibly inactivated by reduction of their disulfide group to sulfhydryl. There are other activation and inactivation processes not involving sulfhydryl groups. The chemical stimuli that induce these reversible changes are known as *effectors* or *modifiers,* which are substances that positively or negatively modify the affinity of an enzyme for its substrate, presumably by changing it from the inactive to the active form, or vice versa. They are of two main kinds: (1) *active site effectors,* which produce their effect directly on the active sites of the enzyme molecule as previously described and (2) *allosteric effectors,* which attach to some part of the enzyme molecule other than the active site and produce their effect by a conformational change of the protein molecule. Due to the conformational change, allosteric effectors alter the active site indirectly and therefore alter its reactivity with the substrate (Kelly and Turner, 1970). Melanin appears to be a negative allosteric effector for the enzyme cellulase, bonding to it electrostatically at enzyme groups other than those at the active center (Bull, 1970). It is now obvious that the enzyme cofactors discussed in Chapter 13 are really positive effectors. Some of the coenzymes or prosthetic groups are active site effectors, and the activating ions are usually allosteric effectors.

As in the case of any chemical reaction, the enzymatically controlled reactions may be slowed down or halted by mass action. Malate oxidation to oxaloacetate, for instance, is inhibited by the accumulation of oxaloacetate, and any substance capable of lowering the concentration of oxaloacetate increases the rate of malate oxidation (Hobson, 1970). Conversely, the rate of starch synthesis in the endosperm of developing ears of wheat increases with the concentration of sucrose, from which it is formed (Jenner, 1970). In many cases, however, an apparent mass action effect is actually due to enzyme inhibition or activation, by *feedback control* or *end-product inhibition,* in which an end-product inhibits an enzyme activity in a much earlier reaction. L-malate, for instance, is the end product of carbon dioxide fixation by PEP in the presence of PEP carboxylase (Chapters 14 and 15). The accumulation of L-malate may regulate carbon dioxide fixation in corn root tips by end-product inhibition (Ting, 1968). Similarly, in the cotyledons of *Vicia faba,* synthesis of sucrose-6-phosphate is inhibited by sucrose. This inhibition causes an accumulation of UDPG, which in its turn inhibits further starch breakdown (De Fekete, 1969). The following are the reactions involved:

$$\text{Starch} \underset{P_i}{\overset{\text{Phosphorylase}}{\rightleftharpoons}} \text{G-1-P} \underset{\text{UTP}}{\overset{\text{UDPG pyrophosphorylase}}{\rightleftharpoons}} \text{UDPG} + \text{PP}_i$$

Feedback control

$$\text{UDPG} + \text{F-6-P} \overset{\text{S-6-P synthetase}}{\rightleftharpoons} \text{UDP} + \text{S-6-P} \overset{\text{Sucrose phosphatase}}{\rightleftharpoons} \text{Sucrose}$$

where F-6-P = fructose-6-phosphate
G-1-P = glucose-1-phosphate
S-6-P = sucrose-6-phosphate
UDPG = uridine diphosphoglucose
UTP = uridine triphosphate
P_i = inorganic phosphate

The reverse kind of control, precursor activation, may also occur. These two mechanisms may also involve ATP and ADP. The formation of ATP may inhibit a metabolic process, and, conversely, ADP or AMP may stimulate it (Atkinson, 1966). An example of this control is phosphofructokinase in carrot plants, which is controlled by citrate and ATP concentration (Dennis and Coultate, 1966).

Many other examples of such control have been reported. The ratios of ATP to ADP and of ATP to AMP appear to control the activities of enzymes in germinating lettuce seed (Pradet, 1969). As a result of the different responses of two enzymes of the Calvin cycle to ATP, carbon dioxide fixation may be enhanced by low concentrations of ATP and inhibited by concentrations above 2 mM (Champigny and Bismuth, 1970). The whole course of metabolism may therefore be altered by changes in intracellular concentrations of these substances. A steady state of ATP concentration may, however, be maintained. In rapidly growing algal cells the ATP remains at a rather uniform level, averaging 0.35% of the organic matter, and stays the same whether the cells are in the light or the dark (Holm-Hansen, 1970).

Repression and derepression of enzymes

The second mechanism must be the primary step in enzyme control, since it involves either *repression* or *derepression* of the synthesis of the enzyme protein. For instance, the products of hydrolysis that accumulate in the starchy endosperm of germinating barley seeds control the reaction by regulating the *production* of hydrolytic enzymes by the aleurone layer (Jones and Armstrong, 1971). The importance of this control is obvious from the fact that although each cell contains at least one copy of all of the organism's genetic information, only a small part is usually active in synthesizing proteins at any one time. Most of the DNA in the native chromatin is presumably repressed. The number of genes in man, for instance, is between 6×10^4 and 6×10^6, yet the total number of enzymes is less than 2000 (Pridham and Swain, 1965). Part of the gene population must therefore be concerned with controlling activities, rather than with a simple one gene–one enzyme production. Thus RNA from pea buds does not synthesize pea seed globulin; RNA from developing pea cotyledon does (Bonner et al., 1967). The repressors are believed to be basic proteins known as *histones,* which prevent chromosomal DNA from acting as a template. In agreement with this concept the vegetative nucleus of pollen grains shows high RNA and protein synthesis and no nucleohistone; the generative nucleus shows no detectable synthesis and a high degree of nucleohistone (Sauter and Marquardt, 1967). Furthermore, this protein synthesis controls even the replication of the genetic material itself. In principle, this second mechanism of control may be exerted at different levels: (1) DNA → mRNA, or the transcription level; (2) stability of mRNA; and (3) mRNA → protein, or the translation level. In bacteria, at least, the major control is at the third level—the level of translation (Vogel and Vogel, 1967). The first-described mechanism (inhibition by reversible structural change) can only follow the preceding three control levels and may therefore be considered a fourth level of control.

Both repression and derepression of the synthesis of a single enzyme have been identified. The derepression results in enzyme synthesis. Any such increase in rate of enzyme synthesis in response to a change in a specific environmental parameter is called *enzyme induction* (Filner et al., 1969). Many cases of apparent enzyme induction in response to substrates, hormones, light, dark, air, water, and development have been reported in higher plants, although few have been proved categorically. The most thoroughly investigated is nitrate reductase. In cultured tobacco cells, for instance, the enzyme is induced by the presence of nitrate in the medium but disappears as a result of nitrate starvation (Filner, 1966). Levels as low as 10 μM NO_3 may lead to maximum induction, but if ammonium is included with the nitrate, it partially prevents the induction of nitrate reductase in excised barley roots (Smith and Thompson, 1971). Similarly, the enzyme is not formed in cells grown on casein hydrolysate as the sole nitrogen source. Some amino acids repress synthesis of the enzyme, others derepress it in the presence of the repressor amino acids. Such enzymes, whose formation is selectively provoked by the substrate, are called *adaptive enzymes*. Most of the plant's enzymes are probably *constitutive*—their synthesis is not provoked by the substrate (Bell, 1965).

Amino acids may control other enzymes besides nitrate reductase. It has long been known, in fact, that many amino acids when applied in the culture medium may inhibit growth. In the case of isolated, germinating barley embryos, this seemed to be due to the inhibition of two reactions (Miflin, 1969). In the case of tobacco callus tissue, glycine, aspartic acid, asparagine, and tyrosine inhibited cell growth 25% to 30% (Koiwai et al., 1970). Glycine, methionine, lysine, and ornithine repressed glutamic acid decarboxylase; aspartic acid and tyrosine were negative effectors for the same enzyme. Several of the methods available to the plant for controlling its metabolism by enzyme control are indicated in Fig. 16-7.

Although, as previously mentioned, proteins are continuously being synthesized and broken down, the cessation of synthesis of an enzyme does not normally lead to its rapid disappearance. Another control mechanism is needed—the presence of proteins capable of inhibiting or degrading enzymes that are no longer useful to the cell (Filner et al., 1969). This has been shown in the case of nitrate reductase. The amount present in barley leaves may be controlled by a balance between activating and inactivating systems (Travis et al., 1969). In some cases only specific isoenzymes are controlled in this way. Of nine phosphorylase isoenzymes present in potato tubers during starch synthesis, only two remained during starch breakdown (Gerbrandy and Verleur, 1971). The existence of such multiple forms of the same enzyme provides another control mechanism, since each isoenzyme may be controlled by a different metabolite (Datta, 1969). In any one metabolic process, such as in photosynthetic C-assimilation, several of these control mechanisms may operate (Bassham, 1971).

There are still other control mechanisms that are not fully understood. A basic mechanism is the formation and breakdown of polyribosomes. In 10-day-old corn seedlings, the apparent requirement of light for the induction of NO_3^- reductase is the result of the dependence of polyribosome formation on light (Travis and Key, 1971).

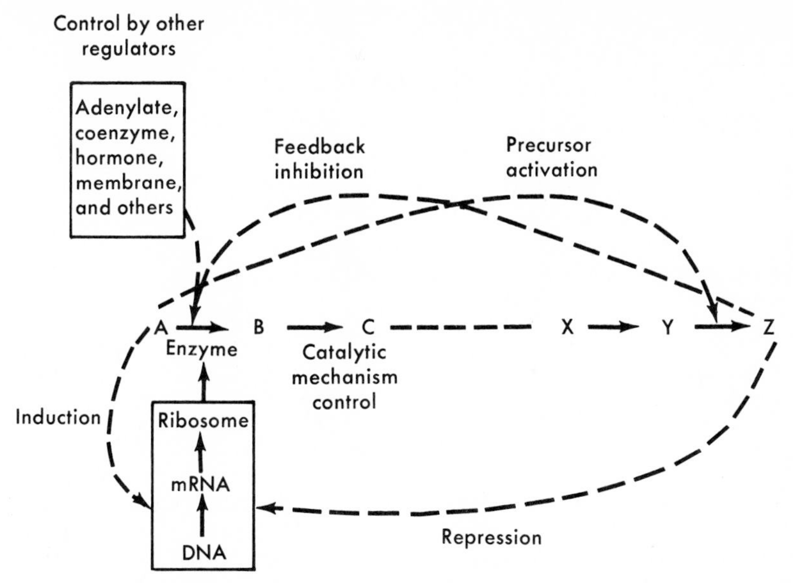

Fig. 16-7. Metabolic control by control of enzyme activity. (From Hammes, G. G., and C.-W. Wu. 1971. Science **172**:1205-1211.)

A newly discovered factor in metabolic control is cyclic 3,5-AMP, which is involved in the induction of the enzymes of barley endosperm (Nickells et al., 1971). Although its role in plant metabolism is only beginning to unfold, its importance is indicated by its accumulation just before cell activation, for instance, in germinating lettuce seed after 2 hr of imbibition (Narayanan et al., 1970). Hormonal control of metabolism will be considered in Chapter 19.

Artificial control is also possible by supplying antibiotics. Actinomycin D prevents the formation of mRNA from DNA (i.e., it inhibits transcription), puromycin appears to destroy the activated amino acid-tRNA complex, and chloramphenicol inhibits the transfer of activated amino acids to mRNA (i.e., they inhibit the translation process). Other antibiotics inhibit other steps in protein synthesis.

It must be realized, of course, that the control of enzyme activity or synthesis is merely a means to an end, not an end in itself. The actual goal of regulatory control is twofold (Atkinson, 1966): (1) to maintain an essentially constant level of building blocks (ATP, NADPH, etc.), despite wide fluctuation in demand; and (2) to facilitate future maintenance by storage of excess energy.

SECONDARY METABOLITES

The aforementioned control obviously cannot be perfect in all cases, and an escape valve must be available if the plant is to survive. If, for instance, one metabolic chain of reactions is too active, leading to the accumulation of toxic quantities of a substance, the plant must have a mechanism of getting rid of the excess. In the case of simple substances, an enzyme may be present whose function is merely to destroy this toxic substance as rapidly as it is formed. This is apparently the function of the ubiquitous (in aerobic organisms) catalase. It destroys hydrogen peroxide at the rate of 2000 molecules per enzyme molecule per second, thus preventing the accumulation of toxic quantities from aerobic respiration. But most metabolites are too complex to be broken down into such simple substances as water and oxygen. If they are toxic, they must be excreted. Even before excretion they may have to be converted first into a less toxic waste substance. Different species possess different enzymes and metabolic paths and therefore may produce different waste metabolites.

Unlike the animal, the plant does not possess a system for excreting toxic or waste metabolites into the surrounding environment. Instead, each cell has its own internal garbage can—the vacuole. This has one advantage over the other system because if conditions change, so that what was formerly waste or toxic is now needed, it can be reabsorbed from the vacuole. This system is not suited to the animal, which produces so much more waste material than the plant, and unlike the plant, cannot reutilize its waste.

It is usually stated that the function of such waste products is unknown, since many of them remain unaltered in the cell until its death. Some, however, are now known to be reutilized to a degree and therefore may be considered as storage metabolites. But even the vast number of metabolites that are not reutilized may be considered as playing a role in plant metabolism, since in many cases they are the end products of detoxification mechanisms in the plant. It is also possible, however, that many of these waste metabolites are simply the end products of metabolic pathways that may be considered blind alleys. The plant happens to have a series of enzymes leading to these end products but no enzymes to convert them once they are formed.

Since this is a polyglot group of metabolites, including toxic substances, waste substances, substances possibly sometimes serving as reserves, and just accidental end products, they are all grouped together as *secondary metabolites.* They are distinguished from the *primary metabolites,* substances that (1) are found in essentially all plants and (2) play reasonably well understood roles in the physiology of the plant. It is, of course, difficult if not impossible to draw a sharp line between the two groups. For instance, many phenolic substances such as tannins and many terpenoids such as essential oils are definitely secondary metabolites without a known function. Yet combinations of the two (quinone-terpenoids) form the well-known ubiquinones and plastoquinones that are essential components of respiration and photosynthesis, respectively. Also, some of the precursors of terpene synthesis are involved in the synthesis of essential substances such as steroids and gibberellins.

The major secondary metabolites are the alkaloids, terpenes (essential oils, resins, rubber), many phenolic substances (such as tannins), glycosides, etc. Some of the more complex secondary metabolites are found in only one or a few taxonomic groups; the simpler ones are usually more widespread. In the case of the latter, the synthesis may depend on enzymes whose normal function is in association with the primary metabolites. Thus phenolic substances are probably formed from acetyl-coenzyme A in fungi and from aromatic amino acids in higher plants (Neish, 1964). The number of secondary metabolites in plants is large. Some 4000 alkaloids alone are known. Aside from these and the macromolecules, nearly 1000 simpler organic substances have been isolated from plants and chemically identified (Mentzer, 1964). Some of the secondary metabolites accumulate in extraordinarily large amounts; for example, tannins account for 14.5% of the dry matter of some leaves (El Sissi and Abd Alla, 1966). The metabolic pathways involved in the synthesis of these secondary metabolites are gradually being discovered, for example, for rubber (Archer and Audley, 1967). In many cases, however, little is known as yet.

QUESTIONS

1. What chemical change occurs when high protein seeds germinate?
2. What substances accumulate in large

quantity when high protein seeds germinate?

3. What substances accumulate in smaller quantity?
4. When a plant is supplied with inorganic nitrogen, what must it synthesize before proteins?
5. Are any amino acids essential for higher plants?
6. Why are (are not) amino acids essential for higher plants?
7. What is the difference between higher plants and animals in this respect?
8. Where are the amino acids synthesized in the plant?
9. Where are the proteins synthesized?
10. What structures are involved in protein synthesis?
11. Is there any relation between protein synthesis and respiration?
12. Is there any relation between protein synthesis and photosynthesis?
13. When lipids disappear in higher plants, what other substances often disappear?
14. When lipids accumulate, what other substances disappear?
15. What are glycosides?
16. Give an example of a glycoside.
17. On what chemical substances does protein synthesis depend?
18. What basic difference exists between the mechanism of protein synthesis and the synthesis of other substances in the plant?
19. What are effectors?
20. What is repression?
21. What are secondary metabolites?

SPECIFIC REFERENCES

Anfinsen, C. R. 1967. Molecular structure and the function of proteins, p. 1-19. *In* J. M. Allen. Molecular organization and biological function. Harper & Row, Publishers, New York.

Archer, B. L., and B. G. Audley. 1967. Biosynthesis of rubber. Adv. Enzymol. **29**:221-257.

Atkinson, D. E. 1966. Regulation of enzyme activity. Ann. Rev. Biochem. **35**:85-124.

Bassham, J. A. 1971. The control of photosynthetic carbon metabolism. Science **172**:526-534.

Bell, E. 1965. Molecular and cellular aspects of development. Harper & Row, Publishers, New York.

Björn, L. O. 1967. Some effects of light on excised wheat roots with special reference to peroxide metabolism. Physiol. Plant. **20**:149-170.

Bonner, J., M. E. Dahmus, D. Fambrough, R. C. Huang, K. Marushige, and D. Y. H. Tuan. 1967. The biology of isolated chromatin. Science **159**: 47-56.

Brandts, J. F. 1967. Heat effects on proteins and enzymes, p. 25-72. *In* A. H. Rose. Thermobiology. Academic Press, Inc., New York.

Bull, A. T. 1970. Kinetics of cellulase inactivation by melanin. Enzymologia **39**:333-347.

Champigny, M. L., and E. Bismuth. 1970. Effect of ATP and Mg^{2+} on the ribulose-diphosphate carboxylase and ribose-5-phosphate isomerase and kinase activities in an enzymatic preparation of chloroplasts. Canad. J. Bot. **48**:1227-1233.

Datta, P. 1969. Regulation of branched biosynthetic pathways in bacteria. Science **165**: 556-562.

De Fekete, M. A. R. 1969. Zum Stoffwechsel der Staerke. II. Die Unwandlung der Staerke in Saccharose in den Kotyledonen von *Vicia faba*. Planta **87**:324-332.

Dennis, D. T., and T. P. Coultate. 1966. Phosphofructokinase, a regulatory enzyme in plants. Biochem. Biophys. Res. Commun. **25**:187-191.

Dorner, R. W., A. Kahn, and S. G. Wildman. 1957. The proteins of green leaves. VII. Synthesis and decay of the cytoplasmic proteins during the life of the tobacco leaf. J. Biol. Chem. **229**: 945-952.

El Sissi, H. I., and M. F. Abd Alla. 1966. Polyphenolics of the leaves of *Catha edulis* (L.). Planta Med. **14**:76-83.

Ellis, R. J., and M. R. Hartley. 1971. The sites of synthesis of chloroplast proteins. Biochem. J. **124**:11P-12P.

Epstein, C. J., R. F. Goldberger, and C. B. Anfinsen. 1963. The genetic control of tertiary protein structure; studies with model systems. Sympos. Quant. Biol. **28**:439-449.

Filner, P. 1966. Regulation of nitrate reductase in cultured tobacco cells. Biochim. Biophys. Acta **118**:299-310.

Filner, P., J. L. Wray, and J. E. Varner. 1969. Enzyme induction in higher plants: environmental or developmental changes cause many enzyme activities of higher plants to rise or fall. Science **165**:358-367.

Fowden, L. 1965. The chemical approach to plants. Sci. Progr. **53**:583-599.

Gerbrandy, S. J., and J. D. Verleur. 1971. Phosphorylase isoenzymes: localization and occurrence in different plant organs in relation to starch metabolism. Phytochem. **10**:261-266.

Gibbs, S. P. 1967. Synthesis of chloroplast RNA at the site of chloroplast DNA. Biochem. Biophys. Res. Commun. **28**:653-657.

Goodenough, U. W., and R. P. Levine. 1971. The effects of inhibitors of RNA and protein synthesis on the recovery of chloroplast ribosomes, membrane organization, and photosynthetic electron transport in the ac-20 strain of *Chlamydomonas reinhardtii.* J. Cell Biol. **50**: 50-62.

Green, B. R., and M. P. Gordon. 1966. Replication of chloroplast DNA of tobacco. Science **152**: 1071-1074.

Gutfreund, H. 1967. The dynamic behavior of proteins during catalysis. Naturwissenschaften **54**:402-406.

Gyldenholm, A. D. 1968. Macromolecular physiology of plastids. V. On the nucleic acid metabolism during chloroplast development. Hereditas **59**:142-168.

Hammes, G. G., and C.-W. Wu. 1971. Regulation of enzyme activity. Science **172**:1205-1211.

Hess, D. 1969. Versuche zur Transformation an hoheren Pflanzen. Wiederholung der Anthocyan-Induktion bei *Petunia* und erste Charakterisierung des transformierenden Prinzips. Z. Pflanzenphysiol. **61**:286-298.

Hobson, G. E. 1970. The oxidation of malate by mitochondria from normal and abnormal tomato fruit. Phytochemistry **9**:2257-2263.

Holm-Hansen, O. 1970. ATP levels in algal cells as influenced by environmental conditions. Plant Cell Physiol. **11**:689-700.

Holmsen, T. W., and A. L. Koch. 1964. An estimate of protein turnover in growing tobacco plants. Phytochemistry **3**:165-172.

Ingversen, J., and S. Ivanko. 1971. Investigation on the assimilation of nitrogen by maize roots and the transport of some major nitrogen compounds by xylem sap. II. Incorporation of taken-up nitrogen into free amino acids and proteins of maize roots. Physiol. Plant. **24**:199-204.

Jenner, C. F. 1970. Relationship between levels of soluble carbohydrate and starch synthesis in detached ears of wheat. Aust. J. Biol. Sci. **23**:991-1003.

Jones, R. L., and J. E. Armstrong. 1971. Evidence for osmotic regulation of hydrolytic enzyme production in germinating barley seeds. Plant Physiol. **148**:137-142.

Kelly, G. J., and J. F. Turner. 1970. The regulation of pea-seed phosphofructokinase by 6-phosphogluconate, 3-phosphoglycerate, 2-phosphoglycerate and phosphoenolpyruvate. Biochim. Biophys. Acta **208**:360-367.

Kemp, J. D., and D. W. Sutton. 1971. Protein metabolism in cultured plant tissues: calculation of an absolute rate of protein synthesis, accumulation, and degradation in tobacco callus in vivo. Biochemistry **10**:81-88.

Koiwai, A., Y. Tanno, and M. Noguchi. 1970. The effects of amino acids on the growth and glutamic acid decarboxylase activity of tobacco cell cultures. Agr. Biol. Chem. **34**:1754-1756.

Leech, R. M., and P. R. Kirk. 1968. An NADP-dependent L-glutamate dehydrogenase from chloroplasts of *Vicia faba* L. Biochem. Biophys. Res. Commun. **32**:685-690.

Losada, M., and A. Paneque. 1968. Light reduction of nitrate and nitrite by chloroplasts. An. Edajol. Agrobiol. **26**:335-349.

Machold, O. 1971. Lamellar proteins of green and chlorotic chloroplasts as affected by iron deficiency and antibiotics. Biochim. Biophys. Acta **238**:324-331.

Marcus, A., and J. Feeley. 1965. Protein synthesis in imbibed seeds. II. Polysome formation during imbibition. J. Biol. Chem. **240**:1675-1680.

Marcus, A., D. P. Weeks, J. P. Leis, and E. B. Keller. 1970. Protein chain initiation by methionyl-tRNA in wheat embryo. Proc. Nat. Acad. Sci. USA **67**:1681-1687.

Miflin, B. J., 1969. The inhibitory effects of various amino acids on the growth of barley seedlings. J. Exp. Bot. **20**:810-819.

Mothes, K. 1966. Zur Problematik der metabolischen Exkretion bei Pflanzen. Naturwissenschaften **53**:317-323.

Narayanan, A., J. Vermeersch, and A. Pradet. 1970. Enzymatic determination of cyclic 3′,5′-AMP in seeds of lettuce cv. 'Reine de mai.' Compt. Rend. Acad. Sci. (Paris) **217**:2406-2407.

Neupert, W., and G. D. Ludwig. 1971. Sites of biosynthesis of outer and inner membrane proteins of *Neurospora crassa* mitochondria. Eur. J. Biochem. **19**:523-532.

Nickells, M. W., G. M. Schaefer, and A. G. Galsky. 1971. The action of cyclic-AMP on GA_3 controlled responses. I. Induction of barley endosperm protease and acid phosphatase activity by cyclic -3′,5′-adenosine monophosphate. Plant & Cell Physiol. **12**:717-725.

Novelli, G. D. 1967. Amino acid activation for protein synthesis. Ann. Rev. Biochem. **36**:449-484.

Oritani, T., and R. Yoshida. 1970. Studies on nitrogen metabolism in crop plants. VII. The nitrogenous compounds in the bleeding sap and various organs of the crop plants. Proc. Crop Sci. Soc. Jap. **39**:355-362.

Perutz, M. F. 1967. Some molecular controls in biology. Endeavour **26**:3-8.
Pradet, A. 1969. Étude des adénosine 5'mono, di, et triphosphates dans les tissus végétaux. V. Physiol. Vég. **7**:261-275.
Pridham, J. B., and T. Swain (Eds.). 1965. Biosynthetic pathways in higher plants. Academic Press Inc. (London) Ltd.
Rich, A. 1967. The structural basis of protein synthesis, p. 20-36. *In* J. M. Allen. Molecular organization and biological function. Harper & Row, Publishers, New York.
Sauter, J. J., and H. Marquardt. 1967. Die Rolle des Nukleohistons bei der RNS -und Proteinsynthese während der Mikrosporogenese von *Paeonia tenuifolia* L. Z. Pflanzenphysiol. **58**:126-137.
Schachman, H. K. 1963. Considerations on the tertiary structure of proteins. Sympos. Quant. Biol. **28**:409-430.
Shah, S. P. J., and L. J. Rogers. 1969. Compartmentation of terpenoid biosynthesis in green plants: a proposed route of acetyl-coenzyme A synthesis in maize chloroplasts. Biochem. J. **114**:395-405.
Shephard, D. C. 1965. An autoradiographic comparison of the effects of enucleation and actinomycin D on the incorporation of nucleic acid and protein precursors by *Acetabularia* chloroplasts. Biochim. Biophys. Acta **108**:635-643.
Smillie, R. M., N. S. Scott, and D. Graham. 1968. Biogenesis of chloroplasts: roles of chloroplast DNA and chloroplast ribosomes *(Euglena gracilis)*, p. 332-353. *In* Comparative biochemistry and biophysics of photosynthesis. University of Tokyo Press, Tokyo, and University Park Press, State College, Pa.
Smith, F. W., and J. F. Thompson. 1971. Regulation of nitrate reductase in excised barley roots. Plant Physiol. **48**:219-223.
Smith, M. 1966. The amino acid composition of proteins. J. Theor. Biol. **13**:261-282.
Spencer, D., and P. R. Whitfeld. 1967. DNA synthesis in isolated chloroplasts. Biochem. Biophys. Res. Commun. **28**:538-542.
Spiegelman, S., and M. Hayashi, 1963. The present status of the transfer of genetic information and its control. Sympos. Quant. Biol. **28**:161-181.
Taylor, W. I. 1966. The source of indole alkaloids. Science **153**:954-956.
Tewari, K. K., and S. G. Wildman. 1966. Chloroplast DNA from tobacco leaves. Science **153**: 1269-1271.
Teipel, J. W., and D. E. Koshland. 1971. Kinetic aspects of conformational changes in proteins. II. Structural changes in renaturation of denatured proteins. Biochemistry **10**:798-805.
Ting, I. P. 1968. CO_2 metabolism in corn roots. III. Inhibition of *p*-enolpyruvate carboxylase by L-malate. Plant Physiol. **43**:1919-1924.
Travis, R. L., W. R. Jordan, and R. C. Huffaker. 1969. Evidence for an inactivating system of nitrate reductase in *Hordeum vulgare* L. during darkness that requires protein synthesis. Plant Physiol. **44**:1150-1156.
Travis, R. L., and J. L. Key. 1971. Correlation between polyribosome level and the ability to induce nitrate reductase in dark-grown corn seedlings. Plant Physiol. **48**:617-620.
Tromp, J., and J. C. Ovaa. 1971. Spring mobilization of storage nitrogen in isolated shoot sections of apple. Physiol. Plant. **25**:16-22.
Vogel, H. J., and R. H. Vogel. 1967. Regulation of protein synthesis. Ann. Rev. Biochem. **36**:519-538.
Wallace, W., and J. S. Pate. 1967. Nitrate assimilation in higher plants with special reference to the cocklebur (*Xanthium pennsylvanicum* Wallr.). Ann. Bot. **31**:212-228.
Willaman, J. J., and H. L. Li. 1963. General relationships among plants and their alkaloids. Econ. Bot. **17**:180-185.
Wollgiehn, R., M. Ruess, and D. Munsche. 1966. Ribonucleinsäuren in Chloroplasten. Flora **157**: 92-108.

GENERAL REFERENCES

Bonner, J. F., and J. E. Varner. 1965. Plant biochemistry. Academic Press, Inc., New York.
Chibnall, A. C. 1939. Protein metabolism in the plant. Yale University Press, New Haven, Conn.
Cohen, N. R. 1966. The control of protein biosynthesis. Biol. Rev. **41**:503-560.
Davies, D. D., J. Giovanelli, and T. ap Rees. 1964. Plant biochemistry. F. A. Davis Co., Philadelphia.
Goodwin, T. W. 1965. Chemistry and biochemistry of plant pigments. Academic Press, Inc., New York.
Haurowitz, F. 1963. The chemistry and functions of proteins. Academic Press, Inc., New York.
Leete, E. 1967. Alkaloid biosynthesis. Ann. Rev. Plant Physiol. **18**:179-196.
Lehninger, A. L. 1965. Bioenergetics; the molecular basis of biological energy transformations. W. A. Benjamin, Inc., New York.
Lienier, I. E. 1966. Organic and biological chemistry. The Ronald Press Co., New York.
McKee, H. S. 1962. Nitrogen metabolism in plants. Clarendon Press. London.

Mentzer, C. 1964. Actualités de phytochimie fondamentale. Masson et Cie, Paris.

Morton, R. A. 1965. Biochemistry of quinones. Academic Press, Inc., New York.

Neish, A. C. 1964. Major pathways of biosynthesis of phenols, p. 293-359. *In* J. B. Harborne. Biochemistry of phenolic compounds. Academic Press, Inc., New York.

Pridham, J. B. 1967. Terpenoids in plants. Academic Press, Inc., New York.

Schubert, W. J. 1965. Lignin biochemistry. Academic Press, Inc., New York.

PART FOUR

Growth and development

17 Growth

The preceding treatment of the fundamental biophysics and biochemistry of the plant is, in a sense, merely a preliminary to the remaining aspects of the physiology of the plant. The ultimate goal of the plant physiologist is to discover and explain in terms of physics and chemistry the complete mechanism of plant growth and development. Today, this requires a utilization of the most modern principles of molecular biology as a basis for experimental attack. The result is an explosion of evidence that is daily flooding the literature of plant physiology. Although it is impossible to do justice to the available experimental evidence, a slightly different approach will be necessary from that in the preceding chapters. Because of the many different lines of attack and today's far from complete understanding of the physiology of growth and development, it is necessary to adopt a kind of "recent advances" approach (e.g., Chapters 19 and 20). In this way the student will be given a glimpse of the tremendous ferment of activity in these fields of plant physiology and of some of the principles that are slowly but surely beginning to appear. At the same time the student must realize that when he has mastered the following information, it will be already out of date, and he will have to supplement it with the more current publications.

QUANTITATIVE RELATIONS

Of all the phases of physiology, growth and development are the most complex, since they depend on all other physiological processes. From the chemical point of view, growth has been defined as the transformation of simple, inorganic foodstuffs into new chemical entities that form the organized protoplasm of the plant (Robertson, 1923). However, this cannot be a complete definition of growth, since there are also physical aspects of the process. It is therefore impossible to understand plant growth without a knowledge of the basic facts of cell physiology and of the physiology of the plant as a whole—transfer of substances, mineral nutrition, and metabolism. Besides these processes, other factors specific to growth and development must be introduced. Foremost among these are the substances known as growth regulators (Chapter 19). Yet it must be remembered that these substances are essentially switch mechanisms for turning growth or development on or off. No understanding of the switch can substitute for an understanding of the growth mechanism itself.

The first requirement for studying growth is a suitable method of measuring it. The simplest definition of growth is an increase in size. Since size is a synonym for volume, this means that.

$$\text{Growth rate} = V_t - V_o \quad (1)$$

$$\text{Growth} = \frac{V_t - V_o}{t} \quad (2)$$

where V_t = volume at end of time t
V_o = original volume at time zero

$$\text{Relative growth rate} = \frac{V_t - V_o}{V_o t} \quad (3)$$

If enough measurements are made, the more useful differential equations may be applied. If growth rate is represented by R:

$$R = \frac{dv}{dt} \quad (4)$$

And if relative growth rate is represented by r:

$$r = \frac{dv}{dt} \cdot \frac{1}{V_o} \quad (5)$$

The ideal method of measuring growth would therefore be to determine the volume of the plant or plant part, but this is usually difficult to do with any degree of accuracy. In the case of growing leaves the area is more readily determined:

$$\text{Growth} = S_t - S_o$$

where S_t = surface area at end of time t (6)
S_o = original surface area

In the case of stems and roots, a simpler measurement is length:

$$\text{Growth} = L_t - L_o$$

where L_t = length at end of time t (7)
L_o = original length

The easiest and most accurate measurement, however, is usually the weight of the plant or plant part and:

$$\text{Growth} = W_t - W_o$$

where W_t = weight at end of time t (8)
W_o = original weight

This leads to some complicating factors, however. It must be decided whether fresh or dry weight is a more nearly correct measurement of growth. In the case of germinating seeds, growth occurs rapidly, yet dry weight decreases steadily for some time because of rapid loss of reserves. During this period only the fresh weight reveals the amount of growth. The same is true in the case of a growing tree in the spring. On the other hand, during midsummer when the tree has stopped growing, there is a steady increase in dry weight because of an accumulation of reserves. Dry weight can therefore serve as a measurement of growth only when it does not include any significant changes (increases or decreases) in the plant's food reserves. It is a satisfactory measurement of growth of seedlings only if the endosperm or seed leaves are not included. In other words, weight is a measure of growth only if $W \propto V$.

CELLULAR BASIS OF GROWTH

The previous quantitative relations are based on the definition of growth as an increase in size. However, this definition does not differentiate between plant growth and the growth, for instance, of a nonliving crystal in a supersaturated solution. Similarly, the first uptake of water by dry seeds is purely an imbibition process and may occur equally in living and dead seeds. Since true plant growth occurs only in living cells, volume or fresh weight measurements in this case would reveal a spurious growth, and increased volume is therefore not an adequate definition of growth. Growth must be defined in another way, preferably on a cellular basis. Since the growth of a plant is initiated in special tissues (meristems), the cellular changes in these tissues should throw some light on the process. The three main changes in the complete development of a cell are cell division, cell enlargement (or elongation), and cell differentiation (or maturation). That growth of a plant part may result

Table 17-1. Growth of isolated wheat roots with low and high phosphate and nitrate*

Nutrient solution	Increase in root length (mm)	Cell length (μm)	Increase in cell number (longitudinally)
Low P, low N	13.7 ± 1.7	177 ± 3	35 ± 4
High P, low N	25.5 ± 2.0	170 ± 3	120 ± 5
Low P, high N	25.2 ± 1.3	269 ± 4	51 ± 3

*From Burström, H. 1951. *In* F. Skoog (Ed.). Plant growth substances. University of Wisconsin Press, Madison, Wis.

from either cell division or cell enlargement is shown in Table 17-1. In this case both phosphorus and nitrogen increased root growth—phosphorus because of increased cell division, nitrogen mainly because of increased cell enlargement.

However, even growth resulting from cell division is possible only because the daughter cells enlarge to the size of the mother cell before they again divide. Furthermore, regardless of whether cell enlargement goes only this far or much farther, it must not include any simple turgor-induced change in cell volume if it is a true measure of cell growth. Consequently, *the growth of a plant or plant part is the increase in volume caused by cell enlargement when measured at constant positive cell turgor.*

According to the classical concept of cell growth, the motivating force is turgor pressure. As evidence in support of this concept, a reduction of turgor pressure inhibited the growth in area of the cell wall of root hairs of *Tradescantia fluminensis* (Schroter and Sievers, 1971).

Yet it is a fact that mature, nongrowing cells usually have higher turgor (and wall) pressures than growing cells (Table 17-2). Therefore the difference between growing and nongrowing cells does not result from a difference in turgor pressure. There is another factor that must be considered: the resistance to the force. The mature cells have more rigid cell walls than the growing cells. The latter stretch more readily; and during cell growth, the decrease in rigidity (i.e., the increase in elastic and plastic stretch) of the cell walls is directly related to the rate of growth (Fig. 17-1). Consequently, a small turgor pressure is sufficient to stretch the walls of young, growing cells. The relation between the turgor force and the resistance to it has been expressed as follows (Green et al., 1971):

$$r = m\,(P - Y)$$

where r = growth rate
P = turgor pressure
Y = yield value of cell wall (or threshold turgor for growth)
m = apparent fluidity $\left(\frac{1}{\text{viscosity}}\right)$ of cell wall

In the case of *Nitella*, P and Y usually have values of 5 to 6 atm, and P–Y is

Table 17-2. Wall (or turgor) pressures of developing cells in the hypocotyl of *Helianthus* seedlings; maximum rate of cell enlargement in zone 2, and beyond this, increasing zone number corresponds to increasing maturity*

Zone	Distance from cotyledons (mm)	Wall pressure (atm)	
		Epidermis	Cortical cells
1	0-5	2.0	0.235
2	5-10	1.44	0.210
3	10-15	1.39	0.120
4	15-20	1.14	0.120
5	20-25	2.05	0.42
6	25-30	2.28	
7	30-35	2.58	0.42

*Modified from Beck, W. A., and B. Andrus. 1943. Bull. Torrey Bot. Club **70**:563-598.

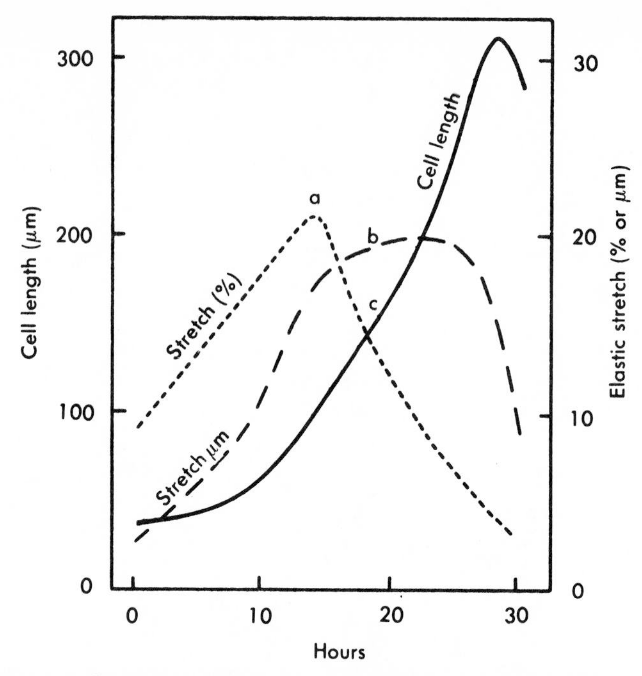

Fig. 17-1. Change in the elastic stretch of root epidermis during cell elongation. Between 10 and 25 hr the ability to stretch elastically is at a maximum (curves **a** and **b**). During this same period, the growth rate (curve **c**) is most rapid. (From Burström, H. 1951. *In* F. Skoog [Ed.]. Plant growth substances. University of Wisconsin Press, Madison, Wis.)

roughly 0.2 atm. The growth rate of the cell is therefore extremely sensitive to changes in P, so that a small decrease in turgor (e.g., 0.7 atm) stops growth completely, and a similar rise greatly accelerates it. The value of Y, however, changes rapidly (within 15 min). Such changes act as a governor, tending to maintain a constant value of P–Y in spite of changes in P. However, when P drops from 5 to 2 atm, growth stops because this is the lowest attainable Y value.

For cells to continue growth, other changes besides the stretching of cell walls must take place. In the simplest cases of cell growth there is nothing more than an uptake of water, producing a turgor pressure sufficient to overcome the force of attraction between the particles in the cell wall. This may be called *osmotic growth*—the growth due to osmosis (or diffusion) of water into the growing body. As a result, the wall stretches and becomes thinner, and the osmotic pressure drops because of uptake of water, which dilutes the cell contents. The consequent increase in the cell's water potential soon brings water uptake (and therefore growth) to a stop. This simplest type of growth rarely if ever occurs in nature, but it can be induced artificially in some cases. Yet under normal conditions, cells are able to increase in size more than fifteenfold (Fig. 17-1). In these cases, as water enters the cell during enlargement (accounting for about 90% of the increase in cell size), solutes are also absorbed (Table 17-3). In this way the cell maintains a high enough turgor pressure to continue stretching its wall.

Besides water and solute absorption, other changes occur during cell growth. New wall material must be laid down by *intussusception,* an insertion of new particles between the old wall particles that

Table 17-3. Growth and solute content of epidermal cells of *Helianthus annuus;* explanation of zones as in Table 17-2*

Zone	Average distance from cotyledons (mm)	Average cell length (μm)	Average equivalent concentration (moles sucrose)	Total solutes per cell (arbitrary units)
1	2.5	25.6	0.270	7.18
2	7.5	96.1	0.248	23.90
3	12.5	144.3	0.243	34.60
4	17.5	173.2	0.228	39.60
5	22.5	203.6	0.225	44.90
6	27.5	212.8	0.226	48.00
7	32.5	221.4	0.226	50.00

*Adapted from Beck, W. A. 1941. Plant Physiol. **16**:637-642.

have become separated. The exact mechanism is not fully understood, although it has been suggested that the cell wall material is synthesized in the Golgi bodies and ejected through the plasma membrane into the cell wall (Wilson, 1964). Wall thickening by *apposition* (a deposit of new particles on the inner surface of the wall) occurs during cell maturation (after enlargement has ceased). The cellulose deposited may differ at various stages of cell development. In pea epicotyls, it increases in molecular weight from less than 500 in young tissues to more than 7000 in mature tissues (Spencer and MacLachlan, 1972).

As mentioned earlier, under artificial conditions it is possible for cell enlargement to occur without cell wall synthesis. In *Avena* coleoptiles, rapid extension of over 30% is induced by low pHs of 3.0 to 3.6 (Rayle et al., 1970) in the absence of cell wall synthesis. In pea stem segments, longitudinal growth is not correlated with the formation of cellulose, yet the segments may become long (Winter, 1966). It may therefore be concluded that during this longitudinal growth there is neither intussusception nor apposition of cellulose microfibrils. Lateral growth, on the other hand, was positively correlated with the formation of cellulose. During normal plant cell growth, however, intussusception probably always occurs.

Recent evidence indicates a possible role of cell wall protein in cell enlargement, and it has therefore been given the name *extensin.* It is rich in hydroxyproline, which is an amino acid found in only a few proteins, such as animal collagen, and which occurs in all species of plants that have been tested (Lamport and Miller, 1971). Free hydroxyproline inhibits cell enlargement at as low a concentration as 1 mM, supposedly because it blocks the utilization of this protein (Cleland, 1967). On the other hand, incorporation of radioactive proline indicates that a stable protein moiety is associated with the cell wall (Sadava and Chrispeels, 1969). Twenty percent of the total hydroxyproline isolated from cell walls of tomato was accounted for by five glycoproteins (Lamport, 1969). The sugar of the glycoprotein apparently formed a glycosidic link with the OH group of the hydroxyproline. According to Cleland (1968), hemicelluloses that are incorporated into the cell wall by intussusception must be attached to a hydroxyproline protein.

It is obvious that phospholipids must also increase during cell enlargement, since the plasma membranes increase tremendously in area, due, apparently, to coalescence of the membranes of vesicles of the Golgi apparatus with the plasma membrane (Risueno et al., 1968).

The meristematic cell is filled with cytoplasm and has few small vacuoles. As enlargement progresses, the vacuole occupies a larger and larger fraction of the cell, and the cytoplasm becomes a thin and almost invisible layer around it. Plant cell enlargement is therefore mostly vacuole enlargement. This might lead one to suspect that no new cytoplasm is formed during cell growth. Such is not the case. Despite the smaller fraction of the cell occupied by the cytoplasm, as a result of cell enlargement, the actual amount per cell increases as much as fourfold, as judged by the protein-nitrogen content of the cell (Blank and Frey-Wyssling, 1944). More detailed analyses have since revealed nearly tenfold increases in protein and twentyfold increases in RNA during cell enlargement in growing fruit (Fig. 17-2). There was, however, no increase in DNA. The protein content of the nucleus also increases during cell enlargement (Lyndon, 1963), but nuclear RNA decreases. The nucleus even increases in volume and hydration.

Nevertheless, the nucleus is apparently not essential per se for growth, since both nucleate and enucleate cells of the alga *Acetabularia* grow equally rapidly, due apparently to the predominant extranuclear RNA synthesis (Dillard and Schweiger, 1969). In the large-celled alga *Chara,* when the cell size increased from 10 to 200 μl, the nitrogen content of all the fractions increased (Mercer and Mercer, 1971). Similarly, in maize the activity of most enzymes increased with cell size (Zeleneva and Khavkin, 1971). Certain specific proteins may, however, be more closely associated with growth than others. For instance, in the initiation of adventitious roots on mung bean hypocotyls, cell division was paralleled by the formation of four new peroxidase isoenzymes (Chandra et al., 1971). That protein synthesis is an essential component of cell growth is further supported by evidence that inhibitors of protein synthesis (e.g., chloramphenicol) also stop cell enlargement (Chapter 19).

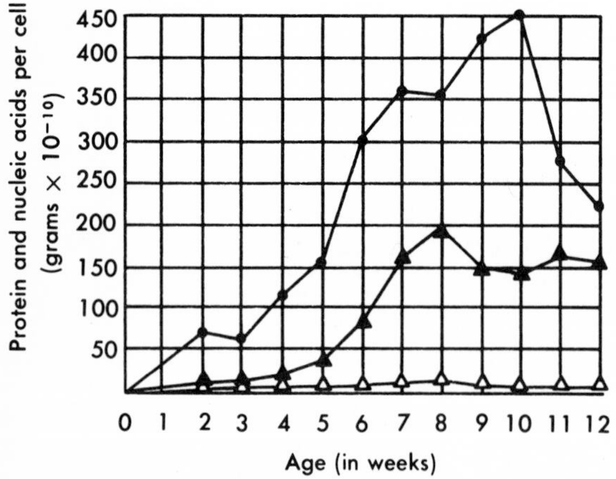

Fig. 17-2. Changes in level of protein, ribonucleic acid, and deoxyribonucleic acid per cell. November fruit (tomato), natural set. ● = protein per cell; ▲ = RNA per cell; △ = DNA per cell. (From Davies, J. W., and E. C. Cocking. 1965. Planta **67**:242-253.)

The increase in RNA per cell during cell enlargement is easy to understand, since the RNA is required for protein synthesis, and the increased metabolic rate accompanying cell enlargement is possible only if there is an increase in rate of protein (i.e., enzyme) synthesis. Despite the lack of increase during cell enlargement, changes in DNA content have been associated with the growth process as is to be expected if cell division is involved. But even before significant cell division or enlargement, changes apparently occur in preparation for growth. For instance, in cell suspensions of *Acer pseudoplatanus,* both DNA and RNA contents rose sharply early in the period of culture and before any significant increase in cell number (Short et al., 1969). In the case of germinating peas, cell elongation consists of two phases: a lag phase during which the DNA content doubles, and a true elongation phase, which begins only after the DNA doubling (Van Parijs and Vandendriessche, 1966*a*). Even during this true cell elongation phase, a 20% increase in DNA occurs (Van Parijs and Vandendriessche, 1966*b*).

In agreement with this result, internode cells of etiolated plantlets of *Kalanchoe daigremontiana* show reduced elongation when DNA is inhibited by 5-FUDR (fluoro-2′-deoxyuridine) or by mitomycin (Bopp, 1970), and the effect of FUDR is reversed only by thymidine. On the other hand, root cells of *Vicia faba* were not inhibited by FUDR.

The maximum amount of RNA and DNA is in the meristematic zone in maize. The zone of differentiation has a high DNA and a low RNA content compared to other zones. But the highest ratio of RNA to DNA is in the region of elongation, although this ratio drops with growth (Sytnyk and Musatenko, 1968). In the seedling root, however, the requirement for RNA synthesis during cell elongation appears to be restricted to the AMP-rich species of RNA, presumed to be mRNA (Lin and Key, 1968). Conversely, when expressed as a percent of total RNA, tRNA was lower in dividing cells than in nondividing cells of pea root (Vanderhoeff and Key, 1971).

It is apparent that during normal cell elongation all parts of the cell increase; but all the increases do not occur at the same rate (Williams and Rijven, 1965). Therefore the *proportions* change as the cell expands. During 5 days prior to leaf emergence in wheat plants, proteins decreased from 69% to 45% of the dry weight, and cell wall materials increased from 12% to 40%. Yet the protein per cell increased eightfold, and the RNA threefold, reaching a maximum 7 days after emergence.

The increase in cytoplasm during cell enlargement has now been shown by a more direct method—electron microscopy. It has been found that all the cell organelles in the cytoplasm increase in number per cell (Table 17-4).

Table 17-4. Increase in protoplasmic organelles per cell during cell enlargement in a root cap*

	Endoplasmic reticulum (area)	**Mitochondria (no.)**	**Golgi bodies (no.)**	**Plastids (no.)**
Original quantity	300	200	30	15
Final quantity	10,000	2000	300	25

*From Juniper, B. E., and F. A. L. Clowes. 1965. Nature (London) **208**:864-865.

Thus although the simplest type of growth involves nothing more than a stretching of the wall caused by an uptake of water, nevertheless, under normal conditions the following processes all take place:

1. Absorption of water
2. Absorption of solutes, maintaining the osmotic potential and turgor pressure at an almost constant level
3. Deposit of new wall material within the wall (intussusception)
4. Formation of new protoplasm (synthesis of proteins, nucleic acids, phospholipids, etc.) and multiplication of organelles

The bases of cellular growth can therefore be represented as shown on p. 270.

Growth is thus a complex process, involving all or nearly all the other physiological processes occurring in the plant. It requires large quantities of energy, since (1) absorption of solutes is probably active and therefore dependent on respiratory energy and (2) syntheses of cell wall and protoplasmic and vacuole substances are endergonic processes. It is therefore dependent on respiration or photosynthesis or both, since these are the plant's only sources of reserve energy (carbohydrates) and available energy (ATP). As a result, the highest rate of metabolism in the roots of *Vicia faba* appears to occur in the zone of elongation (Hadacova, 1968).

Although the preceding classical concept of cell growth resulting from turgor pressure is logical enough and is based on considerable sound evidence, it may not always be an exact picture of the order of

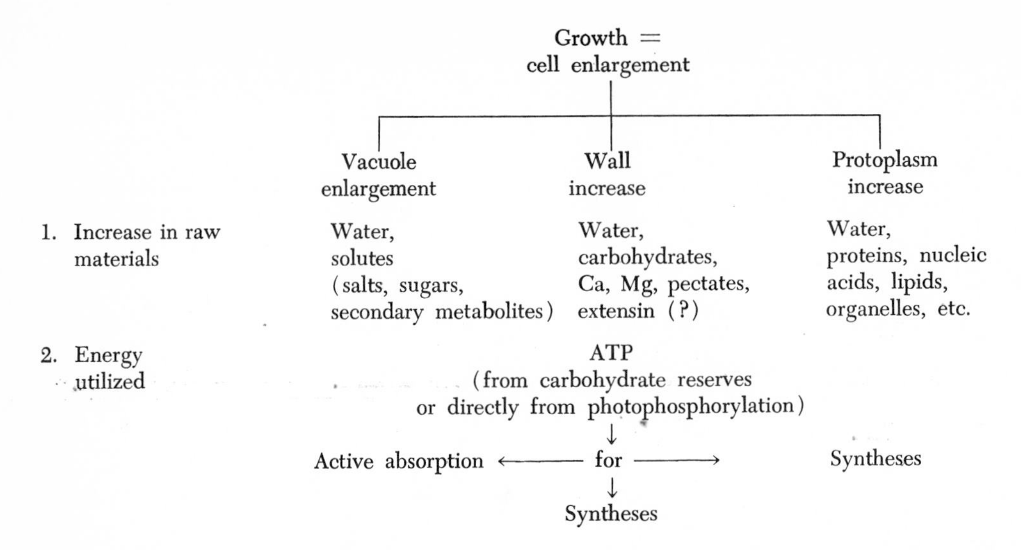

events. That cell enlargement can occur under artificial conditions according to the classical turgor concept is easily proved. When potato slices are aerated in water, their turgor pressure becomes so great that the cell walls are stretched beyond their elastic limits, and a permanent increase in cell volume of as much as 50% occurs. But this is not a normally growing tissue. Roots grow with such a low turgor pressure (about 1.5 atm according to Burström) that the cell wall is not even stretched elastically, let alone plastically. The root cells therefore grow despite the fact that the turgor pressure is too small to produce growth. Instead of being the initiating force in growth, turgor is here only a secondary factor. The initial cause of growth must be the insertion of new material within the cell wall (intussusception), leading to an increase in the perimeter of the cell. The turgor pressure would then serve simply to keep the wall taut and to prevent it from wrinkling as its perimeter increases. This would still be an essential role because in the absence of turgor pressure, the wall particles would soon become too close together to permit any further intussusception. In opposition to Burström's low values, however, in the case of roots of cotton, peas, and peanuts, pressures of 9.4, 13.0 and 11.5 bars, respectively, were measured with a strain gauge (Taylor and Ratliff, 1969). In such cases, turgor pressure presumably provides the initiating force for cell growth.

In the case of many types of cells (e.g., development of vascular cells from cambium cells), growth appears to be uniform. In others (e.g., elongation of epidermal cells) growth seems to be localized at the tips (Frey-Wyssling, 1952). The protoplasm penetrates the loose cellulose framework in their tips and even moves ahead of it, laying down a new framework as the cell elongates. The new cellulose framework is in the form of long strands that comprise the weft (parallel to the elongating cell). When the extending tip pushes further ahead, the warp (the cross strands) is laid down in the wall, reinforcing it. All this represents only the formation of the primary wall. The secondary wall is later formed by apposition.

GRAND PERIOD OF GROWTH

The growth of a plant or plant part characteristically passes through stages represented by an S-shaped curve (Fig. 17-3). The time during which this occurs has been called by Sachs the *grand period of growth.* Several attempts have been made to de-

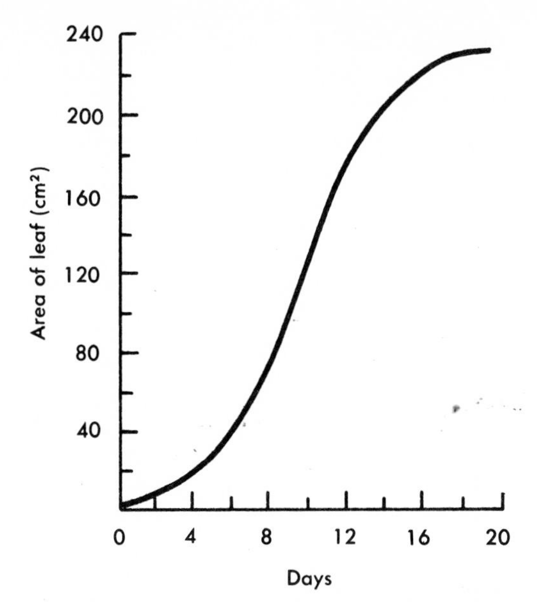

Fig. 17-3. Growth (in area) of cucumber leaf. (From Gregory, F. G. 1921. Ann. Bot. **35**:93-123.)

velop a mathematical expression for the growth curve. Blackman (1919) assumed that the shape of the curve results from the fact that the rate of production of new materials is proportional to the size of the plant. When growth results from cell division (and enlargement of the daughter cells to the size of the mother cell), each cell is replaced by two cells of the same size as the mother cell, and this process is repeated so that there is a geometric progression (2, 4, 8, 16, 32, etc.). Blackman applied the compound interest formula to explain growth:

$$A = ae^{rt}$$

or in logarithmic form:

$$\log_e \frac{A}{a} = rt, \text{ or } 2.3026 \log A/a = rt$$

where A = final size
a = initial size
e = base of natural logarithm
r = rate of interest (or growth)
t = time interval

Thus if a plant has doubled itself in a time (t) of 10 days, $A/a = 2$, and since $\log_e 2 = 0.69315$, the rate of increase, or growth, was therefore:

$$r = \frac{\log_e (A/a)}{t} = \frac{0.69315}{10} = 0.0693,$$

or 6.93% per day

If the period of doubling were 5 days, the rate would be 13.8% per day. But this constant rate of growth can apply only up to the end of the increasing slope of the growth curve.

It must be emphasized that only the *relative growth* rate (r) is constant in the kind of growth due to a constant rate of cell division, as visualized by Blackman. The *actual growth* is proportional at every instant to the size at that instant. The size of the plant therefore increases logarithmically. The first, ascending portion of the growth curve during the grand period of growth, to which Blackman's equation applies, is therefore called the *logarithmic* or *exponential* phase. It can be defined as that portion of the growth process in which growth follows the compound interest law. Cells that show this rate of growth are called *exponential cells.*

Robertson assumed that growth is based on a chemical transformation of assimilated or reserve substances into living protoplasm. As soon as it is formed, the new protoplasm begins to participate in these transformations, thus catalyzing its own synthesis. In this respect, growth resembles autocatalyzed chemical reactions (Fig. 17-4)—reactions in which one of the products of the chemical change has the property of accelerating or catalyzing the further progress of the change. Since growth depends on many chemical reactions, Robertson (1923) assumed that the slowest reaction sets the pace for the rest and becomes the "master reaction." Although he admitted that this reaction is probably multimolecular, for purposes of simplification he applied the formula for monomolecular, auto-

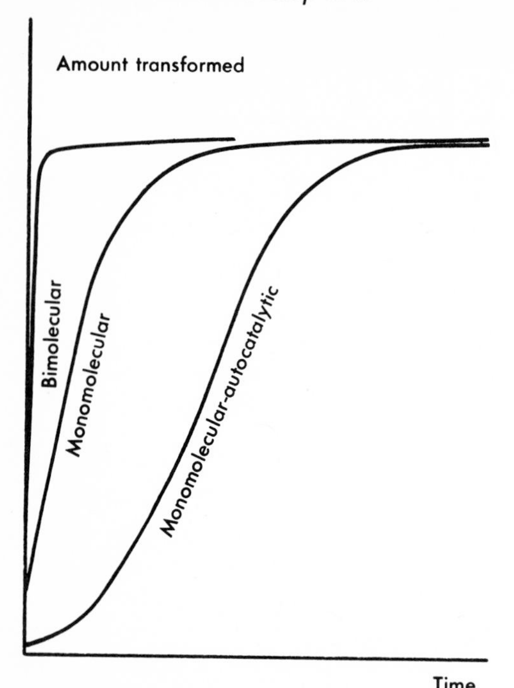

Fig. 17-4. Comparison of the relationship of extent of transformation to time in monomolecular, bimolecular, and autocatalyzed monomolecular reactions. (From Robertson, T. B. 1923. The chemical basis of growth and senescence. J. B. Lippincott Co., Philadelphia.)

catalytic reactions, assuming that a single master reaction limits the growth rate:

$$\frac{dx}{dt} = K(A-x), \text{ or } \log \frac{x}{A-x} = k(t-t_1)$$

where K = empirical constant
x = size reached in t days from beginning of growth
A = final size
t_1 = time to reach one half final size

According to Robertson, there are two possible reasons why the chemical reaction (or the growth process) comes to a stop: (1) by exhaustion of substrate and (2) by accumulation of the products of the reaction, resulting in acceleration of the reverse reaction. A more modern explanation is feedback control. The theoretical curve obtained from Robertson's equation agrees well with the actual growth curve (Fig. 17-5). However, it must be admitted that not all growth curves follow the simple S-shape, and therefore the more complex curves cannot be fully described by either of the preceding equations. Grapes, for instance, show a double S curve (Fig. 17-6), due to two peaks in the rate of growth —the first during cell division, the second during cell enlargement (Coombe, 1960). This double sigmoid growth pattern is apparently characteristic of fruit in general (Kender and Desrochers, 1970). On the other hand, growth may attain a constant rate. The meristem of a root adds about 170,000 cells to the root per day (Clowes, 1971). Finally, even if the S-shaped curve is obtained, and even if it is found to hold for the enlargement of a single cell, this could reflect the result of a combination of concentrations of growth factors and of the capacity of the cell to respond (Lockhart, 1971).

Other formulas have been proposed by various workers (Chen et al., 1968; Radford, 1967; Steward, 1968) but all fail to express adequately the progress of growth, since they do not take into account all the processes and variables involved.

INFLUENCE OF EXTERNAL FACTORS ON GROWTH

Temperature

Growth of higher plants occurs in the range of about 0° to 35° C. Within most (but not all) of this range, raising the temperature 10° C increases the growth rate 2 to 3 times. There are three temperatures known as the *cardinal temperatures* or *points* for growth: the *minimum*, or lowest temperature at which growth can be detected; the *optimum*, or temperature of maximum rate of growth; and the *maximum*, or highest temperature at which growth can be detected (Fig. 17-7). These are not sharp temperatures, and they vary from species to species (Table 17-5).

The optimum temperature for growth as determined in these tests over a short period of time is not necessarily the opti-

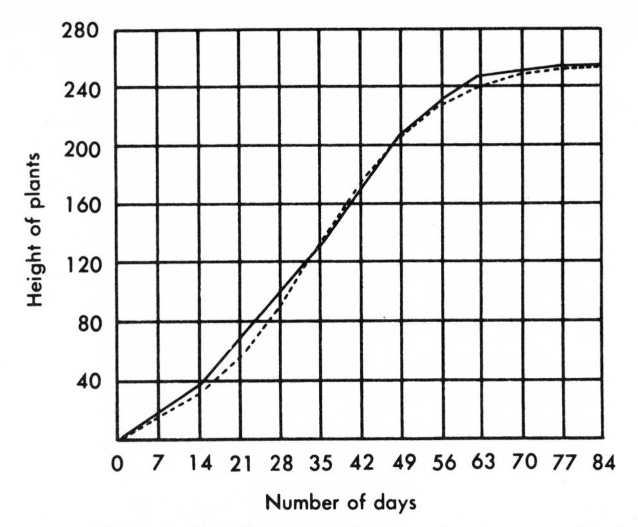

Fig. 17-5. Comparison of observed and calculated values for the mean height of *Helianthus*. Full line indicates the observed heights, and the discontinuous line indicates the calculated heights. (From Robertson, T. B. 1923. The chemical basis of growth and senescence. J. B. Lippincott Co., Philadelphia.)

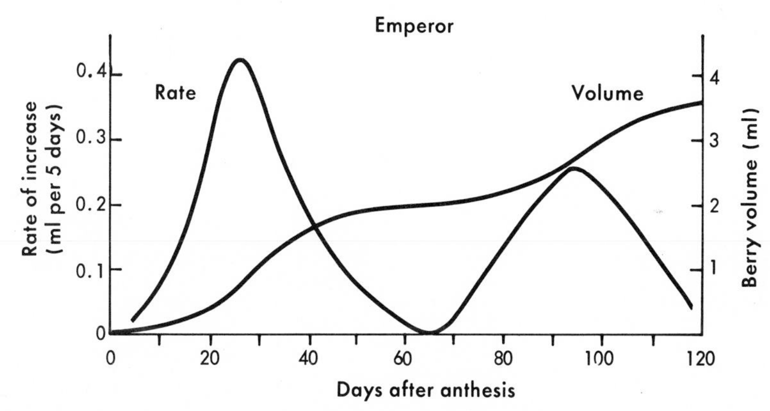

Fig. 17-6. Growth of Emperor berries *(Vitis vinifera)* from anthesis to maturity. (From Coombe, B. G. 1960. Plant Physiol. **35**:241-250.)

mum for the general development and yield of the plant over long periods of time. Furthermore, the cardinal temperatures may change with the stage of development of the plant. In the case of cocksfoot *(Dactylis glomerata)*, relative growth rates increased with temperature from 14° to 26° C during the early stages but not during a later stage (Davidson and Milthorpe, 1965). Furthermore, the growth rate of an organ is dependent not only on its own temperature but also on the temperature of the rest of the plant. When the leaf temperature of corn was kept constant (25° C), for instance, even though it was maintained at a relative humidity of 100%, its growth rate increased with root temperature from below 5° to 15° C (Watts, 1971). It must also be remembered that growth depends on other processes (e.g.,

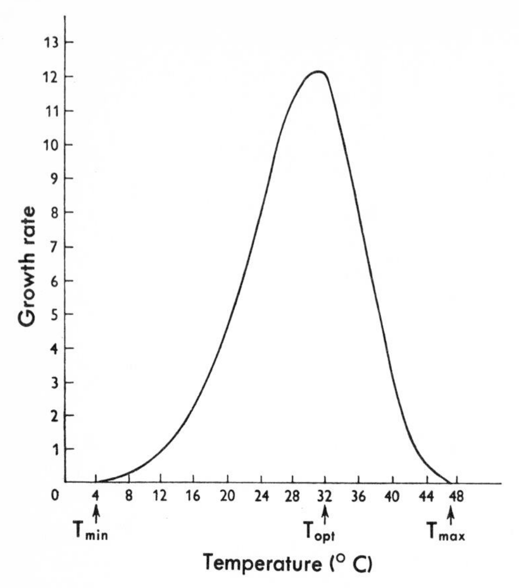

Fig. 17-7. Cardinal temperatures for growth of maize seedlings. T_{min} = minimum temperature for growth. T_{opt} = optimum temperature for growth. T_{max} = maximum temperature for growth. (From Lundegårdh, H. 1957. Klima und Boden, Gustav Fischer Verlag, Jena.)

Table 17-5. Cardinal points (° C) for growth of different plants*

Plant	Minimum	Optimum	Maximum
Barley	5	29	38
White mustard	0	21	28
Scarlet runner bean	10	34	46
Maize	10	34	46
Gourd, squash	14	34	46

*From Palladin, V. I. 1923. Plant physiology. B. E. Livingston (Ed.) Blakiston Division, McGraw-Hill Book Co., New York.

photosynthesis, respiration, etc.), and these processes also possess cardinal temperatures.

The cardinal temperatures for growth are of great practical importance, since maximum growth and yield would require a climate with a maximum time at or near the optimum temperature. But the existence of cardinal temperatures is also of theoretical importance, and an understanding of their cause might also lead to an ability to control them.

Chemical reactions do not show minimum, optimum, and maximum temperatures but yield straight line relations between rates and temperature (Fig. 17-8). Enzymatically controlled reactions, on the other hand, show a similar straight line relation only within a narrow temperature range—between the two temperatures slightly above the minimum and slightly below the optimum (Fig. 17-9). The rate drops off steeply at both extremes. This explains the existence of minimum and maximum temperatures for all plant processes, since all chemical reactions in the plant are enzymatically controlled. A maximum temperature must occur as soon as the enzyme protein is denatured and therefore inactivated. This usually occurs at a temperature slightly above the maximum for a plant process, probably because the denaturation temperature for a protein is lower in vivo than in vitro. The minimum temperature may be explained in the same way. Although an enzyme is not irreversibly denatured at low and moderately high temperatures as it is at very high temperatures, it may be *reversibly* denatured (Brandts, 1967). The cardinal temperatures for growth therefore must be controlled by the denaturation temperatures of a plant's enzymes.

In some cases the effect of temperature on growth may be the opposite of what would be expected. Embryo growth, for instance, is more rapid at low temperature (2° C) than at high temperature (15° C) in stratified seeds (see discussion on dormancy).

Light

Although the growth of higher plants eventually depends on photosynthesis, nevertheless light is not essential for the growth process itself, so long as sufficient quantities of organic substances are available. Some tuberous or bulbous plants (e.g., jack-in-the-pulpit) can complete their whole life cycle in the dark at the expense of their large food reserves, but the kind

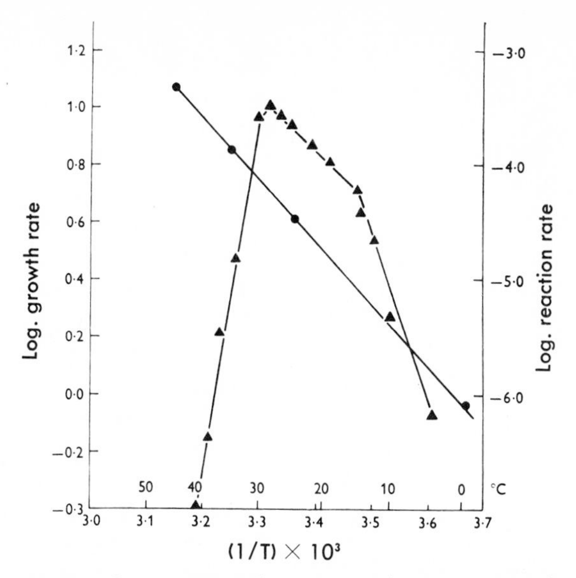

Fig. 17-8. Arrhenius plots for a chemical reaction (decomposition of nitrogen pentoxide, ●—●) and for growth (of roots of *Lepidium sativum,* ▲—▲). (Based on data from Getman, F. H., and F. Daniels. 1937. Outlines of theoretical chemistry. John Wiley & Sons, Inc., New York; and Talma, E. G. C. 1918. Receuil, Trav. Bot. Neerland. **15**:366-422.)

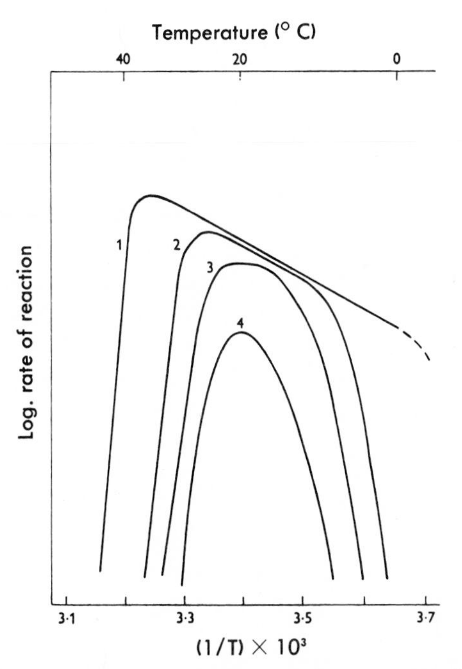

Fig. 17-9. Arrhenius plots for four enzymically controlled reactions. (From Brandts, J. F. 1967. *In* A. H. Rose [Ed.]. Thermobiology. Academic Press, Inc., New York.)

of growth is different when the light is absent. In the dark, higher plants show a weak, spindly growth known as *etiolation.* In the case of most dicotyledons, the stem is excessively elongated, and the leaves are underdeveloped. Little differentiation occurs, the tissue remaining mostly parenchymal. Usually the leaves remain free of chlorophyll, and the color is therefore pale yellow, although exceptions are found among gymnosperm seedlings, some ferns, and many algae. These may form chlorophyll in the dark (Vernon and Seeley, 1966, see Chapter 15). Some nongreen plants show a similar stretching in darkness. Monocotyledons may show excessive elongation of the first internode and normal or excessive leaf development in the dark.

Relatively short, daily exposures to light prevent etiolation. Consequently, light must retard this excessive growth. An extreme case is the dwarfing of alpine plants by the intense light at the high elevations where they grow. This light is richer in violet and ultraviolet radiations, which appear to have more of a stunting effect than other wavelengths.

Despite this commonly observed retardation of growth by light, the effect in many cases is just the opposite. Although the stems of etiolated plants are excessively long, the leaves may fail to grow altogether (e.g., potato sprouts). According to Thomson (1951), the primary effect of light is to *accelerate* growth when cells are in the division stage. It is only after cell enlargement has progressed to near its normal limit that light retards any further excessive enlargement that would occur in the dark.

The relation of root growth to light is somewhat complex (Burström, 1965). Weak illumination may be indispensable for continuous growth of excised roots. It appears to be in some way related to tryptophan requirement—perhaps for synthesis of some growth regulator. Roots of intact plants, however, are comparatively insensitive to light and show only a growth inhibition. Yet if roots respond phototropically at all, they are negatively phototropic (Chapter 18), indicating greater growth on the illuminated side. Even excised roots are inhibited by light if it exceeds a low level. As in the case of the stunting effect (see earlier discussion), the enhancing effect of light on growth may depend on light quality. Short wavelengths (blue light) lead to an increase in protein synthesis in fern gametophytes, making possible the formation of normal, two-dimensional prothallia (Drumm and Mohr, 1967). Long wavelengths (red light) lead to formation of filaments that are much lower in protein content, even when the rate of dry matter accumulation is the same as in the blue light. The RNA content is also higher in the blue light. These results are interpreted to indicate that the blue light initiates differential gene (i.e., DNA) activation.

Other aspects of the effects of light will be discussed in Chapter 20.

Water

Since all growth depends on a hydrostatic turgor pressure, a water deficiency will retard or completely stop it. On the other hand, an excess of water may result in an abnormal type of growth. Thus in a saturated atmosphere the development of leaves is poor and the differentiation of the tissues is retarded. This is undoubtedly the result of excessive stretching of the cell walls because of the abnormally high turgor pressure. Plants adapted to such conditions (e.g., aquatics) have low osmotic pressures and therefore cannot develop such excessive turgor pressures even when their tissues are saturated.

It must be realized that the effect of one environmental factor on growth may actually be due to another environmental factor. For instance, a high temperature may bring growth to a stop due to the increased rate of water loss, leading to loss of turgor. Even the inhibition of growth by low temperature may be sometimes due to its effect on the water factor, in this case a reduced water uptake (Nelson, 1971).

Chemical stimulants and inhibitors

Even nutrient salts required by plants for normal growth may inhibit growth or kill the plants if applied in unbalanced solutions. On the other hand, they stimulate growth when applied in suitable quantities and in balanced solutions. Nonnutrient mineral elements frequently inhibit growth. Salts of the heavy metals (copper, lead, silver, mercury, etc.) are particularly toxic. Even the metabolic products of a plant (e.g., oxalic acid) may be poisonous when supplied to the protoplasm instead of being stored in the vacuole. In weak doses many poisons stimulate growth. Thus phenol is poisonous in 1:1000 concentration but stimulates when used in 4 to 8:100,000; ethyl alcohol checks growth in 25 to 75:1000 and stimulates it in 25 to 75:100,000 (Maximov, 1938). Mercury compounds used to disinfect seeds sometimes stimulate growth. The subject of chemical stimulants and inhibitors will be returned to later when considering so-called growth regulators (Chapter 19). Inhibitors of enzymes and protein synthesis have already been mentioned (Chapters 13 and 16).

DORMANCY OR REST PERIOD

Terminology and phases

Some of the preceding environmental factors may completely stop growth. This is most commonly the case in nature when the plant temperature is below the minimum for growth (e.g., in winter) or when its water content is too low (e.g., in seeds). But even when the environmental conditions are favorable for growth, the plant may still fail to grow because of some internal factors. Twigs collected in the fall, newly harvested potato tubers, or Dutch bulbs (tulips, hyacinths, daffodils, etc.) fail to grow although exposed to temperature, water, and other environmental conditions that are optimum for their later spring growth. In some cases, part of the plant may grow while another part is arrested; for example, in the early growth of grasses the leaves grow but the internodes do not. Similarly, the lateral buds of a plant may fail to grow while the terminal bud is growing; and roots of at least some species may grow appreciably while the aboveground parts are unable to grow (Meyer and Tukey, 1967).

Attempts have been made to differentiate rigorously between these two kinds of suspended growth. Horticulturists have called a nongrowing plant or plant part *dormant* or *quiescent* when growth can be readily induced by transferring it to environmental conditions optimum for growth of nondormant plants. When these conditions failed to induce growth, the plant was said to be in its *rest period.* Plant physiologists, on the other hand, use the term dormancy more commonly than rest period and define it as any phase in the life cycle in which active growth is temporarily suspended (Wareing, 1969). Wareing distinguishes between *imposed dormancy* (or quiescence), which is due to unfavorable environmental conditions, and *innate dormancy* (or rest), which is due to conditions within the dormant plant or organ. In the case of buds there is a special kind of innate dormancy due to *correlative inhibition* (Chapter 19).

It seems best to combine the terminology of horticulturists and physiologists as follows:

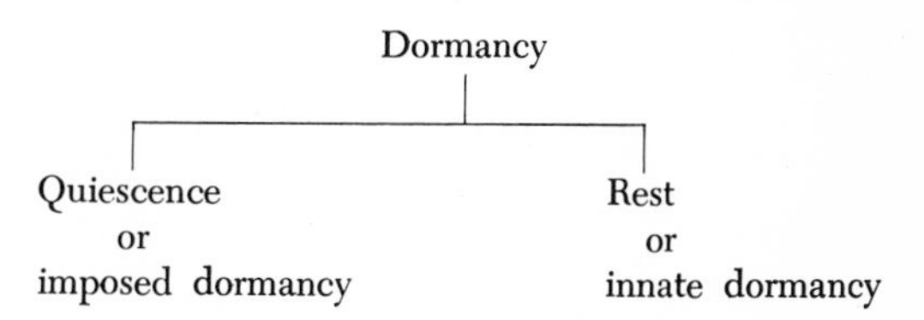

Dormancy, in this usage, would apply to all forms of suspended (or temporary cessation of) growth, whether due solely to unfavorable external conditions or to internal factors.

It was early recognized, however, that the ability to resume growth is a quantitative character (Table 17-6). More recent results have shown that even for a specific degree of suspended growth, the time required for growth to recommence will depend on the particular environmental conditions to which it is exposed. The degree of rest or dormancy therefore depends on an intimate interaction between internal and external factors. Furthermore

Table 17-6. Growth of excised twigs of *Prunus avium* on transfer to the greenhouse*

Date removed from outdoors to greenhouse	Date of first flower opening	Number of days for opening of flowers
Dec. 11	Jan. 10	27
Jan. 10	Jan. 28	18
Feb. 2	Feb. 19	17
March 2	March 14	12
March 11	March 21-22	10½
March 23	March 31	8
April 3	April 8	5

*From Vegis, A. 1964*a*. *In* W. Ruhland. Encyclopedia of plant physiology. Vol. XV. Springer-Verlag, New York, Inc., New York.

Table 17-7. Phases of suspended growth in terms of dormancy and rest period*

Phase	Growth	Dormancy	Rest period
I	Fully normal growth	Nondormancy	Active period
II	Ability to continue normal growth not completely lost	Predormancy	Early rest
III	Immediate normal growth cannot occur	True dormancy	Main rest
IV	Able to start growth within narrow limit of external conditions	Post dormancy	After-rest

*Modified from Vegis, A. 1964. Ann. Rev. Plant Physiol. **15**:185-224.
True dormancy of Vegis is equivalent to innate dormancy of Wareing.

(Table 17-6), the plant or its organ passes successively through phases that differ in degree of suspended growth. It is now common to refer to the degree of suspended growth in the synonymous terms of either innate dormancy or rest period (Table 17-7).

It is only in phase III—the true dormancy, or main rest period—that immediate normal growth does not occur despite the external conditions. But even this phase may not be as rigid as formerly believed. It is now becoming clear that many plants previously believed to have such a theoretical extreme dormancy may actually be induced to grow within a narrow range of environmental conditions (Vegis, 1964*c*). Under natural conditions, of course, the plant may be in the completely nongrowing state, even though it is not in true dormancy, simply because the external conditions do not fall within this narrow range. Many potatoes, for instance, have no true dormancy because even when at their maximum rest, they can sprout at high temperatures (e.g., above 30° C); but since they are normally at temperatures well below this, they remain dormant. Conversely, wheat seed has a high temperature dormancy (George, 1967). At a germination temperature of 20° C, all twelve varieties tested were dormant at harvest, and this dormancy persisted for 20 to 60 days. At 10° C, germination was normal; at 30° C, all varieties showed a persistent dormancy from which only one recovered after 80 days and only two more after a year. Thus in dormant seed the speed and completeness of germination was inversely related to the germination temperature; in nondormant seed the relation was direct.

Dormancy therefore can be best described in terms of the cardinal points for growth. In the actively growing phase a plant possesses the widest range from the minimum to the maximum temperature for growth. As it begins to enter the dormant state (i.e., in the predormant state), the range from minimum to maximum narrows (Fig. 17-10). The smallest range for growth occurs at the phase of true dormancy, or main rest. In some plants this range may be zero (Fig. 17-10, A_2 and B_2). In others it may still have a positive although small value (Fig. 17-10, A_1 and B_1).

Trees go through the four phases of dormancy seasonally, and the transition between them is gradual. Although these transitions depend on the environmental temperature, some processes of after-rest can proceed even though the winter chilling requirement has not been fully completed (Perry, 1970). Different parts of the same tree may pass through the phases of dormancy independently of each other. Thus chilling the shoots of *Taxus* spp. does not substitute for the chilling requirement for breaking the dormancy of the roots (Lathrop and Mecklenburg, 1971). The root regeneration potential of *Taxus* actually increased from a minimum in summer, through an increase in the fall, to a maxi-

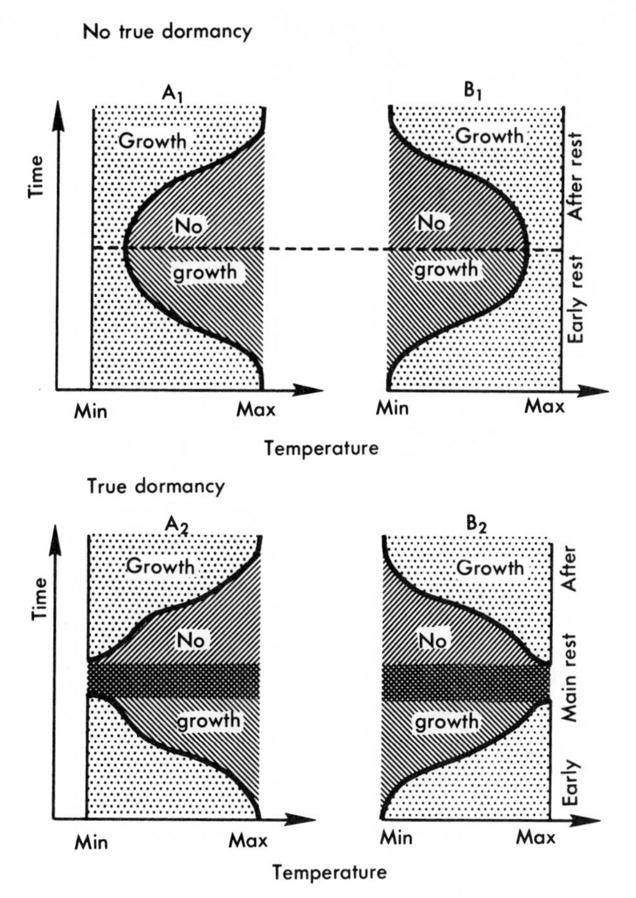

Fig. 17-10. *Top:* No true dormancy occurs, only an early rest and an after-rest. During early rest the temperature range for growth gradually narrows to a minimum; during after-rest it gradually widens until the original maximum range is attained. $\mathbf{A_1}$, During early rest the maximum temperature for growth gradually decreases to a point near the minimum temperature, which remains constant. During after-rest the process is reversed. $\mathbf{B_1}$, During early rest the minimum temperature for growth gradually rises to a point near the maximum temperature, which remains constant. During after-rest the process is reversed. *Bottom:* True dormancy occurs because early rest progresses until the minimum and maximum temperatures for growth are identical. After-rest again mirrors early rest. $\mathbf{A_2}$ and $\mathbf{B_2}$ are the projections of $\mathbf{A_1}$ and $\mathbf{B_1}$. (From Vegis, A. 1964. Ann. Rev. Plant Physiol. **15:**185-224.)

mum in January, followed by a drop in spring and early summer.

All seeds are dormant as long as they are maintained in the air-dried state. Consequently, when a distinction is made between dormant and nondormant seed, this refers to true (innate) dormancy or main rest, versus active (rest) period, or quiescence. It is this true (innate) dormancy or main rest that Crocker (1916—see Crocker, 1948) is referring to when he lists seven kinds of seed dormancy. These may be lumped together, giving two main types: (1) seed coat–induced and (2) embryo-induced. Each of these may, in turn, be divided into two subgroups: (1a) impermeability of the seed coat to water, gases, or both, (1b) mechanical impenetrability of the seed coat by growing embryo, (2a) embryo immaturity, and (2b) dormancy of mature embryo. The relations are shown in Fig. 17-11. There may be

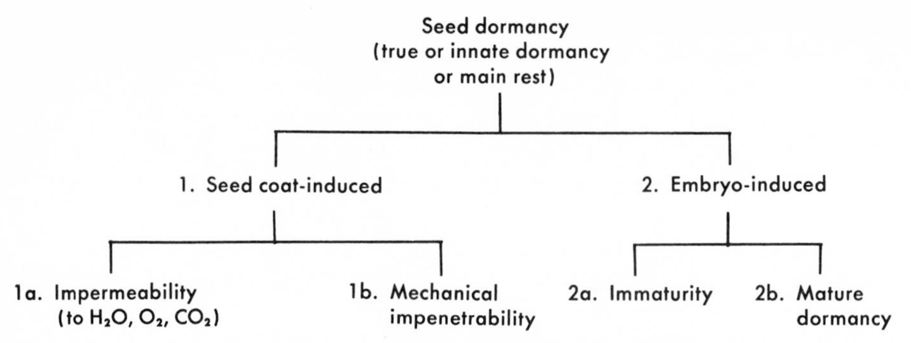

Fig. 17-11. The different kinds of seed dormancy.

combinations of the different kinds of dormancy in a single seed, and the seed may become dormant again after breaking of dormancy, leading to a secondary dormancy.

Breaking of dormancy

Both entrance into true or innate dormancy and passage through it to the active state depend on the external factors. Growth may be "started by such diverse agencies as heat, cold, increased and decreased water content, anesthetics, narcotics, gases and liquids, bases and acids, salts singly or combined, electricity, mechanical injury, and in vacuo" (Howard, 1915). Under natural conditions, dormancy is broken by "after-ripening." This may also be accomplished artificially by *stratification,* for example, by burying the plant with dormant buds (or the dormant seed) in moist soil or sand and by maintaining it at a low temperature (e.g., 0° C) for some days, weeks, or months. For most seed of temperate climates, 0° to 7° C is the optimum range for stratification (Nikolaeva, 1969). In some cases this is due to the slow maturation of immature embryos. A range of 8° to 10° C halts the process, and temperatures of 14° C and above induce secondary dormancy. Temperatures below freezing are relatively ineffective for breaking the rest of pear buds (Westwood and Bjornstad, 1968). In the case of *Taxus,* –2° C (28° F) breaks the dormancy of the buds, but 1.5° C (35° F) is low enough for the roots (Lathrop and Mecklenburg, 1971). The optimum chilling temperature is higher for species from warm climates (7° to 10° C) than for species from cold climates (3° to 5° C).

Some plants show the opposite reaction to temperature—low temperature induces dormancy and high temperature breaks it. Paradoxically, some plants that normally pass through their rest periods at low temperatures may have their rest broken by warm-water baths (30° to 35° C for 9 to 12 hr). Two hours at 43° C gave the best results for astilbes (Lemattre, 1971). Exposure of dry seeds of *Philadelphus lewisii* to 95° C for up to 24 hr can enhance subsequent germination (Uhlinger, 1970); but 100° C is injurious and 90° C is ineffective. This treatment cannot, however, substitute for light during germination or for low-temperature after-ripening. Alternating temperatures (20° and 40° C) in the dark supported better germination than did a constant temperature in the case of longleaf uniola plants (*Uniola sessiliflora*—Wolters, 1970). Similar results were obtained with grapes (Balthazard, 1971). Most of the effects of alternating temperature can be explained by the breaking of dormancy during the low-temperature period in excess of 10 hr, and by the high-temperature acceleration of germination in those seeds which have broken their dormancy (Popay and Roberts, 1970*a*). Different species of the same genus may show markedly different requirements for germination, such as in *Kalmia* (Jaynes, 1971). Moderate temperature and moisture may be the deciding factors in the case of the potato tuber. When stored under moist conditions (20° to 25° C), its dormant period is terminated in 7 days as opposed to 7 weeks in the case of dry-stored tubers (Goodwin, 1966).

In addition to temperature and water, light may also be an important factor. This is particularly true in the case of many seeds. Some (e.g., Poa, tobacco, carrot, lettuce, *Oenothera*) will not germinate under normal conditions unless exposed to light. Others (e.g., *Crataegus*) are prevented from germinating by exposure to light. In the case of many plants, all three factors interact in producing or breaking dormancy. For instance, secondary dormancy caused by high temperature may be reversed by light. As little as 2 mm of soil totally inhibited the germination of light-sensitive *Senecio* and *Capsella* seeds, although this complete suppression is not accounted for solely by the absence of light (Popay and Roberts, 1970*b*). The effect of photoperiod will be considered in Chapter 20.

Interaction between light and water occurs in the case of *Avena fatua*. With low volumes of water, seeds are inhibited by white light, whereas with high volumes they are promoted (Hsiao and Simpson, 1971). Similar interactions occur between temperature and moisture (Woods and MacDonald, 1971). Finally, interactions between light and temperature have been reported. Blue or far-red light, for instance, induced photosprouting of immature tubers of *Begonia evansiana* at higher temperatures and promoted release from dormancy induced by chill of 2° to 5° C in mature tubers but failed to break mature tuber dormancy at room temperature (Esahi, 1969).

The inability of nonripened seed to germinate is often dependent on the presence of intact seed coats, supposedly because they limit the supply of oxygen to the embryo (Vegis, 1964*b*). This relation, however, varies with the species. In the case of light-sensitive seeds of *Phacelia tanacetifolia,* increased oxygen uptake is apparently the result rather than the cause of seed germination, and light does not cause dormancy by inhibiting oxygen uptake (Chen, 1970). Grape seeds, on the other hand, require a 48 hr soaking period for enzyme synthesis, which requires oxygen, but are then stimulated to germinate by total (nitrogen gas) or partial (water) anaerobiosis at 15° to 37° C (Pouget, 1971). The rate of breaking of dormancy under anaerobic conditions was equal to or greater than the rate obtained by stratification at 1° C (2 to 3 months). Some seed coats may restrict germination, even though they do not interfere with either water uptake or RNA formation during the early hours of soaking (Barton et al., 1971). Dormancy resulting from an impermeable seed coat may be overcome by filing the coat or by immersion in high concentrations of sulfuric acid (e.g., 60% to 70% for 24 hr). This exposes the lumens of the macrosclereid cells, permitting imbibition of water (Brant et al., 1971). Even the pressure exerted by such coverings may sometimes be important. However, embryos that are truly dormant are not usually induced to germinate by breaking their coats.

In addition to oxygen, other chemicals have long been known to affect dormancy. Anesthetics such as ether (0.5 ml per liter of air for 24 to 48 hr) are sometimes effective. Nitrates stimulate the germination of light-requiring seeds (Popay and Roberts, 1970). Potassium nitrate increases the respiration rate of bluegrass seed, but less so in the more dormant variety (Maguire and Steen, 1971). Thiourea is often effective both in the case of seeds and buds (e.g., of potato tubers). It may break secondary dormancy induced by high temperatures, or it may substitute for low-temperature after-ripening. It does not seem to be effective against true dormancy. According to Lipp and Ballard (1970), thiourea has both stimulatory and inhibitory effects on germination. Ascorbic acid has been reported to act synergistically with thiourea but also to antagonize the stimulating effect of thiourea in the light (Vyas and Agarwal, 1971). Potassium nitrate also acts synergistically with thiourea in breaking the dormancy of peach buds (Erez et al., 1971). In the case of stylo *(Stylosanthes humilis)* seed, a 1- to 3-day treatment with 0.1 M or 0.2 M thiourea at 30° C may stimulate germination but the seedlings may be abnormal, and sensitivity to thiourea varies with the variety and the age of the seed (Ballard and Buchwald, 1971). Thiocyanates and ethylene chlorohydrin are effective in breaking the rest period of potato tubers. The Boyce Thompson Institute (Denny, 1945) has developed a particularly effective mixture called Rindite (7 parts ethylene chlorohydrin:3 parts ethylene dichloride:1 part carbon tetrachloride) for breaking dormancy of potato tubers (0.8 ml/kg for 48 hr). This mixture profoundly affects the reactions that control the synthesis and degradation of organic acids and free amino acids in the presprouting phase of the tubers (Jolivet, 1968). Over a hundred substances have been reported to have some effect on the breaking of dormancy (Vegis, 1964*b*). Some (e.g., maleic hydrazide and succinic acid 2,2-dimethylhydrazide [Raese, 1971]) are also able to induce dormancy.

It should be emphasized that only the specific part of the plant treated by any of the aforementioned methods has its dormancy broken. The effect is not transmitted to the rest of the plant. Thus if a single branch is exposed to low temperature, immersed in warm water, exposed to ether, etc., only this branch will have its dormancy broken.

Relation to metabolism

In some cases, dormancy is induced by a simple mechanical device such as an impermeable coat in the case of seeds or the formation of dormancy callose in sieve-tubes of trees in the fall, which prevents translocation of needed reserves to the buds. But dormancy occurs in most cases in the complete absence of such mechanical devices. The causes of this kind of dormancy have been sought in metabolic control.

Growth is dependent both on metabolic energy and on the presence of adequate, metabolically produced raw materials for the new cells or cell parts. It is therefore logical to expect a general decrease in metabolism when a plant stops growing and becomes dormant. A plant that did not slow down its metabolism under such conditions would simply be wasting its reserves and would be at a disadvantage in competition with plants that do slow down their metabolism. This is undoubtedly why, as has already been indicated (Chapter 4), dormancy is often associated with minimum surface and therefore with minimum surface energy. It is a mechanism by means of which the plant can conserve its stored energy until conditions are suitable for growth. It is therefore clear that the actively *growing* plant is also *metabolically* more active than the dormant, nongrowing plant. This is seen during the transition of potato discs from the resting to the mitotically active state, which is accompanied by transcription, translation, and increased metabolic activity (Lange et al., 1971). But what about the plant in deep rest (true or innate dormancy), compared to its later phase of after-rest (post dormancy), when its rest is broken but it is still in the quiescent state? In neither case is it actually growing; yet when exposed to favorable environmental conditions, it will immediately start growing if its rest is broken but will show no growth if in the phase of deep rest. Is there any metabolic difference between these two—for instance, between the buds of the deciduous tree in early fall and in spring before bud swelling, or between the truly dormant and nondormant seed when both are allowed to imbibe water under conditions favorable for the growth of the nondormant seed?

The seed that owes its dormancy to a purely physical barrier—a coat impermeable to water or carbon dioxide—will remain at a low level of metabolism, whereas the nondormant seed takes up water and oxygen and quickly increases its metabolism. This is true of the seeds of *Acer saccharum,* which therefore require a long period of stratification (Webb and Dumbroff, 1969). Similarly, the thin inner papery seed coat of the Cape tulip *(Homeria miniata)* is apparently impermeable to water, oxygen, and carbon dioxide and is protected by the outer thick coat, which is not impermeable (Taylor, 1969). But what happens if dormancy is not due to a physical barrier? Even then, low levels of metabolism have been found, for instance in dormant seeds of Douglas fir under conditions favorable for the germination of nondormant seed (Ross, 1969). Nevertheless, it does not necessarily follow that all metabolism is slowed down in the innate dormant state. Some enzymes may be more active, for instance, tyrosinase in dormant potato tubers (Todd, 1953). Ascorbic acid oxidase is apparently also more active in dormant barnyard grass seed because when the seed are treated with inhibitors of this terminal oxidase, their dormancy is broken (Ueki and Shimizu, 1969). Even the general metabolic rate may not be greatly decreased. There is little difference between the respiratory rate of innate dormant and stratified seed or even of their embryos (Nikolaeva, 1969). Dormant seeds of *Avena fatua* that do not germinate when allowed to imbibe water have a respiration rate only 20% less than that of imbibed, nondormant seed (Chen and Varner, 1970). Finally, during embryo maturation of *Fraxinus excelsior,* while they are in the true or innate dormant phase, there occur food interconversions, wall development, and the appearance and proliferation of cell organelles (Villiers, 1971). Similarly, it has been known for some time that an active RNA and protein synthesis develop in trees in the fall, when they are in their true dormant state (Siminovitch et al., 1967).

It has therefore long been assumed that dormancy is due to some specific metabolic block, rather than a general inhibition of all metabolism. Several hypotheses have been proposed as to the nature of this block. Chaudhry et al. (1970) list the following six suggested metabolic changes as possible initiators of the termination of the rest period:

1. Accumulation of soluble nitrogen
2. Resumption of active protein synthesis
3. Accumulation of soluble phosphorus
4. Accumulation of nucleic acids
5. Increase in high energy phosphates
6. Removal of DNA repression

A reversal of any one of the above six changes should act as a metabolic block leading to dormancy. Thus each of the six conditions may be referred to as, for example, either "shift 1" or "block 1" in the following discussion.

Results with potato tubers appear to indicate a combination of shifts 1 and 2. At the height of their dormancy, amino acids and proteins accumulate in the internal storage tissues (Cotrufo and Levitt, 1958). Only when this is reversed, and when (1) amino acids move out of the storage tissues into the buds and (2) proteins are hydrolyzed in the storage tissues, is the tuber capable of rapid growth. As further evidence of shift 1, a deficiency of pool amino acids has been found in dormant grain of winter wheat (Scheffer and Lorenz, 1968). In grapevines this also seemed to be a factor because the amino acids increased during dormancy and reached a maximum just prior to bud burst in the spring (Kliewer, 1967). The bulk of the soluble nitrogen reserves were stored in *Taxus* as arginine in the stems and old needles (Meyer and Splittstoesser, 1971) and therefore had to be converted to other amino acids before use. The hypothetical shift 1 is also indicated by the artificial breaking of the dormancy of achenes of *Ranunculus sceleratus.* Conventional methods, including application of gibberellic acid (Chapter 19) and culture of excised embryos, failed to break dormancy (Sachar et al., 1970). Dormancy was, however, overcome by feeding casamino acids (obtained by hydrolysis of casein) to the achenes in culture. Some seven amino acids were as effective as the total mixture, although subsequent growth was not so good.

In favor of shift 4, mRNA is formed during the imbibition of seeds, prior to germination (Marcus and Feeley, 1964). Presumably, the seeds cannot germinate until this RNA leads to synthesis of the proteins required by the new cells. DNA also increases, both prior to and during cell division (Holdgate and Goodwin, 1965). With increasing length of the cold treatment, the embryos of dormant pear seeds showed a progressively increasing capacity for synthesis of nucleic acids (Khan et al., 1968). Leaf "embryos" of *Bryophyllum crenatum* under short-day conditions showed a progressive transition from an active to a dormant state, during which there was a sharp reduction in synthesis of nucleic acids and proteins and an inhibition of mitosis (Warden, 1970). Inhibitors of the synthesis of nucleic acid (e.g., uracil derivatives) can prevent the release of lettuce seeds from dormancy (Smith and Frankland, 1966). The reverse changes have been found to occur during seed development on the plant. From the time of maximum fresh weight of the seed to the time it is completely air dry, the RNA of the nuclear fraction of castor bean endosperm decreases to about 50% (Cocucci and Sturani, 1965). Similarly, the mitochondrial RNA almost disappears, but the RNA of the soluble fraction increases. The ribosomes completely disappear. Barker and Rieber (1967) have also found that the concentration of functional polysomes is low in dry pea seeds and that this limits their protein synthesizing system. Although most of the evidence has been obtained with seeds, there is mounting evidence for the involvement of NA synthesis in the regulation of bud rest in potato tubers (Shih and Rappaport, 1970) and in the terminal buds of the tubercles of *Stachys sieboldi* (Tort, 1971).

Investigations of the buds of *Pyrus communis,* however, supported only the hypothetical shifts 2, 4, and 6. Chaudry and co-workers (1970) concluded that DNA is liberated from a DNA-phosphoprotein complex and that the released DNA leads to RNA synthesis, which then induces protein synthesis and termination of dormancy. The role of the proteins in this termination is not yet clear. Enzyme activity may be a factor. The activity of nitrate reductase, extracted from embryos of *Agrostemma* seeds, is closely connected with their ability to germinate (Borriss and Schulze, 1966). Dormant embryos showed little or no activity. Embryos from after-ripened (i.e., nondormant) seeds possessed substantial activity even without induction and developed strong activity when induced by potassium nitrate.

Since there was no deficiency of ATP in the dormant pear bud tissues (Chaudhry et al., 1970), the metabolic block must have been related to the release of raw materials for growth. This agrees with earlier observations (Vegis, 1964*b*) that the onset and termination of dormancy are frequently connected (particularly in seeds) with the respective synthesis and decomposition of lipids. More recent evidence suggests that the same metabolic block may operate in seeds. Thus a 1-day stratification of Douglas fir seed removed the block preventing lipid breakdown (Ross, 1969). Presumably this released raw materials for new synthesis and provided substrate for respiration, which increased to the full capacity by the fifth day. Similarly, the treatment of potato tubers with a dormancy-breaking gaseous mixture called Rindite (p. 282) caused an increase in fatty acid content (Cherif, 1971).

From this variety of results, it must be concluded that the block may occur at several different points in the metabolism of the plant. Yet of the six suggested metabolic blocks, five must all lead to the sixth. Thus block 1, the deficiency in pool amino acids found in some dormant plants, must slow down or halt protein synthesis (block 2). This block does not necessarily require a deficiency of total amino acids. In dormant blueberry plants, for instance, the concentrations of total, insoluble, and soluble nitrogen were significantly higher than in nondormant plants (Kocher and Valenzuela, 1971). Only the ratio of total nitrogen to soluble nitrogen was lower. A deficiency of soluble phosphorus (block 3), would lead to block 4, a deficiency of nucleic acids. The remaining three deficiencies, blocks 4, 5, and 6, must also lead to a slow-down or cessation of protein synthesis. Therefore the single metabolic inhibition common to all the above cases of dormancy is of protein synthesis, although the specific metabolic block leading to it may occur at different points in the plant's metabolism. The breaking of dormancy would therefore be due, in all such cases, to an increased protein synthesis. The specific proteins involved are apparently enzymes that convert the reserves in the dormant plant or plant part into the building materials required for new cell synthesis and enlargement.

Mechanism of dormancy control

From the previous results it must be concluded that the plant controls dormancy by developing and removing specific metabolic blocks. How does it accomplish these two opposite controls? There are two possible mechanisms of dormancy (Wareing, 1965). Growth can be prevented (1) by the enclosing structure (seed coats, bud scales, etc.) restricting oxygen uptake and therefore metabolic activity, or (2) by the production of inhibitors. Vegis (1964*b*) favors the first method. He suggests that the high lipid content of the most deeply dormant seeds may be connected with a restricted oxygen supply. In the seeds of charlock *(Sinapis arvensis)*, the diffusion of oxygen into the embryo may be retarded by the mucilages and phenols in the testa (Edwards, 1968*a*). Similarly, excess moisture inhibits the germination of seed of *Spinacea oleracea*, yet there is no evidence of a water-soluble inhibitor, and the result seems to depend on the oxygen supply (Heydecker and Orphanos, 1968). On the other hand, the second mechanism accounts for the 4- to 12-month post-harvest dormancy of seeds of *Anagallis arvensis* (a winter annual). They contain a water-soluble inhibitor that appears to control the length of the dormant period (Singh, 1969). Even the difference between a dormant and a nondormant variety of orchard grass is accounted for by the difference in concentration of a growth inhibitor (Fendall and Canode, 1971). It does not necessarily follow, however, that the two mechanisms are mutually exclusive. In the case of charlock, in fact, an interaction between the two has been suggested (Edwards, 1968*b*): the action of a specific growth-inhibiting substance, which is produced at low oxygen concentration in the interior of the seed and diffuses to the meristem. That the oxygen deficiency is an indirect rather than a direct factor is indicated by the reverse result. If the restriction of the oxygen supply is too extreme, anaerobic respiration may lead to the accumulation of ethyl alcohol and acetaldehyde, which may break the dormancy. Some of the older methods of breaking the rest period may be explained in this way; for example, immersion in warm water may induce a sufficient oxygen deficiency to stimulate an active anaerobic respiration, leading to the accumulation of substances that break dormancy. Dormant buds of grapevine are induced to grow by anaerobiosis (Pouget, 1971). Similarly, when the seeds were allowed to imbibe water for 48 hr and then kept anaerobic for 10 to 12 days, they germinated at 25° C at least as well as seeds stratified at 1° C for 2 to 3 months.

Most of the modern evidence favors the second concept. Many kinds of inhibitors may be involved. In grape seeds, the inhibitors consist of three or more proteins with molecular weights of 10,000 or 20,000 (Ogawa et al., 1971). In seeds of *Aegilops kotschyi* a specific RNA fraction is responsible for dormancy, and germination is dependent on its degradation (Wurzburger and Lesham, 1971). Many investigators (e.g., Bell and Amen, 1970) now support the concept that dormancy may be controlled by a balance between an inhibitor (commonly ABA, see Chapter 19) and a promoter (commonly gibberellic acid). This will be discussed in Chapter 20.

QUESTIONS

1. How is growth measured?
2. Are fresh or dry weights satisfactory measurements?
3. Where does growth occur in the plant?
4. What cellular changes occur in these regions?
5. Which cellular changes really involve growth?
6. Define growth on a cellular basis.
7. What is the motivating force for growth?
8. How does turgor pressure compare in growing as opposed to nongrowing cells?
9. What cell property permits growth of cells at the same turgor as fails to produce it in mature cells?
10. What processes accompany cell enlargement?
11. Why is energy required for cell enlargement?
12. Is it conceivable that cell enlargement may be initiated by something other than turgor pressure?
13. What is the initiating process in this case?
14. Do all cells grow uniformly?
15. Is there any relation between the protoplasm and the growing cell wall?
16. What is the shape of the growth curve?
17. What is the grand period of growth?
18. What concept of growth is used to develop an equation for growth?
19. What is meant by the cardinal points for growth?
20. Are these cardinal points exact?
21. What effect does a 10° C temperature rise have on growth?
22. Is growth possible in the absence of light?
23. What is etiolation?
24. Does light enhance or retard growth?
25. How does a water deficit affect growth?
26. How does excess water affect growth?
27. What is meant by the rest period?
28. What is meant by dormancy?
29. List the phases of the rest period and of dormancy.
30. How may dormancy be overcome artificially?
31. What changes occur in cells entering dormancy?
32. What changes occur in cells emerging from dormancy?
33. What temperature relation changes during the entrance into and emergence from dormancy?

SPECIFIC REFERENCES

Ballard, L. A. T., and T. T. Buchwald. 1971. A viability test for seeds of Townsville stylo using thiourea. Aust. J. Exp. Agr. Anim. Husb. **11:** 207-210.

Balthazard, J. 1971. Apparent and true dormancy of grapevine seeds. Compt. Rend. Acad. Sci. (Paris) **272:**2773-2776.

Barker, G. R., and M. Rieber. 1967. The development of polysomes in the seed of *Pisum arvense*. Biochem. J. **105**:1195-1202.

Barton, K. A., C. H. Roe, and A. A. Kahn. 1971. Imbibition and germination: influence of hard seed coats on RNA metabolism. Physiol. Plant. **25**:402-406.

Beck, W. A. 1941. Production of solutes in growing epidermal cells. Plant Physiol. **16**:637-642.

Beck, W. A., and B. Andrus. 1943. The osmotic quantities of the cells in the hypocotyl of *Helianthus annuus* seedlings. Bull. Torrey Bot. Club **70**:563-598.

Bell, K. L., and R. D. Amen. 1970. Seed dormancy in *Luzula spicarta* and *L. parviflora*. Ecology **51**:492-496.

Blackman, V. H. 1919. The compound interest law and plant growth. Ann. Bot. **33**:353-360.

Blank, F., and A. Frey-Wyssling. 1944. Protoplasmic growth in the hypanthium of *Oenothera acaulis* during cell elongation. Ann. Bot. **8**:71-78.

Bopp, M. 1970. DNA synthesis and cell elongation. Physiol. Vég. **8**:215-230.

Borriss, H., and J. Schulze. 1966. Die Nitratreductase-Aktivität der Embryonen ruhender und nachgereifter *Agrostemma*-Samen. Z. Pflanzenphysiol. **55**:449-457.

Brandts, J. F. 1967. Heat effects on proteins and enzymes, p. 25-72. *In* A. H. Rose. Thermobiology. Academic Press, Inc., New York.

Brant, R. E., G. W. McKee, and R. W. Cleveland. 1971. Effect of chemical and physical treatment on hard seed of Penngift Crownvetch. Crop Sci. **11**:1-6.

Burström, H. 1953. Studies on the growth and metabolism of roots. IX. Cell elongation and water absorption. Physiol. Plant. **6**:262-276.

Burström, H. 1965. Light in the regulation of root growth, p. 45-60. *In* P. R. White and A. R. Grove. Proceedings of the International Conference of Plant Tissue Culture. McCutchan Publishing Corp., Berkeley.

Chandra, G. R., L. E. Gregory, and J. F. Worley. 1971. Studies on the initiation of adventitious roots on mung bean hypocotyl. Plant Cell Physiol. **12**:317-324.

Chaudhry, W. M., T. C. Broyer, and L. C. T. Young. 1970. Chemical changes associated with the breaking of the rest period in vegetative buds of *Pyrus communis*. Physiol. Plant. **23**:1157-1169.

Chen, L. H., B. K. Huang, and W. E. Splinter. 1968. Growth dynamics of small tobacco plants as affected by night temperature and initial plant size. Trans. Asae **11**:126-128.

Chen, S. S. C. 1970. Influence of factors affecting germination on respiration of *Phacelia tanacetifolia* seeds. Planta **95**:330-335.

Chen, S. S. C., and J. E. Varner. 1970. Respiration and protein synthesis in dormant and after-ripened seeds of *Avena fatua*. Plant Physiol. **46**:108-112.

Cherif, A. 1971. Breaking of potato tuber dormancy: effect on lipogenesis. Compt. Rend. Acad. Sci. (Paris) **272**:242-245.

Cleland, R. 1967. Inhibition of formation of protein-bound hydroxyproline by free hydroxyproline in *Avena* coleoptiles. Plant Physiol. **42**:1165-1170.

Cleland, R. 1968. Hydroxyproline formation and its relation to auxin-induced cell elongation in the *Avena* coleoptile. Plant Physiol. **43**:1625-1630.

Clowes, F. A. L. 1971. The proportion of cells that divide in root meristems of *Zea mays* L. Ann. Bot. **35**:249-261.

Cocucci, S., and E. P. Sturani. 1965. Nucleic acids in ripening castor bean seed endosperm. II. Changes of the RNA from different fractions. [Transl. title.] G. Bot. Ital. **72**:357-358.

Coombe, B. G. 1960. Relationship of growth and development to changes in sugars, auxins, and gibberellins in fruit of seeded and seedless varieties of *Vitis vinifera*. Plant Physiol. **35**:241-250.

Cotrufo, C., and J. Levitt. 1958. Investigations of the cytoplasmic particulates and proteins of potato tubers. VI. Nitrogen changes associated with emergence of potato tubers from the rest period. Physiol. Plant. **11**:240-248.

Davidson, J. L., and F. L. Milthorpe. 1965. The effect of temperature on the growth of cocksfoot (*Dactylis glomerata* L.). Ann. Bot. **29**:407-417.

Davies, J. W., and E. C. Cocking. 1965. Changes in carbohydrates, proteins, and nucleic acids during cellular development in tomato fruit locule tissue. Planta **67**:242-253.

Denny, F. E. 1945. Synergistic effects of three chemicals in the treatment of dormant potato tubers to hasten germination. Contrib. Boyce. Thompson Inst. Plant Res. **14**:1-14.

Dillard, W. L., and H. G. Schweiger. 1969. RNA synthesis in *Acetabularia (mediterranea)*. III. The kinetics of RNA synthesis in nucleate and enucleate cells. Protoplasma **67**:87-100.

Drumm, H., and H. Mohr. 1967. Die Regulation der RNS-Synthese in Farngametophyten durch Licht. Planta **72**:232-246.

Edwards, M. M. 1968*a*. Dormancy in seeds of charlock *(Sinapis arvensis)*. I. Development anatomy of the seed. J. Exp. Bot. **19**:575-582.

Edwards, M. M. 1968*b*. Dormancy in seeds of charlock. II. The influence of the seed coat. J. Exp. Bot. **19**:583-600.

Erez, A., S. Lavee, and R. M. Samish. 1971. Improved methods for breaking rest in the peach and other deciduous fruit species. J. Amer. Soc. Hort. Sci. **96**:519-522.

Esashi, Y. 1969. The relation between light and temperature effects in the induction and release of dormancy in the aerial tuber of *Begonia evansiania*. Plant Cell Physiol. **10**:583-595.

Fendall, R. K., and C. L. Canode. 1971. Dormancy-related growth inhibitors in seeds of orchardgrass (*Dactylis glomerata* L.). Crop Sci. **11**:727-730.

Frey-Yyssling, A. 1952. Growth of plant cell walls. Sympos. Soc. Exp. Biol. **6**:320-328.

George, D. W. 1967. High temperature seed dormancy in wheat (*Triticum aestivum* L.). Crop Sci. **7**:249-253.

Goodwin, P. B. 1966. The effect of water on dormancy in the potato. Eur. Potato J. **9**:53-63.

Green, P. B., R. O. Erickson, and J. Buggy. 1971. Metabolic and physical control of cell elongation rate: *in vivo* studies in *Nitella*. Plant Physiol. **47**:423-430.

Gregory, F. G. 1921. The increase of leaves and leaf surface of *Cucumis sativus*. Ann. Bot. **35**:93-123.

Hadacova, V. 1968. The influence of some respiration inhibitors on the respiration of the root zones of *Vicia faba* L. Biol. Plant. **10**:385-397.

Heydecker, W., and P. I. Orphanos. 1968. The effect of excess moisture on the germination of *Spinacea oleracea* L. Planta **83**:237-247.

Hobson, G. E. 1971. A study of mitochondrial complementation in wheat. Biochem. J. **124**: 110P.

Holdgate, D. P., and T. W. Goodwin. 1965. Metabolism of nucleic acids during early stages of the germination process in rye (*Secale cereale*). Phytochemistry **4**:845-850.

Howard, W. L. 1915. An experimental study of the rest period of plants. Res. Bull. Mo. Agric. Exp. Sta. Pub. No. 21. Columbia, Mo.

Hsiao, A. I-H., and G. M. Simpson. 1971. Dormancy studies in seed of *Avena fatua*. VII. The effects of light and variation in water regime on germination. Canad. J. Bot. **49**:1347-1357.

Jaynes, R. A. 1971. Seed germination of six *Kalmia* species. J. Amer. Soc. Hort. Sci. **96**:668-672.

Jolivet, E. 1968. Métabolisme des acides organiques et des acides amines libres dans le tubercule de pomme de terre au cours de la rupture provoquée de son répos végétatif par la Rindite. Physiol. Veg. **6**:221-223.

Juniper, B. E., and F. A. L. Clowes. 1965. Cytoplasmic organelles and cell growth in root caps. Nature (London) **208**:864-865.

Kender, W. J., and J.-C. Desrochers. 1970. Changes in endogenous auxin, gibberellin-like substances, and inhibitors in developing lowbush blueberry fruits. J. Amer. Soc. Hort. Sci. **95**:699-702.

Khan, A. A., C. E. Heit, and P. C. Lippold. 1968. Increase in nucleic acid synthesizing capacity during cold treatment of dormant pear embryos. Biochem. Biophys. Res. Comm. **33**:391-396.

Kliewer, W. M. 1967. Annual cyclic changes in the concentration of free amino acids in grapevines. Amer. J. Enol. Viticult. **18**:126-137.

Kocher, F., and J. Valenzuela. 1971. Nitrogenous constituents in tissues of growing and dormant lowbush blueberry plants. Hort. Science **6**:411-413.

Lamport, D. T. A. 1969. The isolation and partial characterisation of hydroxyproline-rich glycopeptides obtained by enzymic degradation of primary cell walls. Biochemistry **8**:1155-1163.

Lamport, D. T. A., and D. H. Miller. 1971. Hydroxyproline arabinosides in the plant kingdom. Plant Physiol. **48**:454-456.

Lange, H., G. Kahl, and G. Rosenstock. 1971. The glucose metabolism of depressed plant storage parenchyma after inhibition of mitosis activity by Tris-(hydroxymethyl)aminomethane. Physiol. Plant. **24**:1-4.

Lathrop, J. K., and R. A. Mecklenburg. 1971. Root regeneration and root dormancy in *Taxus* spp. J. Amer. Soc. Hort. Sci. **96**:111-114.

Lemattre, P. 1971. Effect of a thermal treatment on the dormancy of astilbes. Ann. Amélior. Plantes **21**:125-139.

Le Page-Degivry, M. T. 1970. Seed dormancy associated with embryo immaturity: in-vitro culture study of *Magnolia soulangeana* Soul. Bod. and *Magnolia grandiflora* L. Planta **90**: 267-271.

Lin, C. Y., and J. L. Key. 1968. Cell elongation in the soybean root: the influence of inhibitors of RNA and protein biosynthesis. Plant Cell Physiol. **9**:553-560.

Lipp, A. E. G., and L. A. T. Ballard. 1970. Thiourea as a stimulator and inhibitor of germination of seed of subterranean clover (*Trifolium subterraneum* L.). Z. Pflanzenphysiol. **62**:83-88.

Lockhart, J. A. 1971. An interpretation of cell growth curves. Plant Physiol. **48**:245-248.

Lyndon, R. F. 1963. Changes in the nucleus during cellular development in the pea seedling. J. Exp. Bot. **14**:418-430.

Maguire, J. D., and K. M. Steen. 1971. Effects of potassium nitrate on germination and respiration of dormant and nondormant Kentucky bluegrass (*Poa pratensis* I.) seed. Crop Sci. **11**:48-50.

Marcus, A., and J. Feeley. 1964. Activation of protein synthesis in the imbibition phase of seed germination. Proc. Nat. Acad. Sci. USA **51**: 1075-1079.

Mercer, M. J., and F. V. Mercer. 1971. Studies on the comparative physiology of *Chara corallina*. III. Nitrogen relations of internodal cell components during internodal cell expansion. Aust. J. Bot. **19**:1-12.

Meyer, M. M., Jr., and H. B. Tukey, Jr. 1967. Influence of root temperature and nutrient applications on root growth and mineral nutrient content of *Taxus* and *Forsythia* plants during the dormant season. Proc. Amer. Soc. Hort. Sci. **90**:440-446.

Meyer, M. M., Jr., and W. E. Splittstoesser. 1971. The utilization of carbohydrate and nitrogen reserves by *Taxus* during its spring growth period. Physiol. Plant. **24**:306-314.

Nelson, L. E. 1971. The effects of root temperature and Ca supply on the growth and transpiration of cotton seedlings (*Gossypium hirsutum* L.). Plant Soil **34**:721-729.

Nikolaeva, M. G. 1969. Physiology of deep dormancy in seeds. Acad. Sci. USSR V. L. Komarov Bot. Inst. Ed. I. N. Konovalov. Israel Program Scientific Translations.

Ogawa, T., T. Higasa, and T. Hata. 1971. Proteinase inhibitors in plant seeds. V. Inhibitors in rape seeds. Mem. Res. Inst. Food Sci. Kyoto Univ. **32**:1-6.

Perry, T. O. 1970. Dormancy of trees in winter. Science **170**:29-36.

Popay, A. I., and E. H. Roberts. 1970*a*. Factors involved in the dormancy and germination of *Capsella bursa-pastoris* (L). Medik. and *Senecio vulgaris* L. J. Ecol. **58**:103-122.

Popay, A. I., and E. H. Roberts. 1970*b*. Ecology of *Capsella bursa-pastoris* (L.) Medik, and *Senecio vulgaris* L. in relation to germination behaviour. J. Ecol. **58**:123-139.

Pouget, R. 1971. The effect of anaerobic conditions, inducing the lifting of dormancy in latent buds, upon embryonic dormancy in grapevine seeds. Compt. Rend. Acad. Sci. (Paris) **272**: 956-959.

Radford, P. J. 1967. Growth analysis formulae: their use and abuse. Crop Sci. **7**:171-175.

Raese, J. T. 1971. Prolonging dormancy of tung trees with spray oil and succinic acid 2,2-dimethyl-hydrazide. Hort. Science **6**:408-410.

Rayle, D. L., P. M. Haughton, and R. Cleland. 1970. An in vitro system that stimulates plant cell extension growth. Proc. Nat. Acad. Sci. USA **67**:1814-1817.

Risueno, M. C., G. Gimenez-Martin, and J. F. Lopez-Saez. 1968. Role of Golgi vesicles in plant cell elongation. Experientia **24**:926.

Ross, S. D. 1969. Gross metabolic activity accompanying the after-ripening of dormant Douglas-fir seeds. Bot. Gaz. **130**:271-275.

Sachar, R. C., S. Guha, and K. Sachar. 1970. Breaking seed dormancy by feeding casamino acids to the unripe achenes of *Ranunculus sceleratus* L. Indian J. Exp. Biol. **8**:129-134.

Sadava, D., and M. J. Chrispeels. 1969. Cell wall protein in plants: autoradiographic evidence. Science **165**:299-300.

Scheffer, F., and H. Lorenz. 1968. Pool-Aminosauren wahrend des Wachstum und der Entwicklung einiger Weizensorten. Phytochemistry **7**:1279-1288.

Schroter, K., and A. Sievers. 1971. Effect of turgor reduction on the Golgi apparatus and cell wall formation in root hairs. Protoplasma **22**:203-211.

Shih, C. Y., and L. Rappaport. 1970. Regulation of bud rest in tubers of potato, *Solanum tuberosum* L. VII. Effect of abscisic and gibberellic acids on nucleic acid synthesis in excised buds. Plant Physiol. **45**:33-36.

Short, K. C., E. G. Brown, and H. E. Street. 1969. Studies on the growth in culture of plant cells. VI. Nucleic acid metabolism of *Acer pseudoplatanus* L. cell suspensions. J. Exp. Bot. **20**:579-590.

Siminovitch, D., B. Rhéaume, and R. Sachar. 1967. Seasonal increases in protoplasm and metabolic capacity in the cells during adaptation to freezing, p. 3-40. *In* Prosser, C. L. (Ed.). "Molecular Mechanisms of Temperature Adaptation." Pub. No. 84. Amer. Soc. Adv. Sci., Washington, D. C.

Singh, K. P. 1969. Seed dormancy and its control by germination inhibitor in *Anagallis arvensis* L. var. Coerulea Gren et Godr. Proc. Nat. Inst. Sci. India Part B. Biol. Sci. **35**:161-171.

Smith, H., and H. Frankland. 1966. Specific inhibition by uracil derivatives of the mechanism of dormancy release in light-sensitive lettuce seeds. Nature (London) **211**:1323-1324.

Spencer, F. S., and G. A. MacLachlan. 1972. Changes in molecular weight of cellulose in the pea epicotyl during growth. Plant Physiol. **49**: 58-63.

Sytnyk, K. M., and L. I. Musatenko. 1968. Nucleic acids in growth zones of maize stalks. Dopov. Akad. Nauk. Ukr. RSR. **30**:760-762.

Talma, E. G. C. 1918. The relation between temperature and growth in the roots of *Lepidium sativum*. Receuil, Trav. Bot. Neerland. **15**: 366-422.

Taylor, B. K. 1969. Physiological studies on

dormancy in Cape tulip. Aust. J. Biol. Sci. **22**: 787-796.

Taylor, H. M., and L. F. Ratliff. 1969. Root growth pressures of cotton, peas, and peanuts. Agron. J. **61**:398-402.

Thomson, B. F. 1951. The effect of light on the rate of development of *Avena* seedlings. Amer. J. Bot. **37**:284-291, 635-638.

Todd, G. W. 1953. Enzyme studies on dormant and active potato tubers. Physiol. Plant. **44**: 168-172.

Tort, M. 1971. Synthèses d'ADN dans le bourgeon terminal des tubercules du Crome du Japon (*Stachys sieboldi* Miq.) dormants ou non dormants (dormance levée par le froid ou par l'acide gibberellique). Compt. Rend. Acad. Sci. (Paris) **272**:1764-1767.

Ueki, K., and N. Shimizu. 1969. Studies on the breaking of dormancy in barnyard grass seed. I. The effects of various chemicals on the breaking of dormancy. Proc. Crop Sci. Soc. Jap. **38**: 261-272.

Uhlinger, R. D. 1970. Germination of heat treated seeds of *Philadelphus lewisi* Pursh. J. Amer. Soc. Hort. Sci. **95**:307-310.

Vanderhoef, L. N., and J. L. Key. 1971. RNA ratios in dividing and non-dividing cells of pea root. Biochim. Biophys. Acta **240**:62-69.

Van Parijs, R., and L. Vandendriessche. 1966*a*. Changes of the DNA content of nuclei during the process of cell elongation in plants *(Pisum sativum). I.* The formation of polytene chromosomes. Arch. Int. Physiol. Biochim. **74**:579-586.

Van Parijs, R., and L. Vandendriessche. 1966*b*. Changes of the DNA content of nuclei during the process of cell elongation in plants *(Pisum sativum).* II. Variations, other than doublings, of the amount of Feulgen stain in maturing plant cells. Arch. Int. Physiol. Biochim. **74**: 587-591.

Villiers, T. A. 1971. Cytological studies in dormancy. I. Embryo maturation during dormancy in *Fraxinus excelsior.* New Phytol. **70**:751-760.

Vyas, L. N., and S. K. Agarwal. 1971. Studies in the seed dormancy of desert plants. II. Effect of thiourea and ascorbic acid on germination of *Verbena bipinnatifida* Nutt. seeds. Phyton. Rev. Int. Bot. Exp. **28**:1-5.

Warden, J. 1969-70. Leaf-embryo dormancy in *Bryophyllum crenatum* under short day conditions. Port. Acta Biol. Ser. A. **11**:319-338.

Wareing, P. E. 1965. Dormancy in plants. Sci. Progr. **53**:529-537.

Wareing, P. W. 1969. Germination and dormancy, p. 605-644. *In* M. B. Wilkins (Ed.). The physiology of plant growth and development. McGraw-Hill Book Co., New York.

Watts, W. R. 1971. Role of temperature in the regulation of leaf extension in *Zea mays.* Nature **229**:46-47.

Webb, D. P., and E. B. Dumbroff. 1969. Factors influencing the stratification process in seeds of *Acer saccharum.* Can. J. Bot. **47**:1555-1563.

Westwood, M. N. and H. O. Bjornstad. 1968. Chilling requirements of dormant seeds of 14 pear *(Pyrus)* species as related to their climatic adaptation. Proc. Amer. Soc. Hort. Sci. **92**:141-149.

Williams, R. F., and A. H. G. C. Rijven. 1965. The physiology of growth in the wheat plant. II. The dynamics of leaf growth. Aust. J. Biol. Sci. **18**:721-743.

Wilson, K. 1964. The growth of plant cell walls. Int. Rev. Cytol. **17**:1-49.

Winter, H. 1966. Effect of auxin and sugar on cell wall synthesis and growth of pea stem segments. Proc. Kon. Nederl. Akad. Wet. (Biol. Med.) **69**:64-72.

Wolters, G. L. 1970. Breaking dormancy of longleaf uniola seeds. J. Range Manage. **23**:178-180.

Woods, L. E., and H. A. MacDonald. 1971. The effects of temperature and osmotic moisture stress on the germination of *Lotus corniculatus.* J. Exp. Bot. **22**:575-585.

Wurzburger, J., and Y. Leshem. 1971. Ribonucleic acid as an inducer of germination inhibition in *Aegilops kotschyi.* Plant Cell Physiol. **12**:211-215.

Zeleneva, I. V., and E. E. Khavkin. 1971. Formation of the respiratory enzyme system in growing cells: glycolytic enzymes in maize sprouts. Ontogenez **2**:311-320.

GENERAL REFERENCES

Burström, H. 1951. Mechanism of cell elongation. *In* F. Skoog (Ed.) Plant growth substances. University of Wisconsin Press, Madison, Wis.

Crocker, W. 1948. Growth of plants. Reinhold Publishing Corp., New York.

Leopold, A. C. 1964. Plant growth and development. McGraw-Hill Book Co., New York.

Maximov, N. A. 1938. Plant physiology. McGraw-Hill Book Co., New York.

Palladin, V. I. 1923. Plant physiology. B. E. Livingston (Ed.). Blakiston Division, McGraw-Hill Book Co., New York.

Robertson, T. B. 1923. The chemical basis of growth and senescence. J. B. Lippincott Co., Philadelphia.

Steward, F. C. 1968. Growth and organization in plants. Addison-Wesley Publishing Co., Inc., Reading, Mass.

Thompson, D. W. 1942. On growth and form. Cambridge University Press, New York.

Vegis, A. 1964*a*. Ruhezustände bei höheren Pflanzen, Induktion, Verlauf und Beendigung; Übersicht, Terminologie, allgemeine Probleme, p. 499-533. *In* W. Ruhland. Encyclopedia of plant physiology. Vol. XV. Springer-Verlag, New York, Inc., New York.

Vegis, A. 1964*b*. Die Bedeutung von physikalichen und chemischen Aussenfaktoren bei der Induktion und Beendigung von Ruhezustanden bei Organen und Geweben höherer Pflanzen, p. 534-668. *In* W. Ruhland. Encyclopedia of plant physiology. Vol. XV. Springer-Verlag, New York, Inc., New York.

Vegis, A. 1964*c*. Dormancy in higher plants. Ann. Rev. Plant Physiol. **15**:185-224.

Wilkins, M. B. 1969. The physiology of plant growth and development. McGraw-Hill Book Co., New York.

18

Irritability and movement

IRRITABILITY

In its broadest sense, *irritability* is the ability to respond physically to an environmental factor. Both irritability and movement are popularly thought of as characteristic of animals and not of plants. This is due to a tendency to recognize only those irritable responses and movements which are rapid enough to observe at a glance. Even these instantaneous responses do, however, occur in some plants—for instance, the rapid folding of the leaves of the sensitive plant *Mimosa pudica* when touched and the rapid closing of the traps of insectivorous plants. In the vast majority of cases, however, irritability and movement in a plant are too slow to be seen at a glance. Its leaves or flowers close or open slowly in response to darkness and light; its stem bends slowly toward the light or grows slowly upward against the force of gravity. Not all plant irritabilities, however, involve movement. An almost human response that has recently been discovered is the change in rate of phloem exudation resulting from massage (Milburn, 1971). Similarly, when an insectivorous plant is stimulated by suitable nitrogenous materials, the glands begin to secrete fluid onto the leaf surface within 1 hr (Heslop-Harrison and Knox, 1971). Even the organelles within the cell may show irritability responses, for example, plastid movements in response to light and streaming movement, which may arise or speed up on exposure to light or because of a rise in temperature. The specific environmental factor that gives rise to the irritability reaction is called the *stimulus,* and the reaction of the plant is called the *response.* The earlier physiologists were fascinated with the phenomenon, but more modern investigators seem to have lost interest in it. It is, in fact, rare nowadays to find reference to the term *plant irritability.* Recently, however, interest in outer space has given rise to attempts to grow plants in the complete absence of polarization of environmental factors, that is, with all the environmental factors either reduced to a minimum or of equal strength on all sides of the plant (Gordon et al., 1964). This tends to remove the stimuli and therefore presumably also the response. In the complete absence of polarization in the environment, the plant has been found to grow as an undifferentiated mass of tissue instead of differentiating into organs. Thus the phenomenon of *polarity* in plant growth (e.g., the formation of a stem tip at one pole, a root tip at the other) is an example of irritability. In fact, the whole process of organ differentiation may be considered to be the result of irritability. This is perhaps one reason why the concept has been dropped by many physiologists because, in its broadest sense, it may include nearly the whole field of growth and development.

MOVEMENTS

Cell movements

When used in its narrower sense, the main irritability reactions of the plant that have been intensively studied are plant movements. There are six distinct kinds of plant movement (Fig. 18-1). But these are all either cell or organ movements and, unlike the animal, the higher plant does not move as a whole but remains anchored in place. Cytoplasmic streaming can also be found in animal cells, but it is usually much more pronounced in plant cells (Ambrose, 1965). Rates as high as 107 μm/sec have been measured (Chapter 10). It occurs in the inner cytoplasm, or *endoplasm,* which is in the sol, or liquid, state. Between the endoplasm and the outer plasma membrane is the *ectoplasm,* which is stationary because of its gel, or solid, state. Although the organelles may be carried along in the streaming cytoplasm, they may move at independent speeds and even in opposite directions within the same stream. Chloroplasts even rotate about their own axes. The direction of the stream may be reversed, and in the case of myxomycete plasmodia, this reversal occurs regularly at about 60 sec intervals (Kamiya, 1942).

The driving force has been sought for years by many investigators. Hydrostatic forces cannot be involved, since the streaming can continue in a *Nitella* cell that has been cut open and is therefore at atmospheric pressure. In this case the cytoplasm may flow right out of one side of the cut surface while flowing up the opposite side (Ambrose, 1965). The evidence indicates that the driving force originates at the contact between the streaming endoplasm and the stationary ectoplasm. The key to the movement may be found in the microtubules, which frequently can be seen under the electron microscope near the plasmalemma (Fig. 2-6). Kamiya found that the movement is dependent on adenosine triphosphate (ATP). He suggests that there

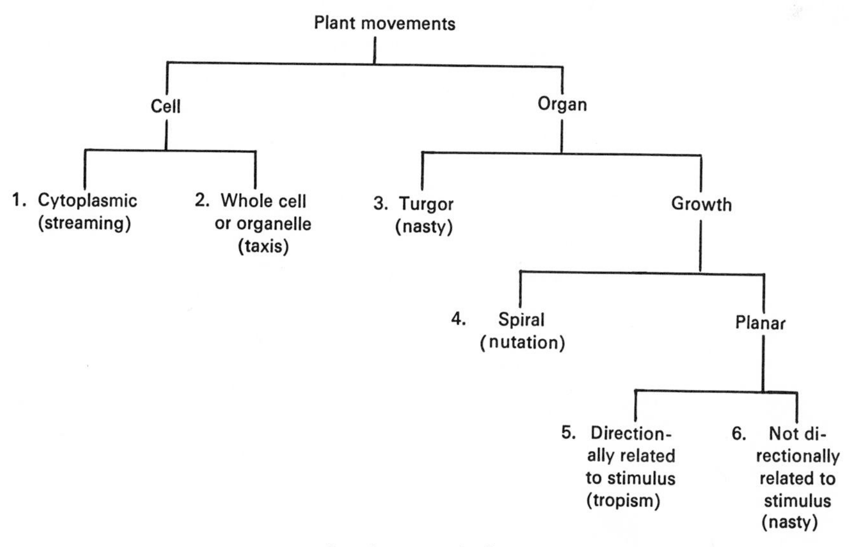

Fig. 18-1. Classification of plant movements.

may be contractile proteins localized near the cell membrane which are responsible for the movement. More direct evidence in favor of this concept has been obtained with thin threads extracted from slime mold (*Physarum polycephalum*) plasmodium. On addition of 5 mM ATP they show a 55% linear contraction (Beck et al., 1970) and a positive reaction for adenosine triphosphatase activity. Electron micrographs reveal a fine structure typical of F-actin. All these results resemble the contractile system of cross-striated muscle. The muscle protein, actomyosin, consists of thin filaments (F-actin) that slide along the thick filaments (myosin) at the expense of ATP energy, due to the adenosine triphosphatase activity of the myosin. It is therefore reasonable to suggest that the contractile proteins of slime mold protoplasm may also contract by a sliding filament mechanism. Results with other plant cells indicate that the contractile proteins occur in the cell as microfilaments 50 Å thick (Wessells et al., 1971). Streaming is reversibly stopped by cytochalasin, a drug that inhibits contractile activity. According to Jahn and Bovee (1969), the contractions drive a thixotropic internal sol that changes reversibly into a gel. The more or less gelated actomyosin may flow as strands or sheets in the case of cyclosis, or in transvacuolar threads.

Although taxis (motility) is more common among lower organisms (blue-green algae, bacteria), the chloroplasts of higher plants are also capable of this kind of movement (Haupt, 1966). In *Vallisneria spiralis* the negative phototaxis of the chloroplasts may be accompanied by a rotation (Seitz, 1967). ATP can promote the movement and can even replace the light (Seitz, 1971). It was therefore concluded that the primary effect of the light is to increase the availability of ATP and that the phototactic movement is due to an ATP gradient built cooperatively by oxidative phosphorylation and photophosphorylation. Other organelles also show mass movements within the cytoplasmic stream but at velocities differing from that of the protoplasmic mass (Mahlberg, 1965). This may also be considered a kind of taxis. The action spectrum for phototaxis in *Euglena* is similar to the absorption spectrum of riboflavin (Diehn and Kent, 1970). Other evidence also supports the conclusion that a flavin, rather than a carotenoid pigment, is the photosensitizer. Chloroplast movement, on the other hand, may be due to light absorption in two different photoreceptors acting together (Gaertner, 1970).

Organ movements

Organ movements are of two main kinds: (1) those caused by turgor changes and (2) those caused by asymmetrical growth. Loss of turgor is frequently a rapid process, leading to rapid movements such as the closing of the leaves of *Mimosa pudica* or the closing of the traps of insectivorous plants (Lloyd, 1942). Such rapid turgor changes are believed to be the result of a rapid loss in semipermeability (see later discussion). Slower turgor changes caused by metabolically produced changes in solute content lead to the slower turgor movements of stomata. The slow sleep movements of plant leaves, however, are believed to be caused by the aforementioned permeability changes. The slowest movements are those resulting from differential (i.e., asymmetrical) rates of growth. Investigations of these so-called growth movements have led to some of the most important information as to the nature of the growth process.

There are three main types of growth movements: (1) *circumnutation,* (2) *tropisms*, and (3) *nastic movements.* Unfortunately, the latter two may be caused by turgor changes as well as by growth, but those caused by growth movements are far more common and have been more intensively studied. All or nearly all plants show circumnutation. This is a spiral type of growth of the plant apex caused by a different rate of growth on opposite sides of the growing tip. The more rapid rate of growth travels around the tip, which, as it grows upward (in the case of the stem), must therefore rotate. The cause of this cyclic change in rates of growth of cells on different sides of the tip is not known. In some plants, circumnutation may come to a stop when the gravitational force has been compensated for on a clinostat (see later discussion) and the plant is kept in the dark; but in other cases the circumnutation has continued under these conditions (Johnsson, 1966). The latter result has led to the conclusion that the movement is autonomous (independent of the environment). If so, circumnutation could not truly be classified as an irritability, since this is a response to the environment. This problem has been circumvented in the past by dividing all plant movements into *autonomic* (controlled by internal stimuli) and *paratonic* (controlled by external stimuli). However, there is no basis for such a classification until an internal stimulus is identified and proved to induce a plant movement. If the relation to gravity is confirmed, circumnutation could conceivably be an overshoot in response to gravity. In support of this explanation the ratio between the periodic time for circumnutation and the reaction time for geotropic curvature (see later discussion) has been found to remain constant with temperature changes from 15° to 40° C (Israelsson and Johnsson, 1967). It has even been possible to predict the amplitude of the circumnutations on the basis of the gravitational response (Johnsson, 1971*a*).

The other two growth movements, however, are definitely a response to an environmental stimulus. The length of time the plant must be exposed to the stimulus in order for a reaction to occur is the *presentation time*. The length of time required for it to react to the stimulus is the *reaction time*. In other words, the plant may be exposed to the stimulus for the presentation time without showing any response. Later, however, in the absence of the stimulus the reaction may appear. The time it takes for the plant to recover its original position after the stimulus has been removed is the *relaxation time*. In the case of tropisms the direction of movement shows a definite relation toward, away from, or at an angle to the direction of application of the stimulus. It is the *plane* of the response that is controlled by the spatial relation between the stimulus and the tropically responding organ (Wilkins, 1966). A movement toward a stimulus is called a positive movement and away from the stimulus, a negative movement. Nastic movements, on the other hand, show no definite directional relation to the stimulus. Since movements can be caused by different stimuli, the name of the movement consists of a prefix that indicates the stimulus and a suffix that states the kind of movement (Table 18-1).

There is some evidence of other growth movements, such as electrotropism and magnetotropism (Pittman, 1963). A greater root concentration on the north and south sides of wheat seedlings has been explained as a geomagnetotropic response (Woolley and Pittman, 1966). Electrotropism has

Table 18-1. Different kinds of tropisms and nasties and their respective stimuli

Stimulus	Movement directionally related to stimulus (tropisms)	Movement not directionally related to stimulus (nasties)
Gravity	Geotropism	
Light (sun)	Phototropism (heliotropism)	Photonasty
Temperature	Thermotropism	Thermonasty
Night		Nyctinasty
Touch	Thigmotropism (haptotropism)	Seismonasty
Chemicals	Chemotropism	
Water	Hydrotropism	

been demonstrated in a number of plants (Murr, 1965). The electrostatic or electrokinetic field can either stimulate or retard their growth, although in all cases leaf damage occurs.

Although chemotropism is usually an artificial phenomenon, it is also found in nature. The pistils of many higher plants contain substances that induce chemotropism of pollen tubes. One substance has been found to induce positive chemotropism and another, negative chemotropism (Miki-Hirosige, 1964). But the only two tropisms that have been studied in considerable detail are geotropism and phototropism.

GEOTROPISM

Primary roots are *positively geotropic* (i.e., bend toward the center of the earth), and primary stems are *negatively geotropic.* Secondary lateral roots and shoots show a weaker response and take up a position at an angle to the gravitational force. They are said to be *plagiogeotropic* as opposed to the primaries, which are *orthogeotropic* (parallel to the gravitational force). Lateral roots and shoots of a higher order are almost insensitive to the stimulus of gravity *(apogeotropic).* Rhizomes grow horizontally; that is, they take up a position at right angles to the gravitational stimulus and are called *diageotropic.* The response of an organ to the gravitational stimulus may change with the stage of development. The peduncle of the poppy bud is positively geotropic, but it gradually becomes negative as the flower opens. A change may also be induced artificially; if the primary root or stem tip is cut off, the lateral nearest to the wound becomes orthogeotropic.

In dicotyledons the first sign of a geotropic response occurs a short distance back from the apex. In grasses and other monocotyledons it occurs near the nodes where the meristems are found. If roots or shoots are *decapitated* (the apex is removed), they fail to show a geotropic response. However, if they are exposed to the stimulus and then decapitated before curvature can occur (after the presentation time but before the reaction time), the stumps will nevertheless show a geotropic response. Thus the tip is necessary for the *perception* of the stimulus (and is called the region of perception) but not for the *response.* The region of response is some distance back from the tip. This control of the responses of one part of a plant by another part is known as *correlation.* A presentation time as short as 30 sec (at least 10 times lower than those usually reported in the literature) has recently been shown to produce a geotropic response (Johnsson, 1971*b*), and an even shorter one of 14 sec has been reported for roots of *Lepidium* (Iversen and Larsen, 1971).

The gravitational stimulus leads to an increased plasticity of the lower side of the hypocotyl. This change may occur as early as 2½ min after commencement of the stimulus (Audus, 1969). The convex side also possesses a higher rate of respiration.

The exact way in which the gravitational stimulus is transmitted to the protoplasm is not known. According to Nemec's *statolith theory,* starch grains fall under the gravitational force, exerting a pressure on the protoplasm that leads in some way to the geotropic response. In the case of the rhizoids of *Chara foetida,* a recent explanation (Sievers, 1967*b*) is that the starch grains block the normal distribution of the Golgi vesicles on the lower side, preventing the growth on this side and stimulating growth on the upper side. On the other hand, in the case of the normal, vertical rhizoid (Sievers, 1967*a*) the statoliths block the central portion of the tip and pack the Golgi vesicles and microvesicles in the region adjacent to the apical wall (Fig. 18-2). This leads to enhanced growth at the tip. The Golgi vesicles appear to resemble the inner part of the cell wall. The plasmalemma is irregular, with invaginations that

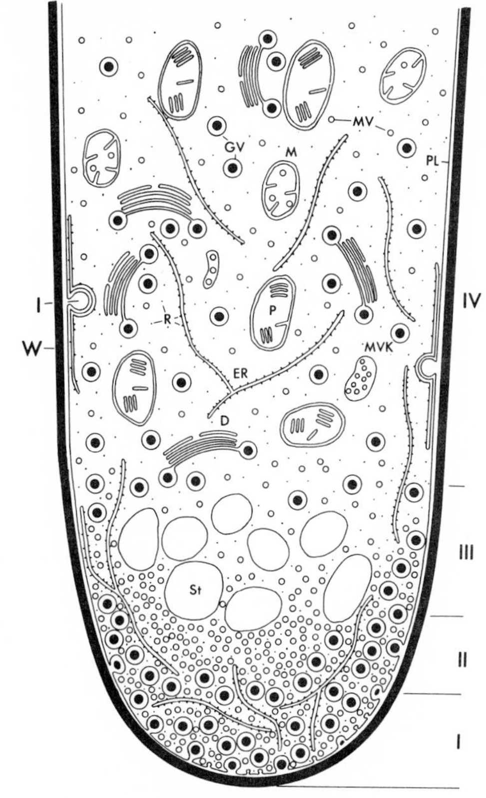

Fig. 18-2. Schematic representation of a longitudinal section through zones I to IV of a normal growing rhizoid of *Chara foetida.* **D** = dictyosome; **ER** = endoplasmic reticulum; **GV** = Golgi vesicle; **I** = invagination of plasmalemma with parallel ER cisterna; **M** = mitochondrion; **MV** = microvesicle; **MVK** = multivesicular body; **P** = plastid; **PL** = plasmalemma; **R** = ribosome; **St** = statolith; **W** = cell wall. (From Sievers, A. 1967. Protoplasma 64:225-253.)

may be caused by the incorporation of Golgi vesicles or microvesicles. Sievers concludes that the Golgi vesicles and microvesicles each supply the apical wall with growth material. During the process, their membranes fuse with the plasmalemma.

In higher plants, storage starch shows no such movement under the influence of gravity. It is only the starch statoliths, which exist in special cells called *statocytes,* that respond to gravity. These statoliths are confined to geotropically sensitive zones such as root caps, coleoptile tips, and the endodermis and starch sheaths surrounding the vascular bundles of hypocotyls (Audus, 1969). The root cap, which contains the georeceptor, is essential for the geotropic response of a typical root, and the coleoptile has statolith cells in a hollow cylinder of parenchyma near its apex. But not all plant parts appear to have such statolith-containing cells (or *georeceptors*). The aerial parts of typical vascular plants show at least some effects of negative geotropism without the mediation of a special georeceptor mechanism (Lyon, 1971). Similarly, evidence has been produced of a response in the absence of starch grains.

Later results, however, indicate that this observation was in error. Iversen (1969) was able to eliminate the geotropic response in roots of cress by removal of statolith starch, although the roots retained their ability to elongate. Other tests of the statolith theory have also supported it (Larsen, 1969; Iversen and Larsen, 1971). On inversion of the root, the amyloplasts in the three youngest of the nonembryonic storeys in the root cap start falling toward the opposite end of the cell at about 72 μm/hr (Iversen et al., 1968). After 6 to 12 min, they virtually come to a stop. It takes 10 to 12 min before any get as close to the ceiling as they were to the floor before inversion. When the root is placed horizontally, they reach the lower, longitudinal wall in 15 min or less. In further agreement with the statolith theory, corn coleoptiles treated with the growth inhibitor morphactin (Chapter 19) show geotropic unresponsiveness and an immobility of the starch grains (Parups, 1970). Similarly, gibberellic acid (Chapter 19) decreases both the amplitude of the geotropic root response and the density of root amyloplasts (Pilet et al., 1971). This is in agreement with the increased amylase activity induced by gibberellic acid in germinating seeds. Finally, in the coleoptiles of the amylomaize mutant, the amyloplasts are much reduced in size compared to the wild type, and geotropic curvature is also significantly less than in the wild type (Hertel et al., 1969). Besides the statolith control, there are aftereffects or *tonic effects* that modify the curvature induced by the statoliths (Iversen and Larsen, 1971). Under natural conditions these apparently counteract or prevent "overshoot" of the curvature.

All these results strongly support the statolith theory. The only other theory has apparently been eliminated by recent results. According to this *geoelectric effect theory,* the gravitational force is converted into an electric charge that leads to the response. Later investigations by the protagonists of this theory (Brauner and Diemer, 1968), however, indicated that the geoelectric effect was the result of a redistribution of auxin (Chapter 19). The electric effect would therefore be the *result* and not the *cause* of an event leading to the curvature. Further evidence indicates that the potential develops in the electrode system itself and not in the plant tissue (Woodcock and Wilkins, 1969) and that even the redistribution of auxin does not occur in roots (Scott and Wilkins, 1968).

The dependence of geotropism on metabolism has been demonstrated by its absence under anaerobic conditions (Wilkins and Shaw, 1967). Since the plant is constantly subjected to the force of gravity, its response to this force must be of fundamental importance in its normal growth and development. Until now, terrestrial plants have been subjected to a constant gravitational force. With the interest in outer space research, it is now becoming important to know the effects on the plant of varying the gravitational force, and experiments have been initiated to determine this. The polarity of the gravitational force can be eliminated by rotating the plant axis while maintaining it in a horizontal position. The instrument that rotates the plant is called a *clinostat.* It restricts the fall of intracellular particles to a quasicircular path such that the position of the particle remains virtually stationary within the plant (Dedolph and Dipert, 1971). A rotation period of 1 to 3 min is best for most plants (Lyon, 1970). Improvements in construction of the clinostat have permitted exposure of the plant to gravitational forces all the way down to zero (Dedolph et al., 1966). No threshold in gravity perception by the plant was detected. In fact, the geotropic curvature response of oat seedlings increased with the gravitational force, reaching a maximum at 0.06 to 0.08 times gravity. In the case of roots of corn *(Zea),* there was no response below 4×10^{-5} g (Lyon, see Audus, 1969).

PHOTOTROPISM

As a rule, stems are positively phototropic and leaves are plagiophototropic. The majority of roots are insensitive. The peduncle of *Lynaria cymbalaria* is positively phototropic when the flower opens and negatively phototropic after fertilization. As a result, the capsule is pressed into cracks between rocks, where the seed may later germinate.

Etiolated coleoptiles of grasses (e.g., oats) have been most commonly used for experimental work on phototropism. As in the case of geotropism, it is easily shown that the region of perception and the region of response to the light stimulus are not the same. The sensitive zone is the first 1 to 5 mm from the tip.

Phototropism is in accord with what has been said earlier about the retarding effect of light on cell elongation. The darkened side grows more rapidly, and, as a result, curvature is toward the light. This, however, is a superficial and not uniformly correct explanation.

As in the case of photosynthesis, the action spectrum for phototropism is known. Maximum sensitivity is in the blue, and essentially no response occurs at the long wave end of the spectrum (Fig. 18-3).

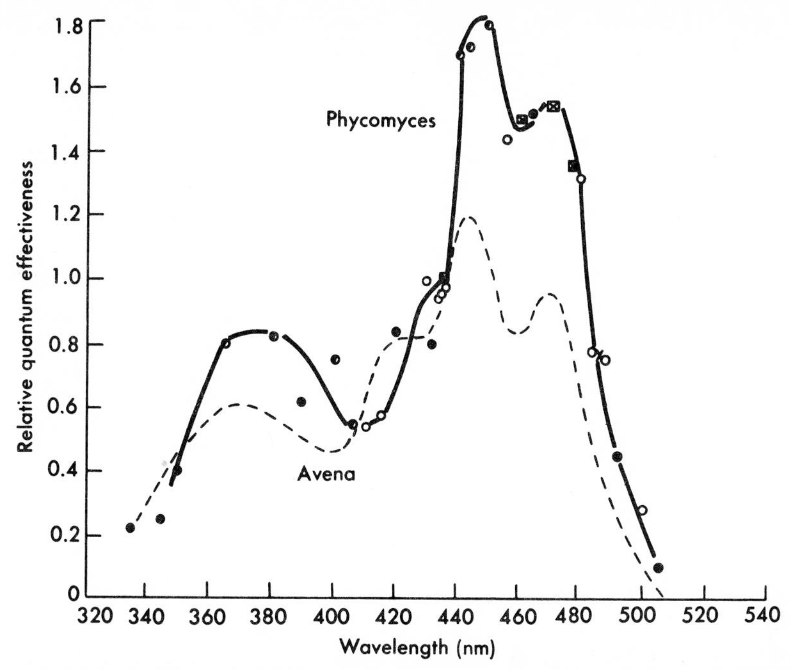

Fig. 18-3. Action spectra for phototropism in *Phycomyces* (a fungus) sporangiophores and *Avena* coleoptiles at low light intensities. (From Curry, G. M., and K. V. Thimann. 1961. *In* B. C. Christensen and B. Buchmann. Progress in photobiology. Elsevier Publishing Co., Amsterdam.)

This action spectrum resembles the absorption spectra of both carotenes and flavins, and there has been considerable controversy as to which is responsible for the absorption of the light that stimulates the phototropic response. Unfortunately, the absorption spectrum of neither pigment matches the action spectrum of phototropism (Thimann, 1964), and analysis of the pigments of *Avena* coleoptiles demonstrates the presence of both flavins and carotenoids. Since light of the wavelength 450 nm was absorbed to only 13% by the carotenoids, 87% by the flavones, Zenk (1967) concluded that carotenoids can play only a minor role in creating a transverse light gradient in the coleoptile.

NASTIC MOVEMENTS

The commonest of nastic movements are the *nyctinastic* movements, the day and night movements of leaves and flowers. Flower perianths and many compound leaves may open during the day and close at night. If these are growth movements, they are possible only while the organ is capable of growth. The amplitude of movement therefore decreases with age, but some retain the ability to open and close even when mature. This is because some nastic movements depend on turgor changes rather than growth differences. Families typically showing nyctinastic leaf movements are *Leguminosae, Oxalidaceae, Euphorbiaceae, Marantaceae.*

Nyctinastic movements may be controlled by temperature, light, etc. The perianths of tulip, crocus, and other similar flowers open at a constant temperature when illuminated and close when darkened. If, on the other hand, the light is kept constant, they open at high temperatures and close at low temperatures. They are therefore both *photonastic* and *thermonastic.* In each case the opening is the result of a more rapid growth of the upper surface *(epinasty).*

Some flowers show a surprisingly high

degree of sensitivity. Thermonastic opening may be induced by a rise of 1° for the tulip and 0.2° C. for the crocus (Ball, 1969). The speed of the temperature change may or may not affect the amplitude of the movement. In the case of the tulip, the mesophyll cells near the outer surface of the perianth segments have an optimum temperature for growth about 10° C lower than that of the cells near the inner surface (Wood, 1953, see Ball, 1969). This difference, together with similar differences between the upper and lower epidermis, explains the physical mechanism of thermonasty. It does not, however, explain the triggering mechanism. A possible factor has been indicated in the case of the photonastic flowers of *Portulaca.* A protein extracted from them, when injected into other plants, caused them to retain their flowers open for 2 hr longer than usual (Cesaire et al., 1968). Nastic movements may also be caused by the presence of certain substances in small quantity. One of the most sensitive tests for ethylene is the epinasty of tomato petioles, which occurs if the ethylene is present in 1:500,000 parts of air. This sensitivity is so great that a doubling of ethylene production by the plants themselves during 2 hr of rotation on a clinostat may also cause epinasty (Leather et al., 1972).

Nastic movements that are due to turgor changes are sometimes called "variation movements," to differentiate them from the above nastic movements due to differential growth (Ball, 1969). Such nyctinastic movements are accompanied by differences in osmotic potentials between the cells in the upper and lower halves of the pulvinus at the base or apex of the petioles or subpetioles. The seismonastic closure of the leaves of *Mimosa pudica* (the "sensitive plant") is accompanied by ejection of liquid into the intercellular spaces of the pulvinus. This was formerly explained by a loss of turgor due to a sudden increase in permeability when stimulated by touch. It is now believed to be due to an increased efflux of K^+ (Allen, 1969). An ATP-ATPase system in association with contractile proteins has also been suggested (Fondeville, 1969). More recently, contractile tannin vacuoles have been observed to undergo conformational changes that parallel the seismonastic leaf movement (Toriyama and Jaffe, 1972). Ca^{++} migrates between the surface of the tannin vacuole and the inside of the central vacuole. Nevertheless, the earlier suggestion of rapid K^+ efflux is believed to apply, the K^+ acting as the osmotic agent.

QUESTIONS

1. What is irritability?
2. What is a stimulus?
3. What is a response?
4. Where exactly does cytoplasmic streaming occur?
5. What is a taxis?
6. Does taxis occur in higher plants?
7. Name the three main classes of growth movements.
8. In each case what is the immediate, direct cause of movement?
9. In what way do tropic and nastic movements differ from circumnutation?
10. What is the difference between tropic and nastic movements?
11. What is the presentation time?
12. What is the reaction time?
13. What is the region of perception?
14. What is the region of response?
15. What is meant by correlation?
16. Name some tropisms.
17. Name some nastic movements.
18. What is epinasty?
19. In what direction does a plant move geotropically?
20. Where is the region of perception for geotropism?
21. Where is the region of response for geotropism?
22. What is a statolith?
23. How may the Golgi apparatus be related to geotropism?
24. What is a clinostat?

25. What is the action spectrum for phototropism?

SPECIFIC REFERENCES

Allen, R. D. 1969. Mechanism of the seismonastic reaction in *Mimosa pudica.* Plant Physiol. **44:** 1101-1107.

Ambrose, E. J. 1965. Cell movements. Endeavour **24:**27-32.

Audus, L. J. 1962. The mechanism of the perception of gravity by plants. Sympos. Soc. Exp. Biol. **16:**197-226.

Audus, L. J. 1969. Geotropism, p. 203-242. *In* M. B. Wilkins (Ed.). Physiology of plant growth and development. McGraw-Hill Book Co., New York.

Audus, L. J., and J. C. Whish. 1964. Magnetotropism, p. 170-182. *In* M. F. Barnothy (Ed.). Biological effects of magnetic fields. Plenum Publishing Corp., New York.

Ball, N. G. 1969. Nastic responses, p. 277-300. *In* M. B. Wilkins (Ed.). Physiology of plant growth and development. McGraw-Hill Book Co., New York.

Beck, R., H. Hinssen, H. Kommick, W. Stockem, and K. E. Wohlfarth-Botterman. 1970. Extensive fibrillar protoplasmic differentiations and their significance for protoplasmic streaming. V. Contraction, ATPase activity and fine structure of actomyosin threads from *Physarum polycephalum.* Cytobiologie **2:**259-274.

Brauner, L., and R. Diemer. 1968. Über das geoelektrische Reaktionsvermögen der Infloreszenzachsen von *Lupinus polyphyllus.* Planta **81:** 113-131.

Cesaire, O. G., Y. Bresson, A. Bellossi, and Y. Le Duc. 1968. Research on the opening and closing movements of the flowers of certain African plants. Med. Afr. Noire **15:**391-394.

Curry, G. M. 1969. Phototropism, p. 245-273. *In* M. B. Wilkins (Ed.). Physiology of plant growth and development. McGraw-Hill Book Co., New York.

Curry, G. M., and K. V. Thimann. 1961. Phototropism; the nature of the photoreceptor in higher and lower plants, p. 127-134. *In* B. C. Christensen and B. Buchmann (Eds.). Progress in photobiology. Proceedings International Congress on Photobiology, 3rd. Elsevier Publishing Co., Amsterdam.

Davies, J. W., and E. C. Cocking. 1965. Changes in carbohydrates, proteins, and nucleic acids during cellular development in tomato fruit locule tissue. Planta **67:**242-253.

Dedolph, R. R., S. A. Gordon, and D. A. Oemick. 1966. Geotropism in simulated low-gravity environments. Amer. J. Bot. **53:**530-533.

Dedolph, R. R., and M. H. Dipert. 1971. The physical basis of gravity stimulus nullification by clinostat rotation. Plant Physiol. **47:**756-764.

Diehn, B., and B. Kent. 1970. The flavin nature of the photoreceptor molecule for phototaxis in *Euglena.* Physiol. Chem. Phys. **2:**483-488.

Fondeville, J.-C. 1969. A botanical guinea pig: the sensitive plant. Nucleus (Paris) **10:**180-190.

Gaertner, R. 1970. The chloroplast movement of *Mesotaenium* in high intensity light. II. Action dichroism and interaction of the photoreceptor with phytochrome. Z. Pflanzenphysiol. **63:**428-443.

Gordon, S. A., et al. 1964. Growth and development in compensated fields: introduction and design. USAEC Argonne Nat. Lab. Rep. **6971:** 166-71.

Graham, L., and C. H. Hertz. 1964. Measurements of the geoelectric effect in coleoptiles. Physiol. Plant. **17:**186-201.

Haupt, W. 1966. Phototaxis in plants. Int. Rev. Cytol. **19:**267-299.

Hertel, R., R. K. De La Fuente, and A. C. Leopold. 1969. Geotropism and the lateral transport of auxin in the corn mutant amylomaize. Planta **88:**204-214.

Heslop-Harrison, Y., and R. B. Knox. 1971. A cytochemical study of the leaf-gland enzymes of insectivorous plants of the genus *Pinguicula.* Planta **96:**183-211.

Israelsson, D., and A. Johnsson. 1967. A theory for circumnutations in *Helianthus annuus.* Physiol. Plant. **20:**957-976.

Iversen, T.-H. 1969. Elimination of geotropic responsiveness in roots of cress *(Lepidium sativum)* by removal of statolith starch. Physiol. Plant. **22:**1251-1262.

Iversen, T.-H., and P. Larsen. 1971. The starch statolith hypothesis and the optimum angle of geotropic stimulation. Physiol. Plant. **25:**23-27.

Iversen, T.-H., K. Pedersen, and P. Larsen. 1968. Movement of amyloplasts in the root cap cells of geotropically sensitive roots. Physiol. Plant. **21:**811-819.

Jahn, T. L., and E. C. Bovee. 1969. Protoplasmic movements within cells. Physiol. Rev. **49:**793-862.

Johnsson, A. 1966. Spontaneous movements in plants studied as a random walk process. Physiol. Plant. **19:**1125-1137.

Johnsson, A. 1971*a.* Geotropic responses in *Helianthus* and their dependence on the auxin ratio: with a refined mathematical description of the course of geotropic movements. Physiol. Plant. **24:**419-425.

Johnsson, A. 1971*b*. Investigations of the geotropic curvature of the *Avena* coleoptile. I. The geotropic response curve. Physiol. Plant. **25**:35-42.

Kamiya, N. 1942. Physical aspects of protoplasmic streaming, p. 199-244. *In* W. Seifriz. The structure of protoplasm. Monogr. Amer. Soc. Plant Physiol. Iowa State University Press, Ames, Iowa.

Konings, H. 1969. The influence of acropetally transported indoleacetic acid on the geotropism of intact pea roots and its modifications by 2,3,5-triiodobenzoic acid. Acta Bot. Neerl. **18**: 528-537.

Larsen, P. 1969. The optimum angle of geotropic stimulation and its relation to the starch statolith hypothesis. Physiol. Plant. **22**:469-488.

Leather, G. R., L. E. Forrence, and F. B. Abeles. 1972. Increased ethylene production during clinostat experiments may cause leaf epinasty. Plant Physiol. **49**:183-186.

Lyon, C. J. 1970. Choice of rotation rate for the horizontal clinostat. Plant Physiol. **46**:355-358.

Lyon, C. J. 1971. Lateral transport of auxin mediated by gravity in the absence of special georeceptor tissue. Plant Physiol. **48**:642-644.

Mahlberg, P. G. 1965. Phase cinemicrographic observations on cultured cells. II. Mass movement of cytoplasm in *Euphorbia marginata*. Amer. J. Bot. **52**:438-443.

Miki-Hirosige, H. 1964. Tropism of pollen tubes to the pistils, p. 152-158. *In* H. F. Linskens. Pollen physiology and fertilization: a symposium. North-Holland Publishing Co., Amsterdam.

Milburn, J. A. 1971. An analysis of the response in phloem exudation on application of massage to *Ricinus*. Planta **100**:143-154.

Murr, L. E. 1965. The biophysics of plant electrotropism. Trans. N. Y. Acad. Sci. **27**:759-771.

Parups, E. V. 1970. Effect of morphactin on the gravimorphism and the uptake, translocation and spatial distribution of indol-3yl-acetic acid in plant tissues in relation to light and gravity. Physiol. Plant. **23**:1176-1186.

Pickard, B. G., and K. V. Thimann. 1966. Geotropic response of wheat coleoptiles in absence of amyloplast starch. J. Gen. Physiol. **49**:1065-1086.

Pilet, P.-E., H. Greppin, and M. Bonzon. 1971. Action de l'acide gibbérellique sur la densité des amyloplastes racinaires. Compt. Rend. Acad. Sci. (Paris) **272**:1760-1763.

Pittman, U. J. 1963. Magnetism and plant growth. Canad. J. Plant Sci. **43**:513-518.

Scott, T. K., and M. B. Wilkins. 1968. Auxin transport in roots. II. Polar flux of IAA in *Zea* roots. Planta **83**:323-334.

Seitz, K. 1967. Eine Analyse der für die lichtabhängigen Bewegungen der Chloroplasten verantwortlichen Photorezeptor-systeme bei *Vallisneria spiralis* ssp. torta. Z. Pflanzenphysiol. **57**: 96-104.

Seitz, K. 1971. Die Ursache der Phototaxis der Chloroplasten: ein ATP-Gradient? Versuch zum Primärprozess der Starklichtbewegung bei *Vallisneria*. Z. Pflanzenphysiol. **64**:241-256.

Sievers, A. 1967*a*. Elektronenmikroskopische Untersuchungen zur geotropischen Reaktion. II. Protoplasma **64**:225-253.

Sievers, A. 1967*b*. Elektronenmikroskopische Untersuchungen zur geotropischen Reaktion. III. Z. Pflanzenphysiol. **57**:462-473.

Sievers, A. 1967*c*. Zum Wirkungsmechanismus der Statolithen in der pflanzlicher Zelle. Naturwissenschaften **54**:252-253.

Thimann, K. V. 1964. Phototropism. Photochem. Photobiol. **3**:463-469.

Toriyama, H., and M. J. Jaffe. 1972. Migration of calcium and its role in the regulation of seismonasty in the motor cell of *Mimosa pudica*. L. Plant Physiol. **49**:72-81.

Wessells, N. K., B. S. Spooner, J. F. Ash, M. O. Bradley, M. A. Luduena, E. L. Taylor, J. T. Wrenn, and K. M. Yamada. 1971. Microfilaments in cellular and developmental processes. Science **171**:135-143.

Wilkins, M. B. 1966. Geotropism. Ann. Rev. Plant. Physiol. **17**:379-408.

Wilkins, M. B., and S. Shaw. 1967. Geotropic response of coleoptiles under anaerobic conditions. Plant Physiol. **42**:1111-1113.

Woodcock, A. E. R., and M. B. Wilkins. 1969. The geoelectric effect in plant shoots. II. Sensitivity of concentration chain electrodes to reorientation. J. Exp. Bot. **20**:687-697.

Woolley, D. G., and U. J. Pittmann. 1966. P^{32} detection of geomagnetotropism in winter wheat roots. Agron. J. **58**:561-562.

Zenk, M. H. 1967. Untersuchungen zum Phototropismus der Avena-Koleptile. II. Pigmente. Z. Pflanzenphysiol. **56**:122-140.

GENERAL REFERENCES

Lloyd, F. E. 1942. The carnivorous plants. Chronica Botanica Co., Waltham, Mass.

Weevers, T. 1949. Fifty years of plant physiology. Scheltema & Holkema, Amsterdam.

19

Growth regulators

TERMINOLOGY

For a plant to grow normally, many substances must be available.

1. The raw materials: all the essential mineral elements, water, carbon dioxide, and oxygen
2. The building blocks for cell production and enlargement: the carbohydrates, proteins, nucleic acids, lipids
3. The sources of energy: both chemical (carbohydrates and high energy phosphates) and physical (water for turgor pressure)
4. The catalysts to maintain adequate rates of metabolism (enzymes plus their cofactors)

Since the green plants are fully autotrophic, the only substances they must be supplied with are the raw materials—the mineral elements, water, carbon dioxide, and oxygen. From these they can synthesize the other three groups of substances. In the presence of two other essential environmental factors—adequate light and suitable temperature—these raw materials are all that need be supplied to the plant. Under certain conditions the plant may not be fully autotrophic, for example, albino mutants or immature embryos. In such cases the plant is heterotrophic and must be supplied with some of the other three groups of substances to grow normally.

If any of the aforementioned substances is present in suboptimal quantities, growth will be decreased. In the broadest sense of the term, all of these substances may therefore be called growth substances, since they can alter the growth rate of the plant. Yet even if all the raw materials are available to the autotrophic green plant in optimum quantities, the plant may fail to grow despite favorable light and temperature. This will occur when the plant is in its rest period. Furthermore, even in the case of a rapidly growing plant, some parts grow continuously (e.g., promeristems), and others grow until a certain point is reached and then stop; yet both the raw materials and the environmental conditions are adequate for the whole plant.

It is evident that the plant must possess some mechanism for regulating its growth. It does this by means of an additional group of substances called *growth regulators.* These are substances that the plant synthesizes in small quantities, which do not serve as raw materials, building blocks, sources of energy, or enzyme components, but which act instead as keys capable of unlocking or locking up the growth process.

Growth regulators are usually *hormones*—regulating substances synthesized in one part of the plant and translocated to another part where they exert their effects. They are therefore frequently called *growth hormones* to distinguish them from other hormones that do not specifically control growth (e.g., flowering hormones, see Chapter 20). However, it is conceivable for a growth regulator to act where it is synthesized, in which case it would not be a hormone. Furthermore, there are many synthetic growth regulators, and since these are not synthesized in the plant, they cannot be hormones if the preceding definition is accepted. Conversely, substances may be growth hormones without belonging to the group of growth regulators. Thiamine and niacin and probably other vitamins may act as growth hormones (Thimann, 1965), yet the mechanism of their activity as enzyme cofactors is fundamentally different from the mechanism of action of growth regulators (see later discussion), and they therefore do not belong in this group. The term *phytohormones* includes all plant hormones—the growth regulators, vitamins, flowering hormones, etc.

The five main groups of naturally occurring growth regulators are the auxins, cytokinins, gibberellins, ethylene, and a miscellaneous group including dormins (or abscisins) and other inhibitors. The high activities of the regulators are illustrated by the low concentrations required to produce an effect on the plant (Table 19-1).

AUXINS

Physiological and chemical properties

Auxins were the first substances recognized to be growth regulators. They were, in fact, believed to be *the* growth regulators, and this led to the conclusion that "without auxin there is no growth" (Went, 1928). This conclusion has been challenged (Mer, 1969), and it is now known that other growth regulators are just as important as auxins. The existence of auxins was first postulated as a result of the discovery of *correlation* (Chapter 18)—the control of the responses in one part of a plant by the region of perception in another part of the same plant. Most of the experiments were performed on oat coleoptiles, which bend toward the light when illuminated unilaterally. If only the region

Table 19-1. Comparison of activities of different kinds of growth regulators*

Regulator	Natural quantities: Quantity found in plant	Natural quantities: Plant	Artificial quantities: Applied concentration	Artificial quantities: Effect
Auxin	6 μg/kg (or ppb)	Pineapple	10^{-2} mg/l (10 ppb)	Increase in length of oat coleoptile
Gibberellin	0.001 g/100 buds	Sunflower	2 x 10^{-11} mole/l (1 ppt)	Initiation of starch digestion
Cytokinin	50-100 ppb in bleeding sap	Grapevine	0.04 mg/l (40 ppb)	Increase in fresh weight of tobacco callus

*Adapted from van Overbeek, J. 1966. Science **152**:721-731.

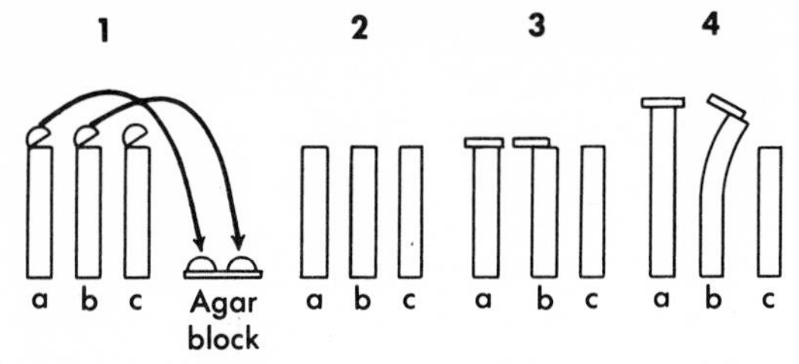

Fig. 19-1. Effects of decapitation and replacement of substances from the tip (in agar blocks) on the growth of oat coleoptiles. **1,** Three coleoptiles decapitated. **2,** Tips from **a** and **b** transferred to agar blocks. **3,** One block placed symmetrically on **a,** one asymmetrically on **b,** none on **c.** **4, a** shows straight growth, **b** growth curvature, **c** no growth.

of perception (the tip) is illuminated, the bend occurs in the same region of response, some distance below, as when the whole coleoptile is illuminated. This means that the illuminated tip is able to transmit to the darkened part the message that it has "seen" unilateral light. The question asked by the early investigators was how this message is transmitted—electrically or chemically. The bend did not occur, and therefore the message was not transmitted, if the tip was separated from the stump by an electrically conducting strip of metal that failed to permit diffusion of solutes. It was transmitted across a layer of agar gel that permits diffusion of water-soluble substances (Fig. 19-1). An electric wave also travels down the coleoptile at speeds up to 14 mm/hr (Newman, 1963), but this appears to be caused by the stream of translocated substance.

The substances responsible for this correlation are known as *auxins,* a term originally coined for any chemical capable of inducing a growth curvature in the oat coleoptile when applied unilaterally in low concentrations. A more sophisticated definition has been attempted (Thimann, 1969), but it leads to difficulties in relation to other growth regulators. Three auxins were originally reported to have been isolated from plant and animal materials, and their chemical structures were published. Two of these were nonindolic substances and were called "auxins *a* and *b*." The third was an indolic substance (Fig. 19-2) and was called heteroauxin. Later thorough and repeated investigations have failed to detect the first two, and even so-called authentic samples left by the "discoverers" of auxins *a* and *b* have proved to be completely different, well-known chemical substances with no auxin activity (Vendrig, 1971). It has therefore been concluded that no such substances were ever found and that the published results were simply fabricated (personal communications). The third substance, heteroauxin, has been isolated from many plant tissues by many investigators and is now generally believed to be the only auxin. It is the chemical substance β-indoleacetic acid, or indole-3-acetic acid (IAA), which is closely related to, and therefore assumed to be formed from, the ubiquitous amino acid tryptophan (Fig. 19-2).

Many attempts have been made to identify the biosynthetic pathway for IAA. Direct evidence is now available. ^{14}C-labeled tryptophan (as well as tryptamine and indole ethanol) was converted into IAA in cucumber seedlings under sterile conditions (Sherwin and Purves, 1969).

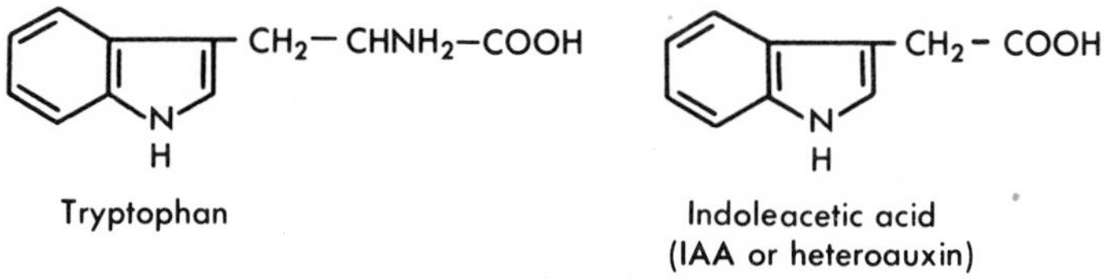

Fig. 19-2. Chemical structure of indoleacetic acid (IAA) and its chemical relation to tryptophan.

The same was true of sterile pea sections, provided that inhibitors of IAA oxidase were included (Libbert et al., 1969); but the amounts produced were small, and the authors suggest that a more important pathway leading to IAA may bypass tryptophan. The plant enzymes are able to produce six indole substances from tryptophan (Libbert et al., 1970*a*). The pathway is apparently as follows (Libbert et al., 1970*b*):

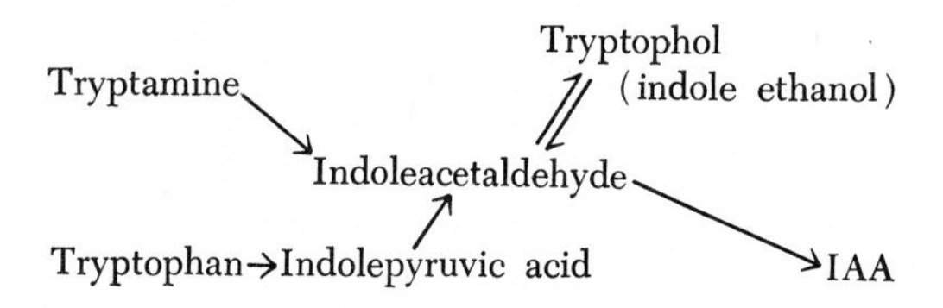

Labeling with ^{3}H serine and ^{14}C indole has also supported the conclusion that tryptophan is the native precursor of IAA (Erdmann and Schiewer, 1971).

Other organisms besides the higher plants can also produce IAA from tryptophan—for instance, epiphytic bacteria normally present on higher plants (Libbert et al., 1969) and the blue-green alga *Chlorogloea fritschii* (Ahmad and Winter, 1970). Although tryptophan itself has no auxin activity, some presumed intermediates between it and IAA (e.g., indoleacetaldehyde, indoleacetonitrile) are active. When applied uniformly to decapitated coleoptiles, these substances cause increased growth. When applied to one side, they cause curvature toward the other side.

It is always a question, however, whether these are active auxins per se, or simply because they can be converted by the plant into IAA which then produces the auxin effect. The latter explanation appears to be true in the case of indolacetonitrile (Evans and Rayle, 1970). Conversely, when IAA was applied to intact pea seedlings, a large proportion was converted to indole-3-acetyl aspartic acid (Morris et al., 1969); but this substance was not transported and therefore cannot act as a true auxin (see later discussion). IAA-myoinositol esters have been found in maize kernels (Veda et al., 1970), and a new chlorinated form has been found in immature seeds of *Pisum sativum*—4-chloroindolyl-3-acetic acid (Marumo et al., 1971). There is thus a large variety of indole substances in plants, many of which may account for the existence of so-called "bound" (or inactive) auxin in the plant (Thimann, 1969). It is still doubtful, however, whether any of these other than IAA can act as auxins. Similarly, an unidentified but nonindolic "citrus auxin" was reported (Lewis et al., 1965), but it now appears to be clearly established that the auxin of citrus plants is IAA as in other plants (Goldschmidt et al., 1971; Igoshi et al., 1971), and the term "citrus auxin" should be dropped.

Nevertheless, native, nonindolic auxins that are steroid in nature have been reported in *Coleus* (Vendrig, 1967). p-Hydroxybenzoic acid has been isolated from woody cuttings of *Ribes rubrum* (Vieitez et al., 1966) and has been found to act as an auxin in the oat coleoptile straight growth test, producing both growth inhibition at high concentration and growth stimulation at low concentration. It is therefore not surprising that among the synthetic substances with auxin activity, some may be found which are much different chemically from IAA (Kefford and Caso, 1966).

However, the effects of these other substances may still be secondary, either by acting as antiauxins (antagonizing the effect of IAA) or as synergists (enhancing the effect of IAA). Thus monophenols (p-coumaric acid, dichlorophenol, kaempferol) may decrease auxin action by promoting its oxidation; diphenols (caffeic, protocatechuic, and chlorogenic acids, catechol) may increase the auxin action by preventing its oxidative destruction (Thimann, 1965). Even hydroxyproline, which has been suggested to play a direct role in wall expansion and therefore in cell enlargement, owes at least part of its effect to control of auxin levels by IAA oxidase (Whitmore, 1971). Simple phenolic glucosides also regulate the IAA level by influencing the activity of IAA oxidase, and this effect is mediated by free phenols (Psenak et al., 1970). Even the alkane lipids may act by enhancing the effect of the IAA (Stowe and Dotts, 1971). Other nonnative lipids such as estradiol may have the same effect (Kopgewicz, 1970). In some of these cases, for instance the nitrophenols, the substances may be as active as the most active auxins (Harper and Wain, 1969).

In addition to the natural auxins, there are many synthetic substances capable of producing the same effects on plant growth. By definition they are also auxins (since they induce curvature of oat coleoptiles), although not naturally occurring. Examples of these are indolebutyric acid, naphthaleneacetic acid, and the important weed killer 2,4-D (2,4-dichlorophenoxyacetic acid). Quantitatively, some of these substances are more effective than the naturally occurring IAA and some are less effective, depending on the test that is used, since there are many effects of both the natural and the synthetic substances. The substance that is most effective for one may be least effective for another of these plant responses (Avery et al., 1947). Some may be more effective than the native auxin because they are more persistent in the plant, perhaps because they resist oxidation by IAA oxidase.

Effects on growth

Since auxin was discovered as a result of investigations of phototropism, its relation to this asymmetric kind of growth should be clearly understood. It was soon demonstrated that phototropic curvature is caused by a higher concentration of auxin on the side of the coleoptile further from the light. Two theories have been proposed to explain this asymmetry. According to the Went-Cholodny theory, it is caused by translocation of auxin from the illuminated to the dark side of the coleoptile. Other investigators have shown that IAA can be destroyed in vitro by light and explain the lower concentration on the illuminated side by this destruction. More recent and thorough investigations appear to have conclusively established the Went-Cholodny explanation (Thimann, 1964, see Chapter 18). This means that phototropism is not simply caused by an inhibition of growth by light, since the redistribution of auxin must result in both a reduced rate on the lighted side and an enhanced rate on the darker side. According to Hager and Schmidt (1968), the phototropic curvature is triggered by a photooxidative formation of 3-methyleneoxindole from IAA on the side adjacent to the light. This substance partly blocks the flow of growth regulator downward and directs it to the shaded side. This explanation seems to vindicate both theories.

So much of the fundamental work on phototropism is based on the oat coleoptile that there is some danger in generalizing from these results. In the case of sunflower seedlings (Lam and Leopold, 1966) an asymmetry of auxin content can be produced without the involvement of lateral auxin transport from the lighted side of the stem to the shaded side. Darkening one cotyledon will cause curvature of the stem toward the lighted cotyledon. More diffusible auxin is obtained from the stem below the darkened cotyledon than below the lighted one. The relationship is basically similar to that in the oat coleoptile, but in this case the *cotyledons* induce the stem curvature by producing an asymmetrical auxin concentration in the stem. The cotyledons are also the region of perception in this case.

Geotropism can be similarly explained. The gravitational stimulus causes a greater concentration of auxin on the lower side of all organs (even in leaves—Lyon, 1971). But it must be remembered that this causes negative geotropism in stems and positive geotropism in roots. This has been explained by the difference in sensitivity of roots and stems to the concentration of auxin (Fig. 19-3). All cells may be either stimulated or inhibited, depending on the concentration of auxin; but the concentration (10^{-6}M) causing increased growth of stem cells decreases the growth of root cells. Consequently, if the gravitational stimulus increased the auxin concentration on the lower side to 10^{-6}M, this would lead to an upward curvature of the stem and a downward curvature of the root. Although this is an adequate explanation of geotropism, other investigations point to the involvement of gibberellins (see later discussion).

Even nastic growth movements have been shown to be controlled by auxins (Brauner and Arslan, 1951).

The dual effect of auxin gives the plant a switch mechanism for turning the growth process on or off. The inhibition, for instance, has been proposed to explain the *polarity* of growth, which refers to the specialized kind and rate of growth at the top and bottom of the plant. *Apical dominance* in the shoot is the growth of the apical bud at a far greater rate than the lateral buds or with complete dormancy of the lateral buds. This is also called *correlative inhibition*—the inhibition of the lateral buds by the apical bud. It is explained as follows. The auxin is synthesized or released in the terminal bud; and since the movement of IAA is strongly polar in a basipetal direction, it is translocated down the stem. The concentration in the terminal bud apparently remains optimum for its growth. The concentration produced in the lateral buds by the downward translocation of the excess is high enough to inhibit growth, and they are kept in the dormant state. This apical dominance also prevents suckering from the roots of aspen (Eliasson, 1971*a*), particularly at the time of the year when shoot growth is occurring and auxin is at its highest level (Eliasson, 1971*b*). This relation is used in practice with potato tubers, which can be prevented from sprouting by soaking them in a concentrated solution of auxin. However, the difference in sensitivities of different organs is also apparent here. Auxin can promote the growth of inhibited pea *shoots*, although it cannot promote the growth of the inhibited pea *buds* (Sachs and Thimann, 1967). Similarly, once a lateral bud is induced to begin growing, it becomes less sensitive to inhibition by the terminal apex. Finally, the downward flow of auxin leads

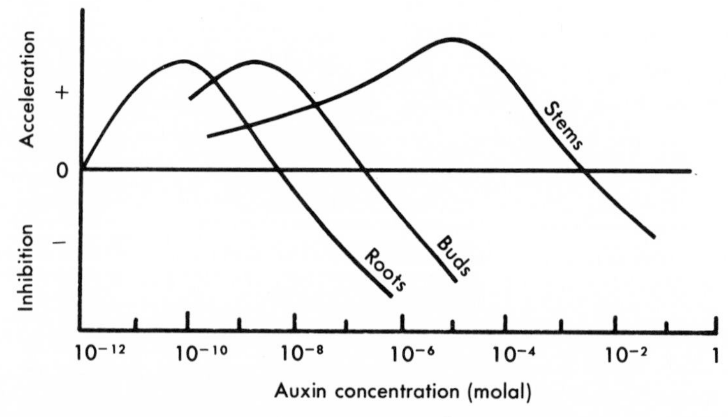

Fig. 19-3. Inhibition and acceleration of growth of different organs as a function of auxin concentration. (From Thimann, K. V. 1937. Amer. J. Bot. **24**:407-412.)

to a sufficient accumulation at the base of cuttings to attain the high concentrations required for the initiation of new roots (Greenwood and Goldsmith, 1970).

Although the movement of IAA in the stem is strongly polar in a basipetal direction, in the root it is polar in an acropetal direction (Wilkins and Scott, 1968) due to a dependence on metabolic energy. This occurs at all temperatures from 1° to 50° C in roots of *Zea mays* (Wilkins and Crane, 1970), with a maximum value of 8 mm/hr at 31° C. This difference between stem and root ensures that auxin moving basipetally, down the stem, can continue to move acropetally into the root and to its tip. The movement is through the sieve-tubes (Lepp and Peel, 1971*a*) in willow in a predominantly basipetal direction (Hoad, et al., 1971). Even in the oat coleoptile, diffusion does not seem to play a major role in the normal transport of auxin (Newman, 1970). Auxin may also enhance the translocation of sugars in the phloem of willow, either due to a stimulation of sugar loading from the storage parenchyma into the sieve elements or to a direct effect on longitudinal transport (Lepp and Peel, 1971*b*). This effect is complicated and may not occur in the same way in all plants. IAA greatly enhanced the upward translocation of phosphate and sucrose in bean and pea plants but not in coleus or sunflower (Bowen and Wareing, 1971).

As pointed out earlier, to have full control over its metabolism, the plant must be able to increase or decrease the concentrations of its metabolic regulators (e.g., the enzymes), which control specific metabolic paths. In the same way, to control its growth, the plant must be able to increase or decrease the concentration of its growth regulators (Fig. 19-4, *A*). Therefore the plant must have a mechanism not only for synthesizing IAA but also for removing it. Removal may be (1) by inactivation due to chemical change or (2) by translocation to other parts of the plant. The simplest method of removal is by oxidation of IAA in the presence of an enzyme IAA oxidase, which is commonly present in plant tissues. This is demonstrated by the metabolism of applied IAA—in the willow to the extent of 80% within 24 hr of application (Field and Peel, 1971). The quantity of IAA oxidase in cabbage roots is so high that variations in level of the enzyme do not seem essential for regulation of the endogenous IAA concentration (Raa, 1971). The IAA oxidase may consist of a peroxidase or a nonperoxidative enzyme (van der Mast, 1970). There are also IAA protectors that inhibit the peroxidase-catalyzed oxidation of IAA (Stonier et al., 1970*a*) and therefore presumably prevent its rapid breakdown, for instance in peas, in which the only metabolite is IAA-aspartate (Kendall et al., 1971) or IAA-methanol (Magnus et al., 1971). The oxidation of IAA in the presence of IAA oxidase can presumably begin only after these protectors themselves are oxidized. The protectors have no effect by themselves but may have a profound effect in the presence of IAA, for instance on rooting (Jackson and Harney, 1970). NADH may mimic the native auxin protectors, thus demonstrating that their effect is due to an oxidation-reduction reaction. In the absence of the peroxidase complex the protectors may actually accelerate the oxidation of NADH; in the presence of the peroxidase complex they slow it down.

Because of this delicate balance between protectors and IAA oxidase, it has been suggested that the IAA oxidase system may function as part of an IAA-activating mechanism rather than an IAA-destroying system (Meudt, 1970). It must therefore be concluded that one method used by the plant to control its growth is a redox poising (or balancing) system, which can reversibly control the oxidation state and therefore the activity of IAA.

What, then, is the nature of the protectors? The involvement of peroxidase in IAA oxidation immediately points to phenolic

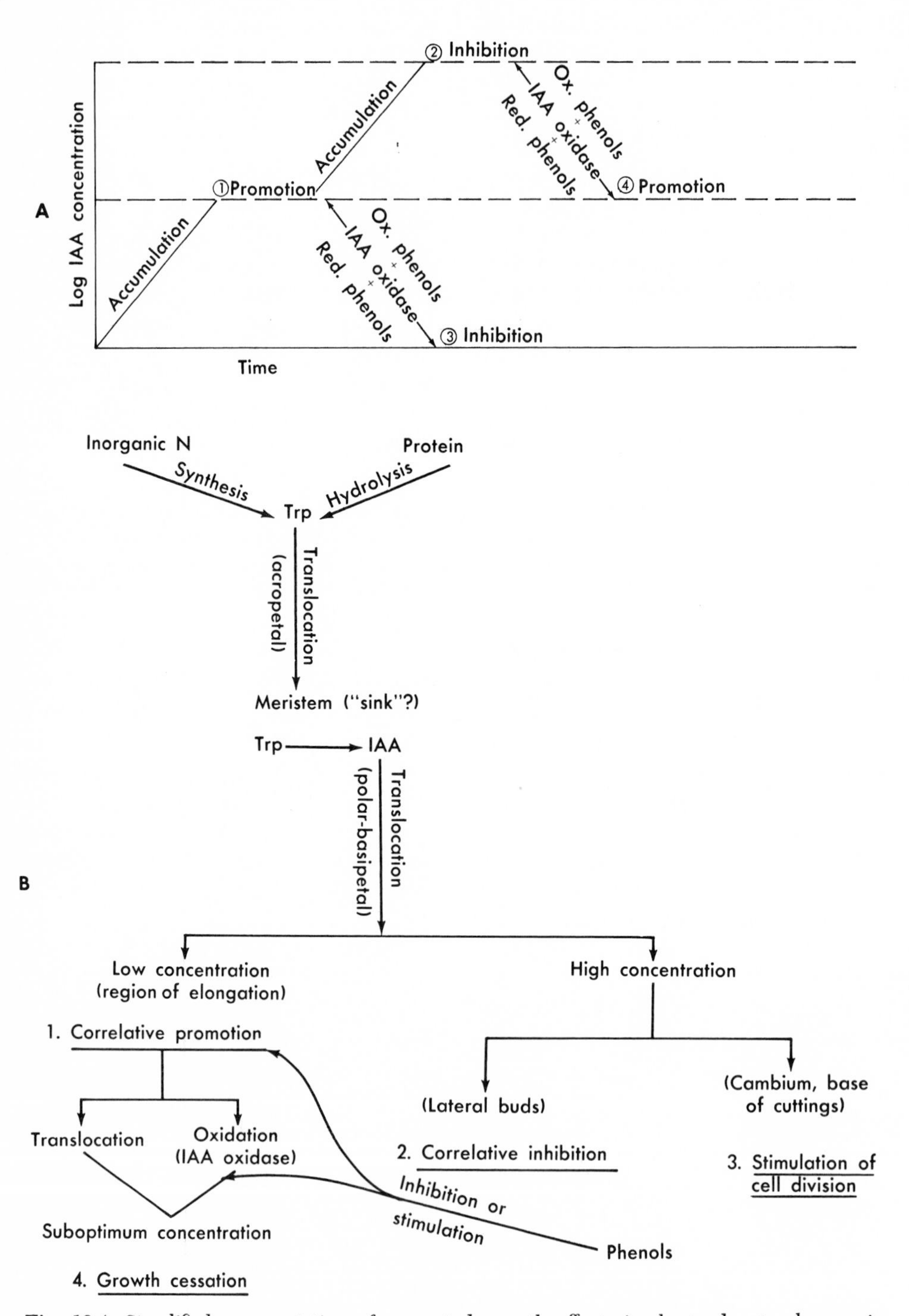

Fig. 19-4. Simplified representation of suggested growth effects in shoots due to changes in IAA concentration brought about by IAA oxidase and phenols, **A,** and by these together with translocation and interrelations with Trp (tryptophan), **B.**

substances, which are the normal substrates of peroxidase (Chapter 13). It is therefore not surprising that phenols may either accelerate or inhibit IAA oxidation (Thimann, 1965), and that auxin protectors contain o-dihydroxyphenolic groups at their active sites (Stonier et al., 1970*b*). There are as many as thirty-three substances that inhibit the destruction of IAA by an extract from pea roots (Janssen, 1970). Furthermore, the effect may depend on the concentration. The phenol scopoletin inhibits IAA oxidase activity at high concentrations (12.5 to 250 nmol/ml) but stimulates it at low concentrations (0.25 to 10 nmol/ml, Imbert and Wilson, 1970). All these interactions between IAA concentration and its effects on growth are summarized in Fig. 19-4, *B*.

The second method used by the plant to control its auxin-regulated growth is by translocation of IAA to or away from the region to be controlled (see previous discussion). It is not always easy to prove which of the two methods is responsible for the control of growth.

In all the cases just mentioned, the effects of auxins on growth result from an increased or decreased cell enlargement at low or high concentration, respectively. However, auxins may also affect cell division. Callus tissue may be formed when IAA is applied to a cut stem or petiole surface. Similarly, root and fruit initiation (Fig. 19-5) is favored by the same substance, and fruit abscission can be prevented or induced. It should be emphasized, however, that root initiation (and therefore cell division) requires much higher concentrations of IAA, sometimes high enough to injure (Huisinga, 1967). To obtain subsequent growth, the excess IAA must therefore be removed. Another example of a relation to cell division is the renewed activity of the cambium in the spring, which is caused by auxin travelling down the stem at about 1 ft a day and controlling cell division (Thimann, 1969; Skene, 1971). The amount of auxin is highest in the xylem, less in the cambium, and least in the phloem, suggesting that it is actually formed in differentiating xylem tissue (Sheldrake, 1971) or that it may induce xylem differentiation rather than cell division (Shininger, 1971).

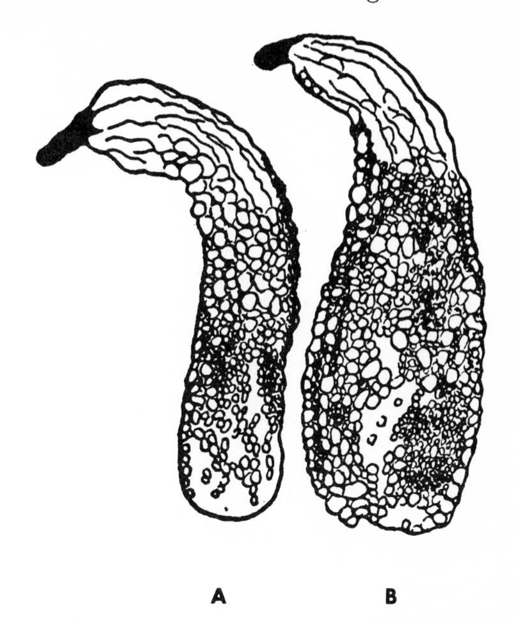

Fig. 19-5. Fruit formation in squash by treatment with 2% indolebutyric acid in lanolin, **A,** as compared with normal fruit by pollination, **B.** (From Gustafson, F. G. 1951. *In* F. Skoog. Plant growth substances. University of Wisconsin Press, Madison, Wis.)

In summary, IAA may produce the following effects on the plant (Linser, 1966): (1) promotion or inhibition of cell enlargement, (2) promotion of cell division, (3) promotion of root formation, (4) inhibition of root growth, (5) inhibition of bud growth or formation, (6) enhancement or inhibition of abscission (Chapter 20), (7) enhancement or inhibition of flowering (Chapter 20), (8) formation of parthenocarpic fruit, (9) promotion of respiration, and (10) promotion of protein synthesis (see later discussion). Optimum concentrations for growth are 10^{-10} M for roots, 10^{-8} M for buds, 10^{-5} M for stems, and 10^{-4} M for flowers.

Mechanism of growth regulation

Any complete explanation of auxin action must include the effects on cell division and cell enlargement; yet most theories have attempted primarily to explain the effect on cell enlargement. This is perhaps understandable, since auxins apparently control the growth of only those organisms that show a manifold cell enlargement—the higher plants. They generally have no effect on the growth of bacteria or fungi (Leelavathy, 1969), although a mutant of *Saccharomyces ellipsoideus* responds to auxin (Yanagishima and Shimoda, 1968), but its free protoplasts do not. If present at all in marine algae, the level is below 0.1 μg IAA/gfw (Buggeln and Craigie, 1971). IAA failed to stimulate the growth of the green alga *Chlorella* (Lien et al., 1971), although it inhibited growth at 6×10^{-5} M. Weak auxins promoted the growth of blue-green algae (Ahmad and Winter, 1970). They also fail to induce any growth response in organisms without cellulose cell walls, as for example, animals. These facts favor the view that auxins alter the cellulose cell walls in such a way as to lead to cell enlargement.

As already shown (Chapter 17), cells enlarge due to a decrease in resistance to the force of turgor pressure. This decrease in resistance is induced by a softening of the cell wall leading to an increase in wall plasticity. IAA produces such a softening (Heyn, 1940), and many recent investigations have corroborated this fact, for instance, by demonstrating a decrease in Young's modulus in both stems and roots (Burström et al., 1970). As a result, the extension rate is markedly greater in auxin-pretreated walls (Cleland, 1971). As further evidence of an intimate relation to the cell wall, ^{3}H-labeled IAA has been localized specifically in the cell walls of the youngest xylem elements (Sabnis et al., 1969).

The cell wall component involved has not been established. Although the logical substance to suspect is cellulose, much of the evidence favors other substances in the cell wall. For a time it was thought that the effect might be on the pectins that are present in small quantities in the walls of growing plant cells; but more recent evidence has eliminated this possibility (Cleland, 1963). Other noncellulosic polysaccharides have also been suggested (Yamamoto and Masuda, 1971). Lamport (1965) has proposed that the cell wall protein that he calls "extensin" can be alternately split and rejoined. Many results suggest that IAA activates the formation of one or more new enzymes that act on cell walls to increase plasticity (Nooden and Thimann, 1965).

In agreement with this hypothesis, cell expansion of a strain of yeast that responds to IAA was enhanced even more by the polysaccharide-hydrolyzing enzyme β-1,3-glucanase than by IAA, but was not enhanced by it in the strain that fails to respond to IAA (Shimoda and Yanagishima, 1968). Furthermore, the auxin-induced expansion was inhibited by an inhibitor of β-1,3-glucanase (Shimoda and Yanagishima, 1971). These results support the idea that auxin induces the expansion of yeast cells in the responsive strain by enhancing the activity of enzymes that degrade the cell wall. The above enzyme was also able to induce as much elongation of oat coleoptiles as by IAA, at least during the first 1 to 3 hr (Tanimoto and Masuda, 1968). The results with oat coleoptiles suggest that auxin primarily causes a partial degradation of the noncellulosic polysaccharide components of the cell wall (Yamamoto and Masuda, 1971). In agreement with the emphasis on noncellulosic polysaccharides, a weakening of the cell walls by the action of cellulase was not sufficient to promote the growth of oat coleoptiles (Ruesink, 1969), even though wall extensibility was increased.

Auxins have a second effect on the cell walls. In the case of sycamore cells in culture, the auxin 2,4-D increased the incorporation of arabinose into cell wall polysaccharides (Rubery and Northcote, 1970). In pea stem tissue, β-glucan synthetase (which catalyzes polysaccharide synthesis) was the only one of the eight enzymes tested whose activity was increased by treatment with IAA.

On the basis of these results, it now appears that IAA has a twofold effect on the cell wall—a loosening of the wall structure plus an addition of new cell wall substance. These two effects are apparently due to the increase in activity of a cell wall hydrolyzing enzyme and a cell wall synthesizing enzyme, respectively. This conclusion is in agreement with the earlier-described cell wall changes during cell enlargement due to a combination of increased plasticity and intussusception (Chapter 17).

This, then, is at least part of the explanation for the increased cell enlargement. Such an explanation, however, is by no means complete. The mode of action of the auxin on the cell wall is not at all clear. It has been shown that auxins greatly affect respiration rate. Those concentrations of auxin that increase growth usually cause increased respiration and those that decrease growth usually cause decreased respiration, although there are exceptions. The way in which this effect on respiration is linked to the effects on the cell wall is impossible, as yet, to say. Perhaps the auxin effect on respiration releases the energy needed to break the bonds between the micelles of the cell wall and to synthesize cellulose molecules (Ray and Baker, 1962), thus both increasing wall plasticity and permitting the intussusception of new wall material. Unfortunately, there is no evidence that the increase in respiration rate is a *direct* effect of the auxin. Since growth is dependent on respiration (Chapter 17), the auxin-induced increase in respiration rate may be a *result* of the auxin-induced increase in growth rate. Similarly, besides the wall-loosening and wall-synthesizing enzymes, many others are apparently activated by IAA (Table 19-2). An enzyme that forms an aspartate conjugate with the auxin is absolutely auxin-specific (Venis, 1972).

It is usually assumed that an auxin-induced increase in enzyme activity is due to de novo synthesis, and this assumption has been supported by experimental results (Evans et al., 1971). The enzymes involved in wall loosening are apparently unstable. Continued protein synthesis is therefore necessary to maintain the wall-loosening process (Cleland, 1970; Penny, 1971). Cleland (1971) suggests three ways by means of which auxin might expand the pool of

Table 19-2. Enzymes reported to increase in activity due to auxin

Enzyme	Plant	Reference
Cellulase	Pea seedlings	Fan and Maclachlan, 1966
Dextranase	*Avena* coleoptile	Heyn, 1970
Lipase	Aleurone layer of germinating wheat	Eastwood et al., 1969
β-1,3-Glucanase	Oat coleoptile	Tanimoto and Masuda, 1968
Peroxidase isoenzyme	Oat coleoptile	Stuber and Levings, 1969
Invertase	Chicory root	Flood et al., 1967
Aldolase	*Atropa belladonna* (NAA)	Simola and Sopanen, 1971
Aminopeptidase	*Atropa belladonna* (NAA)	Simola and Sopanen, 1971
Glutamate: oxaloacetate transaminase	*Atropa belladonna* (NAA)	Simola and Sopanen, 1971
β-Glucan synthetase	Pea stem	Abdul-Baki and Ray, 1971
IAA oxidase	Tobacco callus	Lee, 1971*a*

these "growth-limiting proteins." Nevertheless, a threefold increase in phosphatase activity due to IAA was not prevented by cycloheximide, an inhibitor of protein synthesis, and therefore must be due to activation of the enzyme rather than de novo synthesis (Palmer, 1970).

The postulated relation of auxin action to protein synthesis has stimulated an intensive search for evidence of a relation to nucleic acids, which control protein synthesis (Chapter 16). Growth-promoting concentrations of IAA were soon found to enhance RNA synthesis and growth-inhibitory concentrations, to decrease it (Key, 1964; Key and Shannon, 1964). Some evidence seemed able to separate this from the growth effect, since cell wall extensibility was apparently increased even by auxin concentrations that inhibited RNA synthesis more than 90% (Cleland, 1965). However, later investigations of a number of different plants have demonstrated an inhibition of auxin-induced cell wall extensibility in all cases when RNA synthesis was inhibited (Coartney et al., 1967). Similarly, specific inhibitors of both RNA and protein synthesis suppress IAA-induced growth. On the basis of these results, Knypl (1966) suggests that IAA is a specific inductor of messenger RNA (m-RNA) synthesis. This conclusion is supported by the incorporation of labeled uracil and orthophosphate into RNA in the presence of IAA (Masuda et al., 1967). Trewavas (1968), however, has concluded that ribosomal RNA is the major species whose synthesis is increased by IAA, although small changes occurred also in mRNA.

The RNA synthesized as a result of auxin treatment is not a qualitatively new species (Thompson and Cleland, 1971), and the particular RNA fraction that increases depends on the length of treatment (Miassod et al., 1970). No complex between ^{3}H-labeled IAA and tRNA could be detected. DNA synthesis appears to be required for auxin induction of cell enlargement in tobacco pith explants (Maheshwari and Nooden, 1971). Auxin also induced DNA synthesis in soybean hypocotyls (Holm and Key, 1971). According to Fellenberg (1969*b*), during root formation in pea epicotyls IAA apparently loosens the histone bound to DNA. In this way it acts as a true initiator of RNA synthesis. In agreement with this conclusion, applied histone inhibited IAA-induced root formation in cuttings of dark-grown pea seedlings and light-grown tomato plants (Fellenberg, 1971*b*). Conversely, washing sugar beet roots with auxin enhanced template availability of the isolated chromatin (Duda and Cherry, 1971).

All these results have given rise to the gene-activation hypothesis, according to which auxins regulate growth by activating specific genes, which lead to the synthesis of specific enzymes required for cell growth. This hypothesis would require auxin action at the DNA level. There are many experimental results, however, that fail to support this kind of action. Thus there is a lag period (45 to 60 min) before auxin-induced growth can be depressed by an inhibitor of protein synthesis (cycloheximide) but not by inhibitors of RNA synthesis (8-azaguanine or actinomycin D—Nelson et al., 1969). These results are compatible with the action of auxin at the level of protein synthesis but indicate that mRNA synthesis is not involved. Similarly, gougerotin, an inhibitor of protein synthesis, inhibits auxin-induced elongation and protein synthesis in *Avena* coleoptiles (Burkett et al., 1970). In agreement with these results, auxin-regulated xylem differentiation in stem segments of *Coleus* was not due to a control of DNA synthesis, although DNA synthesis, which occurred in the absence as well as the presence of auxin, was necessary for the auxin action (Fosket, 1970). These results point to auxin effects at the translation level rather than at the duplication or transcription level, and therefore do not support the gene activation hypothesis.

Other auxin effects are also impossible

to explain by the gene activation hypothesis, and it has been suggested that auxin acts as an allosteric effector (Dalheim, 1969). Auxin may either inhibit or enhance coleoptile growth within 1 to 3 min of application (Rayle et al., 1970). This is too short a period to permit the complete series of gene activation followed by transcription, protein synthesis, and enzyme action. There are so many more equally rapid or more rapid effects of auxin that the primary interaction of auxin has been suggested to involve the plasma membrane rather than the genes (Rayle et al., 1970). Thus IAA enhances the movement of tritiated water into and out of pea stem segments well within 1 min after application of the auxin (Kang and Burg, 1971), and it produces a rapid bursting response in isolated *Avena* coleoptile protoplasts (Hall and Cocking, 1971). In agreement with these results, a specific, reversible binding of auxin to plasma membrane fractions from maize coleoptiles has been demonstrated (Lwmbi et al., 1971; Hertel et al., 1972). IAA also stimulated the incorporation of ^{32}P-labeled orthophosphate by sterile pea stems in 5 min and before the onset of stimulated growth (Neumann, 1971), as well as the synthesis of cyclic AMP by 100% (Azhar and Murti, 1971). Finally, auxin acted cooperatively with GTP (ITP) as an effector of a membrane-bound anisotropic adenosine triphosphatase or proton pump (Hager et al., 1971). According to them, this auxin-activated pump utilizes respiratory energy to raise the proton concentration in a compartment at

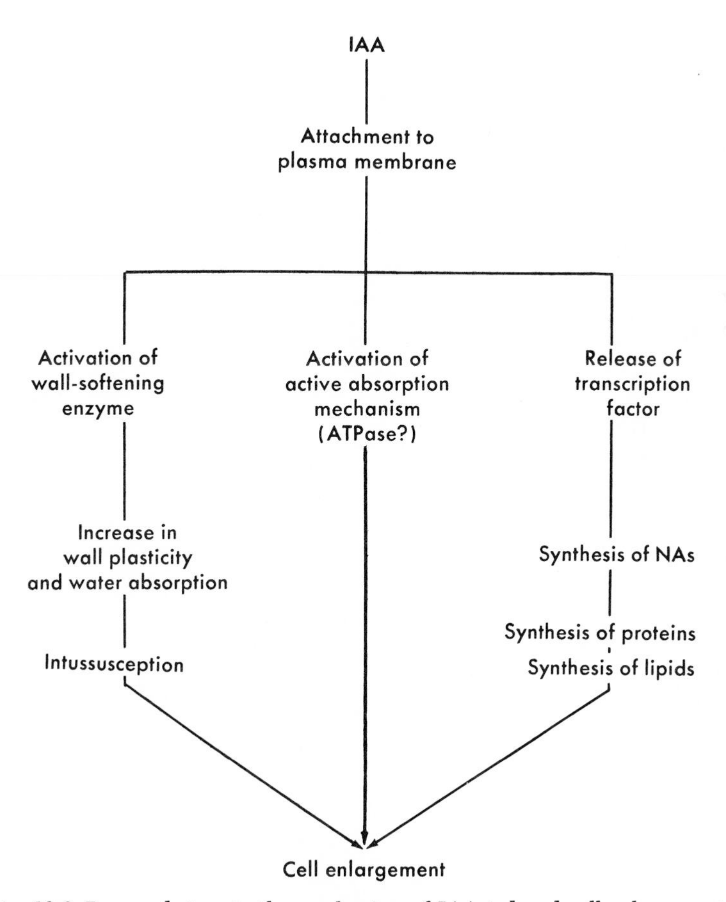

Fig. 19-6. Proposed steps in the mechanism of IAA-induced cell enlargement.

the cell wall. This event leads to an increase in activity of enzyme softening cell walls and thus triggers cell elongation. This transport or secretion of protons into the cell wall compartment should be compensated by a flow of cations into the interior of the cytoplasm or by flow of anions to the cell periphery, causing secondary auxin effects.

Although these rapid membrane effects of IAA occur before the effects on protein synthesis, recent results have indicated a possible interrelation between the two. Auxin treatment of isolated plasma membrane fractions of soybean hypocotyl led to a release from the membrane of a transcriptional factor that significantly stimulated the activity of soybean RNA polymerase (Hardin et al., 1972).

All these membrane results establish the inadequacy of the gene activation hypothesis as an explanation of the mechanism of auxin action. Nevertheless, the effects of auxin on nucleic acid (NA) and protein synthesis are so striking that they must play some role in the effect of auxin on cell enlargement. Unfortunately, so much emphasis has been placed on the cell wall that investigators have completely lost sight of the fact that cell enlargement also includes vacuole and protoplasm enlargement. Cell enlargement involves active absorption of solutes, as well as NA, protein, and lipid synthesis. When IAA induces cell enlargement, it must stimulate all of these processes. It is therefore probable that the rapid effects of IAA are due to activation of already present enzyme systems, leading to rapid wall softening and cell enlargement. This must be followed by a slower synthesis of NA, protein, and lipids. The total process is represented by Fig. 19-6.

The exact mechanism of each step in Fig. 19-6 has yet to be worked out. Two or more steps may perhaps be affected simultaneously. For instance, an (H^+, K^+)-activated adenosine triphosphatase may conceivably soften the cell wall by the high acidity resulting from the efflux of H^+ ions while simultaneously inducing active absorption of K^+ ions. However, even this broader concept of the mechanism of auxin action is inadequate. It has become apparent that the effects of auxin on growth cannot be fully understood except in conjunction with the effects of the other growth regulators.

GIBBERELLINS

Physiological and chemical properties

At the same time as the European (especially the Dutch) investigators were proving the existence and functions of auxins, Japanese investigators were discovering gibberellins in rice plants affected by the "foolish disease." However, it was not until about 30 years later that these substances were found to be present in normal plants and their importance was recognized by scientists throughout the world. Just as auxins are defined by the method used to measure them—the ability to induce a growth curvature in oat coleoptiles—so,

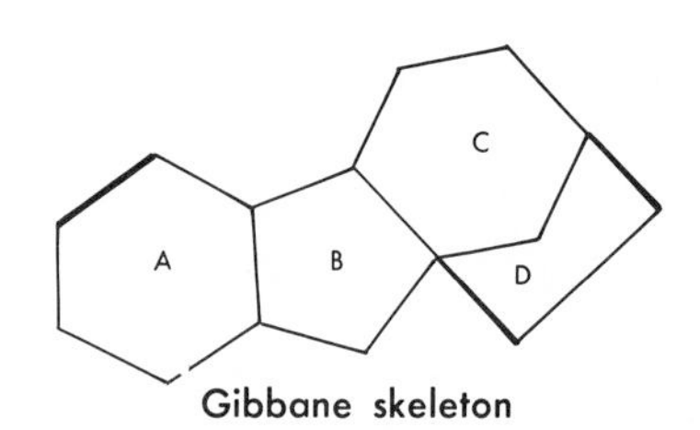

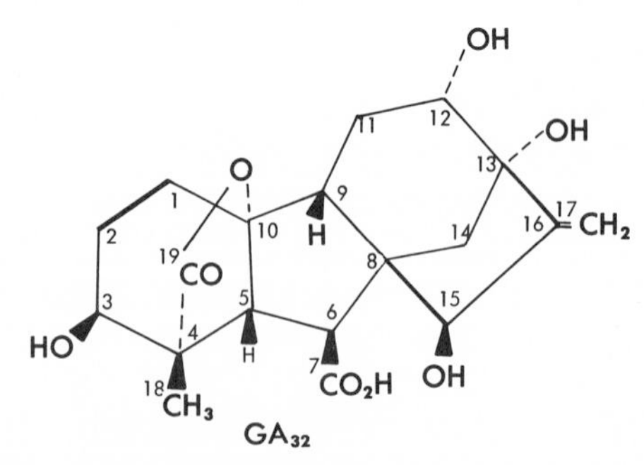

Fig. 19-7. Chemical structure of gibberellins. (From Coombe, B. G. 1971. Science **172**:856-857. Copyright 1971 by the American Association for the Advancement of Science.)

also, gibberellins may be defined physiologically as compounds that are active in gibberellin assays (see following discussion). They may also be defined chemically as possessing a gibbane structure (Fig. 19-7) consisting of two cyclohexane and two cyclopentane rings. By this definition, even those gibbanes that are inactive in gibberellin bioassays are still considered to be gibberellins. On the other hand, kaurene is active in some gibberellin bioassays, but it does not have the gibbane structure and is therefore not considered to be a gibberellin. The gibbane structure resembles but it differs from the structure of steroids such as cholesterol, which consist of three cyclohexane and one cyclopentane ring. The abbreviation GA was originally coined for gibberellic acid but is now used for any gibberellin, the ones of known structure being numbered (e.g., GA_3 for gibberellic acid). Unlike auxins, of which only one (IAA) appears to be the normal hormone in most plants, there are thirty-four known gibberellins (Coombe, 1971). Not all of these, however, are physiologically active, and many are synthesized in organisms in which they are not growth regulators. Gibberellic acid (GA_3) is the most common and GA_{32} the most polar (Coombe, 1971), differing from GA_3 by possession of two OH groups at carbons 12 and 15 (Fig.

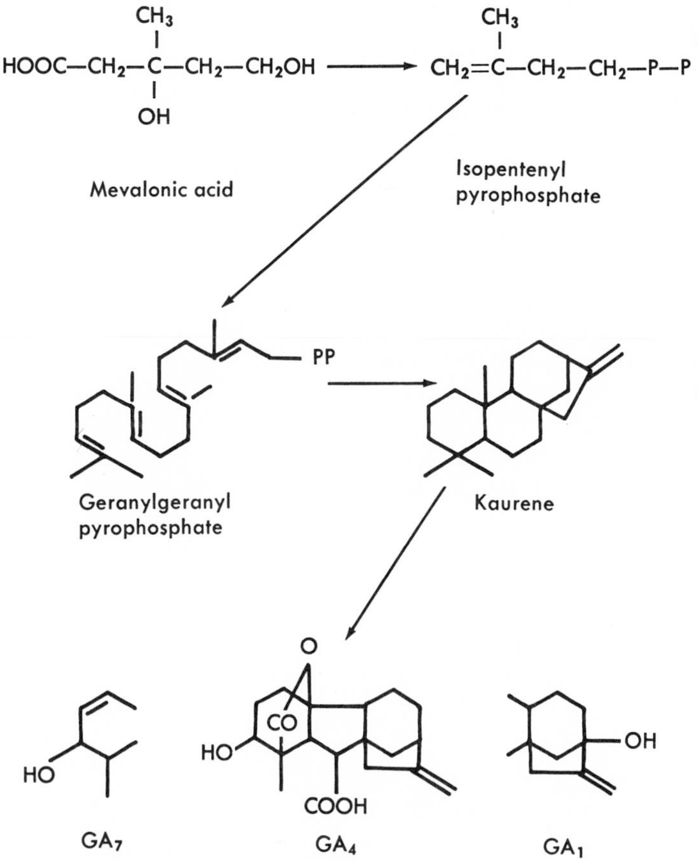

$$HOOC{-}CH_2{-}C(CH_3)(OH){-}CH_2{-}CH_2OH \longrightarrow CH_2{=}C(CH_3){-}CH_2{-}CH_2{-}P{-}P$$

Mevalonic acid

Isopentenyl pyrophosphate

PP

Geranylgeranyl pyrophosphate

Kaurene

O

CO

HO

HO

COOH

OH

GA_7

GA_4

GA_1

Fig. 19-8. Pathway of gibberellin biosynthesis in *Fusarium moniliforme*. (From Cleland, R. E. 1969. *In* M. B. Wilkins [Ed.]. The physiology of plant growth and development. Copyright 1969 McGraw-Hill Book Co. [UK] Ltd., Maidenhead, England. Used with permission.)

19-7). All the gibberellins are chemically related to, and therefore probably synthesized from, diterpenes ($C_{20}H_{39}$). The biosynthetic pathway of gibberellins is probably as represented in Fig. 19-8. The highest levels of gibberellins are found in seeds, as high as 470 μg/gfw of endosperm. In vegetative regions the values range from 1 to 10 μg and are higher in young than old tissues. In general, they tend to be more concentrated in more rapidly growing and developing regions. Unlike auxins, gibberellins are translocated nonpolarly in both directions. It has been suggested that they may be synthesized in the roots. According to recent results, however, the shoot is the primary site of GA biosynthesis and the root is merely a site of GA interconversion (Crozier and Reid, 1971).

Effects on growth

Although they do not all have the same effects on growth (Crozier et al., 1970), gibberellins, like auxins, are best known for their enhancement of cell elongation such as occurs in the disease of rice that led to their discovery. For instance, they induce internode elongation in plants during their rosette stage of growth and therefore lead to the phenomenon called "shooting" (or "bolting")—the significant stem elongation that precedes flowering in many plants (e.g., biennials). Natural differences in GA content apparently lead to differences in internode length, for instance between tall and dwarf forms of peas (Broughton and McComb, 1967) and rice (Harada and Wada, 1968). In the Alaska pea the growth rate is correlated with the synthesis of gibberellic acid (Moore, 1967). This internode elongation is frequently accompanied by a slight decrease in root growth, probably because of a competition for raw materials. Excised roots, on the other hand, show increased growth in the presence of gibberellin. In fact, van Overbeek believes that root growth is promoted only by gibberellin, not by auxin as indicated earlier (Fig. 19-3).

Even the rate of root hair elongation was increased by GA concentrations of 10^{-7} to 10^{-12} M (Devlin and Brown, 1969). Similar low concentrations promoted root formation by pea cuttings, although higher concentrations (10^{-6} to 10^{-3} M) inhibited root formation (Eriksen, 1971). At concentrations of 10^{-6} M, however, GA had no effect on root growth of *Lens culinaris,* although it significantly reduced the roots' geotropic reaction.

Gibberellins stimulate the growth of a yeast (Makarem and Alldridge, 1969) and a unicellular green alga (Paster and Abbott, 1970). They are involved in the growth regulation of red algae (Jennings and McComb, 1967). Gibberellins have, in fact, been found in all the groups of lower as well as higher plants (Cleland, 1969).

The enhancement of cell elongation by gibberellins is, at least partly, a secondary effect due to a reversal of the light-induced inhibition of cell expansion in the internodes (Cleland, 1969). The induction of bolting in rosette plants, which is observed due to the excessive cell enlargement of the internode cells, is actually initiated by an activation of cell division. This division occurs in a normally inactive subapical meristem and is believed to be the primary effect of the gibberellin (Cleland, 1969). The apical meristem seems to function independently of the gibberellins (Cleland, 1969). Similarly, in the case of *Eucalyptus camaldulensis* the major response to GA seemed to be an increase in transverse cell divisions (Bachelard, 1969). On the other hand, GA_3 inhibited regeneration in *Begonia* leaves, probably by blocking the organized cell divisions initiating the formation of buds and root primordia (Heide, 1969*a*).

When applied artificially, gibberellins may also stimulate cell division in meristems, leading to flowering (Bernier et al., 1964). They may also stimulate seed germination. Leaf growth responds to gibberellin, although not to auxin; yet gibberellins may promote fruit growth, resembling the effect of auxin. It has even been suggested that gibberellins that are produced in the apple ovule after fertilization are responsible for fruit set (Dennis, 1967). In agreement with this suggestion, spraying grapes with GA increased the movement of ^{14}C assimilates into the fruit (Weaver et al., 1969).

The grape berry responds with maximum increase in berry size when treated with GA just after bloom or at the stage of early fruit set, and therefore the stage of cell division (Weaver and Pool, 1971). After fruit set there is a gradual decrease in response to GA.

The response to gibberellin may be altered by an environmental factor: The dwarf pea fails to show its dwarfness in the dark when supplied with gibberellin; light, however, renders it less responsive to gibberellic acid, and it fails to grow to full size.

In summary, gibberellins (0.05 to 0.5 ppm) can promote (1) internode growth, (2) hypocotyl growth, (3) germination, (4) sexual development of flowers, (5) flower formation, (6) parthenocarpy, (7) breaking of the rest period of buds and underground organs, (8) flower formation, (9) cell division in the cambial zone, (10) change in shape and size of leaves, and (11) activity of enzymes (Linser, 1966). They may also participate in the expression of apical dominance (Ruddat and Pharis, 1966) and may increase the storage life of fruit (Kitagawa et al., 1966). Some of these effects will be discussed in Chapter 20. Any attempt to explain the mechanism of gibberellin action must be capable of explaining all of these results.

Mechanism of growth regulation

As in the case of auxins, many of the effects of gibberellins can be explained by their control of enzyme activity, although the enzymes involved are different. Stimulation of seed germination is apparently the result of an activation of α-amylase, which then digests the endosperm starch.

Although α-amylase is the enzyme most commonly reported to increase in activity as a result of GA treatment, the activation of many other enzymes has been reported (Table 19-3). At least in some cases, however, α-amylase increases first and the others follow later (Pollard, 1969). Furthermore, the increase in α-amylase activity is not confined to seeds. Even in isolated leaves of *Begonia rex*, applied GA (10^{-6}g/ml) increased the activity of starch dissolving enzymes 2.3 times (Muenzel, 1970). In the leaf sheath of dwarf maize seedlings, GA always induced both elongation and an increase in α-amylase activity, but the two processes are apparently independent of one another (Katsumi, 1970).

Many lines of evidence indicate that GA stimulates growth by increasing metabolism. The first observed metabolic effect in cereal grains is the increase in soluble carbohydrate (Pollard, 1969) as a result of the above-mentioned activation of α-amylase. This was followed by increased oxygen consumption and increased secretion of many other enzymes (Table 19-3). Potato tuber slices showed similar metabolic changes (Coutrez, 1968). It was even possible to mimic the effect of GA by simply injecting glucose or its derivatives into pea internodes (Broughton and McComb, 1971). Finally, the effect on shoot growth may be due to a stimulation of translocation of photosynthate from the leaf to the growing region (Lovell, 1971).

Just as the effect of auxins on enzyme activity is commonly due to protein synthesis, so also the activation by gibberellin is apparently the result of *synthesis* of the α-amylase protein because when barley endosperm halves were incubated with phenylalanine-^{14}C, the α-amylase protein became labeled in the presence of gibberellin but not in its absence. In the case of dwarf peas, internode expansion can be induced by GA treatment. This results in a considerable increase in protein synthesis, which apparently increases the rate of cell

Table 19-3. Some reports of enzyme stimulation by GA

Enzyme	Plant	Reference
Stimulation		
Lipase	Cottonseed	Jones et al., 1967
Protease	Barley	Jacobsen and Varner, 1967
Adenosine triphosphatase		
GTPase		
Phytase		
Phosphomonoesterase	Grains	Pollard, 1969
Phosphodiesterase		
Alcohol dehydrogenase		
Isocitrate lyase		
β-Glycerophosphatase	*Cyamopsis tetragonoloba*	Kathju et al., 1971
Pyrophosphatase		
Amylase		
β-Fructofuranosidase	Pea internodes	Broughton and McComb, 1971
Starch phosphorylase (slight stimulation)		
PEP carboxylase		
RuDP carboxylase	Dwarf pea	Broughton et al., 1970
α-Amylase	*Avena*	Naylor, 1969
α-Amylase	Corn	Katsumi, 1970
α-Amylase, four forms	Barley aleurone	Jacobsen et al., 1970
β-1,3-glucanase (release but not synthesis)	Barley aleurone	Jones, 1971
α-Amylase	Tobacco leaves	Lee and Rosa, 1969
Invertase	Tobacco leaves	Lee and Rosa, 1969
Invertase	Pea internodes	Moll, 1971
No effect		
Cellulase	Pea internodes	Broughton and McComb, 1971
Pectinesterase	Pea internodes	

wall synthesis (Broughton and McComb, 1967).

The evidence for growth regulation by NA activation is stronger in the case of gibberellins than in the case of auxins. Gibberellin is prevented from inducing α-amylase synthesis by treatment with actinomycin D, an inhibitor that forms complexes with DNA but not with RNA. Presumably, it inhibits DNA-dependent but not RNA-dependent synthesis (Khan, 1967*b*). The scheme is as follows:

$$\text{DNA} \xrightarrow[\text{gibberellic acid}]{\text{Action of}} \text{mRNA} \longrightarrow \alpha\text{-Amylase}$$

On this basis, GA may conceivably lead to α-amylase synthesis by uncovering the genes responsible for synthesis of the α-amylase molecule. The same explanation may apply to the stimulation of the RNA-controlled protein synthesis in yeast, leading to sporulation (Kamisaka et al., 1967). But the foregoing scheme does not seem to apply to all tissues. In the case of lentil epicotyl, DNA synthesis was promoted by gibberellin (Nitsan and Lang, 1966). This stimulating effect on DNA has since been corroborated for many plants by many investigators (pea: Soleimani, 1968; hazel seeds: Pinfield and Stobart, 1969; pea epicotyl: Nakamura et al., 1970; potato tuber: Shih and Rappaport, 1970; *Chrysanthemum* shoots: Mahmoud and Steponkus, 1970; cucumber hypocotyl: Degane and Atsmon, 1970). The synthesis of many other NA fractions (e.g., mRNA: Aramaki and Kuroiva, 1967) has also been reported to increase. Such increases as well as the increased enzyme synthesis must follow any increased synthesis or derepression of DNA. In some cases, however, GA apparently enhances growth without increasing DNA synthesis or, at least, before any increase in DNA synthesis (pea internodes: Broughton, 1969; soybean hypocotyl: Holm and Key, 1969; dwarf pea plants: McComb et al., 1970). It has been suggested (McComb et al., 1970) that these negative results may be due to small changes not detectable by the technique used. On the other hand, the possibility of enhancing DNA activity without having to increase its synthesis was not supported (Spiker and Chalkley, 1971), for GA had no effect on the histones, which are believed to repress DNA activity. Nevertheless, washing sugar beet tissue in solutions of GA (and auxin) enhanced the template availability of isolated chromatin (Duda and Cherry, 1971).

Interactions with IAA

Many apparent discrepancies in the gibberellin literature are due to interactions between gibberellins and auxins. Because of these interactions, and because there are always auxins and gibberellins present in the plant, it is difficult to interpret the results when a tissue is treated with a single growth regulator.

The interactions may be synergistic or antagonistic (Fig. 19-9). GA by itself does not induce coleoptile curvature, but it may increase the curvature induced by IAA; this is a synergistic effect. On the other hand, GA antagonizes the correlative inhibition of axillary buds (apical dominance) induced by IAA (Phillips, 1971*a*). The opposite effect, enhancement of IAA-induced apical dominance, has also been reported by Tomaszewski (1970). These opposite effects of GA on IAA-induced growth changes can be explained by changes in IAA concentration. There have been many reports of GA-induced increases in auxin concentration (Katsumi and Sano, 1968). In many cases this has been accompanied by an inhibitory effect of GA on IAA oxidases and therefore presumably a decrease in IAA breakdown (Katsumi and Sano, 1968). There are, however, several different IAA oxidases (see previous discussion), some of which may be inhibited, others stimulated by GA. Thus GA promotes IAA oxidation in pea buds by increasing the activity of some (but not other) IAA oxidases by as much as 50 times (Ockerse and Waber, 1970). Even this increase in IAA oxidation may either enhance or decrease IAA-induced growth, depending on whether the original IAA concentration is inhibitory or stimulatory. In the case of tobacco callus cultures, GA promoted the development of three IAA oxidase isoenzymes (Lee, 1971*a*). Still another reported effect of gibberellic acid on IAA is an enhancement of the metabolism of tryptophan to IAA in *Avena* coleoptile tips (Valdovinos and Sastry,

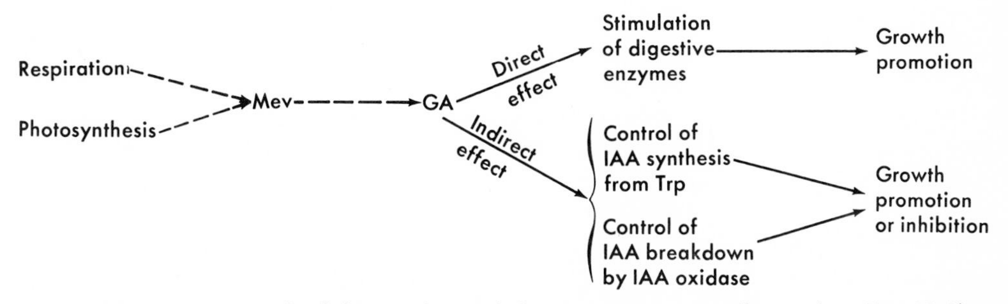

Fig. 19-9. Promotion and inhibition of growth by GA. Mev = mevalonate (see Fig. 19-8).

1968) and in dwarf peas (Lantican and Muir, 1969), and when apical dominance is enhanced (Tomaszewski, 1970).

On the other hand, IAA antagonizes the GA-induced extension growth of internodes (Phillips, 1971*b*). All these interactions have led to the suggestion that the sole effect of GA is an indirect one due to its control of IAA, which is suggested to be the direct growth regulator. However, there are some GA-induced effects such as the enhancement of α-amylase that are independent of IAA (Cleland, 1969). Furthermore, despite their interactions, the sites of induced elongation are different—the apical region of pea stems for GA, a region well below the apex for IAA. The primary effect of GA is now believed to be on cell division, the primary effect of IAA on cell enlargement. There are still other differences.

On the basis of the previous evidence, it appears that GA produces its initial effect at the DNA level, either by increasing DNA synthesis or by derepressing DNA. Whether the effect is the same in all plants is not yet clear. In any case, GA, as well as IAA, regulates growth by the subsequent enhancement of protein synthesis and therefore of enzyme activity. Whether it also induces a more rapid membrane effect, as in the case of IAA, has not yet been determined. The two growth regulators differ basically in the enzyme systems they activate. IAA appears to enhance primarily the activity of enzymes involved in cell wall hydrolysis and synthesis. GA appears to enhance primarily the activity of digestive enzymes that convert the insoluble reserves (starch, fats) into readily metabolizable soluble substances as well as oxidative enzymes, which control IAA concentration. Nevertheless, enhancement by GA of wall-hydrolyzing enzymes has also been reported. There may be a difference between the two growth regulators with respect to the specific point of attack in the protein synthesizing system. Most of the evidence points to an effect of GA on DNA duplication or transcription, and to an effect of IAA on translation.

CYTOKININS

Physiological and chemical properties

In opposition to auxins, which primarily affect cell enlargement, substances have been found that primarily affect cell division and have been called *kinins* (from kinesis, meaning division). Unfortunately, the name has priority for a chemically different (peptide) group of substances in animal physiology. The terms *phytokinin* (Osborne, 1963, see Chapter 20) and *cytokinin* (Skoog et al., 1965) have therefore been suggested, and the latter is now most commonly used and is abbreviated CK.

Cytokinins have been defined as substances, that, regardless of other activities, promote cytokinesis in cells of various plant origins (Skoog et al., 1965). This activity appears to be confined, in nature, to 6-substituted purine derivatives, the purine being normally adenine. Many bioassays have been used to test for CK activity—cell division, chlorophyll preservation, cell enlargement, germination, and differentiation (Fox, 1969).

Unlike auxins and gibberellins, CK activity was not discovered in plant extracts. Adenine sulfate was the first CK discovered (Skoog, 1951). It greatly alters the kind of growth in tissue cultures (e.g., tissues cut from apical or cambial meristems and grown in suitable culture media). It may convert the undifferentiated tissue so formed into one that develops buds and even mature organs (Fig. 19-10). Later investigations resulted in the isolation of a more active substance formed by the breakdown of nucleic acid. It was named *kinetin* and is now known to be 6-furfuryl-aminopurine.

Many extracts of plants are now known to have cytokinin activity. This may result from the presence of *zeatin,* which has been isolated from corn grains. It is a hydroxy

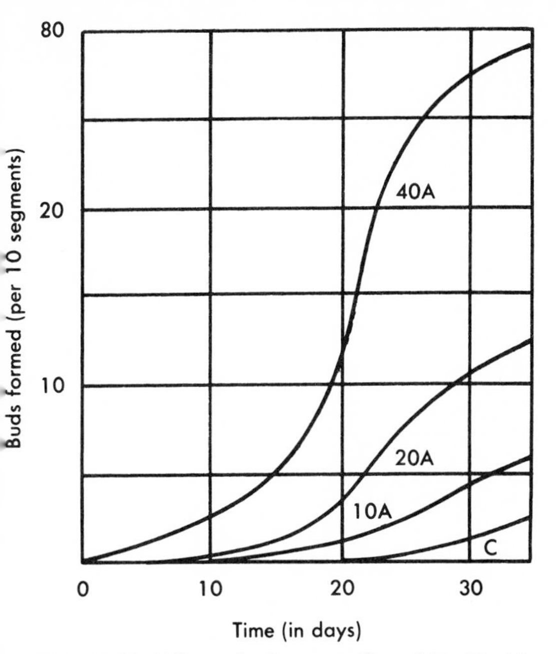

Fig. 19-10. Effect of adenine sulfate (40, 20, 10, and 0 mg per liter) on bud formation in tobacco stem segments. (From Skoog, F., and C. Tsui. 1951. *In* F. Skoog. Plant growth substances. University of Wisconsin Press, Madison, Wis.)

derivative of 6-dimethylallyl adenine (Fig. 19-11). CKs commonly occur in seeds or fruit and are concentrated in meristematic regions such as root tips, although they are not measurable in mature regions (Short and Torrey, 1972). Three chromatographically distinct CKs have been found in seeds of the pumpkin (Gupta and Maheshwari, 1970) and watermelon (Prakash and Maheshwari, 1970) as well as in apple fruit (Letham and Williams, 1969). In the apple, they appear to be zeatin, zeatin riboside, and zeatin ribotide. When ^{14}C-labeled zeatin was taken up by bean axes, the ^{14}C was found mainly in zeatin, zeatin riboside, zeatin ribotide, and dihydrozeatin derivatives (Sondheimer and Tzou, 1971). So far, all naturally occurring cytokinins appear to be purine derivatives. Transfer RNA possesses CK activity due to the presence of a new nucleoside (Hecht et al., 1969). It has even been suggested that tRNA may be involved in regulatory activities by serving as a source of CKs (Dyson et al., 1970).

The best known synthetic CK is benzyladenine. Even urea derivatives have CK activity although not related to them chemically. Phenylurea is the simplest active compound (Bruce and Zwar 1966), although over 200 ureas were found to be active. Another group of cell division–promoting substances are nicotinamide derivatives (Wood et al., 1969). It was even suggested that they, rather than the CKs, are the naturally occurring factors, promoting cell division in higher plants. Other substances with CK activity have also been reported (Isogai et al., 1970; Sassa et al., 1970).

Earlier observations seemed to indicate that CKs were not translocated but acted in situ, where they were synthesized. This, if true, would eliminate them from the class of hormones. More recent results, however, indicate that they must be translocated under normal conditions. When applied to the base of the stem of entire, etiolated *Cicer arietinum* plants, CKs move acropetally toward the terminal buds and accumulate there (Hugon, 1967). When applied to the leaf blade of several plants, kinetin is immobile (Lagerstedt and Langston, 1967), but it is not immobile when applied to petiole, vein, or root. Basipetal as well as acropetal translocation occurred in tobacco leaves. Kinetin was not taken up by cotton roots but was both absorbed

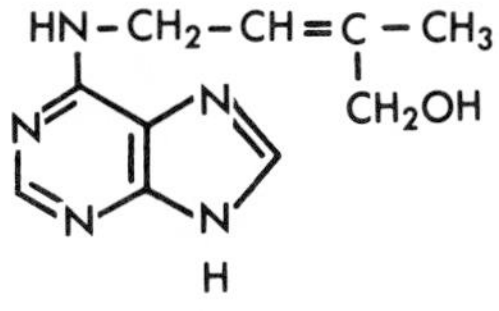

HN−CH_2−CH=C−CH_3 (C bearing CH_2OH), on a purine ring (N, N, N, N, H)

Zeatin

Fig. 19-11. Zeatin. (From van Overbeek, J. 1966. Science **152**:721-731. Copyright 1966 by the American Association for the Advancement of Science.)

by and translocated from tobacco roots.

It has, in fact, been established in many cases (Burrows and Carr 1969; Robbins and Hervey, 1971; Radin and Loomis, 1971) that CK is synthesized in the growing root apices and transported to the shoot whose growth and development it then regulates, for instance, in cereals (Dezsi and Farkas, 1964). CK activity is therefore found in root exudates (of rice —Yoshida et al., 1971) or in root extracts (of radish—Radin and Loomis, 1971), due apparently to all three CKs—zeatin, its nucleoside, and its nucleotide. Ribosylzeatin has, in fact, been crystallized from chicory root (2 mg from 250 kg fresh weight—Bui-Dang-Ha and Nitsch, 1970). The formation of adventitious roots on a leaf (Engelbrecht, 1964) or on a hypocotyl (Wheeler, 1971) was required for survival due to a supply of CK from these roots. This, in fact, is apparently the sole function of nonabsorbing roots of *Scrofularia arguta* (Miginiac, 1971), which ensure the vegetative development of cotyledonary buds as opposed to their floral development in the absence of these roots. Roots of submerged hydrophytes, which, again, are not required for absorption, also supply the shoot with CK (Waisel and Shapira, 1971). Even the growth of the oat coleoptile is apparently regulated by CKs of the root (Jordan and Skoog, 1971). The specific effects produced by the CKs of the root on arrival in the shoot are not always distinguished. The normal branching of the inflorescence of *Carex* seems to depend on an adequate supply of CKs from the roots (Smith, 1969).

Effects on growth

The two most striking effects of CKs are the induction of cell division and morphogenesis (organ differentiation) in cultures of excised tissues (Fig. 19-10). The result is a conversion of an amorphous mass of undifferentiated tissue into buds and finally into a whole plant. Zeatin induces cell division at concentrations of less than 5×10^{-11} M (Letham, 1969). The only organic compounds required for active growth of some tissues in vitro are sucrose, thiamine, myoinositol, auxin, and a cytokinin. CK acts as a "trigger" for mitosis (Fox, 1969) and may actually be used to synchronize mitosis in tobacco cells (Jouanneau, 1971). For differentiation, on the other hand, it does not act simply as a trigger, but must be present during a critical period until differentiation is stabilized (Vrandes and Kende, 1968). It appears to be involved in the normal growth even of algae, at least of the Phaeophyta and Rhodophyta (Jennings, 1969), and has been reported to increase the rate of cell division in several microorganisms, including protozoans, phytoflagellates, and bacteria (Fox, 1969). There is even some positive evidence in the case of higher animals. CKs are produced by fungi (Phillips and Torrey, 1970; Sassa et al., 1970), and they increase or inhibit mycelial growth in low and high concentrations, respectively (Nada, 1970). The CK (a zeatin-like compound) released by *Rhizobium* is apparently sufficient to initiate the cortical cell division necessary to produce root nodules (Phillips and Torrey, 1970).

There are some reports of unusual effects of CKs on growth. High concentrations of kinetin may inhibit growth (Simpkins and Street, 1970; Nudel and Bamberger, 1971). Although the primary effects of CK are on cell division and morphogenesis, it may also markedly increase cell enlargement (Fox, 1969), and this response may even be used as a bioassay for CK (Letham, 1971).

Cytokinins have the capacity to direct the flow of substances within the plant. This may be partly a result of the high ratio of protein nitrogen to soluble nitrogen that they produce (Banerji and Laloraya, 1967). This "antagonism" between kinetin and amino acids has been explained by an inhibition of proteolysis rather than a promotion of protein synthesis (Kuraishi, 1968; Shibaoka and Thimann, 1970; Mizrahi et al., 1970). The consequent low concentration of amino acids could induce translocation of amino acids from regions of higher concentration. However, this flow occurs even in the case of an amino acid not synthesized into a protein. The movement is therefore not simply a "source and sink" relationship (i.e., a movement to a region where the substance is used). CKs also control the flow of assimilates. The nearly expanded leaves of *Vitis vinifera* imported only trace amounts of ^{14}C-labeled assimilates after assimilation of $^{14}CO_2$ by a lower leaf. Benzyladenine (4.4×10^{-3} M) caused a marked increase in this movement (Quinlan and Weaver, 1969). Similarly, kinetin mobilized the starch reserves and increased the flow of sugars in Chinese cabbage (Berridge and Ralph, 1971). CKs are able to inhibit senescence (Chapter 20), perhaps because of this directional flow of usable materials. This effect is demonstrated by its inhibition of the chlorophyll breakdown in higher plants, and also in blue-green algae (Hoffmann and Rathsack, 1970). The maintenance of chlorophyll in darkened leaves may be used as a bioassay for CK (Shibaoka and Thimann, 1970). Kinetin may also induce greening in nongreen tissue (Kaul and Sabharwal, 1971).

In summary, as in the case of the other two groups of growth regulators, CKs have many effects on the plant (Linser, 1966 and Fig. 19-12). They may accelerate or promote (1) cell division and the related DNA and RNA synthesis (see below); (2) cell enlargement in leaves (they may also inhibit it); (3) leaf bud formation (eliminating the related polarity); (4) root formation and growth (inhibition occasionally); (5) germination; (6) breaking of rest; (7) respiration; and (8) translocation of low molecular weight nitrogen compounds. They may also (9) inhibit protein degradation and (10) influence leaf shape and pigments. Kinetin has also been reported to (11) stimulate fruit set (Varga, 1969) and (12) induce tuber formation (Palmer and Smith, 1972).

Mechanism of growth regulation

As in the case of auxins and gibberellins, CKs affect the activity of some enzymes. Kinetin suppressed the increase in ribonuclease level under illumination but maintained the levels of lipase and esterase in detached leaves in which these activities normally decrease (Sodek and Wright, 1969). They also activated thiamine biosynthesis in tobacco callus cultures

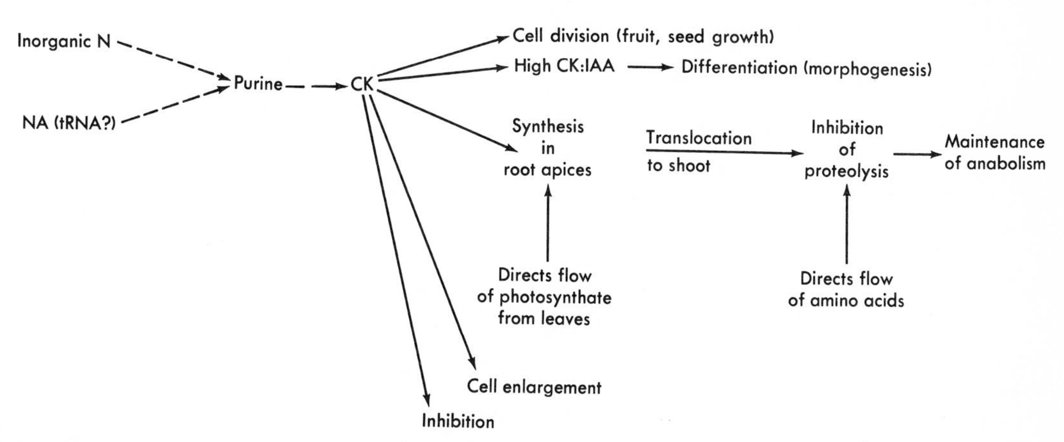

Fig. 19-12. The many effects of cytokinins on growth.

(Dravnieks et al., 1969). Kinetin stimulated protein synthesis in mitochondria isolated from seedlings of *Vigna sinensis* (Bhattacharyya and Roy, 1969). Similarly, the rate of synthesis of photosynthetic enzymes in the primary leaves of dark grown rye seedlings is determined by the level of CK (Feierabend, 1969). At the beginning of germination only the cytoplasmic enzymes are promoted by application of kinetin. In the cotyledons of *Phaseolus vulgaris,* amylase activity is promoted by CK and not by GA (in opposition to the endosperm of cereals), so that CK and not GA controls starch degradation in this seed (Gepstain and Ilan, 1970). On the other hand, 6-benzylaminopurine was found to decrease amylase activity in weakly growing plants of *Stellaria media* (Häggman and Haapala, 1971) and to increase it only at high concentrations. Benzyladenine enhanced nitrate reduction activity in *Agrostemma githago* (Kende et al., 1971).

As mentioned previously, the favorable effect of CKs on the leaf protein level is apparently due to a retardation of protein breakdown rather than to a stimulation of protein synthesis. Nevertheless, there is also evidence of a stimulatory effect.

As might be expected from its chemical relation to nucleic acids, kinetin has been found to promote RNA metabolism (e.g., in moss protonemata, Brandes, 1967). Aminoacyl-transfer-RNA synthetase activity increased in discs of tobacco leaf after treatment for 7 days with kinetin (Anderson and Rowan, 1966). A labeled cytokinin (N,6-benzyladenine), when supplied to tobacco and soybean, was incorporated into several RNA components (Fox, 1966). Similarly, kinetin increased the incorporation of labeled orotate into the RNA of radish leaf discs (Burdett and Wareing, 1966). Kinetin has also been found to inhibit the breakdown of adenine in *Pelargonium* leaves (Schlee et al., 1966). A highly purified soluble RNA from roots and shoots of germinating corn *(Zea mays)* seeds has been found to possess CK activity (Letham and Ralph, 1967), leading to the conclusion that it induces growth by allowing the synthesis of specific transfer RNAs (tRNA). CKs can also inhibit RNA synthesis (Srivastava, 1967); but this is to be expected, since they can also inhibit growth (e.g., of roots, Gaspar and Xhaufflaire, 1967).

Although many attempts have been made to explain cytokinin-induced effects by a change in metabolism of mRNA (Rijven and Parkash, 1970), rRNA (Trewavas, 1970; Nudel and Bamberger, 1971) or tRNA (Rijven and Parkash, 1971), most investigators seem to conclude that the effect is an unspecific stimulation of total RNA (Frankhauser and Erismann, 1961; Schneider et al., 1969). Nevertheless, tRNAs are known to include four compounds that promote cell division and growth and therefore possess CK activity (Hecht et al., 1970; Bezemer-Sybrandy and Veldstra, 1971*a*). In the presence of an unnatural CK (6-benzylaminopurine or BAP), tobacco callus tissue generates these natural CK-active tRNA components (Burrows et al., 1971). The BAP is not itself incorporated into tRNA (Bezemer-Sybrandy and Veldstra, 1971*b*). Kinetin, however, is incorporated, but only into the purine (not the pyrimidine) nucleotides (Onckelen and Verbeek, 1971). On the other hand, despite earlier positive results, kinetin failed to increase DNA synthesis in tobacco pith tissue although IAA did (Simard, 1971), and both the initiation and replication of DNA were controlled by the balance between the two growth regulators (Giles, 1971). Even when CK does induce growth of the nucleus, nucleolus, and cytoplasm in isolated mature (epidermal) cells, its primary effect is on RNA metabolism (Kohlenbach, 1970). It does this apparently by removing a limitation that prevents the synthesis of RNA and genome expression.

Interactions with IAA and GA

The most striking effect of CKs is their interaction with other hormones. Alone, they usually have little effect. One of the earliest discovered characteristics of CKs was their interaction with auxins. The aforementioned ability of tissues to differentiate in culture and to form buds (Fig. 19-10) was found to depend on the presence of a high enough ratio of CK to auxin, and the CKs were active only in the presence of auxins. Similarly, when lateral buds of peas are inhibited by the growing apex (because of the excess auxin transported to the lateral buds from the apex), this inhibition is overcome by applying kinetin directly to the lateral buds. This release from apical dominance must be followed by an application of auxin to the bud apices in order to obtain normal elongation (Sachs and Thimann, 1967). The fact that auxins sometimes induce cell division (although their normal effect is on cell enlargement) and that CKs may accelerate elongation (although their normal effect is on division) may perhaps be a result of this interrelationship. In the presence of auxins, CKs strongly promote the formation of lignin at the expense of polysaccharide synthesis (Koblitz, 1967). Even the oat coleoptile may show a slight increase in elongation due to applied CK, provided that the roots are removed (Jordan and Skoog, 1971). One cause of this interaction is the effect of CK on IAA oxidase activity. The optimal concentration of kinetin or zeatin for producing its activation in tobacco callus cultures is 0.2 μM (Lee, 1971*c*), but only some of the isoenzymes are affected. Concentrations above this actually inhibit IAA oxidase activity. This would explain the importance of the ratio of CK to auxin in controlling growth. Kinetin-inhibited root tips also contain more peroxidase and destroy more IAA. In lentil roots this is due to a lowering of the level of indoleacrylic acid, which is apparently the main auxin in this plant (Darimont et al., 1971). The two may also reinforce each other. In the case of willow, both IAA and kinetin increased the rate of loading of sugars into sieve elements (Lepp and Peel, 1970).

Interactions between GA and CK also occur, and they may be either competitive or cooperative. GA stimulates germination during the first three days and kinetin inhibits after the third day (Verbeek et al., 1969). The kinetin is apparently required during the first 24 hr before the GA, to increase the α-amylase activity in wheat (Eastwood et al., 1969). Other interactions between CK and GA have been shown by several investigators. As in the case of IAA-CK interrelations the effect of GA on the growth of tobacco callus tissue depends on the concentration of CK applied (Helgeson and Upper, 1970). In *Phaseolus vulgaris*, applications of CK and GA stimulated the activity of carboxydismutase, although this appeared to be an activation rather than a synthesis (Treharn et al., 1970). Suitable concentrations of the two permit the induction of nitrate reductase in leaves of tobacco in the dark, eliminating the requirement of light (Lips and Roth-Bejerano, 1969); but the concentration of GA required for optimal induction varied with the concentration of kinetin supplied. Adding increasing concentrations of kinetin reduced the concentration of GA_3 required (Roth-Bejerano and Lips, 1970); but this was true only in summer. In winter, the higher the concentration of kinetin applied, the higher the concentration of GA required for maximum nitrate reductase activity. Kinetin at physiological concentrations causes a significant reduction of the GA_3-promoted growth in excised *Avena* stem segments. Conversely, kinetin negated the blocking effect of GA_3 on cell division in the intercalary meristem. It is therefore considered to be a noncompetitive gibberellin antagonist, at least in this system (Jones and Kaufman, 1971).

All three groups of growth regulators are apparently interrelated. They sometimes act in sequence. CKs and gibberellins appear to dominate the early phase of development from seeds, and auxins dominate the later phases. The three interacted in directing the transport of organic compounds in apple seedlings, provided that root competition was removed (Hatch and Powell, 1971*b*). They also induced each other's translocation.

Gibberellic acid and kinetin are known to inhibit the auxin-induced root initiation. In the case of pea epicotyls, this inhibition occurs only when the two substances are applied during the first 24 to 72 hr (Fellenberg, 1969*a*). All three can bind to the nucleoprotein of the pea epicotyl in vitro at pH 8. Similarly, they all appear to affect the cucumber chromatin; for when intact, etiolated, 2-day embryos were treated with IAA, GA, or CK, the chromatin from the embryonic axis had an increased capacity to support RNA synthesis (Johnson and Purves, 1970). Interactions between the three growth regulators may also control xylem promotion (Torrey et al., 1971). The normal pattern is for each regulator to rise in concentration quickly to a peak, then to decline quickly. When the concentration is kept artificially high for long periods, this normal rise and fall is interfered with and the normal interactions are disturbed. This may be the reason why 2,4-D acts as a weed killer (van Overbeek, 1966). It does not necessarily follow that these three groups include all the growth regulators. Isolated potato buds required both GA and IAA for their growth but were found to cease growth when the stem was 6 to 10 mm long unless 8-hydroxyquinoline was present (Goodwin, 1966).

These interactions are reminiscent of Went's earlier classification of growth hormones into *rhizocalines* and *caulocalines.* The former were supposed to be formed in the root and translocated to the stem where they controlled the growth, and conversely the latter were formed in the stem and translocated to the roots where they controlled the growth. Thus formation of enzymes in the shoot was strongly decreased by root excision early in development, which is known to lower the supply of cytokinins (Feierabend, 1969). The high rate of enzyme formation was restored by feeding kinetin to the rootless seedlings. Not only does CK fail to produce these effects in the root, it may even be deleterious to the initiation and elongation of the main axis of roots (Fox, 1969). On the other hand, a root-promoting substance moves slowly from soybean leaves to the stem after cutting (Usami, 1968). In the case of peas, however, grafting of tall to dwarf varieties indicated that the roots have no direct control over stem growth (Lochard and Grunwald, 1970). CK differs in one respect from Went's rhizocaline: it also regulates the cambial activity of the root itself (Radin and Loomis, 1971), and it is required for growth of excised roots (Robbins and Hervey, 1971).

The interactions between the above three groups of growth regulators are most readily explained on the basis of gene activation and protein synthesis (Fig. 19-13). For instance, it has been suggested that the growth regulators may help to maintain DNA as a functional template for RNA synthesis (Fletcher and Osborne, 1965). Kinetin-treated tissues do, in fact, show a net synthesis of both RNA and protein, which is totally suppressed by the RNA inhibitor thiouracil (Wallgiehn and Parthier, 1964). All three groups of growth regulators have indeed been found to stimulate RNA synthesis in coconut milk nuclei. This stimulation is inhibited by actinomycin D, and the inhibition is removed by higher concentrations of growth regulators (Roychaudhury and Sen, 1965).

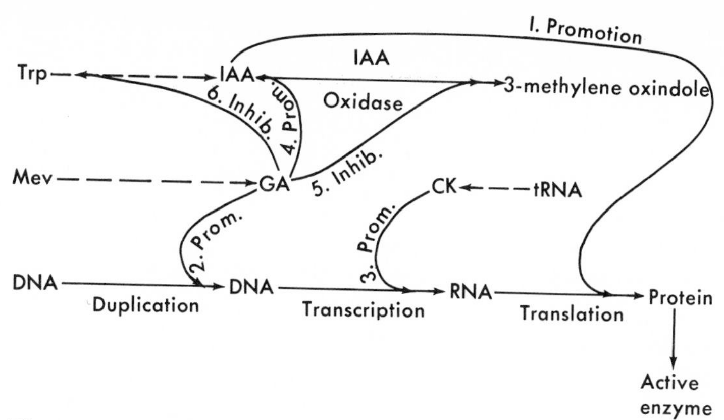

Fig. 19-13. The gene-activation concept and the interactions between IAA, GA, CK, NA, proteins, and enzymes. Trp = tryptophan; Mev = mevalonic acid; Inhib. = growth inhibition; Prom. = growth promotion.

Each of the three growth regulators could conceivably control the concentration of the others by controlling the synthesis of enzymes involved in the synthesis or breakdown of the other growth regulators. In support of this explanation, all three are bound to DNA by unstable H-bonds at pH > 7 (Fellenberg, 1971*a*). IAA and GA were able to loosen the H-bonds of the double helix; CK was not. Nevertheless, an interaction between kinetin and gibberellin affects the activity of the enzyme nitrate reductase without apparently stimulating general protein synthesis (Roth-Bejerano et al., 1970). Another exception is the enhancement of elongation in cucumber hypocotyls by IAA without any effect on DNA synthesis, although GA enhanced both (Degane and Atsmon, 1970).

The gene activation and protein synthesis hypothesis is tempting but is not the only logical explanation of the mechanism of growth control by growth regulators. All three groups may also affect growth by directing the transport of other substances through the phloem, such as ^{32}P- (Hatch and Powell, 1971*a*) and ^{14}C-labeled photosynthate (Mullins, 1970). Furthermore, some of the effects of the growth regulators appear to occur too rapidly to be accounted for by protein synthesis, although this may occur within 90 min of CK application (Usciati and Codaccioni, 1971). Some investigators have therefore sought an explanation in permeability changes. The newer evidence points to this not only in the case of IAA (see previous discussion) but also as an effect of CK. Kinetin stimulates the uptake of K^+ and inhibits Na^+ absorption by leaf discs of *Helianthus annuus* (Ilan, 1971). The K^+:Na^+ ratio was in this way increased by 80% to 100%. Benzyladenine had a similar effect. On the basis of such results, both auxins and cytokinins are believed to regulate K^+ absorption (Ilan et al., 1971). Similarly, experiments with isolated chloroplasts led to the conclusion that the major effect of kinetin on chloroplasts may be related primarily to an effect on hydration and permeability of the chloroplasts and their membranes and not directly on their machinery for protein synthesis (Richmond et al., 1971). Unfortunately, there is no suggestion as to how such changes can control growth or what their relation is to the apparently slower NA and protein changes. A relation between the two may be involved in the suggestion of an enhanced synthesis of membrane components in the presence of CK (Schaeffer and Sharp, 1971).

ETHYLENE

Although its ability to induce ripening, epinasty, and tropisms has long been known (Chapters 14 and 18), it is only in recent years that ethylene (abbreviation: ETH) has been recognized as a growth regulator (Galston and Davies, 1970). It is the only one that is a gas throughout the normal temperature and pressure range of the plant. Due to its gaseous state, it can readily diffuse to cells other than those in which it is produced. If it produces its growth effect on the cells to which it diffuses, it must be considered a growth hormone. It is difficult, however, to think of ethylene as producing a localized or asymmetrical effect, since it might be expected to diffuse uniformly and rapidly in all directions. Ethylene has by far the simplest chemical structure of all growth regulators ($CH_2 = CH_2$), and it is apparently synthesized from simple amino acids—from methionine in apple, tomato, and cauliflower (Mapson et al., 1970), or its aldehyde methional (Ku et al., 1970), from β-alanine in bean cotyledons (Stinson and Spencer, 1969) or from isoamyl alcohol in conifer tissues (Lona and Raffi, 1970).

The most characteristic growth effect of ethylene is an inhibition of cell elongation in both roots and shoots and an enhancement of lateral cell growth. Therefore it induces shorter and wider cells. Nevertheless, it may result in either a clear-cut inhibition or a stimulation of growth. Low concentrations (2 ppm) of ethylene can completely suppress the growth of buds on decapitated stems (Letham, 1969). Cucumber plants at the eighth, tenth, or twelfth leaf stage were severely checked by an ethylene-releasing substance called *Ethrel* (2-chloroethyl-phosphonic acid) at a concentration of 120 ppm (Lower et al., 1970).

Ethylene may either stimulate or inhibit the growth of fig fruit, depending on whether it is applied during cell enlargement or cell division, respectively (Marei and Romani, 1971; Marei and Crane, 1971). The growth rate of rice coleoptiles is increased by low concentrations (0 to 100 ppm) of ethylene, especially at oxygen concentrations lower than in air (Ku et al., 1970). This effect is specific to ethylene and is not induced by ethane, propane, methane, propylene, or butane. Ethrel also increased the number of roots on mung bean cuttings but decreased their length (Krishnamoorthy, 1970). Both of these effects of ethylene on root growth may be expected to occur in nature. On the one hand, the extension of the root axes of barley was severely inhibited by concentrations of ethylene considerably lower than those occurring in anaerobic soil (Smith and Robertson, 1971). On the other hand, submerging softwood cuttings of willow in water stimulated root formation, due presumably to the increased ethylene production (Kawase, 1971). Growth of pea pollen is stimulated by ethylene (Search and Stanley, 1970), although a high concentration (1000 ppm) was required to obtain this effect. Aged seed of rape produced less ethylene than fresh seed during early stages of germination, and the application of exogenous ethylene accelerated their rate of germination although not significantly increasing their percent of germination (Takayanagi and Harrington, 1971). Ethylene was able partially to replace the hypocotyl hook by influencing greening in excised cotyledons, although no other growth regulator was able to do this (Hardy et al., 1971). In the case of sycamore cells in suspension culture, ethylene had no effect (Mackenzie and Street, 1970). The same was true of suspension cultures from many other plants (LaRue and Gamborg, 1971).

There are so many similarities between the effects of auxin and ethylene that the effects of one have frequently been ascribed to the other. This is because the interactions between the two are so marked that it is difficult to separate the effects of auxin from those of ethylene (Fig. 19-14). Growth promotion by ethylene, for instance, did not occur in the absence of exogenous auxin (Imaseki and Pjon, 1970) or in the absence of the synthesis of endogenous auxin (Kamienska and Chrominski, 1971). On the other hand, at least some of the auxin-induced growth effects are actually due to ethylene. Ethylene production is stimulated by applied auxin, and when ethylene is applied alone, it may produce the same effect as the auxin (Letham, 1969). The ethylene-producing system apparently has a rapid turnover rate, and therefore a constant supply of auxin is necessary to maintain the ethylene levels (Sakai and Imaseki, 1971). When applied auxin *inhibits* growth, this may be due to the induced ethylene formation. Thus a large portion of the IAA-induced inhibition of excised pea root tips and virtually all such inhibitions of intact roots are the result of IAA-dependent ethylene production (Chadwick and Burg, 1970). The ethylene production in response to applied IAA is governed by the level of auxin already present in the root.

The similarities between the effects of auxin and ethylene, however, are never complete. Ethylene, for instance, produced the same effect on pea root growth as auxin, except that lateral root development was normal as opposed to the proliferation of massed lateral roots obtained with auxin (Scott and Norris, 1970). In the case of mung bean cuttings, IAA inhibited both the number and length of the roots, yet in combination with ethylene-producing Ethrel, both the number and length of the roots were promoted synergistically (Krishnamoorthy, 1970). Similarly, although ethylene mimics auxin by causing swelling of stems of intact seedlings, in contrast to the effect of auxin no swelling occurs when decapitated pea seedlings are exposed to ethylene, and ethylene never increases the activity of cellulase or cellobiase (Ridge and Osborne, 1969). It was therefore concluded that the effects of auxin on cell growth and enzyme activities cannot be attributed solely to regulation by ethylene. In opposition to

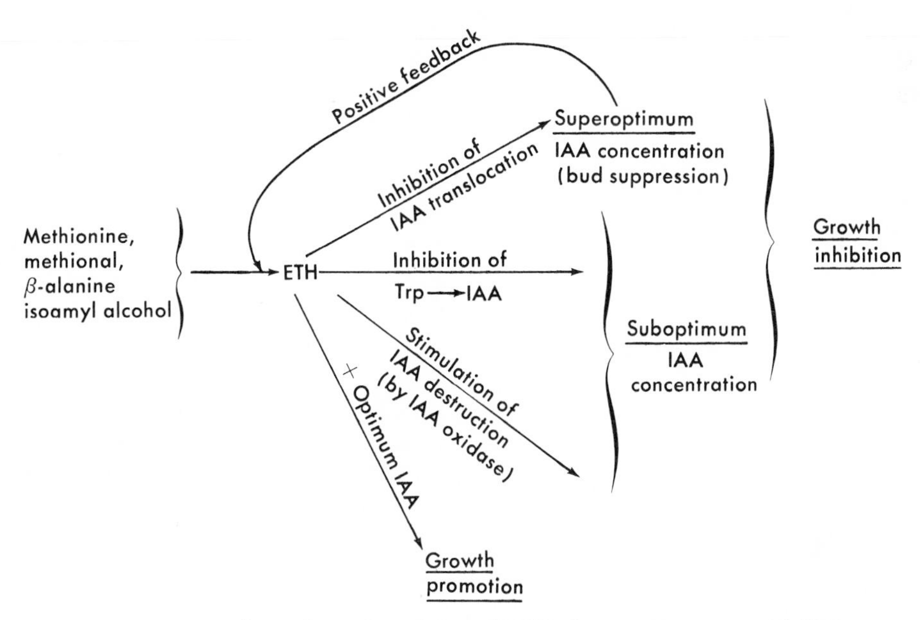

Fig. 19-14. Growth regulation by ethylene (ETH) due to interactions with IAA.

whole pea seedlings, cellulase activity in the separation zones of citrus fruit was associated with the ethylene concentration in their internal atmosphere (Rasmussen and Jones, 1971). There is further evidence, however, against the concept that auxin effects are due to the ethylene produced by the auxin.

In segments of etiolated pea shoots, the rate of ethylene production saturates at a higher concentration of auxin than the rate of growth (Kang et al., 1971). This seems to prove that ethylene can account for the growth inhibition by high auxin concentrations but not the growth stimulation by low concentrations. In further agreement with this conclusion, ethylene cannot replace auxin in tissue cultures (LaRue and Gamborg, 1971).

The similarities that do exist between auxin and ethylene effects on growth may be partly due to an interference by ethylene in auxin transport (Morgan and Gausman, 1966); such an interference leads to production of tropisms by ethylene (Burg and Burg, 1966, 1967). Cotton stem sections excised from plants exposed to ethylene for 3 hr show a reduced capacity to transport auxin basipetally (Beyer and Morgan, 1969). Although ethylene also increased the release of carbon dioxide from IAA, this accelerated metabolism of auxin cannot explain the disrupted transport but is the effect rather than the cause (Beyer and Morgan, 1970). In agreement with this conclusion, IAA transport was inhibited 50% by ethylene in the absence of decarboxylation of IAA (Ernest and Valdovinos, 1971). The rate of formation of IAA from tryptophan in the presence of enzyme was, however, decreased 50%, and the conjugation of IAA as indoleaspartic acid (a nontranslocated auxin) was increased 100%. It has been suggested that ethylene also increases the sensitivity of tissue to auxin, but this was not observed in the auxin-induced growth of excised rice coleoptile segments (Imaseki and Pjon, 1970).

Ethylene did not appear to mediate the auxin effects on nucleic acid synthesis in soybean hypocotyl with the possible exception of inhibition in the apical tissue (Holm et al., 1970). In agreement with these results, ethylene had either no effect or an inhibitory effect on the synthesis of RNA and protein in sections of avocado and banana fruit (Sacher and Salminen, 1969). Nevertheless, there is evidence that ethylene may control protein synthesis. It may induce de novo synthesis of peroxidase, increasing the content nearly 100 times in sweet potato slices after 84 hr in air containing 1 μl ethylene per liter air (Shannon et al., 1971). At 0.1 ppm it increased both peroxidase activity and hydroxyproline levels in the cell walls of pea apices, and these effects suggest the implication of ethylene in the regulation of cell wall growth (Ridge and Osborne, 1970*a*, *b*). In further evidence of this conclusion, supraoptimal concentrations of IAA as well as ethylene cause newly formed microfibrils to be deposited at the inner wall surface in a predominantly longitudinal direction (Apelbaum and Burg, 1971). Ethylene also promoted the synthesis of β-1,3-glucanase in *Phaseolus vulgaris* (Abeles and Forrence, 1970). In ripening apples it may stimulate both RNA and protein synthesis (Hulme et al., 1971). In the case of peroxidase, however, ethylene regulation apparently occurred at the translation level rather than the transcription level (Ridge and Osborne, 1970*a*). It has been suggested by Ridge and Osborne (1971) that ethylene increases cytoplasmic hydroxylation of proline, leading to enrichment in a specific, hydroxyproline-rich, wall peroxidase. Another enzyme, phenylalanine ammonia lyase (PAL), is also enhanced by ethylene in pea seedlings (Hyodo and Yang, 1971) and in swedes and turnips (Rhodes and Wooltorton, 1971). The induction period is about 6 hr and the peak in activity is reached at 30 hr, then declines. Both cases apparently involved de novo synthesis. In gherkin hypocotyl tissue, ethylene played a role in PAL synthesis only in combination with the formation or disappearance of another factor (Englesma and van Bruggen, 1971).

An effect of ethylene on cell permeability has been suggested, since it induces seismonasty (Jaffe, 1970). Other evidence, however, indicates that its effect is not on cell permeability (Sacher and Salminen, 1969; Mehard and Lyons, 1970; Mehard et al., 1970). On the basis of all the evidence, both the formation and the growth effects of ethylene appear to be closely related to auxins but not to gibberellins (Sankhla and Shurla, 1970) or cytokinins. In leaf sections of wheat, however, ethylene production was stimulated not only by IAA but also by GA_3 and kinetin (Loverys and Wareing, 1971*a*, *b*), yet ethylene played no important role in the stimulation of leaf unrolling by GA_3 and kinetin. Furthermore, ethylene (presumably produced from Ethrel) and GA have essentially opposite effects on the growth of pea plants, the ethylene reducing growth and the GA increasing shoot growth (Andersen, 1971). The Ethrel can also nullify the effects of a previous GA treatment during the first week. It will be seen below that other growth regulators may also have pronounced effects on ethylene formation in the plant.

GROWTH INHIBITORS

Antagonists of growth promoters

Auxins, gibberellins, and cytokinins most commonly stimulate or promote growth. As seen previously, however, they may also inhibit growth. This usually occurs at high ("unphysiological") concentrations not commonly found in the plant; and in the case of auxin this effect is apparently due to its stimulation of ethylene synthesis. Therefore in the normal plant these growth regulators may be considered growth stimulators or promoters. There are also naturally occurring growth regulators that inhibit growth at concentrations normally found in the plant. These may be called *growth inhibitors,* even though they are sometimes found to stimulate growth. In many cases these growth inhibitors appear to act by opposing one or another of the three groups of growth promoters and may therefore be called antagonists of growth promoters. If the antagonists are competitive, they are called antipromoters. *Antiauxins,* for instance, counteract auxins (competitively), and *antigibberellins* counteract gibberellins. No natural growth regulator has been called an *anticytokinin,* although synthetic ones do exist. Although antagonists of growth promoters normally inhibit growth, they may also promote growth. For instance, if the auxin is originally present at a level capable of causing growth inhibition, a competitive antiauxin can counteract part of the auxin and can reduce its effective concentration to the level of growth stimulation. In this case the antiauxin indirectly stimulates growth. The same result may be brought about by an auxin-destroying enzyme, indoleacetic oxidase (IAA oxidase), as by a competitive inhibitor.

The kind as well as the quantity of growth may be affected. Differences in IAA oxidase activity between varieties give rise to differences in IAA content and therefore to differences in growth habit (i.e., in tillering ability—Hanada et al., 1969). In the growing plant it is believed that IAA is constantly being destroyed by IAA oxidase. Diphenols such as caffeic acid have growth-promoting activities because they inhibit this oxidation (i.e., destruction) of IAA. They may therefore be thought of as antagonists of auxin antagonists. Monophenols, on the other hand, activate IAA oxidation and are therefore auxin antagonists. At the more common physiological concentrations of auxin they will inhibit growth.

Coumarin is a growth regulator chemically related to the above phenols. It is a

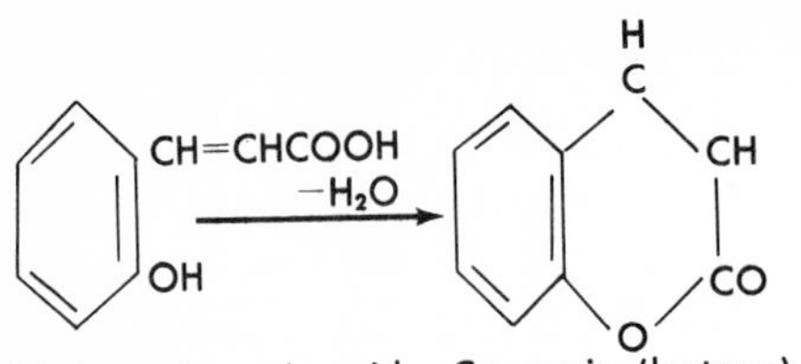

CH=CHCOOH —H_2O→ OH; H C CH CO O

o-Hydroxycinnamic acid Coumarin (lactone)

Fig. 19-15. Relation of the lactone coumarin to the phenol o-hydroxycinnamic acid.

lactone that can be formed by dehydration of the phenol o-hydroxycinnamic acid (Fig. 19-15). Coumarin derivatives inhibit the germination of light-sensitive seed such as lettuce. The seed do not germinate until the coumarin derivative is destroyed in the light. Only those derivatives that are more reduced than coumarin show this inhibitory activity (Berrie et al., 1968). *Seselin* is a natural coumarin derivative isolated from citrus roots that inhibits the radicle growth of several plants in the dark. This growth retardation is accompanied by increased activity of peroxidase and IAA oxidase (Goren and Tomer, 1971). It would therefore seem reasonable to classify coumarin or its derivatives with the phenols as anti-auxins. More recently, however, they have been found to act as antigibberellins, presumably competing with them for the same active site due to a chemical similarity to part of the gibberellin molecule (Berrie et al., 1968). In agreement with this conclusion, coumarin reversed the GA-induced stimulation of germination in light-requiring seeds of *Verbena bipinnitifida* (Vyas and Garg, 1971) and in lettuce seed (Harada and Koizumi, 1971). There are further interactions. The inhibitory effect of coumarin on the germination of lettuce seed in the dark can be reversed by a CK as well as by GA, and even more effectively by a combination of the two (Bryant and Skinner, 1968). Coumarin, and one of its derivatives *scopolin,* may also result in dwarfing of Engelmann spruce *(Picea glauca),* since high-altitude dwarf forms have higher concentrations than low-altitude normal forms (Love et al., 1970). It appears to inhibit longitudinally directed processes (transverse cell divisions and cell elongation) and to stimulate radially directed processes (longitudinal divisions and lateral cell enlargement—Svensson, 1971). Coumarin also accelerates protein and chlorophyll loss in the light (but not in the dark) at concentrations too low to affect growth (Knypl, 1971), and this effect is not prevented by either GA or CK (Knipl and Kulaeva, 1970). Coumarin (0.1 to 1 mM) also inhibits the light-induced opening of the hypocotyl hook of etiolated beans. It does this by stimulating ethylene synthesis, the ethylene mediating the inhibitory actions of coumarin (Morgan and Powell, 1970). On the other hand, in the same concentration range it stimulated the growth of the straight portion of the hypocotyl, and this action was not due to the ethylene. The growth-regulating effect of coumarin and its derivatives obviously involves interactions with all the above growth regulators.

Dormins and abscisic acid

The most important growth inhibitors are those which, like the growth stimulators, are hormones. They are therefore involved in correlative inhibition (Arney and Mitchell, 1969). Wareing showed that one such substance is produced in leaves and exerts its inhibitory effect in the apical meristems. He called such growth-inhibiting hormones *dormins,* since they occurred in the buds of dormant woody plants in winter. In the presence of dormins, vegetative buds change to dormant winter buds by converting the developing leaf primordia into bud scales. One of these dormins is particularly potent, and its inhibitory effect is overcome by gibberellin. It is identical with an independently discovered substance, abscisin II (now called *abscisic acid*—Fig. 19-16), which stimulates abscission (e.g., of cotton).

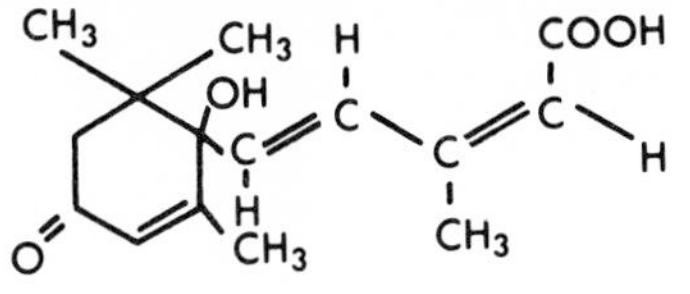

Abscisic acid

Fig. 19-16. Dormin or abscisin II or abscisic acid. (From van Overbeek, J. 1966. Science **152**:721-731. Copyright 1966 by the American Association for the Advance of Science.)

Abscisic acid (or ABA) is a sesquiterpene of exceptionally high optical activity that was characterized and synthesized in 1965 (Milborrow, 1969). The synthetic and the natural ABAs are not identical; they have opposite optical activities and their boiling points are different (Wareing and Ryback, 1970). Yet they are equally inhibitory. ABA has been isolated in crystalline form from five species of angiosperms and has been detected in more than thirty other species of plants. It usually occurs at a concentration of 0.01 to 1 ppm up to 4 ppm. It has even been detected in roots of peas (Tietz, 1971). It is light-sensitive and is slowly transformed into trans-ABA by the ultraviolet radiations of sunlight. It is transported in cotyledonary petioles of cotton seedlings at a rate of 22.4 mm/hr, independently of the length of the petiole or the concentration of ABA applied (Ingersoll and Smith, 1970, 1971). This rate is much greater than IAA velocities in a similar system (6 to 7 mm/hr), and is also much more rapid than the rate of diffusion. Abscisic acid showed no polarity of transport, and inhibition of metabolism slowed down the transport 80% to 98%.

Because of its sesquiterpene (and therefore isoprenoid) structure, ABA is related to the half carotene vitamin A (Fig. 19-16). Light appears to stimulate its formation from the products of photooxidation of the carotenoid xanthophyll (Taylor and Smith, 1967; Taylor and Burden, 1970*a*). This relation as well as its relation to gibberellins is indicated by the biosynthetic pathway proposed for all three substances (Graebe, 1967, and personal communication) (Fig. 19-17).

Besides inducing bud dormancy, ABA is responsible for the dormancy of some seed, for instance, of apple and several species of Gramineae (Oritani and Oritani, 1971). It decreases during stratification of the seed until the end of 3 weeks, when none is detectable (Rudnicki, 1969). It also retards the growth of tissue cultures (Letham, 1969). In the case of the small aquatic plant *Spirodela oligorrhiza,* ABA completely arrests the growth of the whole plant at concentrations down to 10^{-1} mg/l, although it stimulates growth at 10^{-8} mg/l (van Staden

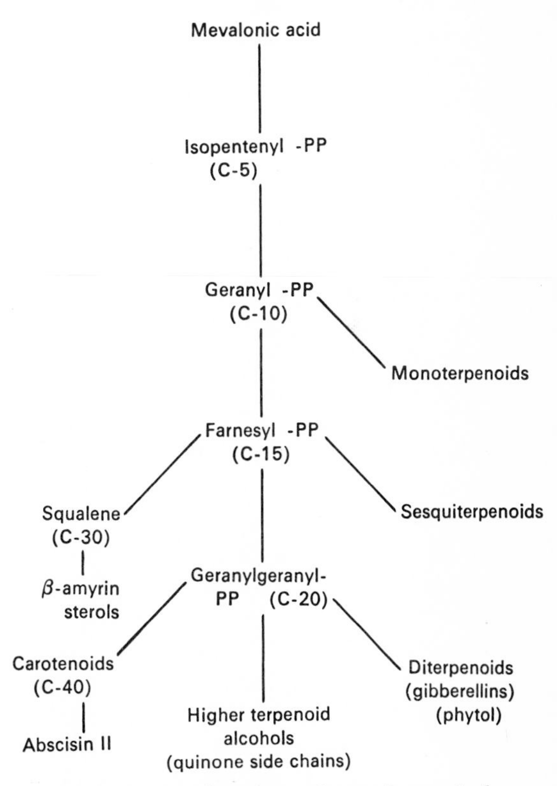

Fig. 19-17. Proposed pathway for synthesis of gibberellins and abscisins (abscisic acid). (From Graebe, J. E. 1967. Science **157**:73-75, and personal communication.)

and Bornman, 1969). In the case of wheat seedlings, the inhibition of the coleoptile and first leaves is greater than in the roots (Belhanafi and Collet, 1970; Isaia, 1971). In fact, ABA may stimulate the rooting of cuttings and may overcome the inhibitory effect of GA on root formation (Chin et al., 1969). This effect may conceivably occur in nature, since ABA has been found in roots (Tietz, 1971).

Since ABA inhibits the outgrowth of lateral buds when applied to decapitated pea plants, it has been suggested that it may be responsible for the natural correlative inhibition of the lateral buds by the apical bud, which was formerly ascribed to excess auxin (Arney and Mitchell, 1969). On the other hand, it has been found to stimulate the formation of callus in bud cultures (Altman and Goren, 1971) and to inhibit the production of ethylene by cell suspension cultures (Gamborg and LaRue, 1971).

One of the most striking effects of ABA is its inhibition of GA-induced formation of α-amylase in the aleurone layer of barley (Chrispeels and Varner, 1967). Therefore it acts as a GA antagonist or an antigibberellin. Nevertheless, it requires a combination of kinetin and GA for nearly complete reversal of this ABA-inhibition of α-amylase synthesis. Furthermore, these two growth promoters fail to reverse the inhibition of growth by ABA (Khan and Downing, 1968). The antigibberellin effect of ABA is, in fact, far from a general one. ABA does not act as an antigibberellin in relation to potato tuberization (Claver, 1970). Its inhibition of the first wheat leaf is, in fact, enhanced by GA (Isaia, 1971). On the other hand, it can suppress the growth-promoting activities of both auxins and cytokinins (Letham, 1969). ABA inhibits the basipetal transport of IAA in seedling epicotyls (Pilet, 1971). Its inhibitory effect on wheat coleoptiles can, in fact, be overcome to a greater degree by kinetin than by GA, and neither of these growth promoters has any effect on the ABA-induced inhibition of root growth (Belhanafi and Collet, 1970). According to Khan and Downing (1968), cytokinins may possibly remove the ABA inhibition of enzyme-specific sites, thereby allowing GA to produce α-amylase.

The complexity of the ABA effects is indicated by its promotion of fruit set in *Rosa* (Jackson and Blundell, 1966) and by its depression of light-induced formation of vitamin K, chlorophylls, and carotenoids, but not the synthesis of these isoprenoids in the dark (Lichtenthaler and Becker, 1970). This was interpreted as an interference with induction of thylakoid formation. In the case of two strains of lettuce, the inhibitory effect of ABA on seed germination and seedling growth was completely overcome by kinetin in both the light and the dark (Sankhla and Sankhla, 1968, see Chapter 20). Neither GA nor IAA was effective, but GA affected the subsequent seedling growth as usual.

Although ABA is the only dormin whose chemistry and growth effects have been so thoroughly investigated, other dormins do exist. Xanthoxin is similar chemically to ABA and has been isolated from dwarf beans and wheat (Taylor and Burden, 1970*b*). Phaseic acid is another relative of ABA from *Phaseolus multiflorus* (MacMillan and Pryce, 1969*a*, *b*). Lunularic acid resembles ABA chemically and probably replaces ABA as the normal growth inhibitor in liverworts (Pryce, 1971). Naringenin is a flavone that inhibits α-amylase and catalase synthesis (Correa, 1969).

One of the growth inhibitors found earlier was called *inhibitor* β by Bennet-Clark and Kefford in 1953 (see Holst, 1971). It was identical to an acid inhibitor extracted from potato peel by Hemburg (1952). The occurrence of inhibitor β in plants is widespread. Several investigators have shown it to be a mixture of substances, the most active component being ABA (Milborrow, 1969). According to Holst (1971), it consists of three components: (1) a phenolic, weakly inhibiting substance, (2) a substance that is probably ABA, and (3) a substance that is equally inhibitory. Inhibitor β produced total growth inhibition in oat coleoptiles, an effect unobtainable by ABA alone, even in unphysiological concentrations.

Inhibitor-promoter concept of dormancy

As was mentioned in Chapter 17, there are two major theories that attempt to explain dormancy. According to the more recent one, the *inhibitor-promoter concept* (Cornforth et. al., 1966), dormancy occurs when the ratio of inhibitor to promoter is high (the higher the ratio, the deeper the dormancy). Release from dormancy (growth) occurs when the ratio is low. We are now in a position to examine the evidence for and against this theory.

The existence of large amounts of growth inhibitors (inhibitor β) in growing stem tissues of aspen, in the presence of high concentrations of a growth promoter (probably IAA—Eliasson, 1969) shows that the growth or dormancy of the plant depends not on the absolute amounts of inhibitors or promoters but on the amounts relative to each other. Nevertheless, growth promoters tend to be present in high concentrations in growing tissues and in low concentrations in dormant tissue. IAA, for instance, was found during the whole vegetative period in buds and sprouting shoots of *Pinus sylvestris* but not during winter dormancy (Alden, 1971).

On the other hand, the breaking of dormancy has frequently been related to a rise in concentration of gibberellin without any evidence of an inhibitor. In the case of lettuce seed, not enough gibberellin is produced in the dark for normal germination; on illumination with near red there is a sharp increase in gibberellin within an hour (Köhler, 1966). GA may also overcome the dormancy of nonlight-requiring seed. In dormant apple seed the level of GA_4 rose during the fourth week of stratification to 10,000-fold higher than in the dormant seeds, then decreased to the initial level between the fiftieth and sixtieth days of stratification (Sinska and Lewak, 1970). Stratification of seeds of *Ginkgo biloba* leads to an increase in GA_3 to 100 times that of nonstratified dormant seed, and exogenously applied GA_3 largely substitutes for stratification (West et al., 1970). GA also broke the dormancy of freshly harvested barley seed (Khan et al., 1970). However, not all plants respond to GA (Vegis, 1964*a*, see Chapter 17). In some seeds (e.g., oats, Barralis, 1965) a double dormancy mechanism appears to exist—the caryopsis dormancy and that caused by the hulls. GA can break the dormancy of some seeds and of excised embryos, but it has no effect on intact seeds containing pericarp and endosperm inhibitors (e.g., *Fraxinus excelsior,* Szalai, 1965*b*) and therefore seems to be inactive on the second mechanism. Furthermore, if short of oxygen, such seeds may enter a second dormancy. Seed dormancy may also be overcome by CK, and Khan (1971) lists eight different combinations of these three growth regulators that may control dormancy. Thus a seed will remain dormant (1) if GA is absent regardless of whether CK or inhibitor is present or (2) in the presence of GA when inhibitor is present but CK is lacking. Germination will occur (3) in the presence of GA and absence of inhibitor regardless of whether CK is present, or (4) in the presence of inhibitor and GA and CK. However, in the case of dormant buds of *Syringa vulgaris,* GA_3 and kinetin were unable to replace the chilling that breaks the bud dormancy under natural conditions. They promoted bud growth only during predormancy and postdormancy (Leike and Lau, 1967).

Apparently contradictory results have also been reported for tuber dormancy, which is inversely related to GA content, both in potatoes and in the Jerusalem artichoke (Bradshaw and Edelman, 1971). As the nondormant new tubers enter their rest period, the gibberellin level drops (Szalai 1965*a*), and during sprouting, it increases to 3 times the dormant quantity. Nevertheless, when artifically stimulated to sprout with Rindite, the gibberellin level drops. Perhaps the Rindite destroys ABA. This may also indicate that dormins were possibly involved in the other cases in which gibberellins overcame dormancy.

Therefore in some cases the dormancy seems to depend on the quantity of promoter in the absence of inhibitor. In the case of correlative inhibition, for instance, the balance may be between different promoters rather than between promoters and inhibitors, and even simple nutrition may be involved. Thus the two most popular theories of the correlative inhibition of axillary buds by apical buds (so-called apical dominance) are (1) the nutrition theory (Phillips, 1969), which ascribes it to the competition for nutrients translocated mainly to the growing apical bud (the metabolic sink), and (2) the direct theory of Thimann and Skoog, which ascribes the inhibition to auxin diffusing from the apical bud to the lateral buds, attaining a high enough concentration there to inhibit them, due to their greater sensitivity to auxin.

More recent evidence has established the existence of other factors, such as the ability of CK to release the axillary buds from correlative inhibition. This CK-induced growth, however, soon comes to a stop, and it appears that both CK and auxin have to be present in sufficient quantity before the bud can grow normally. GA, on the other hand, can often lead to enhanced apical dominance, possibly due to increased auxin synthesis.

Two stages in this interaction between IAA and CK have been suggested in the case of the plantlets that arise on the margin of Bryophyllum leaves (Viana and Novais, 1969-70). (1) Due to an excess of CK over IAA, the embryos are freed from the first (IAA) inhibition. (2) The growth may still be inhibited by inhibitor β unless this, too, is overcome by CK.

As in the case of the other growth regulators, the dormins appear to control nucleic acid metabolism, Cultures of *Lemna minor* become nearly completely dormant when ABA is added (3.8×10^{-6} M). In this case only the cytokinin benzyladenine (not auxin or gibberellin) was able to counteract the effect of ABA (van Overbeek et al., 1967); and even this was possible only if the ABA concentration did not exceed a critical level. The ABA suppressed nucleic acid synthesis. Inhibition of DNA synthesis seemed to precede that of RNA. Cytokinins reversed this process. It is therefore suggested that ABA is an inhibitor of DNA synthesis and that cytokinin is an activator of the process.

Similar NA effects have been reported for the GA-ABA interactions. Ribosomal preparations from GA-treated segments of barley leaf had a greater percentage of polysomes and a greater capacity for amino acid incorporation into proteins than similar preparations from ABA-treated segments (Poulson and Beevers, 1970). Similarly, the breaking of the rest of barley seeds by GA stimulated RNA synthesis in the mitochondria and the ribosome fraction (Rejowski and Kulka, 1970). In conformity with its reported antigibberellin nature, ABA essentially blocked the GA-promoted synthesis of DNA and RNA in potato tubers, as well as their synthesis in the absence of added gibberellic acid (Shih and Rappaport, 1970).

Other results oppose an effect of ABA on DNA synthesis (Walton et al., 1970), and suggest an inhibition of RNA synthesis or transcription (Villiers, 1968; Wareing and Ryback, 1970). A marked decrease in RNA has, indeed, been established (Poulson and Beevers, 1970). This has been explained by an increase in ribonuclease activity leading to an increased breakdown of RNA rather than a decreased synthesis (Belhanafi and Collet, 1970; Pilet, 1970; Leshem, 1971). Its stimulation of abscission, however, is due to an increase in cellulase activity and ethylene production (Abeles and Craker, 1969) apparently requiring an increase in at least some nucleic acids. As in the case of the growth promoters, ABA has also been shown to raise the permeability to water markedly, at least in carrot storage tissue (Glinka and Reinhold, 1971). It has also been found to produce a localized effect on the cell wall, which was irregularly thickened with protuberances formed by the plasmalemma (Roland and Pilet, 1971).

Synthetic growth inhibitors

A number of synthetic substances have been produced that shorten the internodes of plants. They are called *growth retardants.* Amo 1618, chlorocholine chloride (CCC), B995, and Phosfon D (Fig. 19-18) are the best known; they also inhibit α-amylase production in germinating barley seed. The evidence indicates that they do this by inhibiting the biosynthesis of gibberellins (Paleg et al., 1965; Khan and Faust, 1967). This inhibition is almost complete in the fungus *Gibberella fujikuroi* (Cross and Myers, 1969), although there is no effect on the biosynthesis of the metabolically related sterols (Barnes et al., 1969). The chemical similarity be-

Chlorocholine chloride or CCC

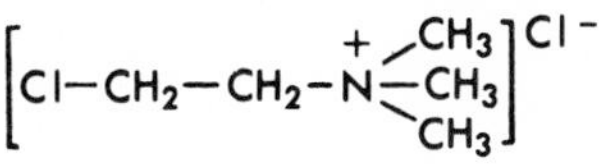

Amo 1618

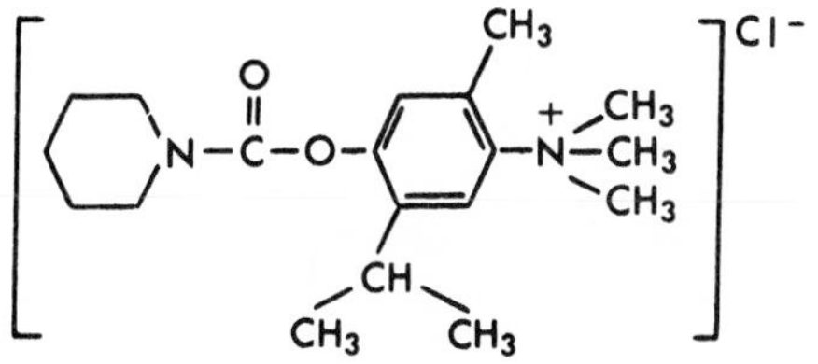

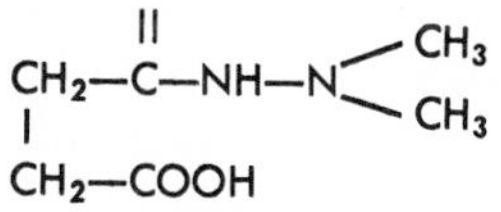

Phosfon

B995

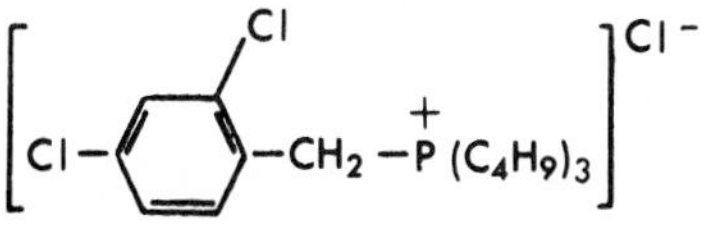

Morphactins

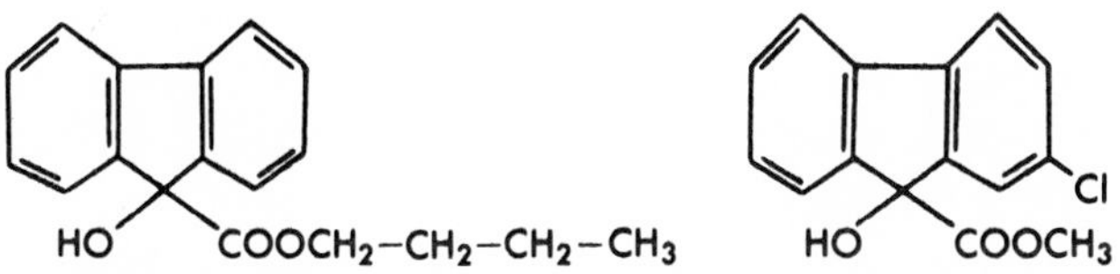

IT 3233

IT 3456

Fig. 19-18. Synthetic growth retardants. (From Jung, J. 1967. Naturwissenschaften **54**:356-360, and Khan, A. A. 1967. Physiol. Plant. **20**:306-313.)

tween these synthetic growth retardants and some natural metabolites led to the suggestion that choline, hordenine, and other basic substances may regulate the endogenous synthesis of gibberellins in germinating barley. Other gibberellin antagonists may act competitively rather than by controlling biosynthesis (Ziegler et al., 1966). Even roots (of grape) become shorter and thicker in the presence of CCC and this is counteracted by GA (Skene and Mullins, 1967). However, the retarding effect of CCC may also be overcome by kinetin (Knypl, 1967) so that classifying it as a gibberellin antagonist may be an oversimplification. Similarly, some investigators have concluded that these growth retardants are antagonists of GA but not true antigibberellins, since they can stimulate growth at low doses (Chailakyan and Kochankov, 1967). Furthermore, despite the inhibition of GA biosynthesis in the fungus, CCC apparently does not have this effect in tomato plants (van Bragt, 1969). It actually increases the GA-content of pea seedlings despite its inhibitory effect (Reid and Grozier, 1970), and of Gladiolus, which it stimulates (Halevy and Shijo, 1970). On the other hand, growth retardants definitely disrupted GA biosynthesis in root tips of peas (Wylie et al., 1970).

CCC also lowers the level of auxin in *Begonia* leaves and may actually stimulate bud formation and growth at certain concentrations (Heide, 1969*a*). Both CCC and Amo 1618 appear to exert their effects on the growth of barley by acting on auxin catabolism (Gaspar and Lacoppe, 1968). Amo 1618 seems to have a similar auxin effect, since it enhances a peroxidase that is unaffected by GA (Gaspar et al., 1971). In agreement with this conclusion, it fails to inhibit red-light–stimulated GA production in etiolated leaves (Loverys and Wareing, 1971*b*). CCC, in fact, may act either as an auxin antagonist or an auxin synergist (Mishra and Paul, 1967). It has also been reported to act synergistically with ascorbic acid (Chinoy et al., 1968). The growth retardant β-9(N-dimethylamino succinamic acid) actually reversed the growth inhibition of cowpea produced by cytokinin and IAA (Mishra and Mohanty, 1968). By retarding vegetative growth these substances often promote flowering and fruiting. Furthermore, even though their general effect is to retard growth, they may result in *increased* yields. In the case of potato plants, B995 decreased stem weight but increased tuber weight and therefore yield (Bodlaender and Algera, 1966).

Maleic hydrazide is another growth inhibitor whose effect may be antagonized by gibberellic acid (Saha, 1966). Another group of synthetic growth inhibitors is called *morphactins* (Fig. 19-19—see Ziegler, 1970). They induce internode shortening and loss of dormancy of lateral buds in citrus seedlings (Mann et al., 1966); but they neither replace nor antagonize gibberellins. Seedlings grown in their presence lose their capacity to respond to gravity or to unilateral light (Khan, 1967*c*). IAA and GA did not modify their effect on geotropism. Nevertheless, their most characteristic effect appears to be as auxin antagonists, by reducing the capacity of the plant to transport IAA (Krelle and Libbert, 1968; Linke and Marinos, 1970). The morphactin-induced changes in geotropic and phototropic responses are explained by an inhibition of basipetal and promotion of acropetal and lateral transport of IAA, equalizing the levels in horizontally placed tissues (Parups, 1970). On the other hand, they may even enhance growth. Morphactins act synergistically with the synthetic auxin 2,4-D in stimulating cotyledon expansion (Sankhla, 1970). Similarly, p-fluorophenylalanine enhances elongation of the *Avena* coleoptile, apparently by depressed deamination of phenylalanine and consequent lowering of potentially inhibitory low molecular weight phenolic constituents such as chlorogenic acid (Hopkins and Orkwiszewski, 1971). This effect may again be due to interactions with auxins, since phenols are known to interact with them.

As in the case of other growth inhibitors, however, the effects of the morphactins cannot all be explained by their antagonism toward a single group of growth promoters. They have been reported to act partially as gibberellin antagonists in leaves cultured in vitro (Besemer et al., 1969), although not in whole plants or rootings of cuttings (Krelle, 1970). Two other morphactins converted CK-requiring tobacco tissue cultures into CK-independent, tumor-like tissues (Bednar and Linsmaer-Bednar, 1971). Apparently a directed and heritable change is induced that regularly and persistently activates the CK-biosynthetic system.

There is also an interaction with auxin when another inhibitor, *malformin,* stimulates the abscission of primary leaves of *Phaseolus aureus* because this effect is inhibited by indoleacetic acid (Curtis, 1968). The growth disturbances produced by malformin are apparently caused by the stimulation of ethylene production, at least in some cases. Again, however, the numerous effects of malformin on plant growth and development cannot be explained by a single effect, in this case enhanced ethylene production (Curtis, 1971). Again, too, it may stimulate as well as inhibit. It appears to inhibit elongation, leading to root curvature (Izhar et al., 1969), but it may also stimulate root hair and lateral root formation. Other synthetic growth regulators (Amchem 66-329 and picloram) also appear to produce their effects as a result of the stimulation of ethylene production (Warner and Leopold, 1967; Morgan and Baur, 1970). Studies of air pollutants have revealed the presence of oxidants (e.g., peroxyacetyl nitrate) that inhibit growth. Auxin can induce recovery from such inhibitors (Ordin et al., 1966).

INTERACTIONS BETWEEN GROWTH REGULATORS

All the growth regulators discussed previously may be classified into nine main groups (Fig. 19-19), three groups of regulators that are primarily growth promoters, five groups that are primarily growth inhibitors, and one transition substance that is difficult to place squarely into one or another of these two main categories.

The different groups of growth regulators have specific effects on growth. For instance, the elongation of seedlings of the grass family is controlled almost exclusively by auxin, but the leaves and leaf sheaths elongate under the exclusive control of GA (Thimann, 1965). It must al-

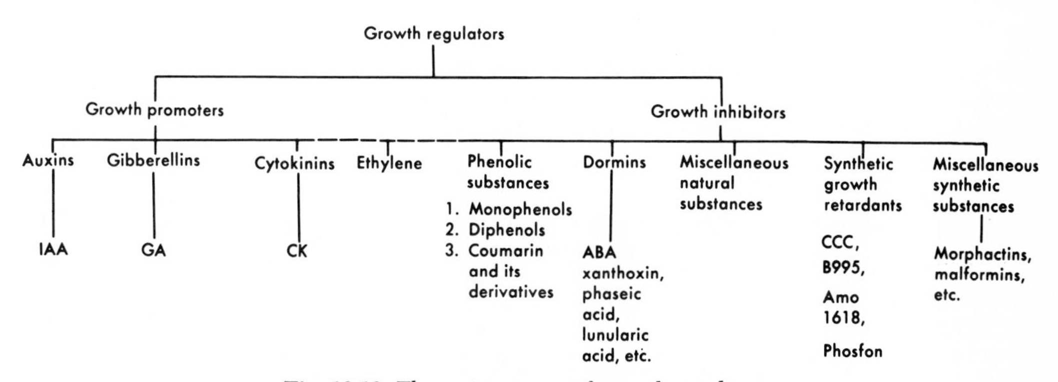

Fig. 19-19. The major groups of growth regulators.

ways be remembered that the promoters may inhibit growth, although far less commonly than they promote it; and the inhibitors may promote growth, although far less commonly than they inhibit it. Also, the frequent similarity in effects of the three groups of growth promoters on plant growth is surprising in view of the pronounced differences in their chemical structures. Auxins and gibberellins induce cell division, although their more characteristic effect is on cell elongation (Digby and Wareing, 1966). Conversely, CKs may affect elongation, although they commonly control division. All three may induce parthenocarpy (Crane and van Overbeek, 1965) and retard senescence (Fletcher and Osborne, 1965). However, such effects are judged by artificial applications to plant parts that have unknown quantities of all three regulators. Thus the observed effect is more likely to depend on the altered *balance* between the three than on the single addition of the one.

In any specific case, therefore, the effect of a growth regulator will depend not only on its concentration inside the growing cells, but also on the concentrations of all the other growth regulators that happen to be there. This is true even in the case of the synthetic growth regulators. The clear-cut prevention by a growth inhibitor of GA synthesis in a fungus as opposed to the absence of such a clear-cut effect in higher plants can be explained by the greater complexity of interactions of growth regulators in the higher plants. GA, for instance, is known to affect auxin activity, and therefore the growth retardant effect may show up more as an auxin than as a GA effect. The opposite actions of a single growth regulator may be similarly explained. The synthetic auxin 2,4-D is most commonly used as a weed killer. Nevertheless, it is frequently required in media used for the growth of tissue cultures. It may promote both cell proliferation and tracheary element differentiation, which in its absence require both an auxin and a CK (Foskett and Torrey, 1969). From all these results, it is evident that no generalization as to the effects of a single growth regulator can be valid unless the quantities of all the other growth regulators can be given. Since adequate methods have not yet been evolved for quantitative determinations of all the growth regulators present in the plant, this goal cannot yet be attained. On the other hand, the fact that all three groups of regulators interact with one another (Table 19-4) may indicate that they all affect the same metabolic process. The chemical structure of only one of these groups of regulators—the CKs—gives a clue as to which metabolic process may be involved. The CKs are chemically related to the purine component of nucleic acids and can logically be expected to affect nucleic acid and, therefore, protein metabolism. Most of the evidence supports this expectation, which is illustrated in Fig. 19-20.

Table 19-4. Interactions between growth regulators*

Regulators	Phenomenon	Effect
1. Auxin as opposed to cytokinin	Apical dominance	Auxin promotes; cytokinin overcomes
2. Auxins plus gibberellins	Activation of cambium	Act together to promote it
3. Auxins, gibberellins, cytokinins	Formation of wood	Interaction of all three
4. Auxin and gibberellin	Internode elongation of dicots	Interdependent
5. Auxin and abscisin	Abscission	Interaction to promote or inhibit
6. Gibberellin, auxin, or cytokinin as opposed to abscisic acid	Dormancy	Abscisic acid promotes; gibberellin, auxin, or cytokinin overcomes its promotion of dormancy

*Modified from Thimann, K. V. (1965). Recent Progr. Hormone Res. **21**:579-596.

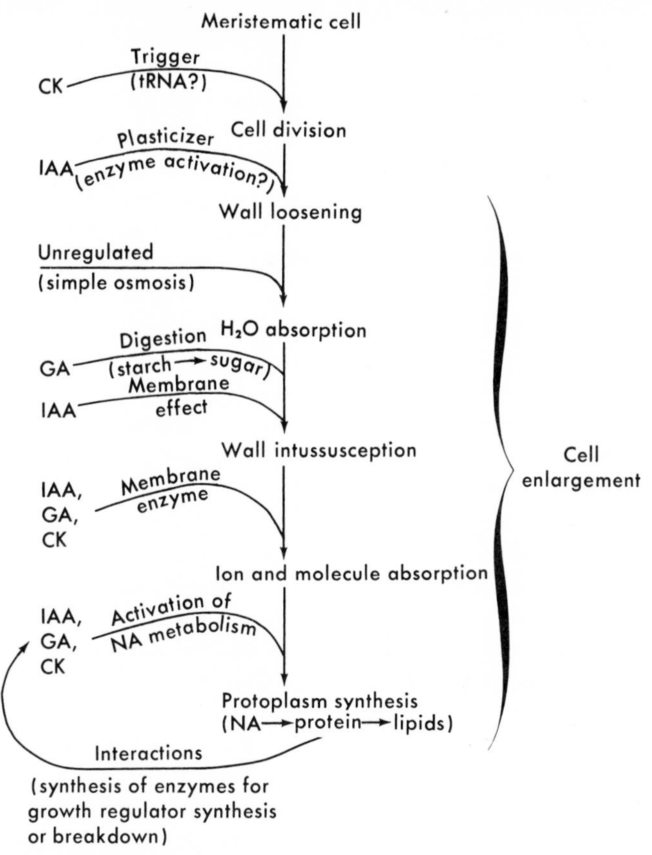

Fig. 19-20. Concept of rapid (top) and slow (bottom) effects of growth promoters on different stages of cell development.

However, van Overbeek (1966) believes that this relation between growth regulators and nucleic acids may not be a direct one, since it is possible to detect the regulator's action long before the nucleic acid response. The growth process is, of course, dependent on nucleic acid metabolism; therefore any growth response to the regulators is sure to be accompanied by a stimulation of nucleic acid metabolism, even if this is not the immediate cause of the growth response. In the case of the growth inhibitor ABA, the lowering of the protein and RNA contents is significant only after a few days (Pilet and Roland, 1971). ABA does, however, induce other effects far more rapidly. For instance, it increases stomatal resistance within 10 min (Mittelheuser and van Steveninck, 1971).

Therefore it seems reasonable to suggest that all the growth regulators (both the promoters and the inhibitors) produce their rapid effect on the membrane, leading to permeability change, enzyme activation, wall softening and synthesis, and ion absorption. Their slower effect would then be on nucleic acid and protein synthesis, leading to the synthesis of new protoplasm (Fig. 19-20). This slower effect also explains the many interactions between the growth regulators; if each growth regulator controls one step in the series of reactions leading to protein synthesis, it may also control the synthesis of one or more enzymes required for the formation of another growth regulator.

OTHER GROWTH REGULATORS

Other growth regulators that do not fit into the preceding major classes have been found in the plant. Traumatin, for instance, is a wound hormone that accumulates at the cut surface of potatoes. Unfortunately, there has been no confirmation of the existence of this growth regulator since its report some time ago. Apparently, others have attempted to detect it without success (personal communication).

Sometimes an unknown mixture may greatly stimulate growth. A good example is coconut milk (van Overbeek, 1942). Its growth-regulating activity is now believed to be caused by auxin, CK, hexitols (e.g., myoinositol), and reduced N-compounds (Kefford, 1963), all of which act synergistically. Myoinositol, for instance, acts synergistically with CK to promote cell division in carrots but not with IAA (Letham, 1966). There must also be some natural growth regulators that have not yet been discovered. In favor of this conclusion is the fact that the tissues of some plants cannot be grown indefinitely in tissue culture. Several other substances, for example, glucobrassicin and helminthosporol (Schantz, 1966), may also stimulate growth. As mentioned earlier, some growth regulators of the auxin type may be used as herbicides. Other substances used to kill plants are not truly growth regulators. The quaternary salts of bipyridyl react with the normal photosynthetic process to produce hydrogen peroxide in concentrations too high for catalase to deal with and therefore kill the plant (Boon, 1967).

In all the preceding discussions, growth regulators have been considered only from the point of view of their effects on the plant that produces them. There are, however, many substances produced by one plant that may regulate the growth of other plants. Normally, this is an inhibition and is then called *allelopathy*. Volatile substances, such as terpenes, produced by one species of plant have in this way inhibited the growth of other species (Muller and Del Moral, 1966; Muller et al., 1968). This may sometimes lead to a pure stand of the former (e.g., *Centaurea*, Fletcher and Renney, 1963). Leachates may be similarly effective (Wilson and Rice, 1968). Lower concentrations of volatile oils from several species were rarely inhibitory to germination of moistened lettuce seed, but had a marked effect on dry seed (Barbalic, 1967). The longitudinal growth was inhibited more than the germination.

QUESTIONS

1. What is a growth regulator?
2. What is a hormone?
3. What is an auxin?
4. What is the chemical substance in the plant responsible for auxin activity?
5. Have any other auxins been found?
6. What effect does auxin have on growth?
7. What effects do different concentrations of auxin have?
8. Are all plant parts equally sensitive to auxins?
9. What phase(s) of cell development is (are) affected?
10. How is cell enlargement induced?
11. What effect does IAA have on metabolism?
12. How is phototropism explained in terms of auxins?
13. How is geotropism explained in terms of auxins?
14. What is an antiauxin?
15. Does an antiauxin increase or decrease growth?
16. What other effects do auxins have?
17. Name two other kinds of naturally occurring growth regulators.
18. Is their chemistry known?
19. How do they differ from auxins?
20. What organisms are affected by auxins?
21. What amino acid is IAA probably formed from?
22. What practical use are auxins put to?

23. What chemical substances are gibberellins related to?
24. What effects do gibberellins have on shoot growth?
25. What effects do gibberellins have on root growth?
26. What effects do gibberellins have on seed germination?
27. What effects do gibberellins have on enzyme activity?
28. What effects do gibberellins have on leaf growth?
29. What effects do gibberellins have on oat coleoptiles?
30. What phase of cell development are cytokinins believed to affect?
31. Which of the three groups of growth regulators interact with other regulators?
32. Which of the three groups affect senescence?
33. Which of the three groups are hormones?
34. What is the major group of natural growth inhibitors?
35. Which growth regulators do they react with?
36. What is believed to be the mechanism of action of growth regulators?
37. Are there any other natural substances capable of controlling growth beside the four main groups of growth regulators?
38. Are there any synthetic substances?

SPECIFIC REFERENCES

Abdul-Baki, A. A., and P. M. Ray. 1971. Regulation by auxin of carbohydrate metabolism involved in cell wall synthesis by pea stem tissue. Plant Physiol. **47**:537-544.

Abeles, F. B., and L. E. Craker, 1969. Abscission: role of abscisic acid. Abst. XI. Vol. 1. Int. Bot. Cong. Seattle.

Abeles, F. B., and L. E. Forrence. 1970. Temporal and hormonal control of β-1,3-glucanase in *Phaseolus vulgaris* L. Plant Physiol. **45**:395-400.

Ahmad, M. R., and A. Winter. 1970. The effect of weak auxins on the growth of blue-green algae. Hydrobiologia **36**:305-316.

Alden, T. 1971. Seasonal variations in the occurrence of indole-3-acetic acid in buds of *Pinus sylvestris*. Physiol. Plant. **25**:54-57.

Altman, A., and R. Goren. 1971. Promotion of callus formation by abscisic acid in citrus bud cultures. Plant Physiol. **47**:844-846.

Andersen, A. S. 1971. Plant growth modification by 2-chlorethyl phosphonic acid (Ethrel). II. Interactions between Ethrel and gibberellin in the growth and fruiting of pea plants. Kong. El. Vet. Landbohojsk Arsskr. **1971**:11-19.

Anderson, J. W., and K. S. Rowan. 1966. Activity of aminoacyl-transfer-ribonucleic acid synthetases in tobacco-leaf tissue in relation to senescence and to the action of 6-furfuryl-aminopurine. Biochem. J. **101**:15-18.

Apelbaum, A., and S. P. Burg. 1971. Altered cell microfibrillar orientation in ethylene-treated *Pisum sativum* stems. Plant Physiol. **48**:648-652.

Aramaki, K., and Y. Kuroiva. 1967. Effect of gibberellic acid on nucleic acid metabolism of germinating barley and barley endosperms. Rep. Res. Lab. Kirin Brew. Co. Ltd. **10**:29-37.

Arney, S. E., and D. L. Mitchell. 1969. The effect of abscisic acid on stem elongation and correlative inhibition. New Phytol. **68**:1001-1015.

Aspinall, D., L. G. Paleg, and F. D. Addicott. 1967. Abscisin II and some hormone-regulated plant responses. Aust. J. Biol. Sci. **20**:869-882.

Azhar, S., and C. R. Krishna Murti. 1971. Effect of indole-3-acetic acid on the synthesis of cyclic 3′-5′ adenosine phosphate by Bengal grain seeds. Biochem. Biophys. Res. Com. **43**:58-64.

Bachelard, E. P. 1969. Effects of gibberellic acid on internode growth and starch contents of *Eucalyptus camaldulensis* seedlings. New Phytol. **68**:1017-1022.

Bagni, N., E. Corsini, and D. S. Fracassini. 1971. Growth-factors and nucleic acid synthesis in *Helianthus tuberosus*. I. Reversal of actinomycin D inhibition by spermidine. Physiol. Plant. **24**: 112-117.

Banerji, D., and M. M. Laloraya. 1967. Correlative studies on plant growth and metabolism. III. Metabolic changes accompanying inhibition of the longitudinal growth of stem and root by kinetin. Plant Physiol. **42**:623-627.

Barbalic, L. 1967. Über die Einwirkung einiger ätherischer Öle auf höhere Pflanzen. Qual. Plant Mat. Veg. **15**:135-153.

Barnes, M. F., E. N. Light, and A. Lang. 1969. The action of plant growth retardants on terpenoid biosynthesis: inhibition of gibberellic-acid production in *Fusarium moniliforme* by CCC and AMO-1618; action of these retardants on sterol biosynthesis. Planta **88**:172-182.

Barralis, G. 1965. La germination des folles avoines. Ann. Épiphyt. **6**:295-314.

Bednar, T. W., and E. M. Linsmaier-Bednar. 1971. Induction of cytokinin-independent tobacco tissues by substituted fluorenes. Proc. Nat. Acad. Sci. USA **68**:1178-1179.

Belhanafi, A., and G. F. Collet. 1970. Types of inhibition of growth and nucleic acid synthesis in wheat seedlings by abscisic acid. Physiol. Plant. **23**:859-870.

Bernier, G., R. Bronchart, and A. Jacqmard. 1964. Action of gibberellic acid on the mitotic activity of the different zones of the shoot apex of *Rudbeckia bicolor* and *Perilla nankinensis*. Planta **61**:236-244.

Berridge, M. V., and R. K. Ralph. 1971. Kinetin and carbohydrate metabolism in Chinese cabbage. Plant Physiol. **47**:562-567.

Berrie, A. M. M., W. Parker, B. A. Knights, and M. R. Hendrie. 1968. Studies on lettuce seed germination. I. Coumarin induced dormancy. Phytochemistry **7**:567-573.

Besemer, J., U. Harden, and J. Reinert. 1969. The influence of kinetin and gibberellic acid on organ formation on leaves of *Cichorium intybus* cultured in vitro. Z. Pflanzenphysiol. **60**:123-134.

Beyer, E. M., Jr., and P. W. Morgan. 1969. Time sequence of the effect of ethylene on transport uptake and decarboxylation of auxin. Plant. Cell Physiol. **10**:787-799.

Beyer, E. M., Jr., and P. W. Morgan. 1970. Effect of ethylene on the uptake, distribution, and metabolism of indoleacetic acid-1-^{14}C and -2-^{14}C and naphthaleneacetic acid -1-^{14}C. Plant Physiol. **46**:157-162.

Bezemer-Sybrandy, S. M., and H. Veldstra. 1971*a*. Investigations on cytokinins. III. Cytokinin activity in *Lemna minor* tRNA hydrolysates. Physiol. Plant. **24**:369-373.

Bezemer-Sybrandy, S. M., and H. Veldstra. 1971*b*. Investigations on cytokinins. IV. The metabolism of 6-benzylaminopurine in *Lemna minor*. Physiol. Plant. **25**:1-7.

Bhattacharyya, J., and S. C. Roy. 1969. Growth promoters and the synthesis of protein in plant mitochondria. I. Effects of kinetin on the incorporation of amino acids into mitochondrial protein. Biochem. Biophys. Res. Comm. **35**:606-610.

Bodlaender, K. B. A., and S. Algera. 1966. Influence of the growth retardant B995 on growth and yield of potatoes. Eur. Potato J. **9**:242-258.

Boon, W. R. 1967. The quaternary salts of bipyridyl—a new agricultural tool. Endeavour **26**: 27-32.

Bowen, M. R., and P. F. Wareing. 1971. Further investigations into hormone-directed transport in stems. Planta **99**:120-132.

Bradshaw, M. J., and J. Edelman. 1971. The effects of growth substances and retardants on renewed processes of Jerusalem artichoke tuber tissues. J. Exp. Bot. **22**:391-399.

Brandes, H. 1967. Die Wirkungmechanismus des Kinetins bei der Induktion von Knospen am Protonema der Laubmoose. Planta **74**:55-71.

Brauner, L., and N. Arslan. 1951. Experiments on the auxin reactions of the pulvinus of *Phaseolus multiflorus*. Rev. Fac. Sci. Univ. Instanbul 16B **3**:257-300.

Broughton, W. J. 1969. Relations between DNA, RNA, and protein synthesis, and the cellular basis of the growth response in gibberellic acid-treated pea internodes. Ann. Bot. **33**:227-243.

Broughton, W. J., E. O. Helmuth, and D. Yeung. 1970. Role of glucose in development of the gibberellin response in peas. Biochim. Biophys. Acta **222**:491-500.

Broughton, W. J., and A. J. McComb. 1967. The relation between cell-wall protein synthesis in dwarf pea plants treated with gibberellic acid. Ann. Bot. **31**:359-366.

Broughton, W. J., and A. J. McComb. 1971. Changes in the pattern of enzyme development in gibberellin-treated pea internodes. Ann. Bot. **35**:213-228.

Bruce, M. I., and J. A. Zwar. 1966. Cytokinin activity of some substituted ureas and thioureas. Proc. Roy. Soc. Ser. B. Biol. Sci. **165**:245-265.

Bryant, S., and C. G. Skinner. 1968. Reversal of chemically-inhibited germination by 6-benzylaminopurine-gibberellic acid combinations. Phyton. Rev. Int. Bot. Exp. **25**:69-73.

Buggeln, R. G., and J. S. Craigie. 1971. Evaluation of evidence for the presence of indole-3-acetic acid in marine algae. Planta **97**:173-178.

Bui-Dang-Ha, D., and J. P. Nitsch. 1970. Isolation of zeatin riboside from the chicory root. Planta **95**:119-126.

Burdett, A. N., and P. F. Wareing. 1966. The effect of kinetin on the incorporation of labelled orotate into various fractions of ribonucleic acid of excised radish leaf discs. Planta **71**:20-26.

Burkett, A. R., K. K. Schlender, and H. M. Sell. 1970. Gougerotin: an inhibitor of protein synthesis and auxin-induced elongation in plants. Phytochemistry **9**:545-547.

Burg, S. P., and E. A. Burg. 1966. The interaction between auxin and ethylene and its role in plant growth. Proc. Nat. Acad. Sci. USA **55**:262-269.

Burg, S. P., and E. A. Burg. 1967. Lateral auxin transport in stems and roots. Plant Physiol. **42**: 891-893.

Burrows, W. J., and D. J. Carr. 1969. Effects of flooding the root system of sunflower plants on the cytokinin content in the xylem sap. Physiol. Plant. **22**:1105-1112.

Burrows, W. J., F. Skoog, and N. J. Leonard. 1971. Isolation and identification of cytokinins located in the transfer ribonucleic acid of tobacco callus grown in the presence of 6-benzylaminopurine. Biochemistry **10**:2189-2194.

Burström, H. G., I. Uhrstrom, and B. Olausson. 1970. Influence of auxin on Young's modulus in stems and roots of Pisum and the theory of changing the modulus in tissues. Physiol. Plant. **23**:1223-1233.

Cameron, R. J. 1970. Translocation of carbon-14-labelled assimilates in shoots of *Pinus radiata*: the effects of girdling and indolebutyric acid. J. Exp. Bot. **21**:943-950.

Chadwick, A. V., and S. P. Burg. 1970. Regulation of root growth by auxin-ethylene interaction. Plant Physiol. **45**:192-200.

Chailakhyan, M. K., and V. G. Kochankov. 1967. Effect of retardants on the growth and flowering of plants. Fiziol. Rast. **14**:773-784.

Chin, T.-Y., M. M. Meyer, Jr., and L. Beevers. 1969. Abscisic-acid-stimulated rooting of stem cuttings. Planta **88**:192-196.

Chinoy, J. J., K. Gurumuti, K. Shastri, P. G. Abraham, I. C. Dave, P. N. Shah, R. B. Pandya, and O. P. Saxena. 1968. Effect of ascorbic acid, CCC and their interaction on germination and metabolism in peanut. Indian J. Plant Physiol. **11**:216-225.

Chrispeels, M. J., and J. E. Varner. 1967. Hormonal control of enzyme synthesis; on the mode of action of gibberellic acid and abscisin in aleurone layers of barley. Plant Physiol. **42**: 1008-1016.

Claver, F. K. 1970. The effects of abscisic acid on tuberization of potato sprouts in vitro. Phyton. Rev. Int. Bot. Exp. **27**:25-29.

Cleland, R. 1963. Independence of the effects of auxin on cell wall methylation and elongation. Plant Physiol. **38**:12-18.

Cleland, R. 1965. Auxin-induced cell wall loosening in the presence of actinomycin D. Plant Physiol. **40**:595-600.

Cleland, R. 1967. A dual role of turgor pressure in auxin-induced cell elongation in *Avena* coleoptiles. Planta **77**:182-191.

Cleland, R. E. 1969. The gibberellins, p. 49-81. *In* M. B. Wilkins (Ed.). The physiology of plant growth and development. McGraw-Hill Book Co., New York.

Cleland, R. 1970. Protein synthesis and wall extensibility in the *Avena* coleoptile. Planta **95**: 218-226.

Cleland, R. 1971. Instability of the growth-limiting proteins of the *Avena* coleoptile and their pool size in relation to auxin. Planta **99**:1-11.

Coartney, J. S., D. J. Morré, and J. L. Key. 1967. Inhibition of RNA synthesis and auxin-induced cell wall extensibility and growth by actinomycin D. Plant Physiol. **42**:434-439.

Coombe, B. G. 1971. GA_{32}: a polar gibberellin with high biological potency. Science **172**:856-857.

Cornforth, J. W., B. V. Milborrow, and G. Ryback. 1966. Identification and estimation of (+) abscisin II ("dormin") in plant extracts by spectropolarimetry. Nature (London) **210**:627-628.

Correa, N. S. 1969. Effect of naringenin (4′,5,7 trihydroxyflavenone) on growth of wheat coleoptiles and α-amylase in barley endosperm, and its interaction with gibberellic acid and IAA. Phyton. Rev. Int. Bot. Exp. **26**:125-134.

Coupland, D., and A. J. Peel. 1971. Uptake and incorporation of ^{14}C-labelled maleic hydrazide in the roots of *Salix viminalis*. Physiol. Plant. **25**:141-144.

Coutrez, D. 1968. The effect of some inhibitors on the respiration of plant tissues treated with gibberellic acid. Ann. Physiol. Veg. Univ. Bruxelles **13**:58-81.

Crane, J., and J. van Overbeek. 1965. Kinin-induced parthenocarpy in the fig, *Ficus carica* L. Science **147**:1468-1469.

Cross, B. E., and P. L. Myers. 1969. The effect of plant growth retardants on the biosynthesis of diterpenes by *Gibberella fujikoroi*. Phytochem. **8**:79-83.

Crozier, A., C. C. Kuo, R. C. Durley, and R. P. Pharis. 1970. The biological activities of 26 gibberellins in nine plant bioassays. Can. J. Bot. **48**:867-877.

Crozier, A., and D. M. Reid. 1971. Do roots synthesize gibberellins? Can. J. Bot. **49**:967-975.

Curtis, R. W. 1968. Mediation of a plant response to malformin by ethylene. Plant Physiol. **43**:76-80.

Curtis, R. W. 1971. Potentiation and inhibition of the effects of 2-chloroethylphosphonic acid by malformin. Plant Physiol. **47**:478-482.

Dalheim, H. 1969. Untersuchungen zum Wirkungsmechanismus von Indolyl-3-Essigsäure mit Hilfe von schwerem Wasser. Planta **86**:224-234.

Darimont, E., T. Gaspar, and M. Hofinger. 1971. Auxin-kinetin interaction on the lentil root growth in relation to indole-acrylic acid metabolism. Z. Pflanzenphysiol. **64**:232-240.

Davies, P. J. 1971. Further evidence against an IAA-tRNA complex. Plant Cell Physiol. **12**: 785-789.

Degane, Y., and D. Atsmon. 1970. DNA synthesis

and hormone-induced elongation in the cucumber hypocotyl. Nature **228**:554-555.

Dennis, F. G., Jr. 1967. Apple fruit-set; evidence for a specific role of seeds. Science **156**:71-73.

Devlin, R. M., and D. P. Brown. 1969. Effect of gibberellic acid on the elongation rate of *Agrostis alba* root hairs. Physiol. Plant. **22**:759-763.

Dezsi, L., and G. L. Farkas. 1964. Effect of kinetin on enzymes of glycolic acid metabolism in cereal leaves. Acta Biol. Hung. **14**:325-332.

Digby, J., and P. F. Wareing. 1966. The effect of applied growth hormones on cambial division and the differentiation of the cambial derivatives. Ann. Bot. **30**:539-548.

Digby, J., and P. F. Wareing. 1966. The effect of growth hormones on cell division and expansion in liquid suspension cultures of *Acer pseudoplatanus*. J. Exp. Bot. **17**:718-725.

Dravnieks, D. E., F. Skoog, and R. H. Burris. 1969. Cytokinin activation of de novo thiamine biosynthesis in tobacco callus cultures. Plant Physiol. **44**:866-870.

Duda, C. T., and J. H. Cherry. 1971. Chromatin- and nuclei-directed ribonucleic acid synthesis in sugar beet root. Plant Physiol. **47**:262-268.

Dyson, W. H., C. M. Chen, S. N. Alam, R. H. Hall, C. I. Hong, and G. B. Chheda. 1970. Cytokinin activity of ureidopurine derivatives related to a modified nucleoside found in transfer RNA. Science **170**:328-330.

Eastwood, D., R. J. A. Tavener, and D. L. Laidman. 1969. Sequential action of cytokinin and gibberellic acid in wheat aleurone tissue. Nature **221**:1267.

Eliasson, L. 1969. Growth regulators in *Populus tremula*. I. Distribution of auxin and growth inhibitors. Physiol. Plant. **22**:1288-1301.

Eliasson, L. 1971*a*. Growth regulators in *Populus tremula*. IV. Apical dominance and suckering in young plants. Physiol. Plant. **25**:263-267.

Eliasson, L. 1971*b*. Growth regulators in *Populus tremula*. III. Variation of auxin and inhibitor level in roots in relation to root sucker formation. Physiol. Plant. **25**:118-121.

Engelbrecht, L. 1964. Über Kinetinwirkungen bei intakten Blattern von *Nicotiana rustica*. Flora 154:57-69.

Englesma, G., and J. M. H. van Bruggen. 1971. Ethylene production and enzyme induction in excised plant tissues. Plant Physiol. **48**:94-96.

Erdmann, N., and U. Schiewer. 1971. Tryptophan-dependent indoleacetic acid biosynthesis from indole, demonstrated by double-labelling experiments. Planta **97**:135-141.

Eriksen, E. N. 1971. Promotion of root initiation by gibberellin. Kong. El. Vet. Landbohojsk. Arsskr. **1971**:50-59.

Ernest, L. C., and J. G. Valdovinos. 1971. Regulation of auxin levels in *Coleus blumei* by ethylene. Plant Physiol. **48**:402-406.

Evans, M. L., and D. L. Rayle. 1970. The timing of growth promotion and conversion to indole-3-acetic acid for auxin precursors. Plant Physiol. **45**:240-243.

Evans, M. L., P. M. Ray, and L. Reinhold. 1971. Induction of coleoptile elongation by carbon dioxide. Plant Physiol. **47**:335-341.

Fan, D. F., and G. A. MacLachlan. 1966. Control of cellulase activity by indoleacetic acid. Canad. J. Bot. **44**:1025-1034.

Feierabend, J. 1969. Der Einfluss von Cytokinen auf die Bildung von Photosynthese enzyme in Roggenkeimlingen. Planta **84**:11-29.

Fellenberg, G. 1969*a*. Influence of gibberellic acid and kinetin on auxin-induced root initiation and on the nucleoproteid of pea epicotyls. Z. Pflanzenphysiol. **60**:457-466.

Fellenberg, G. 1969*b*. Veränderungen des Nucleoproteids unter dem Einfluss von Auxin und Ascorbinsäure bei der Wurzelneubildung an Erbsenepikotylen. Planta **84**:324-338.

Fellenberg, G. 1971*a*. Untersuchungen über die Bindung pflanzlicher Wuchsstoffe an verschiedene komponenten des Chromatins in vitro. I. Bindung an DNS. Z. Naturforsch. **26b**:607-612.

Fellenberg, G. 1971*b*. Inhibition of some growth and differentiation processes in relation to the concentration of applied histone. Z. Pflanzenphysiol. **64**:427-436.

Field, R. J., and A. J. Peel. 1971. The metabolism and radial movement of growth regulators and herbicides in willow stems. New Phytol. **70**:743-749.

Fletcher, R. A., and A. J. Renney. 1963. A growth inhibitor found in *Centaurea* spp. Canad. J. Plant Sci. **43**:475-481.

Fletcher, R. A., and D. J. Osborne. 1965. Regulation of protein and nucleic acid synthesis by gibberellin during leaf senescence. Nature (London) **207**:1176-1177.

Flood, A. E., P. P. Rutherford, and E. W. Weston. 1967. Effects of 2,4-dichlorophenoxyacetic acid on enzyme systems in Jerusalem artichoke tubers and chicory roots. Nature (London) **214**:1049-1050.

Fosket, D. E. 1970. The time course of xylem differentiation and its relation to deoxyribonucleic acid synthesis in cultured *Coleus* stem segments. Plant Physiol. **46**:64-68.

Fosket, D. E., and J. G. Torrey. 1969. Hormonal control of cell proliferation and xylem differentiation in cultured tissues of *Glycine max* var. Biloxi. Plant Physiol. **44**:871-880.

Fox, J. E. 1966. Incorporation of a kinin, N,6-benzyladenine into soluble RNA. Plant Physiol. **41**: 75-82.

Fox, J. E. 1969. The cytokinins, p. 85-123. *In* M. B. Wilkins (Ed.). The physiology of plant growth and development. McGraw-Hill Book Co., New York.

Frankhauser, M., and K. H. Erismann. 1969. The effect of kinetin on protein, amino acid and RNA metabolism in *Lemna minor* L. Planta **88**: 332-343.

Galston, A. W., and P. J. Davies. 1970. Control mechanisms in plant development. Prentice-Hall, Inc., Englewood Cliffs, N. J.

Gamborg, O. L., and T. A. G. LaRue. 1971. Ethylene production by plant cell cultures: the effect of auxins, abscisic acid, and kinetin on ethylene production in suspension cultures of rose and *Ruta* cells. Plant Physiol. **48**:399-401.

Gaspar, T., and J. Lacoppe. 1968. The effect of CCC and Amo-1618 on growth, catalase, peroxidase and indoleacetic acid oxidase activity of young barley seedlings. Physiol. Plant. **21**: 1104-1109.

Gaspar, T., R. Verbeek, and A. A. Khan. 1971. Some effects of Amo-1618 on growth, peroxidase and α-amylase which cannot be easily explained by inhibition of gibberellin biosynthesis. Physiol. Plant. **24**:552-555.

Gaspar, T., and A. Xhaufflaire. 1967. Effect of kinetin on growth, auxin catabolism, peroxidase and catalase activities. Planta **72**:252-257.

Gepstain, S., and I. Ilan. 1970. A promotive action of kinetin on amylase activity in cotyledons of *Phaseolus vulgaris*. Plant Cell Physiol. **11**: 819-822.

Giles, K. L. 1971. The control of chloroplast division in *Funaria hygrometrica*. II. The effects of kinetin and indole-acetic acid on nucleic acids. Plant Cell Physiol. **12**:447-450.

Glinka, Z., and L. Reinhold. 1971. Abscisic acid raises the permeability of plant cells to water. Plant Physiol. **48**:103-105.

Gogala, N. 1970. Effect of the natural cytokinins of *Pinus sylvestris* L. and other growth substances on the growth of the mycelium of *Boletus edulis* var. pinicolus Vitt. Oesterr. Bot. Z. **118**:321-333.

Goldschmidt, E. E., R. Goren, S. P. Monselise, N. Takahashi, H. Igoshi, I. Yamaguchi, and K. Hirose. 1971. Auxins in citrus: a reappraisal. Science **174**:1256-1257.

Goldschmidt, E. E., S. P. Monselise, and R. Goren. 1971. On the identification of native auxins in citrus tissues. Canad. J. Bot. **49**:241-245.

Goodwin, P. B. 1966. An improved medium for the rapid growth of isolated potato buds. J. Exp. Bot. **17**:590-595.

Goren, R., and E. Tomer. 1971. Effects of seselin and coumarin on growth, indoleacetic acid oxidase, and peroxidase, with special reference to cucumber (*Cucumis sativa* L.) radicles. Plant Physiol. **47**:312-316.

Graebe, J. E. 1967. Isoprenoid biosynthesis in a cell-free system from pea shoots. Science **157**: 73-75.

Greenwood, M. S., and M. H. M. Goldsmith. 1970. Polar transport and accumulation of indole-3-acetic acid during root regeneration by *Pinus lambertiana* embryos. Planta **95**:297-313.

Gupta, G. R. P., and S. C. Maheshwari. 1970. Cytokinins in seeds of pumpkin. Plant Physiol. **45**:14-18.

Hager, A., and R. Schmidt. 1968. Auxintransport und Phototropismus. I. Die lichbedingte Bildung eines Hemmstoffes für den Transport von Wuchsstoffen in Koleoptilen. Planta **83**: 347-371.

Hager, A., H. Menzel, and A. Krauss. 1971. Experiments and hypothesis concerning the primary action of auxin in elongation growth. Planta **100**:47-75.

Häggman, J., and H. Haapala. 1971. The effect of 6-benzylaminopurine on the starch metabolism of *Stellaria media*. Physiol. Plant. **24**:548-551.

Halevy, A. H., and R. Shilo. 1970. Promotion of growth and flowering and increase in content of endogenous gibberellins in *Gladiolus* plants treated with the growth retardant CCC. Physiol. Plant. **23**:820-827.

Hall, M. D., and E. C. Cocking. 1971. The bursting response of isolated *Avena* coleoptile protoplasts to indol-3-yl acetic acid. Biochem. J. **124**: 33P.

Hanada, K., K. Akutsu, and A. Gozawa. 1969. Studies on branching habits in crop plants. VII. Proc. Crop Sci. Soc. Jap. **38**:67-73.

Harada, H., and T. Koizumi. 1971. Effects of cinnamic acid derivatives on the geo- and phototropism and the germination of *Lactuca sativa* L. cv. Grand Rapids. Z. Pflanzephysiol. **64**:350-357.

Harada, J., and K. Wada. 1968. Leaf characters of tall and short lines of rice and their response to gibberellin. Tahaku J. Agr. Res. **19**:27-31.

Hardin, J. W., J. H. Cherry, D. J. Morré, and C. A. Lembi. 1972. Enhancement of RNA polymerase activity by a factor released by auxin from plasma membrane. Proc. Nat. Acad. Sci. USA **69**:3146-3150.

Hardy, S. I., P. A. Castelfranco, and C. A. Rebeiz. 1971. Effect of the hypocotyl hook on chlorophyll accumulation in excised cotyledons of *Cucumis sativus* L. Plant Physiol. **47**:705-708.

Harper, D. B., and R. L. Wain. 1969. Studies on plant growth-regulating substances. XXX. The plant growth-regulating activity of substituted phenols. Ann. Appl. Biol. **64**:395-407.

Hatch, A. H., and L. E. Powell. 1971*a*. Hormone-directed transport of ^{32}P in *Malus sylvestris* seedlings. J. Amer. Soc. Hort. Sci. **96**:230-234.

Hatch, A. H., and L. E. Powell. 1971*b*. Hormone-directed transport of certain organic compounds in *Malus sylvestris* seedlings. J. Amer. Soc. Hort. Sci. **96**:399-400.

Hecht, S. M., R. M. Bock, N. J. Leonard, R. Y. Schmitz, and F. Skoog. 1970. Cytokinin activity in $tRNA^{Phe}$. Biochem. Biophys. Res. Com. **41**:435-440.

Hecht, S. M., N. J. Leonard, W. J. Burrows, D. J. Armstrong, F. Skoog, and J. Occlowitz. 1969. Cytokinin of wheat germ transfer RNA: 6-(4-hydroxy-3-methyl-2-butenylamine)-2-methylthio-9-β-D-ribofuranosylpurine. Science **166**:1272-1274.

Heide, O. M. 1969*a*. Interaction of growth retardants and temperature in growth, flowering, regeneration, and auxin activity of *Begonia* x *cheimantha* Everett. Physiol. Plant. **22**:1001-1012.

Heide, O. M. 1969*b*. Non-reversibility of gibberellin-induced inhibition of regeneration in *Begonia* leaves. Physiol. Plant. **22**:671-679.

Helgeson, J. P., and C. D. Upper. 1970. Modification of logarithmic growth rates of tobacco callus tissue by gibberellic acid. Plant Physiol. **46**:113-117.

Hemberg, T. 1952. The significance of the acid growth–inhibiting substance for the rest period of the potato tuber. Physiol. Plant. **5**:115-129.

Hertel, R., K.-St. Thomson, and V. E. A. Russo. 1972. *In vitro* auxin binding to particulate cell fractions from corn coleoptiles. Planta **107**: 325-340.

Heyn, A. N. J. 1970. Dextranase activity in coleoptiles of *Avena*. Science **167**:874-875.

Hoad, G. V., S. K. Hillman, and P. F. Wareing. 1971. Studies on the movement of indole auxins in willow (*Salix viminalis* L.). Planta **99**:73-88.

Hoffman, P., and P. Rathsack. 1970. On the effect of cytokinin in blue-green algae. Biochem. Physiol. Pflanz. **161**:95-96.

Holm, R. E., and J. L. Key. 1969. Hormonal regulation of cell elongation in the hypocotyl of rootless soybean: an evaluation of the role of DNA synthesis. Plant Physiol. **44**:1295-1302.

Holm, R. E., and J. L. Key. 1971. Inhibition of auxin-induced deoxyribonucleic acid synthesis and chromatin activity by 5-fluordeoxyuridine in soybean hypocotyl. Plant Physiol. **47**:606-608.

Holm, R. E., T. J. O'Brien, J. L. Key, and J. H. Cherry. 1970. The influence of auxin and ethylene on chromatin-directed ribonucleic acid synthesis in soybean hypocotyl. Plant Physiol. **45**:41-45.

Hopkins, W. G., and J. A. J. Orkwiszewski. 1971. Inhibition of phenylalanine ammonia lyase and enhancement of *Avena* coleoptile segment elongation by fluorophenylalanine. Canad. J. Bot. **49**:129-135.

Holst, V. B. 1971. Some properties of inhibitor B from *Solanum tuberosum* compared to abscisic acid. Physiol. Plant. **24**:382-395.

Hugon, E. 1967. Influence du froid sur l'effet des cytokinines et de l'adénine appliquée sur la tige de plantes étiolées entières de *Cicer arietinum* L. Ann. Sci. Nat. Bot. Biol. Veg. **8**:479-486.

Huisinga, B. 1967. Root induction in *Avena* mesocotyls by indoleacetic acid. Acta Bot. Néerl. **16**:123-124.

Hulme, A. C., M. J. C. Rhodes, and L. S. C. Wooltorton. 1971. The relationship between ethylene and the synthesis of RNA and protein in ripening apples. Phytochem. **10**:749-756.

Hyodo, H., and S. F. Yang. 1971. Ethylene-enhanced synthesis of phenylalanine ammonia-lyase in pea seedlings. Plant Physiol. **47**:765-770.

Igoshi, M., I. Yamaguchi, N. Takahashi, and K. Hirose. 1971. Plant growth substances in the young fruit of *Citrus unshiu*. Agr. Biol. Chem. **35**:629-631.

Ilan, I. 1971. Evidence for hormonal regulation of the selectivity of ion uptake by plant cells. Physiol. Plant. **25**:230-233.

Ilan, I., T. Gilad, and L. Reinhold. 1971. Specific effects of kinetin on the uptake of monovalent cations by sunflower cotyledons. Physiol. Plant. **24**:337-341.

Imaseki, H., and C.-J. Pjon. 1970. The effect of ethylene on auxin-induced growth of excised rice coleoptile segments. Plant Cell Physiol. **11**:827-829.

Imbert, M. P., and L. A. Wilson. 1970. Stimulatory and inhibitory effects of scopoletin on IAA oxidase preparations from sweet potato. Phytochemistry **9**:1787-1794.

Ingersoll, R. B., and O. E. Smith. 1970. Movement of (RS)-abscisic acid in the cotton explant. Plant Physiol. **45**:576-578.

Ingersoll, R. B., and O. E. Smith. 1971. Transport of abscisic acid. Plant Cell Physiol. **12**:301-309.

Isaia, A. 1971. Comparative action of (±)-abscisic acid, coumarin, p-coumaric acid and some of their derivatives on the growth of the first wheat leaf. Planta **96**:175-182.

Isogai, Y., Y. Komoda, and T. Okamoto. 1970. Plant growth regulators in the pea plant (*Pisum sativum* L.). Chem. Pharm. Bull. (Tokyo) **18:** 1872-1879.

Izhar, S., J. M. Bevington, and R. W. Curtis. 1969. Effect of malformin on root growth. Plant Cell Physiol. **10:**687-698.

Jackson, G. A. D., and J. B. Blundell. 1966. Effect of dormin on fruit set in *Rosa.* Nature (London) **212:**1470-1471.

Jackson, M. B., and P. M. Harney. 1970. Rooting cofactors, indoleacetic acid, and adventitious root initiation in mung bean cuttings *(Phaseolus aureus).* Canad. J. Bot. **48:**943-946.

Jacobsen, J. V., J. G. Scandalios, and J. E. Varner. 1970. Multiple forms of amylase induced by gibberellic acid in isolated barley aleurone layers. Plant Physiol. **45:**367-371.

Jacobsen, J. V., and J. E. Varner. 1967. Gibberellic acid–induced synthesis of protease by isolated aleurone layers of barley. Plant Physiol. **42:**1596-1600.

Jaffe, M. J. 1970. Physiological studies on pea tendrils. VII. Evaluation of a technique for the asymmetrical application of ethylene. Plant Physiol. **46:**631-633.

Jakubowska, J., and M. Wlodarczyk. 1970. Observations on growth and metabolism of *Saccharomyces cerevisiae* influenced by beta-indole acetic acid (IAA). Ser. B. Microbiol. Appl. Acta Microbiol. Pol. **2:**75-81.

Janssen, M. G. H. 1970. Inhibitors of pea root indoleacetic acid oxidase. Acta Bot. Neerl. **19:** 109-111.

Jennings, R. C. 1969. Cytokinins as endogenous growth regulators in the algae *Ecklonia* (Phaeophyta) and *Hypnea* (Rhodophyta). Aust. J. Biol. Sci. **22:**621-627.

Jennings, R. C., and A. J. McComb. 1967. Gibberellins in the red alga, *Hypnea muaciformis* (Wulf.) Lamour. Nature (London) **215:**872-873.

Johnson, K. D., and W. K. Purves. 1970. Ribonucleic acid synthesis by cucumber chromatin: development and hormone-induced changes. Plant Physiol. **46:**581-585.

Jones, H. C., H. S. Black, and A. M. Altschul. 1967. Comparison of the effects of gibberellic acid and aflatoxin in germinating seeds. Nature (London) **214:**171-172.

Jones, R. A., and P. B. Kaufman. 1971. Regulation of growth in *Avena* stem segments by gibberellic acid and kinetin. Physiol. Plant. **24:**491-497.

Jones, R. L. 1971. Gibberellic acid–enhanced release of β-1,3-glucanase from barley aleurone cells. Plant Physiol. **47:**412-416.

Jordan, W., and F. Skoog. 1971. Effect of cytokinins on growth and auxin in coleoptiles of derooted *Avena* seedlings. Plant Physiol. **48:** 97-99.

Jouanneau, J. P. 1971. Control of synchronization of mitosis in tobacco cells by cytokinins. Exp. Cell Res. **67:**329-337.

Jung, J. 1967. Synthetische Wachstums regulatoren, insbesondere Chlorcholinchlorid. Naturwissenschaften **54:**356-360.

Kamisaka, S., Y. Masuda, and N. Yanagishima. 1967. Gibberellin-induced yeast sporulation in relation to RNA and protein metabolism. Physiol. Plant. **20:**98-105.

Kamienska, H., and A. Chrominski. 1971. Auxin-like activity of (2-chloroethyl) phosphonic acid. Bot. Gaz. **132:**229-232.

Kang, B. G., and S. P. Burg. 1971. Rapid change in water flux induced by auxins. Proc. Nat. Acad. Sci. USA **68:**1730-1733.

Kang, B. G., W. Newcomb, and S. P. Burg. 1971. Mechanism of auxin-induced ethylene production. Plant Physiol. **47:**504-509.

Kathju, S., M. N. Tewari, and U. Chatterji. 1971. Study of the effect of gibberellic acid and cycocel on the activity of phosphatases. Z. Pflanzenphysiol. **64:**169-174.

Katsumi, M. 1970. Effect of gibberellin A_3 on the elongation, α-amylase activity, reducing sugar content and oxygen-uptake of the leaf sheath on dwarf seedlings. Physiol. Plant. **23:** 1077-1084.

Katsumi, M., and H. Sano. 1968. Relationship of IAA-oxidase activity to gibberellin- and IAA-induced elongation of light-grown cucumber seedlings. Physiol. Plant. **21:**1348-1355.

Kaul, K., and P. S. Sabharwal. 1971. Effects of sucrose and kinetin on growth and chlorophyll synthesis in tobacco tissue cultures. Plant Physiol. **47:**691-695.

Kawase, M. 1971. Causes of centrifugal root promotion. Physiol. Plant. **25:**64-70.

Kefford, N. P. 1963. Natural growth regulators; report on 25th Int. Conf. on Plant Growth Regulation. Science **142:**1495-1505.

Kefford, N. P., and O. H. Caso. 1966. A potent auxin with unique chemical structure—4-amino-3,5,6-trichloropicolinic acid. Bot. Gaz. **127:**159-163.

Kendall, F. H., C. K. Park, and C. L. Mer. 1971. Indole-3-acetic acid metabolism in pea seedlings: a comparative study using carboxyl- and ring-labelled isomers. Ann. Bot. (London) **35:**565-579.

Kende, H., H. Hahn, and S. E. Kays. 1971. Enhancement of nitrate reductase activity by benzyladenine in *Agrostemma githago.* Plant Physiol. **48:**702-706.

Key, J. L. 1964. RNA and protein synthesis as

essential processes for cell elongation. Plant Physiol. **39**:365-370.

Key, J. L., and J. C. Shannon. 1964. Enhancement by auxin of RNA synthesis in excised soybean hypocotyl tissue. Plant Physiol. **39**:360-364.

Khan, A. A. 1967*a*. Antagonism between cytokinins and germination inhibitors. Nature (London) **216**:166-167.

Khan, A. A. 1967*b*. Dependence of gibberellic acid–induced dark germination of lettuce seed on RNA synthesis. Planta **72**:284-288.

Khan, A. A. 1967*c*. Physiology of morphactins: effect of gravi- and photo-response. Physiol. Plant. **20**:306-313.

Khan, A. A. 1971. Cytokinins: permissive role in seed germination. Science **171**:853-859.

Khan, A. A., and R. D. Downing. 1968. Cytokinin reversal of abscisic acid inhibition of growth and α-amylase synthesis in barley seed. Physiol. Plant. **21**:1301-1307.

Khan, A. A., and M. A. Faust. 1967. Effect of growth retardants on α-amylase production in germinating barley seed. Physiol. Plant. **20**:673-681.

Khan, R. A., S. H. Hashmi, and S. Ahmad. 1970. Some applied aspects of gibberellic acid in barley and wheat. Pak. J. Sci. Ind. Res. **13**:294-298.

Kitagawa, H., A. Sugiura, and M. Sugiyama. 1966. Effects of gibberellin spray on storage quality of kaki. Hort. Sci. **1**:59-60.

Knipl, Y. S., and O. N. Kulaeva. 1970. Effect of coumarin and synthetic growth retardants on the content of chlorophyll and protein in barley leaf discs in darkness and light. Fiziol. Rast. **17**:13-22.

Knypl, J. S. 1966. Specific inhibitors of RNA and protein synthesis as suppressors of the IAA- and coumarin-induced growth responses. Acta Soc. Bot. Pol. **35**:357-373.

Knypl, J. S. 1967. Synergistic inhibition of kale seed germination by coumarin and (2-chloroethyl)-trimethylammonium chloride and its reversal by kinetin and gibberellic acid. Planta **72**:292-296.

Knypl, J. S. 1971. Control of chlorophyll synthesis by coumarin and plant growth retarding chemicals. Acta Soc. Bot. Pol. **39**:321-332.

Koblitz, H. 1967. Beiträge zur Zellwandforschung in der pflanzlichen Gewebekultur. II. Biochemische Aspekte der Zellwandontogenese in vitro kultivierter Pflanzengewebe. Qualitas Plant. Mater. Veg. **14**:37-69.

Kohlenbach, H. W. 1970. The kinetin-induced growth of nucleus, nucleolus and cytoplasm of isolated epidermal cells of *Rhoeo spathacea*. Z. Pflanzenphysiol. **63**:297-307.

Köhler, D. 1966. Veränderungen des Gibberellingehaltes von Salatsamen nach Belichtung. Planta **70**:42-45.

Kopgewicz, J. 1970. Effect of estradiol-17B, estrone and estriol on the endogenous auxins content in plants. Acta Soc. Bot. Pol. **39**:339-346.

Krelle, E. 1970. Interaction of morphactin with gibberellic acid in whole plants and in rooting of cuttings. Biol. Plant (Praha) **12**:256-264.

Krelle, E., and E. Libbert. 1968. Inhibition of the polar auxin transport by a morphactin. Planta **80**:317-320.

Krishnamoorthy, H. N. 1970. Promotion of rooting in mung bean hypocotyl cuttings with ethrel, an ethylene releasing compound. Plant Cell Physiol. **11**:979-982.

Ku, H. S., H. Suge, L. Rappaport, and H. K. Pratt. 1970. Stimulation of rice coleoptile growth by ethylene. Planta **90**:333-339.

Ku, H. S., S. F. Yang, and H. K. Pratt. 1970. Inactivity of apoperoxidase in indoleacetic acid oxidation and in ethylene formation. Plant Physiol. **45**:358-359.

Kuraishi, S. 1968. The effect of kinetin on protein level of *Brassica* leaf disks. Physiol. Plant. **21**:78-83.

Lagerstedt, H. B., and R. G. Langston. 1967. Translocation of radioactive kinetin. Plant Physiol. **42**:611-622.

Lam, S. L., and A. C. Leopold. 1966. Role of leaves in phototropism. Plant Physiol. **41**:847-851.

Lamport, D. T. A. 1965. The protein component of primary cell walls. Adv. Bot. Res. **2**:151-218.

Lantican, B. P., and R. M. Muir. 1969. Auxin physiology of dwarfism in *Pisum sativum*. Physiol. Plant. **22**:412-423.

LaRue, T. A. G., and O. L. Gamborg. 1971. Ethylene production by plant cell cultures: variations in production during growing cycle and in different plant species. Plant Physiol. **48**:394-398.

Lee, T. T. 1971*a*. Increase of indoleacetic acid oxidase isoenzymes by gibberellic acid in tobacco callus cultures. Canad. J. Bot. **49**:687-693.

Lee, T. T. 1971*b*. Promotion of indoleacetic acid oxidase isoenzymes in tobacco callus cultures by indoleacetic acid. Plant Physiol. **48**:56-59.

Lee, T. T. 1971*c*. Cytokinin-controlled indoleacetic acid oxidase isoenzymes in tobacco callus cultures. Plant Physiol. **47**:181-185.

Lee, T. T., and N. Rosa. 1969. Regulation of starch and sugar levels in tobacco leaves by gibberellic acid. Canad. J. Bot. **47**:1595-1598.

Leelavathy, K. M. 1969. Effects of growth regulating substances on fungi. Canad. J. Microbiol. **15**:713-721.

Leike, H., and R. Lau. 1967. Wirkung von Gibberellinsäure und Kinetin auf ruhende Terminalknospen von *Syringa vulgaris* L. Flora **157**:467-470.

Lepp, N. W., and A. J. Peel. 1970. Some effects of IAA and kinetin upon the movement of sugars in the phloem of willow. Planta **90**:230-235.

Lepp, N. W., and A. J. Peel. 1971*a*. Patterns of translocation and metabolism of ^{14}C-labelled IAA in the phloem of willow. Planta **96**:62-73.

Lepp, N. W., and A. J. Peel. 1971*b*. Influence of IAA upon the longitudinal and tangential movement of labelled sugars in the phloem of willow. Planta **97**:50-61.

Leshem, Y. 1971. Abscisic acid as a ribonuclease promoter. Physiol. Plant. **24**:85-89.

Letham, D. S. 1963. Zeatin; a factor inducing cell division isolated from *Zea mays*. Life Sci. **8**:569-573.

Letham, D. S. 1966. Regulators of cell division in plant tissues. II. A cytokinin in plant extracts; isolation and interaction with other growth regulators. Phytochemistry **5**:269-286.

Letham, D. S. 1969. Cytokinins and their relation to other hormones. BioScience **19**:309-316.

Letham, D. S. 1971. Regulators of cell division in plant tissues. XII. A cytokinin bioassay using excised radish cotyledons. Physiol. Plant. **25**: 391-396.

Letham, D. S., and R. K. Ralph. 1967. A cytokinin in soluble RNA from a higher plant. Life Sci. **6**:387-394.

Letham, D. S., and M. W. Williams. 1969. Regulators of cell division in plant tissues. VIII. The cytokinins of the apple fruit. Physiol. Plant. **22**:925-936.

Lewis, L. N., R. A. Khalifah, and C. W. Coggins, Jr. 1965. The existence of the non-indolic citrus auxin in several plant families. Phytochemistry **4**:203-205.

Libbert, E. 1964. Kontrolliert Auxin die Apikaldominanz? Physiol. Plant. **17**:371-378.

Libbert, E., A. Drawert, and R. Schroder. 1969. Pathways of IAA production from tryptophan by plants and by their epiphytic bacteria: a comparison. I. IAA formation by sterile pea sections in vivo as influenced by IAA oxidase inhibitors and by transaminase coenzyme. Physiol. Plant. **22**:1217-1225.

Libbert, E., E. Fischer, A. Drawert, and R. Schroder. 1970*a*. Pathways of IAA production from tryptophan by plants and by their epiphytic bacteria: a comparison. II. Establishment of the tryptophan metabolites, effects of a native inhibitor. Physiol. Plant. **23**:278-286.

Libbert, E., R. Schroder, A. Drawert, and E. Fischer. 1970*b*. Pathways of IAA production from tryptophan by plants and by their epiphytic bacteria: a comparison. III. Metabolism of tryptamine indolacetaldehyde, indoleethanol and indoleacetamide effects of a native inhibitor. Physiol. Plant. **23**:287-293.

Lichtenthaler, H. K., and K. Becker. 1970. Inhibition of the light-induced vitamin K_1 and pigment synthesis by abscisic acid. Phytochemistry **9**:2109-2113.

Lien, T., R. Pettersen, and G. Knutsen. 1971. Effects of indole-3-acetic acid and gibberellin on synchronous cultures of *Chlorella fusca*. Physiol. Plant. **24**:185-190.

Linke, R. D., and N. G. Marinos. 1970. Effects of pregermination pulse treatment with morphactin on *Pisum sativum*. Aust. J. Biol. Sci. **23**:1125-1131.

Linser, H. 1966. The hormonal systems of plants. Angew. Chem. Int. Ed. English **5**:776-784.

Lips, S. H., and N. Roth-Bejerano. 1969. Light and hormones: interchangeability in the induction of nitrate reductase. Science **166**:109-110.

Lockard, R. G., and C. Grunwald. 1970. Grafting and gibberellin effects on the growth of tall and dwarf peas. Plant Physiol. **45**:160-162.

Lona, F., and F. Raffi. 1970. Ethylene production from isoamylalcohol and other compounds associated with essential oils. Ateneo Parmense Sez. II. Acta Natur. **6**:89-96.

Love, D., C. McLellan, and I. Gamow. 1970. Coumarin and coumarin derivates in various growth-types of Engelmann spruce. Sv. Bot. Tidskr. **64**:284-296.

Lovell, P. H. 1971. Translocation of photosynthates in tall and dwarf varieties of pea, *Pisum sativum*. Physiol. Plant. **25**:382-385.

Loverys, B. R., and P. F. Wareing. 1971*a*. The hormonal control of wheat leaf unrolling. Planta **96**:117-127.

Loverys, B. R., and P. F. Wareing. 1971*b*. The red light controlled production of gibberellin in etiolated wheat leaves. Planta **98**:109-116.

Lower, R. L., C. H. Miller, F. H. Baker, and C. L. McCombs. 1970. Effects of 2-chloroethylphosphonic acid treatment at various stages of cucumber development. Hortscience **5**:433-434.

Lwmbi, C. A., D. J. Morré, K. St. Thomson, and R. Hertel. 1971. N-1-Naphthylphthalamic-acid-binding activity of a plasma membrane-rich fraction from maize coleoptiles. Planta **99**:37-45.

Lyon, C. J. 1971. Lateral transport of auxin mediated by gravity in the absence of special georeceptor tissue. Plant Physiol. **48**:642-644.

Mackenzie, I. A., and H. E. Street. 1970. Studies on the growth in culture of plant cells. VIII. The production of ethylene by suspension cultures of *Acer pseudoplatanus* L. J. Exp. Bot. **21**: 824-834.

MacMillan, J., and R. J. Pryce. 1969*a*. Plant hormones. X. The constitution of phaseic acid, a relative of abscisic acid from *Phaseolus multiflorus:* an interpretation of the mass spectrum of phaseic acid and a probable structure. Tetrahedron **25**:5903-5914.

MacMillan, J., and R. J. Pryce. 1969*b*. Plant hormones. IX. Phaseic acid, a relative of abscisic acid from seed of *Phaseolus multiflorus:* possible structures. Tetrahedron **25**:5893-5901.

MacMillan, J., and N. Takahashi. 1968. Proposed procedure for the allocation of trivial names to the gibberellins. Nature (London) **217**:170-171.

Magnus, V., S. Iskric, and S. Kveder. 1971. Indole-3-methanol: a metabolite of indole-3-acetic acid in pea seedlings. Planta **97**:116-125.

Maheshwari, M. G., and L. D. Nooden. 1971. A requirement for DNA synthesis during auxin induction of cell enlargement in tobacco pith tissue. Physiol. Plant. **24**:282-287.

Mahmoud, T. A., and P. L. Steponkus. 1970. Effects of phosfon on shoot histogenesis of *Chrysanthemum morifolium*. J. Amer. Soc. Hort. Sci. **95**:295-299.

Makarem, E. H., and N. Alldridge. 1969. The effect of gibberellic acid on *Hansenula wingei*. Canad. J. Microbiol. **15**:1225-1230.

Mann, J. D., H. Hield, K. H. Yung, and D. Johnson. 1966. Independence of morphactin and gibberellin effects upon higher plants. Plant Physiol. **41**:1751-1752.

Mapson, L. W., F. March, M. J. C. Rhodes, and L. S. C. Wooltorton. 1970. A comparative study of the ability of methionine or linoleic acid to act as precursors of ethylene in plant tissues. Biochem. J. **117**:473-479.

Marei, N., and J. C. Crane. 1971. Growth and respiratory response of fig (*Ficus carica* L. cv. Mission) fruits to ethylene. Plant Physiol. **48**: 249-254.

Marei, N., and R. Romani. 1971. Ethylene-stimulated synthesis of ribosomes, ribonucleic acid, and protein in developing fig fruits. Plant Physiol. **48**:806-808.

Marumo, S., H. Hattori, and H. Abe. 1971. Chromatography of a new natural auxin, 4-chlorindolyl-3-acetic acid and related chloro derivatives. Analyt. Biochem. **40**:488-490.

Masuda, Y., E. Tanimoto, and S. Wada. 1967. Auxin-stimulated RNA synthesis in oat coleoptile cells. Physiol. Plant. **20**:713-719.

McComb, A. J., J. A. McComb, and C. T. Duda. 1970. Increased ribonucleic acid polymerase activity associated with chromatin from internodes of dwarf pea plants treated with gibberellic acid. Plant Physiol. **46**:221-223.

Mehard, C. W., and J. M. Lyons. 1970. A lack of specificity for ethylene-induced mitochondrial changes. Plant Physiol. **46**:36-39.

Mehard, C. W., J. M. Lyons, and J. Kumamoto. 1970. Utilization of model membranes in a test for the mechanism of ethylene action. J. Membrane Biol. **3**:173-179.

Mer, C. L. 1969. Plant growth in relation to endogenous auxin, with special reference to cereal seedlings. New Phyt. **68**:275-294.

Meudt, W. J. 1970. Indole-3-acetic acid oxidase in a *Nicotiana* hybrid and its parental types. Physiol. Plant. **23**:841-849.

Miassod, R., P. Penon, M. Teissere, J. Ricard, and J.-P. Cecchini. 1970. Contribution to the study of rapidly labeled RNA from lentil roots: isolation and identification, effect of a hormone on their synthesis and on the synthesis of peroxidases. Biochim. Biophys. Acta **224**:423-440.

Miginiac, E. 1971. Influence des racines sur le développement vegetatif ou floral des bourgeons cotylédonaire chez le *Scrofularia arguta*: rôle possible des cytokinines. Physiol. Plant. **25**: 234-239.

Milborrow, B. V. 1969. The occurrence and function of abscisic acid in plants. Sci. Prog. **57**: 533-558.

Mishra, D., and B. Mohanty. 1968. β-9(N-dimethylaminosuccinamic acid) reversal of benzimidazole- or IAA-induced inhibition of growth and yield of cowpea plants. J. Exp. Bot. **19**: 567-574.

Mishra, D., and S. C. Paul. 1967. Interaction of CCC and coumarin or IAA on seedling growth of rice. Curr. Sci. (India) **36**:550-551.

Mitsuhashi, M., H. Shibaoka, and M. Shimokoriyama. 1969. Portulal: a rooting promoting substance in *Portulaca* leaves. Plant Cell Physiol. **10**:715-723.

Mittelheuser, C. J., and R. F. M. van Steveninck. 1971. Rapid action of abscisic acid on photosynthesis and stomatal resistance. Planta **97**: 83-86.

Mizrahi, Y., J. Amir, and A. E. Richmond. 1970. The mode of action of kinetin in maintaining the protein content of detached *Tropaeolum majus* leaves. New Phytol. **69**:355-361.

Moll, A. 1971. Soluble invertase in dwarf pea internodes during normal and gibberellin-stimulated growth. Biochem. Physiol. Pflanz. **162**:334-344.

Moore, T. C. 1967. Gibberellin relationship in the

"Alaska" pea *(Pisum sativum)*. Amer. J. Bot. **54:** 262-269.

Morgan, P. W., and J. R. Baur. 1970. Involvement of ethylene in picloram-induced leaf movement response. Plant Physiol. **46:**655-659.

Morgan, P. W., and H. W. Gausman. 1966. Effects of ethylene on auxin transport. Plant Physiol. **41:**45-52.

Morgan, P. W., and R. D. Powell. 1970. Involvement of ethylene in responses of etiolated bean hypocotyl hook to coumarin. Plant Physiol. **45:** 553-557.

Morris, D. A., R. E. Briant, and P. G. Thomson. 1969. The transport and metabolism of ^{14}C-labelled indoleacetic acid in intact pea seedlings. Planta **89:**178-197.

Muenzel, E. 1970. Photosynthesis and hydrolase activity in isolated leaves of *Begonia rex* Putzey under the effect of gibberellic acid. Beitr. Biol. Pflanz. **47:**117-125.

Muller, C. H., and R. Del Moral. 1966. Soil toxicity induced by terpenes from *Salvia leucophylla*. Bull. Torrey Bot. Club **93:**130-137, 332-351.

Muller, W. H., P. Lorber, and B. Haley. 1968. Volatile growth inhibitors produced by *Salvia leucophylla:* effect on seedling growth and respiration. Bull. Torrey Bot. Club **95:**415-422.

Mullins, M. G. 1970. Hormone-directed transport of assimilates in decapitated internodes of *Phaseolus vulgaris* L. Ann. Bot. **34:**897-909.

Nakamura, T., Y. Yabuuchi, and N. Takahashi. 1970. Effect of gibberellic acid on elongation and nucleic acid synthesis in etiolated Alaska pea epicotyl. Develop. Growth Different. **12:** 189-197.

Naylor, J. M. 1969. Regulation of enzyme synthesis in aleurone tissue of *Avena* species. Canad. J. Bot. **47:**2069-2072.

Nelson, H., I. Ilan, and L. Reinhold. 1969. The effect of inhibitors of protein and RNA synthesis on auxin-induced growth. Isr. J. Bot. **18:** 129-134.

Neumann, P. M. 1971. Possible involvement of a glycerophosphate compound in auxin induced growth. Planta **99:**56-62.

Newman, I. A. 1963. Electric potentials and auxin translocation in *Avena.* Aust. J. Biol. Sci. **16:** 629-646.

Newman, I. A. 1970. Auxin transport in *Avena.* I. Indoleacetic acid-^{14}C distributions and speeds. Plant Physiol. **46:**263-272.

Ninnemann, H., J. A. D. Zeevaart, H. Kende, and A. Lang. 1964. The plant growth retardant CCC as inhibitor of gibberellin biosynthesis in *Fusarium moniliforme.* Planta **61:**229-235.

Nitsan, J., and A. Lang. 1966. DNA synthesis in the elongating, non-dividing cells of the lentil epicotyl and its promotion by gibberellin. Plant Physiol. **41:**965-970.

Nooden, L. D., and K. V. Thimann. 1965. Inhibition of protein synthesis and of auxin-induced growth by chloramphenicol. Plant Physiol. **40:** 193-201.

Nudel, V., and E. S. Bamberger. 1971. Kinetin inhibition of ^{3}H-uracil and ^{14}C-leucine incorporation by tobacco cells in suspension culture. Plant Physiol. **47:**400-403.

Ockerse, R., and J. Waber. 1970. The promotion of indole-3-acetic acid oxidation in pea buds by gibberellic acid and treatment. Plant Physiol. **46:**821-824.

Ordin, L., M. J. Garber, B. P. Skee, and G. Rolle. 1966. Role of auxin in growth of inhibitor-treated oat coleoptile tissue. Physiol. Plant. **19:** 937-945.

Oritani, T., and T. Oritani. 1971. Studies on nitrogen metabolism in crop plants. IX. Identification of (+)-abscisic acid in the immature seeds of several species of *Gramineae.* Proc. Crop. Sci. Soc. Jap. **40:**34-39.

Paleg, L., H. Kende, H. Ninnemann, and A. Lang. 1965. Physiologic effects of gibberellic acid. VIII. Growth retardants on barley endosperm. Plant Physiol. **40:**165-169.

Palmer, C. E., and O. E. Smith. 1970. Effect of kinetin on tuber formation on isolated stolons of *Solanum tuberosum* L. cultured in vitro. Plant Cell Physiol. **11:**303-314.

Palmer, J. M. 1970. The induction of phosphatase activity in the slices of Jerusalem artichoke tissue by treatment with indoleacetic acid. Planta **93:**53-59.

Parups, E. V. 1970. Effect of morphactin on the gravimorphism and the uptake, translocation and spatial distribution of indol-3yl-acetic acid in plant tissues in relation to light and gravity. Physiol. Plant. **23:**1176-1186.

Paster, Z., and B. C. Abbott. 1970. Gibberellic acid: a growth factor in the unicellular alga *Gymnodinium breve.* Science **169:**601-602.

Penny, P. 1971. Growth-limiting proteins in relation to auxin-induced elongation in lupin hypocotyls. Plant Physiol. **48:**720-723.

Phillips, D. A., and J. G. Torrey. 1970. Cytokinin production by *Rhizobium japonicum.* Physiol. Plant. **23:**1057-1063.

Phillips, I. D. J. 1969. Apical dominance, p. 165-202. *In* M. B. Wilkins (Ed.). Physiology of plant growth and development. McGraw-Hill Book Co., New York.

Phillips, I. D. J. 1971*a.* Effect of relative hormone concentration on auxin-gibberellin interaction in correlative inhibition of axillary buds. Planta **96**:27-34.

Phillips, I. D. J. 1971*b*. Factors influencing the distribution of growth between stem and axillary buds in decapitated bean plants. J. Exp. Bot. **22**:465-471.

Pilet, P.-E. 1970. The effect of auxin and abscisic acid on the catabolism of RNA. J. Exp. Bot. **21**:446-451.

Pilet, P.-E. 1971. Abscisic acid action on basipetal auxin transport. Physiol. Plant. **25**:28-31.

Pilet, P.-E., and J.-C. Roland. 1971. Effects of abscisic acid on the growth and ultrastructure of tissues cultivated in vitro. Cytobiologie **4**:41-61.

Pinfield, N. J., and A. K. Stobart. 1969. Gibberellin-stimulated nucleic acid metabolism in the cotyledons and embryonic axes of *Corylus avellana* (L.) seeds. New Phytol. **68**:993-999.

Pollard, C. J. 1969. A survey of the sequences of some effects of gibberellic acid in the metabolism of cereal grains. Plant Physiol. **44**:1227-1232.

Poulson, R., and L. Beevers. 1970. Effects of growth regulators on ribonucleic acid metabolism of barley leaf segments. Plant Physiol. **46**:782-785.

Prakash, R., and S. C. Maheshwari. 1970. Studies on cytokinins in watermelon seeds. Physiol. Plant. **23**:792-799.

Pryce, R. J. 1971. Lunularic acid, a common endogenous growth inhibitor of liverworts. Planta **97**:354-357.

Psenak, M., A. Jindra, and P. Kovacs. 1970. Simple phenolic glycosides in potential regulators of the IAA-oxidase system. Biol. Plant. (Praha) **12**:241-245.

Quinlan, J. D., and R. J. Weaver. 1969. Influence of benzyladenine, leaf darkening, and ringing on movement of C-labeled assimilates into expanded leaves of *Vitis vinifera* L. Plant Physiol. **44**:1247-1252.

Raa, J. 1971. Degradation of indol-3yl-acetic acid in homogenates and segments of cabbage roots. Physiol. Plant. **24**:498-505.

Radin, J. W., and R. S. Loomis. 1971. Changes in the cytokinins of radish roots during maturation. Physiol. Plant. **25**:240-244.

Rasmussen, G. K., and J. W. Jones. 1971. Cellulase activity and ethylene concentration in citrus fruit sprayed with abscission-inducing chemicals. Hortscience **6**:402-403.

Ray, P. M., and D. B. Baker. 1962. Promotion of cell wall synthesis by indolylacetic acid. Nature (London) **195**:1322.

Rayle, D. L., M. L. Evans, and R. Hertel. 1970. Action of auxin on cell elongation. Proc. Nat. Acad. Sci. USA **65**:184-191.

Reid, D. M., and A. Grozier. 1970. CCC-induced increase of gibberellin levels in pea seedlings. Planta **94**:95-106.

Rejowski, A., and K. Kulka. 1970. Influence of gibberellic acid on nucleic acids synthesis in resting spring barley seeds. Acta Soc. Bot. Pol. **39**:243-250.

Rhodes, M. J. C., and L. S. C. Wooltorton. 1971. The effect of ethylene on the respiration and on the activity of phenylalanine ammonia lyase in swede and parsnip. Phytochemistry **10**:1989-1997.

Richmond, A. E., B. Sachs, and D. J. Osborne. 1971. Chloroplasts, kinetin and protein synthesis. Physiol. Plant. **24**:176-180.

Ridge, I., and D. J. Osborne. 1969. Cell growth and cellulases: regulation by ethylene and indole-3-acetic acid in shoots of *Pisum sativum*. Nature **223**:318-319.

Ridge, I., and D. J. Osborne. 1970*a*. Regulation of peroxidase activity by ethylene in *Pisum sativum:* requirements for protein and RNA synthesis. J. Exp. Bot. **21**:720-734.

Ridge, I., and D. J. Osborne. 1970*b*. Hydroxyproline and peroxidases in cell walls in *Pisum sativum:* regulation by ethylene. J. Exp. Bot. **21**:843-856.

Ridge, I., and D. J. Osborne. 1971. Role of peroxidase when hydroxyproline-rich protein in plant cell walls is increased by ethylene. Nature **229**: 205-208.

Rijven, A. H. G. C., and V. Parkash. 1970. Cytokinin-induced growth responses by fenugreek cotyledons. Plant Physiol. **45**:638-640.

Rijven, A. H. G. C., and V. Parkash. 1971. Action of kinetin on cotyledons of fenugreek. Plant Physiol. **47**:59-64.

Robbins, W. J., and A. Hervey. 1971. Cytokinin and growth of excised roots of *Bryophyllum calycinum*. Proc. Nat. Acad. Sci. USA **68**:347-348.

Roland, J.-C., and P.-E. Pilet. 1971. Ultrastructural changes of the cell walls of the swamp bramble under the influence of abscisic acid. Compt. Rend. Acad. Sci. (Paris) **272**:72-75.

Roth-Bejerano, N., and S. H. Lips. 1970. Seasonal characteristics of interaction between kinetin and gibberellic acid with regard to level of nitrate reductase activity in tobacco leaves. Isr. J. Bot. **19**:30-36.

Roychoudhury, R., and S. P. Sen. 1965. The effect of gibberellic acid on nucleic acid metabolism in coconut milk nuclei. Plant Cell Physiol. **6**:761-765.

Rubery, P. H., and D. H. Northcote. 1970. The

effect of auxin (2,4-dichlorophenoxyacetic acid) on the synthesis of cell wall polysaccharides in cultured sycamore cells. Biochim. Biophys. Acta **222**:95-108.

Ruddat, F., and R. P. Pharis. 1966. Participation of gibberellin in the control of apical dominance in soybean and redwood. Planta **71**:222-228.

Rudnicki, R. 1969. Studies on abscisic acid in apple seeds. Planta **86**:63-68.

Ruesink, A. W. 1969. Polysaccharidases and the control of cell wall elongation. Planta **89**:95-107.

Russell, D. W., and A. W. Galston. 1969. Blockage by gibberellic acid of phytochrome effects on growth, auxin responses, and flavonoid synthesis in etiolated pea internodes. Plant Physiol. **44**:1211-1216.

Sabnis, D. D., G. Hirschberg, and W. P. Jacobs. 1969. Radioautographic analysis of the distribution of label from ^{3}H-indoleacetic acid supplied to isolated *Coleus* internodes. Plant Physiol. **44**: 27-36.

Sacher, J. A., and S. O. Salminen. 1969. Comparative studies on effect of auxin and ethylene on permeability and synthesis of RNA and protein. Plant Physiol. **44**:1371-1377.

Sachs, T., and K. V. Thimann. 1967. The role of auxins and cytokinins in the release of buds from dominance. Amer. J. Bot. **54**:136-144.

Saha, A. K. 1966. Antagonism between gibberellic acid and maleic hydrazide on the growth of Zinnia *(Zinnia elegans)* plants. Sci. Cult. **32**: 548-549.

Sakai, S., and H. Imaseki. 1971. Auxin-induced ethylene production by mungbean hypocotyl segments. Plant Cell Physiol. **12**:349-359.

Sankhla, N. 1970. Expansion of isolated cotyledons of *Ipomoea pentaphylla* in response to added auxins, cytokinins and morphactin. Biochem. Physiol. Pflanz. **161**:183-186.

Sankhla, N., and S. N. Shukla. 1970. Observations on hypocotyl coiling: effect of ethrel and gibberellin on seedling growth of *Phaseolus radiatus*. Z. Pflanzenphysiol. **63**:284-287.

Sassa, T., T. Tojyo, and K. Munakata. 1970. Isolation of a new plant growth substance with cytokinin-like activity. Nature **227**:379.

Schaeffer, G. W., and F. T. Sharpe, Jr. 1971. Cytokinin function: increase in [methyl ^{14}C] metabolism to phosphatidylcholine and a decrease in oxidation to carbon dioxide–^{14}C. Physiol. Plant. **25**:456-460.

Schlee, D., H. Reinbothe, and K. Mothes. 1966. Wirkungen von Kinetin auf den Adeninabbau in Chlorophyll defekten Blattern von *Pelargonium zonale*. Z. Pflanzenphysiol. **54**:223-236.

Scott, P. C., and L. A. Norris. 1970. Separation of effects of auxin and ethylene in pea roots. Nature **227**:1366-1367.

Search, R. W., and R. G. Stanley. 1970. Stimulation of pollen growth in vitro by ethylene. Phyton. Rev. Int. Bot. Exp. **27**:35-39.

Shannon,, L. M., I. Uritani, and H. Imaseki. 1971. De novo synthesis of peroxidase isozymes in sweet potato slices. Plant Physiol. **47**:493-498.

Sheldrake, A. R. 1971. Auxin in the cambium and its differentiating derivatives. J. Exp. Bot. **22**: 735-740.

Sherwin, J. E., and W. K. Purves. 1969. Tryptophan as an auxin precursor in cucumber seedlings. Plant Physiol. **44**:1303-1309.

Shibaoka, H., and K. V. Thimann. 1970. Antagonisms between kinetin and amino acids: experiments on the mode of action of cytokinins. Plant Physiol. **46**:212-220.

Shih, C. Y., and L. Rappaport. 1970. Regulation of bud rest in tubers of potato, *Solanum tuberosum* L. VII. Effect of abscisic and gibberellic acids on nucleic acid synthesis in excised buds. Plant Physiol. **45**:33-36.

Shimoda, C., and N. Yanagishima. 1968. Strain dependence of the cell-expanding effect of β-1, 3-glucanase in yeast. Physiol. Plant. **21**:1163-1169.

Shimoda, C., and N. Yanagishima. 1971. Role of cell wall-degrading enzymes in auxin-induced cell expansion in yeast. Physiol. Plant. **24**:46-50.

Short, K. C., and J. G. Torrey. 1972. Cytokinins in seedling roots of pea. Plant Physiol. **49**:155-160.

Shininger, T. L. 1971. The regulation of cambial division and secondary xylem differentiation in *Xanthium* by auxins and gibberellin. Plant Physiol. **47**:417-422.

Simard, A. 1971. Initiation of DNA synthesis by kinetin and experimental factors in tobacco pith tissues in vitro. Canad. J. Bot. **49**:1541-1549.

Simpkins, I., and H. E. Street. 1970. Studies on the growth in culture of plant cells. VII. Effects of kinetin on the carbohydrate and nitrogen metabolism of *Acer pseudoplatanus* L. cells grown in suspension culture. J. Exp. Bot. **21**: 170-185.

Simola, L. K., and T. Sopanen. 1971. Effect of α-naphthalene- and α-naphthoxyacetic acid on the activity of certain enzymes of *Atropa belladonna* cv. lutea cells in suspension culture. Physiol. Plant. **25**:8-15.

Sinska, I., and S. Lewak. 1970. Apple seeds gibberellins. Physiol. Veg. **8**:661-667.

Skene, K. G. M. 1971. Hormonal effects on sugar release from vine canes. Ann. Bot. (London) **35**:277-286.

Skene, K. G. M., and M. G. Mullins. 1967. Effect

of CCC on the growth of roots of *Vitus vinifera* L. Planta **77**:157-163.

Skoog, F., F. M. Strong, and C. O. Miller, 1965. Cytokinins. Science **148**:532.

Smith, D. L. 1969. The role of leaves and roots in the control of inflorescence development in *Carex*. Ann. Bot. **33**:505-514.

Smith, K. A., and P. D. Robertson. 1971. Effect of ethylene on root extension of cereal. Nature **234**:148-149.

Sodek, L., and S. T. C. Wright. 1969. The effect of kinetin on ribonuclease, acid phosphatase, lipase and esterase levels in detached wheat leaves. Phytochem. **8**:1629-1640.

Soleimani, A. 1968. Effect of gibberellin A_3 on concentration of DNA, RNA, and protein in pea plant *(Pisum sativum* L.). Hortscience **3**: 276.

Sondheimer, E., and D.-S. Tzou. 1971. The metabolism of hormones during seed germination and dormancy. II. The metabolism of 8-^{14}C zeatin in bean axes. Plant Physiol. **47**:516-520.

Spiker, S., and R. Chalkley. 1971. Electrophoretic analysis of histones from gibberellic acid–treated dwarf peas. Plant Physiol. **47**:342-345.

Srivastava, B. I. S. 1967. Effect of kinetin on nucleic acid synthesis in barley leaf segments. Biochim. Biophys. Acta **145**:166-169.

Stinson, R. A., and M. Spencer. 1969. β-Alanine as an ethylene precursor: investigations towards preparation and properties of a soluble enzyme system from a subcellular particulate fraction of bean cotyledons. Plant Physiol. **44**:1217-1226.

Stonier, T., J. Hudek, R. Vande-Stowe, and H.-M. Yang. 1970. Studies of auxin protectors. VIII. Evidence that auxin protectors act as cellular poisers. Physiol. Plant. **23**:775-783.

Stonier, T., R. W. Singer, and H.-M. Yang. 1970. Studies on auxin protectors. IX. Inactivation of certain protectors by polyphenol oxidase. Plant Physiol. **46**:454-457.

Stowe, B. B., and M. A. Dotts. 1971. Probing a membrane matrix regulating hormone action. I. The molecular length of effective hormone lipids. Plant Physiol. **48**:559-565.

Stuber, C. W., and C. S. Levings, III. 1969. Auxin induction and repression of peroxidase isozymes in oats (*Avena sativa* L.). Crop Sci. **9**:415-416.

Svensson, S.-B. 1971. The effect of coumarin on root growth and root histology. Physiol. Plant **24**:446-470.

Szalai, I. 1965*a*. Ueber die quantitative Veränderung der nativen Gibberellin-Antigen Stoff in mit Rindite stimulierten neuen Kartoffelknollen. Acta Univ. Szeged. **11**:101-106.

Szalai, I. 1965*b*. Brechung der Keimruhe von Fraxinus-Samen durch Gibberellinsäure. Acta Univ. Szeged. **11**:93-100.

Takayanagi, K., and J. F. Harrington. 1971. Enhancement of germination rate of aged seeds by ethylene. Plant Physiol. **47**:521-524.

Tanimoto, E., and Y. Masuda. 1968. Effect of auxin on cell wall degrading enzymes. Physiol. Plant. **21**:820-826.

Taylor, H. F., and R. S. Burden. 1970*a*. Identification of plant growth inhibitors produced by photolysis of violaxanthin. Phytochemistry **9**: 2217-2223.

Taylor, H. F., and R. S. Burden. 1970*b*. Xanthoxin, a new naturally occurring plant growth inhibitor. Nature **227**:302-304.

Taylor, H. F., and T. A. Smith. 1967. Production of plant growth inhibitors from xanthophylls; a possible source of dormin. Nature (London) **215**:1513-1514.

Thimann, K. V. 1937. On the nature of inhibitions caused by auxin. Amer. J. Bot. **24**:407-412.

Thimann, K. V. 1969. The auxins, p. 3-48. *In* M. B. Wilkins (Ed.) The physiology of plant growth and development. McGraw-Hill Book Co., New York.

Thompson, W. F., and R. E. Cleland. 1971. Auxin and ribonucleic acid synthesis in pea stem tissue as studied by deoxyribonucleic acid–ribonucleic acid hybridization. Plant Physiol. **48**: 663-670.

Tietz, A. 1971. Identification of abscisic acid in roots. Planta **96**:93-96.

Tomaszewski, M. 1970. Auxin-gibberellin interactions in apical dominance. Bull. Acad. Pol. Sci. Biol. **18**:361-366.

Torrey, J. G., D. E. Fosket, and P. K. Hepler. 1971. Xylem formation: a paradigm of cytodifferentiation in higher plants. Amer. Sci. **59**: 338-352.

Treharne, K. J., J. L. Stoddart, J. Pughe, K. Paranjothy, and P. F. Wareing. 1970. Effects of gibberellin and cytokinins on the activity of photosynthetic enzymes and plastid ribosomal RNA synthesis in *Phaseolus vulgaris*. L. Nature **228**:129-131.

Trewavas, A. J. 1968. Effect of IAA on RNA and protein synthesis. Arch. Biochem. Biophys. **123**: 324-335.

Trewavas, A. J. 1970. The turnover of nucleic acids in *Lemna minor*. Plant Physiol. **45**:742-751.

Usami, K. 1968. Effects of leaves on adventitious root formation in cuttings of *Glycine max* Merrill. Sci. Hum. Life. **8**:66-76.

Usciati, M., and M. Codaccioni. 1971. Research on the early metabolic modifications induced by 6-benzylaminopurine at the time of the lifting of growth inhibition in axillary buds of *Cicer arietinum*. Mem. Soc. Bot. Fr. **1971**:207-219.

Valdovinos, J. G., and K. S. S. Sastry. 1968. The effect of gibberellin on tryptophan conversion and elongation of the *Avena* coleoptile. Physiol. Plant. **21**:1280-1286.

van Bragt, J. 1969. The effect of CCC on growth and gibberellin content of tomato plants. Neth. J. Agr. Sci. **17**:183-188.

van der Mast, C. A. 1970. Separation of IAA degrading enzymes from pea roots on columns of polyvinylpyrrolidone. II. Acta Bot. Neerl. **19**:141-146.

van Onckelen, H. A., and R. Verbeek. 1971. Incorporation of a cytokinin, *N*-(furfurylamino) purine-8-C^{14} in the nucleic acid fraction of barley. Bull. Soc. Roy. Bot. Belg. **104**:161-168.

van Overbeek, J., J. E. Loeffler, and M. I. R. Mason. 1967. Dormin (abscisin II), inhibitor of plant DNA synthesis? Science **156**:1497-1499.

van Staden, J., and C. H. Bornman. 1969. Inhibition and promotion by abscisic acid of growth in *Spirodela*. Planta **85**:157-159.

Varga, A. 1969. Effects of growth regulators on fruit set and June drop of pears and apples. Neth. J. Agr. Sci. **17**:229-233.

Veda, M., A. Ehmann, and R. S. Bandurski. 1970. Gas-liquid chromatographic analysis of indole-3-acetic acid myoinositol esters in maize kernels. Plant Physiol. **46**:715-719.

Vendrig, J. C. 1967. Steroid derivatives as native auxins in *Coleus*. Ann. N. Y. Acad. Sci. **144**:81-93.

Vendrig, J. C. 1971. Further research on the phytohormones in urine. Z. Pflanzenphysiol. **64**:297-308.

Venis, M. A. 1972. Auxin-induced conjugation systems in peas. Plant Physiol. **49**:24-27.

Verbeek, R., H. A. van Onckelen, and T. Gaspar. 1969. Effets de l'acide gibbérellique et de la kinétine sur le developpement de l'activité α-amylasique durant la croissance de l'orge. Physiol. Plant. **22**:1192-1199.

Viana, M. J., and M. C. Novais. 1969-70. Influence of different light conditions on hormonal control of leaf-embryo development in Bryophyllum. Port. Acta Biol. Ser. A. **11**:373-384.

Vieitez, E., E. Seoane, D. V. Gesto, C. Mato, A. Vazquez, and A. Carnicer. 1966. Hydroxybenzoic acid, a growth regulator, isolated from woody cuttings of *Ribes rubrum*. Physiol. Plant. **19**:294-307.

Villiers, T. A. 1968. An autoradiographic study of the effect of the plant hormone abscisic acid on nucleic acid and protein metabolism. Planta **82**:342-354.

Vrandes, H., and H. Kende. 1968. Studies on cytokinin controlled bud formation in moss (*Funaria hygrometrica*) protonemata. Plant Physiol. **43**:827-837.

Vyas, L. N., and R. K. Garg. 1971. Responses to gibberellin of light-requiring seeds of *Verbena bipinnatifida* Nutt. Z. Pflanzenphysiol. **65**:189-194.

Waisel, Y., and Z. Shapira. 1971. Functions performed by roots of some submerged hydrophytes. Isr. J. Bot. **20**:69-77.

Wallgiehn, R., and B. Parthier. 1964. Der Einfluss des Kinetins auf den RNA und Proteinstoffwechsel in abgeschnittenen, mit Hemmstoffen behandelten Tabakblättern. Phytochemistry **3**:241-248.

Walton, D. C., G. S. Soofi, and E. Sondheimer. 1970. The effects of abscisic acid on growth and nucleic acid synthesis in excised embryonic bean axes. Plant Physiol. **45**:37-40.

Wareing, P. F., and G. Ryback. 1970. Abscisic acid: a newly discovered growth regulating substance in plants. Endeavour **29**(107):84-88.

Warner, H. L., and A. C. Leopold. 1967. Plant growth regulation by stimulation of ethylene production. BioScience **17**:722-723.

Weaver, R. J., and R. M. Pool. 1971. Berry response of "Thompson Seedless" and "Perlette" grapes to application of gibberellic acid. J. Amer. Soc. Hort. Sci. **96**:162-166.

Weaver, R. J., W. Shindy, and W. M. Kliewer. 1969. Growth regulator induced movement of photosynthetic products into fruits of Black Corinth grapes (*Vitis vinifera*). Plant Physiol. **44**:183-188.

Went, F. W. 1928. Wuchsstoff und Wachstum. Rec. Trav. Bot. Néerl. **25**:1.

West, W. C., F. J. Frattarelli, and K. J. Russin. 1970. Effect of stratification and gibberellin on seed germination in *Ginkgo biloba*. Bull. Torrey Bot. Club. **97**:380-388.

Wheeler, A. W. 1971. Auxins and cytokinins exuded during formation of roots by detached primary leaves and stems of dwarf French bean (*Phaseolus vulgaris* L.). Planta **96**:128-135.

Whitmore, F. W. 1971. Effect of indoleacetic acid and hydroxyproline on isoenzymes of peroxidase in wheat coleoptiles. Plant Physiol. **47**:169-171.

Wightman, F. 1964. Pathways of tryptophan metabolism in tomato plants. Colloq. Int. Cent. Nat. Rech. Sci. **123**:191-212.

Wilkins, M. B., and A. R. Crane. 1970. Auxin transport in roots. V. Effects of temperature on the movement of IAA in *Zea* roots. J. Exp. Bot. **21**:195-211.

Wilkins, M. B., and T. K. Scott. 1968. Auxin transport in roots. III. Dependence of the polar flux of IAA in *Zea* roots upon metabolism. Planta **83**:335-346.

Wilson, R. E., and E. L. Rice. 1968. Allelopathy as expressed by *Helianthus annuus* and its role in old-field succession. Bull. Torrey Bot. Club. **95**:432-448.

Wood, H. N., A. C. Braun, H. Brandes, and H. Kende. 1969. Studies on the distribution and properties of a new class of cell division-promoting substances from higher plant species. Proc. Nat. Acad. Sci. USA **62**:349-356.

Wylie, A. W., K. Ryugo, and R. M. Sachs. 1970. Effects of growth retardants on biosynthesis of gibberellin precursors in root tips of peas, *Pisum sativum* L. J. Amer. Soc. Hort. Sci. **95**: 627-630.

Yamamoto, R., and Y. Masuda. 1971: Stress-relaxation properties of the *Avena* coleoptile cell wall. Physiol. Plant. **25**:330-335.

Yanagishima, N., and C. Shimoda. 1968. Auxin-induced expansion growth of cells and protoplasts of yeast. Physiol. Plant. **21**:1122-1128.

Yoshida, R., T. Oritani, and A. Nishi. 1971. Kinetin-like factors in the root exudate of rice plants. Plant Cell Physiol. **12**:89-94.

Ziegler, H. 1970. Morphactins. Endeavour **29**: 112-116.

Ziegler, H., D. Köhler, and B. Streitz. 1966. Ist 2-chlor-9-fluorenol-9-carbonsäure ein gibberellin antagonist? Z. Pflanzenphysiol. **54**:118-124.

Zwar, J. A., and M. I. Bruce. 1970. Cytokinins from apple extract and coconut milk. Aust. J. Biol. Sci. **23**:289-297.

GENERAL REFERENCES

Avery, G. S., E. B. Johnson, R. M. Addoms, and B. F. Thomson. 1947. Hormones and horticulture. McGraw-Hill Book Co., New York.

Boysen-Jensen, P. 1936. Growth hormones in plants. McGraw-Hill Book Co., New York.

Heyn, A. N. J. 1940. The physiology of cell elongation. Bot. Rev. **6**:515-574.

Pilet, P.-E. 1961. Les phytohormones de croissance. Masson & Cie, Paris.

Schantz, E. M. 1966. Chemistry of naturally-occurring growth regulating substances. Ann. Rev. Plant Physiol. **17**:409-438.

Skoog, F. (Ed.). 1951. Plant growth substances. University of Wisconsin Press, Madison, Wis.

Thimann, K. V. 1965. Toward an endocrinology of higher plants. Recent Progr. Hormone Res. **21**:579-596.

van Overbeek, J. 1942. Hormonal control of embryo and seedling. Sympos. Quant. Biol. **10**: 126-134.

van Overbeek, J. 1966. Plant hormones and regulators. Science **152**:721-731.

Went, F. W., and K. V. Thimann. 1937. Phytohormones. The Macmillan Co., New York.

Wightman, F., and G. Setterfield, 1968. Biochemistry and physiology of plant growth substances. Runge Press, Ottawa.

20

Development

STAGES OF DEVELOPMENT

Tissues excised (cut) from apical or cambial meristems can be grown in suitable culture media forming so-called callus tissue, which may continue growth indefinitely without forming any specialized, differentiated tissues. It is thus capable of growth without development. The normal plant, however, undergoes both growth and development. The fertilized egg-cell develops into an embryo, the embryo develops into a seedling, and the seedling develops into a flowering and then a fruiting plant. In the case of many annuals (e.g., sunflower), growth ceases with initiation of the reproductive process, and this is followed by senescence and death. Although they are much longer-lived, even perennials eventually senesce and die.

At one stage the plant as a whole, or parts of it, may be either in the active or the resting (dormant) state. Similarly, a single organ of the plant may pass through several developmental subphases. Even different portions of an organ may be at different stages of development. In the case of the stem, for instance, older internodes are completing their development when younger ones are just beginning theirs.

The growing plant and its separate parts thus pass through several stages and phases of development that are associated with definite morphological as well as physiological changes. They may be classified as follows (adapted from Sax, 1962).

1. Vegetative stage
   a. Embryonic phase
   b. Juvenile phase
   c. Adolescent phase
2. Reproductive (adult) stage
   a. Flowering phase (in some preceded by vegetative reproduction)
   b. Fruiting phase
3. Senescent stage

In some cases the characteristics of a specific stage or phase have been investigated; in others the transition from one phase to another has been investigated. The following stages, phases, or transitions have received the most attention:

1. Transition from the embryonic to the juvenile phase, or seed germination
2. The juvenile phase
3. Transition from the vegetative to the reproductive stage
4. Fruit and seed formation
5. Senescence
6. Abscission
7. Longevity

Seed germination

The air-dry seed. Most air-dry seeds normally contain about 12% water, all of which consists of so-called "bound water." This water is held by strong adsorptive forces to the surfaces of the hydrophilic colloids in the seed—starch grains, proteins, cellulose, and other cell wall constituents. There are some exceptions among the larger seeds, whose internal cells may not have come to equilibrium with the vapor pressure of the surrounding air and may therefore contain larger or smaller quantities of free water, such as the "milk" in coconuts. The embryos of oil palm nuts may double their original water content in 9 days at the expense of the endosperm, which is much less hydrated (Rabechault et al., 1970). This transfer of water occurs during storage and appears to be dependent on the initiation of metabolic processes. Normally, however, in the case of truly air-dry seed, the metabolic rate remains low and does not increase until water is imbibed from its environment. This low metabolic rate is largely due to the absence of free water. Other factors, however, may also be deficient, such as cytochrome *c* (Wilson and Bonner, 1971) and ATP (Pradet et al., 1968). The vitamin or cofactor, nicotinic acid, may be present in the dry seed but only in the bound form (Mukherji et al., 1968). Some enzymes, such as those required for fatty acid synthesis, are already present in the dry seed and participate in synthesis as soon as a critical level of water content is achieved (Harwood and Stumpf, 1970). Lipase and phytase, however, are not present in the dry wheat grain but arise within 12 hr of imbibition (Laidman and Tavener, 1971).

Another conceivable impediment to metabolism in the dry seed is the existence of the main reserves in the insoluble and therefore not immediately usable form. Some soluble carbohydrates, however, are also present in the dry seed and are presumably required for rapid germination. In clover seed, galactose, glucose, and fructose were present, as well as ribose, sucrose, raffinose, stachyose and verbascose (Somme, 1971). The latter five were utilized during the first 2 days of germination.

Nearly all air-dry seeds contain phytic acid (inositol hexaphosphoric acid) chiefly as phytin—the calcium, magnesium, and potassium salt (Mayer and Poljakoff-Mayber, 1963). Mature pinto beans contain phytin to the extent of 1% of their weight (Makower, 1969). As much as 76% of the total seed phosphorus occurs as phytin (e.g., in rice—Mukherji et al., 1971). Although many functions have been suggested for phytin, the lack of a change in ATP level during the buildup of phytin in developing seed has led to the conclusion that its role is solely as a store of phosphorus and cations (Williams, 1970). There is no phytase activity in the cotyledons of unsoaked seed (Mandal and Biswas, 1970*a*).

Imbibition. The imbibition of water by air-dry seed is the first requirement for the cell growth that leads to germination. In its strictest sense, imbibition refers solely to water absorption—a diffusion (or osmotic) process that depends on the water potential difference between the imbibing body (e.g., the seed) and its environment. But cell growth requires more than water. All the protoplasmic and cell wall substances must be synthesized before any true cell growth can occur. Therefore it is not surprising that there is an absolute requirement for both protein and nucleic acid synthesis prior to the initiation of cell enlargement in bean seeds (Walton and Soofi, 1969). Some of these syntheses occur as soon as the cells obtain a little free water from their environment (Table 20-1). An increase in respiration must also occur to supply the energy needs for the syntheses. It is therefore not possible to distinguish in time between the water absorption and the associated metabolic and growth processes.

Table 20-1. Changes observed during the early stages of imbibition of some germinating seeds*

Germination stage	Germination time	Condition of embryo cells
1. Air-dry	0 min.	1. ER reduced to few crescents 2. Mitochondria with few cristae 3. Dictyosomes compacted 4. Lipid drops at plasmalemma 5. No chloroplasts (Durzan et al., 1971)
2. Imbibition stage		
Phase I (physical wetting)	0-10 min.	
Phase II (little further imbibition)	10-30 min.	1. Marked increase in ATP (Narayanan and Pradet, 1968) 2. Protein synthesis begins on ribosomes
Phase III (continuous water uptake)	a. 2 hr.	1. Further increase in ATP (Narayanan and Pradet, 1968) 2. Increase in respiration 3. Increase in number of mitochondria and their cristae 4. Changes in all organelles
	b. 6 hr.	1. Circlets fewer, ER associated with nuclear membrane 2. Uridine and thymidine incorporated 3. Large increase in protein synthesis on new ribosomes on ER

*Modified from Hallam, N. D., B. E. Roberts, and D. J. Osborne. 1972. Planta **105**:293-309.

In practice, the water-absorbing seed is said to be in the imbibition stage as long as no germination is visible. The imbibition stage can thus be defined as the first stage of germination up to emergence of the seedling from the seed coat. It is sometimes subdivided into phases.

Seeds germinate at water potentials down to –2 to –8 bars, depending on the species (Kaufmann, 1969). The rate of imbibition depends on other factors such as the presence of inhibiting phenolic substances (Maasch and Ruge, 1970). Other changes may also occur during the initial stages of imbibition. The coats of apple seeds, for instance, enclose adsorbed gases that must be released during imbibition (Come, 1971*a*). This may explain the role of phenolic substances in seed coats, for they may combine with adsorbed oxygen, converting it to water or carbon dioxide, which is then released. It also explains the relatively high respiratory quotients of seeds during imbibition (Come, 1971*b*).

The respiratory increase goes through three phases in mung beans: (1) a sharp rise, (2) a nearly constant rate, and (3) a second rise (Morohashi and Shimokoriyama, 1972). In general, it tends to parallel imbibition.

During imbibition, many fundamental ultrastructural and metabolic changes occur internally. The dry seed is normally deficient in some components of the protein-synthesizing nucleic acid apparatus. Although ribosomes occur in pea seeds with all the general characteristics of normal ribosomes (Gumilevskaya et al., 1971), there are no polysomes (Barker and Rieber, 1967). Polysomes develop early during imbibition, within 4 hr in the case of pine (Saski and Brown, 1971). In rice, the ribosomes are able to dissociate into subunits within 20 min of the initiation of imbibition at 28° C, and this is accompanied by activation of protein synthesis (App et al., 1971). In pea seeds, the cytoplasmic organelles underwent observable changes within 8 hr of soaking, including a proliferation of ER, the appearance of dictyosomes, and an inward migration of lipid bodies (Yoo, 1970). The nucleus, however, showed little change for 48 hr. In the radish there is extensive RNA synthesis during the first 48 hr (Julien et al., 1970). In further contrast to pea seeds, maize seeds showed the most spectacular changes in the nucleus (Deltour and Bronchart, 1971). In the dry seed the chromatin was heavily condensed, but complete dispersion occurred during the first 8 hr of imbibition. The nucleolus also showed profound changes, and RNA synthesis occurred within 4 hr. In wheat embryos the genome is essentially inactive for the first 12 hr of imbibition, and no mRNA is transcribed. Transcription of rRNA, however, begins within 2 hr (Chen et al., 1971). The same is true of beans (Walbot, 1971). Therefore the earliest protein synthesis during germination must occur on ribosomes already present in the dry embryo. The later synthesis of new ribosomes, however, initiates a sharp acceleration of protein synthesis.

The utilization of preformed ribosomes for the initial, relatively slow protein synthesis during imbibition requires the presence of mRNA. Evidence has been produced that preformed mRNA indeed is present in the quiescent wheat embryo (Weeks and Marcus, 1971) and that it converts the messenger-free ribosomes into functional polyribosomes during imbibition. No new mRNA is synthesized during the first 24 hr (Price and Murray, 1969). In the preemergent 1 mm root tip of onion seed, the total RNA per cell began to increase after 18 hr but the DNA did not begin to increase until 20 hr later (Melera, 1971). The total RNA doubles between 24 and 48 hr of germination in barley (Bohmova and Guitton, 1971). A rapid resynthesis of tRNA during germination of wheat seed brought it back to the normal level at about 20 hr (Vold and Sypherd, 1968). Both the heavier 28-S and the lighter 18-S rRNAs are synthesized in approximately linear fashion for up to 6 days in germinating pea cotyledons (Hewish et al., 1971), and tRNA is also synthesized.

Besides the nucleic acid apparatus, enzymes must also be present for metabolism to begin as soon as imbibition has introduced free water into the cells. Some enzymes are, indeed, present in the active state in the air-dry seeds, others are not. Most enzymes increase in activity during the first 1 to 3 days of germination, some by activation of previously inactive proteins, others by de novo protein synthesis. A few enzymes decrease in activity during germination. Phenylalanine ammonia-lyase (PAL) is first detectable in buckwheat seedlings on the second day of germination (Amrhein and Zenk, 1970). New isozymes may also appear during germination, for instance, one of α-amylase (Palmiano and Juliano, 1972). Because of such enzyme deficiencies in the dry seed, the main carbohydrate may not begin to be mobilized until as late as 18 hr after emergence of the radicle (Reid, 1971). Syntheses are even more likely to be delayed than hydrolyses, as was found in the case of sterol (Baisted, 1971).

All of these rapid ultrastructural and metabolic changes during imbibition require the utilization of reserves present in the air-dry seed. Most of these reserves are stored in the insoluble state: the carbohydrates usually as starch, the lipids as fats, the organic nitrogen usually as insoluble protein and nucleic acids. One of the first metabolic processes in imbibing seeds must be the digestion of these insoluble reserves and their conversion into soluble, easily translocated and utilized substances. The endosperm starch, for instance, is converted to glucose, which is transported via the symplast to the scutellum where it is converted to sucrose (Nomura et al., 1969). This conversion to sucrose is essential for transport to the embryonic axis (Chen and Varner, 1969). In the soybean, reducing sugar of the embryonic axis reached a maximum 36 hr after imbibition (Wahab and Burriss, 1971), leading to a doubling of its dry weight. The dry weight of the seedling as a whole decreased at first due to respiration of some of the reserves and did not show a significant increase until the eighth day.

The protein reserves are hydrolyzed to amino acids and translocated to the embryo during the first 48 hr, where they are resynthesized to new proteins (Sircar, 1967). Four proteolytic enzymes capable of producing this hydrolysis were found in pea seeds (Beevers, 1968). The enzymes that hydrolyze the protein and nucleic acid reserves are found in the cotyledon (Beevers and Splittstoesser, 1968). The amino acids produced by hydrolysis of proteins in germinating seeds are commonly largely converted into one or both of the amides asparagine and glutamine (Chibnall, 1938—see Chapter 16). These amides sometimes account for as much as 80% to 90% of the free amino acids. Even in dormant wheat seeds, asparagine is the major component, accounting for as much as 43% of the total amino acids and amides. Within 72 hr of germination this rises to 80% (Wang, 1968). In dormant corn, proline accounts for 40% of the free amino acids and amides, and asparagine accounts for only 13%. On germination, however, the proline is replaced by asparagine. The common feature in the two, therefore, is the predominance of asparagine in the seedling, in both roots and shoot. In the pumpkin, however, most amino acids increased steadily in the cotyledons during germination (Chou and Splittstoesser, 1972). The six exceptions were found in large amounts in the translocation stream.

Nucleic acids are depleted in the coleorhiza and coleoptile regions of germinating wheat embryos during the first 2 days. They are redistributed, particularly to leaf and root tissue (Price and Murray, 1969). More than 50% of the tRNA was lost from wheat seed during the first 10 to 15 hr (Vold and Sypherd, 1968). In *Vicia sativa*, the RNA content of the cotyledons fell from the first to the seventh day (Grellett et al., 1968) while ribonuclease activity increased, but large increases of RNA occurred in the embryo.

Conversion of fat to carbohydrate occurs during the germination of fat-storing seeds. This apparently takes place in specific organelles called glyoxysomes (Cooper and Beevers, 1969). The specific activities of the enzymes in the glyoxysomes of the castor bean increase until the fifth day of germination and then remain constant (Gerhardt and Beevers, 1970). In seeds of other plants the time factors differ somewhat (Longo and Longo, 1970; Schnarrenberger et al., 1971).

As mentioned earlier, a reserve substance characteristic of seeds is phytin. During the germination of rice seed, the phytin is dephosphorylated to inorganic phosphorus (Mukherji et al., 1971) and to free inositol, which increases from 0.04 mg/g seeds before germination to a value of 0.55 mg after germination, or from 2% to 44% of the total inositol. Nevertheless, total free and phytin inositol decreased from 1.885 to 1.245 mg/g seed (Kurasawa et al., 1969). This degradation of phytin is associated with (and dependent on) an increase in activity of the hydrolytic enzyme phytase, which is dependent on the synthesis of new RNA (e.g., in the cotyledon of mung bean) but not of DNA (Mandal and Biswas, 1970*b*). In the case of wheat, however, it was concluded that the induction process probably involves activation of an already present inactive form of the enzyme by glutamine and other nitrogen compounds (Eastwood and Laidman, 1971). The potassium, phosphorus, magnesium, and calcium released from the breakdown of phytin are translocated to the developing seedling (Eastwood and Laidman, 1968).

On the basis of all the above results, the sequence of events during imbibition and therefore the events previous to emergence from the seed can be summarized as follows:

1. Physical absorption of water and release of absorbed gases
2. Diffusion of dissolved substrate to preexisting enzymes, leading to stimulated metabolism. In some cases this is accompanied by polysome formation or activation
3. Active protein synthesis and ultrastructural changes
4. Digestion (by newly formed enzymes) of reserves—starch, fats, proteins, phytin
5. Translocation of soluble products from endosperm or cotyledons to embryo root and shoot
6. Embryo growth

Control of germination by growth regulators. The control of bud and seed dormancy by an inhibitor-promoter complex has been described in Chapter 19. For dormant seed to germinate, the inhibitor must be inactivated and the promoter must be activated. When the inhibitor has been identified, it is usually found to be abscisic acid (ABA) (e.g., Martin et al., 1969). Gibberellic acid (GA) is most commonly at least part of the promoter complex.

Germination can be largely explained by an interaction between gibberellin and auxin. Gibberellin stimulates protease activity, leading to a breakdown of proteins to amino acids, including tryptophan. The tryptophan may, perhaps, then be converted to IAA. According to van Overbeek (1966, see Chapter 19), the sequence of events in barley seed germination is as follows:

1. Embryo is activated by imbibition of water.
2. Gibberellin is produced in the embryo.
3. Gibberellin moves to aleurone layer.
4. Gibberellin induces synthesis of digestive enzymes (hydrolases): α-amylase, protease, (cellulase ?).
5. These enzymes convert starch to sugars, proteins to amino acids, including tryptophan.
6. Gibberellin induces root growth.
7. Tryptophan produced by protease activity is translocated to the coleoptile tip.
8. In the coleoptile tip it is converted to IAA.
9. IAA moves polarly from the coleoptile tip to the base.
10. The coleoptile cell walls become weakened.
11. Water is taken up and the cells enlarge.

This scheme applies in general to cereal seeds with the GA migrating from the embryo to the aleurone layer between 12 and 20 hr after moistening (MacLeod, 1969). Many results with other seeds also agree with this scheme. Seed treatment with GA increased α-amylase and phosphatase activity of rice but ABA depressed both (Sircar, 1967). GA concentration is low in dry bean seeds but high in cotyledons after imbibition for 4 days, leading to development of amylase activity (Dale, 1969). The GA is believed to be released from the bound form in these bean cotyledons; but in barley, it is synthesized in the embryo and in turn initiates α-amylase synthesis in the aleurone layer (Groat and Briggs, 1969). Many marked changes are observable in the organelles of the aleurone layer as a result of this promotion by GA. The mitochondria increase and develop cristae, the aleurone grains and their inclusions swell, dissolve, and vacuolate, spherosomes enlarge and lose their contents, and the ER increases and becomes rough with ribosomes. Since all these changes occur in response to GA, this hormone is believed to lead to a derepression of genes allowing enhanced RNA metabolism and synthesis of enzyme protein at the expense of cell reserves.

According to this concept, the GA causes an increase in DNA-template activity and an increased production of a specific RNA fraction. This explanation, however, is opposed by the ability of GA and ABA to produce their typical effects in lettuce seed prevented from synthesizing DNA by irradiation (Haber et al., 1969). The more plausible explanation is that they both act at the translation level (Chen and Osborne, 1970), controlling the de novo synthesis of hydrolytic enzymes required for the breakdown of food reserves preliminary to germination.

All these results lead to the conclusion that GA is the endogenous, enzyme-iniating and germination-promoting hormone. This does not, of course, eliminate roles for other hormones in the germination process. A role for IAA in the seedling has already been described. In the imbibition stage, however, somewhat contradictory results have been reported. In rice seeds, immediately after hydration, endosperm auxin is translocated to the scutellum, and the auxin content of the embryo increases after the rise in respiration during the first 72 hr of germination (Sircar, 1967). In maize, on the contrary, the total IAA content decreases during 96 hr of germination, declining to about 10% of the original content (Ueda and Bandurski, 1969). Furthermore, IAA and its derivatives completely inhibit the germination of lettuce seed even at relatively low concentrations (Khan, 1971).

In contrast to IAA, cytokinin (CK) has been reported to act as a promoter of germination. CK rather than GA opposes the inhibiting action of ABA in lettuce seed (Khan, 1967). Similarly, in contrast to cereal seed, lettuce seed reveals no major GA-induced changes in protein synthesis before visible germination (Bewley and Black, 1972). Germination was, in fact, promoted by inhibitors of protein synthesis (Black and Richardson, 1968). The synthesis of proteolytic enzymes in the soybean cotyledons appears to be controlled by the embryonic axis, and the hormone involved appears to be a CK (Chen and Lin, 1971). The inhibition of germination of *Arabidopsis* and *Leptadaenia* seed by ABA was also reversed by CK, and not by IAA or GA (Sankhla and Sankhla, 1968*a*). Similar results were obtained with beans (Gepstein and Ilan, 1970) and sunflower (Gilad et al., 1970). These investigators, however, conclude that the regulatory action of the embryo axis on starch degradation in the cotyledons is mediated by CK. Tobacco seeds that were unable to germinate in the dark were induced to germinate by any one of the three growth regulators, GA, IAA, or CK, although best results were obtained by a combination of GA and CK (Spaulding and Steffens, 1969). Even in the case of barley, the GA-promoted seed par excellence, CK may be involved. Although GA stimulates α-amylase activation during the first 3 days of germination, CK inhibits it after the third day (Verbeek et al., 1969). However, GA induces the formation of five α-amylase isozymes, whereas CK has no influence on their formation (van Onckelen and Verbeek, 1969).

Khan (1971) explains the above discrepancies as follows. CK by itself has no profound effect on germination. However, when the GA effect is blocked by ABA, this inhibition is only slightly overcome by excess GA. It is at this point that CK is effective, leading to almost full recovery in the presence of ABA. The inhibitor-promoter control of seed germination therefore depends on a balance between three growth regulators. Of the eight possible

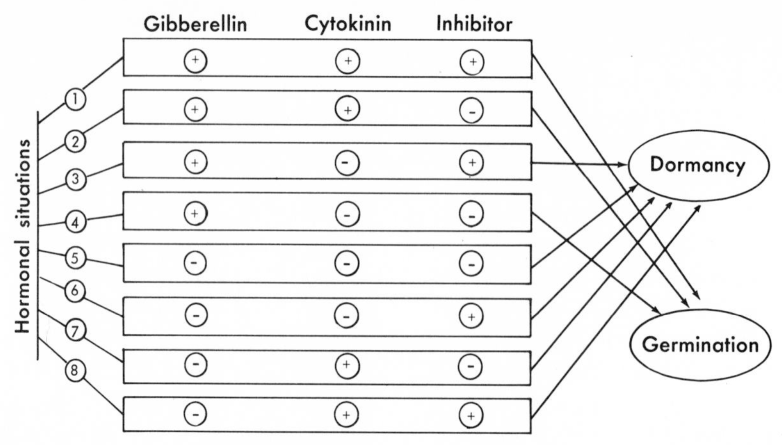

Fig. 20-1. Model for the hormonal mechanisms of seed dormancy and germination using gibberellin, cytokinin, and inhibitor. It shows hormonal or physiological situations likely to be found in seeds. Presence of any one type of hormone at physiologically active concentrations is designated by the plus sign and its absence by the minus sign. (From Khan, A. A. 1971. Science 171:853-859. Copyright 1971 by the American Association for the Advancement of Science.)

combinations, three lead to promotion of germination and five lead to inhibition (Fig. 20-1). The effect of CK is explained by an induction of DNA synthesis.

This concept, however, fails to include the other hormones that have been shown to affect germination—the stimulating or inhibiting effect of IAA (see earlier discussion) and the stimulating effect of ethylene on lettuce seed (Abeles and Lonski, 1969). Furthermore, the balance that favors germination at one stage may not necessarily be optimum for another stage. This may explain the reported inhibition of root and coleoptile growth in barley by CK (Gaspar et al., 1969), possibly by a destruction of IAA.

Whatever the interactions between the growth regulators during seed germination, the mechanism is still not clear. Although the gene-activation and inactivation concept appears to be the most popular, other effects may be important. Each hormone, for instance, has a specific effect on the ability of the endosperm to retain (CK), to release (GA), and to inhibit the release (ABA) of potassium, phosphorus, magnesium, and calcium (Eastwood and Laidman, 1971).

Although GA stimulates α-amylase production, it may also influence germination in other ways. In oats it causes an increased degradation of glucose by way of the pentose phosphate pathway, which is an essential step in the preparation for germination (Simmonds and Simpson, 1971). Furthermore, there are other regulatory mechanisms that may also be involved. For instance, any accumulation of the products of hydrolysis, such as 0.2 to 0.4 M concentrations of maltose or glucose in the starchy endosperm of the germinating barley seed, effectively inhibits production of α-amylase (Jones and Armstrong, 1971).

Juvenile phase

The juvenile phase has been defined in different ways by different investigators on the basis of leaf form, incapacity to produce flowers, degree of thorniness, or leaf abscission in the fall (Schwabe, 1971). Gregory (see Schwabe, 1971) has attempted a quantitative measurement of the juvenile phase by the minimum number of leaves that must be produced before flowering, but this can be markedly reduced by various methods (see following discussion). The emphasis on lack of flowering demonstrates that few investigators attempt to distinguish between the juvenile and the adolescent phases. It is therefore usually contrasted with the adult or reproductive stage. The juvenile phase is readily observed only in plants that persist in this phase for some time. The English ivy *(Hedera helix)* and other species of *Hedera* may persist as a vine in the juvenile phase or may develop into the arborescent plant, which is the adult or reproductive stage.

Besides the morphological differences between the juvenile and adult stages, there are also physiological differences. Eucalyptus seedlings possess juvenile, intermediate, and adult leaf stages. The juvenile leaves possess a number of features that should permit them to function better at a low light intensity (Cameron, 1970). Nevertheless, the rate of apparent photosynthesis was higher in the intermediate leaves. In the case of *Hedera helix*, the rooting response of adult tips to applied IAA was similar in low light intensity to that of juvenile tips in high light intensity.

Grafting of reproductive to juvenile types of shoots of the ivy causes the reproductive shoot to revert to the juvenile state (Schwabe, 1971). That this is probably due to hormone control is indicated by the effects of applied growth regulators, although the results are contradictory. For instance, it is now possible to induce this adult, arborescent form of *Hedera helix* to produce juvenile shoots by spraying with gibberellin (Robbins, 1960) and to revert from a reproductive to a vegetative condition. This gibberellin-induced reversion to juvenile characteristics has been observed also in *Citrus* (Thorpe and Hield, 1970), and in *Centaurea solstitialis* (Feldman and Cutter, 1970*a*). Cycocel, on the other hand, favored the formation of adult leaves. Similarly, the vegetative reproduction of potato plants by tuber formation is inhibited by GA and stimulated by CK or by antigibberellins (Smith and Palmer, 1970; Lezica, 1970). It has therefore been proposed (Feldman and Cutter, 1970*b*) that a gibberellin is at least one of the factors regulating the development of a leaf primordium into a juvenile or adult leaf.

In contrast to the results just mentioned, coastal and giant redwood normally show no flowering until 20 yr and 24 to 70 yr, respectively. At ages of less than 1 yr they were induced to flower precociously by applications of 1500 μg or more of GA (Pharis and Morf, 1969). Arizona cypress responds in the same way to as little as 0.05 μg of GA. Similarly, the change from the juvenile to the adult stage of seedlings of the tea crabapple can be induced by treating the buds with CKs and gibberellins (Zimmerman, 1971). Evans (1971) therefore suggests that juvenility may be due to a low level of endogenous GA.

In the case of the cocklebur, the juvenile cotyledons are characterized by a high level of endogenous auxin, in contrast to the first pair of true leaves (Podol'nyi and Loboda, 1971). A sesquiterpene ([+]–dehydrojuvabione) with high juvenile hormone activity has been isolated from *Abies balsamea* (Mori et al., 1970). Therefore it seems evident that there are regulators of development as well as of growth, and that some substances may play both roles.

Transition to the reproductive stage

In agreement with the concept of juvenility as due to a low level of endogenous GA, there are many cases in the literature of flower promotion by GA (Table 20-2). The endogenous GA content also increases shortly before flower initiation and an inhibitor (possibly ABA) decreases (Badr et al., 1970). In the case of *Blitum capitatum* and *B. virgatum*, however, it was possible to dissociate the well-known induction of stem elongation by GA from an effect on flowering (Jacques, 1969). Evans (1969) concluded that endogenous gibberellins play no direct role in floral induction in *Lolium temulentum* but that

Table 20-2. Promotion of reproduction by applied GA

Plant	Reference
Easter lily	Laiche and Box, 1970
Bryophyllum	Chailakhyan et al., 1970
Ceratopteris thalictroides (fern)	Stein, 1971
Thuja plicata (red cedar)	Owens and Pharis, 1971
Cynara scolymus (globe artichoke)	Snyder et al., 1971
Trifolium pratense	Jones and Stoddart, 1970
Cucumber	Atsmon, 1968
Carnation	Harris et al., 1969
Rudbeckia bicolor	Lezzica, 1969
Olive	Badr et al., 1970

compounds sharing early steps in the biosynthetic pathway to gibberellin may do so, and that their action can be enhanced by applied gibberellins. That GA alone is insufficient to induce flowering is indicated by the results with *Lemna.* Treatment with acetone extracts of flowering cocklebur *(Xanthium)* with or without GA initiated flowering. Treatment with extracts from vegetative plants failed to initiate flowering even when supplemented with GA (Hodson et al., 1970).

A unique role for GA in flower induction is also difficult to reconcile with results obtained with other growth regulators. Chlorocholine chloride (CCC) hastened the flowering of *Hydrangea macrophylla,* although another retardant (B995) retarded flowering (Milletti and DeCapite, 1968). Acetylene, which sometimes can replace ethylene, initiated early flowering in the pineapple (Gifford, 1969). Once flowering is initiated in *Cyperus rotundus,* their normal differentiation and development is promoted by CK (Ram and Batra, 1970).

What, then, is the explanation of these interrelations between growth regulators and reproduction? The similarity between the dependence of both growth and development on growth regulators led to suggestions of a similar control mechanism. According to Heslop-Harrison (1964), flower-forming substances, natural and otherwise, are effectors. They unblock a whole gene system that is quiescent during vegetative growth. This can happen only in active apices possessing an active nucleic acid and protein-synthesizing metabolism. This would explain the flower-inhibiting role of purine and pyrimidine analogs, for instance, the inhibition of flowering in *Anagallis arvensis* by 5-FUDR and by actinomycin D (Taillandier, 1971). This led to the conclusion that a renewal of mitotic activity and an increase in DNA synthesis are indispensable to the processes leading to flowering. Wellensiek (1967) has concluded that flower bud formation depends on the relative ratio of promoting and inhibiting substances within the plant, provided that a sufficient amount of food and energy-producing substances is available. Earlier investigators pointed to another kind of balance. It was long suspected that a high carbon-nitrogen ratio within the plant favors flowering, and a low ratio favors vegetative growth. Although it has been shown that this ratio by itself is incapable of controlling flowering, it nevertheless may have a great effect on the quantity of flowers produced. It may even have an all-or-none effect. Thus *Sinapis alba* is a long-day plant; yet, when apices are grown in vitro, they can all be made to flower in short days if supplied with optimal sucrose and low ammonium nitrate (Deltour, 1967). These concepts will be discussed further in connection with the environmental control of reproduction.

Fruit and seed development

As in the case of other aspects of plant development, fruit set and ripening are controlled by growth regulators. The first one to be involved appears to be CK. When it is deficient, fruit set is impaired. In the case of grapevines, for instance, a rootstock containing less CK in its bleeding sap depressed the yield by reducing the number of berries per bunch (Skene and Antcliff, 1972). The CK content is therefore high during early fruit development, for instance in the avocado, falls off as the rate of fruit growth slows down, and disappears completely by the time the fruit reaches maturity (Blumenfeld and Gazit, 1970). A similar rise occurs in cotton fruit, reaching a peak from the fourth to the ninth day, and declining to little or none by the eighteenth day after anthesis (Sandstedt, 1971).

A correlation between GA concentration and the rate of orange fruit growth led to the suggestion of a cause-and-effect relation between endogenous GA and early stages of fruit growth (Wiltbank and Krezdorn, 1969).

Inhibitors appear to hasten ripening. ABA accumulation in pears was associated with premature ripening (Wang et al., 1972). Similarly, the growth retardant CCC forced the maturation of apples by 10 days (Becka and Sebanek, 1971).

In many cases the effect of one growth regulator depends on the presence of another. The role of ethylene in hastening the ripening of fruit has long been known (Chapter 14). An auxin, on the other hand, delayed the ripening of grapes, leading to the conclusion that an auxin-ethylene antagonism may be involved in the regulation of ripening (Hale et al., 1970). Auxins also delayed climacteric changes and ripening of banana fruit (Vendrell, 1969). This would explain the two physiological stages of apples with regard to their response to exogenous ethylene. During the first stage (within 6 weeks of bloom) ethylene stimulated respiration but did not initiate a climacteric respiratory or color change (Looney, 1969), perhaps due to a high auxin content. In the orange fruit there is a considerable auxin and gibberellin content at all stages of development, leading to the conclusion that development depends on the hormonal balance between these two (Goren and Goldschmidt, 1970).

Most of the biochemical work on fruit development has concentrated on fruit ripening, or the transition from the preclimacteric to the climacteric phase. This would seem to eliminate some fruit that have been reported to lack a climacteric phase. The distinction between these and the climacteric fruit is, however, more apparent than real (Hulme et al., 1969), and therefore they will be considered together. There is some evidence that the pentose phosphate pathway (PPP) operates in the preclimacteric fruit and then decreases with a shift toward glycolysis (Hartmann, 1969). This may perhaps precede the increased production of ethylene, which initiates a sequence of changes in enzyme activity. The ethylene-induced increase in respiration is followed by a peak in the incorporation of amino acids into protein during the early part of the climacteric (Brady et al., 1970*a*, *b*). This is apparently a redirected synthesis, including new enzymes required for ripening (Dilley and Klein, 1969). Phosphofructokinase has been identified as the regulating enzyme that is activated at the climacteric rise in respiration (Chalmers and Rowan, 1971). The simultaneous increase in respiration rate and in sugar content during ripening indicates that something other than carbohydrate is serving as the respiratory substrate. This is at least partly due to the utilization of malic acid as substrate (Wejnar, 1969), due to a large increase in activity of a malate decarboxylating system. A consistent increase in IAA oxidase but not in peroxidase isozymes occurs during the ripening of both climacteric and nonclimacteric fruit (Frenkel, 1972). A net synthesis of ATP occurs during the respiratory climacteric (Rowan et al., 1969). Despite the close relation between the effects of ethylene on respiration and ripening, the two effects are apparently separable (Wang and Hansen, 1970).

No substantial increase in ribosomal material accompanies the increase in rate of protein synthesis early in the climacteric (Wade et al., 1972). There is, in fact, a notable decline in the polymeric forms after the fruit reaches the climacteric peak (Ku and Romani, 1970). No incorporation of radioactive base into ribosomes could be detected once the fruit had reached the climacteric peak. This is in contrast to the young (citrus) fruit. On a dry matter basis there was a decrease in the level of nucleic acids, but the fast growth of the young fruit was apparently accompanied by intensive turnover of nucleic acids rather than their dilution (Goren, 1971).

There are three phases in the development of seeds of *Vicia faba* and other legumes (Payne et al., 1971). (1) 1 to 40 days: Rapid cell division is followed by cell enlargement and the major organs are formed. (2) 40 to 70 days: Intense synthetic activity continues with the accumulation of proteins and starch. There is also a tremendous proliferation of rough ER. (3) 70 to 100 days: Maturation occurs accompanied by dehydration. The separation of the phases can also be made on a water content rather than a time basis. DNA synthesis ceased when the water content was still 75% (Brunori, 1967). Mitotic activity continued as long as the water content was higher than 65%. Accumulation of reserve material lasted as long as the water content was higher than 55%.

Many structural changes occur during seed development. In the case of *Phaseolus vulgaris* (Opik, 1968), the cotyledon parenchyma cells first become highly vacuolate. The vacuoles soon divide and are converted to reserve protein bodies while cell expansion continues. Cytoplasm is synthesized, containing masses of rough-surfaced ER that persist until the cotyledons dry out. The proteins are presumably synthesized here. Starch grains grow within the plastids. The chlorophyll disappears when the seed dries, and the ER is replaced by small vesicles. Most of the organelles, however, remain recognizable in the dry seed, but the ribosomes are dispersed free in the cytoplasm.

The individual growth regulators occur at more or less exact phases in the development of the seed. Neither IAA nor its methyl ester was detected in immature pea seeds (Hattori and Marumo, 1972), although an auxin must have been present at a later phase. GA first appeared in apple seeds about 5 weeks after full bloom, increased to a maximum concentration at 9 weeks, and subsequently decreased again, disappearing completely by the time the seed was mature (Luckwill et al., 1969). The rapid buildup of GA was preceded by another unidentified hormone that reached its peak 6 weeks after bloom. A second unidentified hormone reached its highest concentration 13 weeks after bloom, when GA was first disappearing. The sequence of hormone production in the apple seed is linked with well-defined stages of embryo and endosperm development. The GA changes in developing watermelon seed are similar to those in the apple seed except that time is compressed, and weeks must be replaced by days (Bhalla, 1971). Developing seeds of *Pharbitis nil* accumulate free as well as bound GA until a maximum is reached approximately 25 days after anthesis (Barendse, 1971). The CK content per gram fresh weight of pea seed showed three maxima, coincident with the maximum volume of endosperm per seed and the two maxima in the growth rates of the whole seed and the embryo (Burrows and Carr, 1970).

Depending on the kind of reserves characteristic of the seed, there is an accumulation of oil or carbohydrate or protein or all three. In the case of the castor bean, the rate of oil synthesis rises to a maximum 6 days after fertilization, then decreases until 54 days after fertilization when oil synthesis stops (Agrawal and Canvin, 1971*a*). In the case of rice grain, the dry matter accumulated is mainly starch, but there is also a gradual accumulation of protein throughout seed development, with the highest rate occurring 12 to 16 days after pollination (Paul et al., 1971). Albumins are synthesized early during pea seed development, and globulins are synthesized with increasing maturity. During rapid protein synthesis there is a high percent of polysomes. Older cotyledons with declining capacity had few polysomes and an abundance of monosomes (Beevers and Poulson, 1972).

Phytic acid is another reserve that accumulates in many seeds. It is a reserve of organic phosphorus. In bean seeds it increased rapidly with advancing maturity, whereas total acid-soluble phosphorus decreased. Mature, dry beans contained about 1% phytol, accounting for half the total acid-soluble seed phosphorus (Makower, 1969). Myoinositol-2-phosphate is apparently the early intermediate in formation of phytic acid from myoinositol (Tanaka et al., 1971).

At the same time that all the above anabolic processes are occurring, catabolic activity must also be high. The PPP, for instance, was operative at all stages of castor bean seed development (Agrawal and Canvin, 1971*a*), accounting for 10% to 12% of the metabolized glucose (Agrawal and Canvin, 1971*b*).

DNA forms rapidly during the first 20 days in rice grains and then remains constant although RNA increases steadily until maturity (Paul et al., 1971). During the maturation of maize grains, DNA again remained constant, but RNA content decreased, presumably due to an accompanying increase in ribonuclease activity (Ingle et al., 1965). During the final days of maturation there was an increase in protein and a decrease in nucleic acids and soluble constituents.

Many changes in enzyme activity must accompany all these metabolic changes. Thus the activities of α-amylase, adenosine triphosphatase, and phytase rose to peaks at 20, 32, and 24 days, respectively, in rice grains (Paul et al., 1971). In barley grain, α-amylase increased at first to an early peak, then declined rapidly (MacGregor et al., 1971). Free β-amylase followed the same pattern a little later, and bound β-amylase rose to a maximum at maturity. In wheat grain, both uridine diphosphate glucose (UDPG) pyrophosphorylase and adenosine diphosphate glucose (ADPG) pyrophosphorylase increased in activity during the phase of starch synthesis (Turner, 1969). Many enzymes were active in the earliest stage of developing maize endosperm at 8 days after pollination (Tsai et al., 1970). The activities of all these enzymes except invertase rose to a peak at 22 to 28 days. During the maturation of pea seed, when the relative water content of the cotyledons drops from 65% to 13%, certain mitochondrial enzymes lose part of their activity (Kolloffel, 1970).

Senescence

Senescence is also called the aging of the plant. It is the stage preceding death or, in the case of organs, preceding abscission, which is normally soon followed by death. In the case of artificially cultivated (excised) roots, aging has been defined as a decrease and ultimate cessation of cell division in the apical meristem (Street, 1967). This does not include effects of starvation or the accumulation of an external inhibitor. It can be measured quantitatively by the decrease in growth rate. More generally, senescence implies the "running down" of an organism or one of its organs, involving changes of a degradative or degenerative nature (Wareing and Seth, 1967). The senescence of a whole perennial plant is usually much more gradual than that of organs (e.g., fruit, leaves) or of whole annual or biennial plants. There seems to be a basic qualitative difference between the two (Woolhouse, 1967). In general, the senescence of perennial plants appears to be dependent on factors external to the living cells.

In the case of trees (e.g., *Pinus sylvestris*), aging is apparently caused by competition for available nutrients between the increasing number of branches and also caused by the increasing distance for translocation between shoot apices and roots (Wareing and Seth, 1967). However, because of the time factor, senescence of perennial plants has not been extensively studied. Even in the more intensively investigated annuals and biennials the evidence seems to point to a competition for nutrients as a cause of senescence. Seed formation occurs at the expense of nutrients drained from the vegetative parts of the plant, and this hastens senescence. Thus the prevention of seed formation by growing the plant under environmental conditions unsuited to sexual reproduction prevents or postpones senescence. *Perilla frutescens*, a short-day plant, is normally an annual plant and therefore dies at the end of a single growing season after flowering and fruiting. When grown in long days, it may survive many years in the vegetative state (Woolhouse, 1967). Similarly, cabbage is normally a biennial plant and therefore dies after flowering and fruiting during its second year. Yet it has been kept growing vegetatively for several years by maintaining it at warm temperatures.

Simply removing the seeds as they form may delay leaf senescence by as much as 6 weeks. Decapitation of the shoot has the same effect. Detached leaves, if allowed to root at the petiole end, remain alive much longer than if they are left on the plant. These results all show that the senescence of plant organs is often under the control of the whole plant and is not solely caused by the intrinsic characteristics of its own cells (Wareing and Seth, 1967). In this respect plants differ from animals.

Nevertheless, this draining of nutrients from the vegetative plant cannot be the sole cause of senescence. Male annual plants may show normal senescence in the absence of fruiting, and hermaphroditic plants may show it in the absence even of flowering. Thus in opposition to perennial plants, the senescence of annuals and biennials appears to be programmed by the living cells themselves.

More detailed information as to the nature of senescence has been obtained by following the changes in parts of the plant rather than the plant as a whole, for example, in leaves. Measurements made on individual *Perilla* leaves show that the rate of photosynthesis declines steadily over a period of 20 days from the time of completion of expansion (Woolhouse, 1967). The decline then becomes suddenly more rapid and continues during 7 days prior to abscission, when it can no longer photosynthesize (Fig. 20-2). The decrease in rate of photosynthesis parallels the earlier described export of nutrients out of the senescing leaf. Respiration behaves differently. It remains constant until near the end, when it shows a brief, small climacteric rise, followed by a brief senescent drop similar to the changes in fruit (Chapter 14).

The leaf proteins and pigments also decrease precipitously during senescence (Fig. 20-3). RNA decreases gradually, then more rapidly. DNA remains stable over a long period, followed by a period of rapid loss. The rapid loss in the case of both nucleic acids occurred at the same time as the senescent respiratory drop (Woolhouse, 1967).

The decrease in leaf protein has been shown by many investigators, but it is usually the total leaf protein that is measured. Woolhouse, however, showed that not all the proteins decrease at the same time. Fraction I protein continues to be synthesized up to the time of full leaf expansion, but after this the turnover is slow. Another protein component of low molecular weight, on the other hand, continues to turn over at a relatively high rate. This may explain the difference between the photosynthesis and respiration curves, since the Fraction I protein is from the chloroplast, and the other is not. In soybean cotyledons, there is a decrease in a

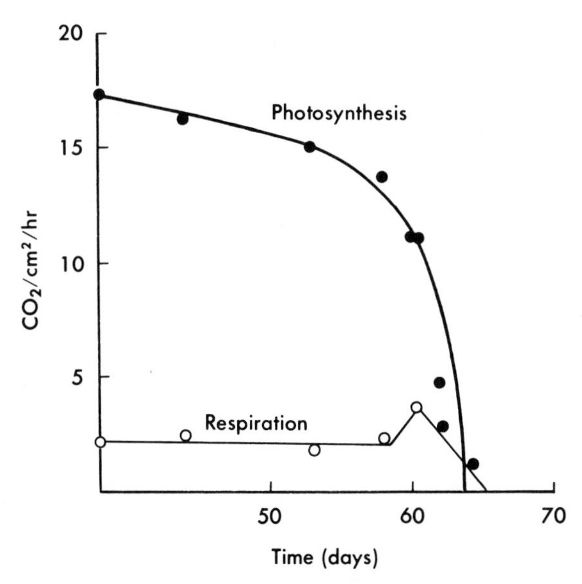

Fig. 20-2. Changes in rates of photosynthesis and respiration of the attached third pair of leaves of *Perilla frutescens* from the time of completion of leaf expansion to abscission. (From Woolhouse, H. W. 1967. Sympos. Soc. Exp. Biol. **21**:179-213.)

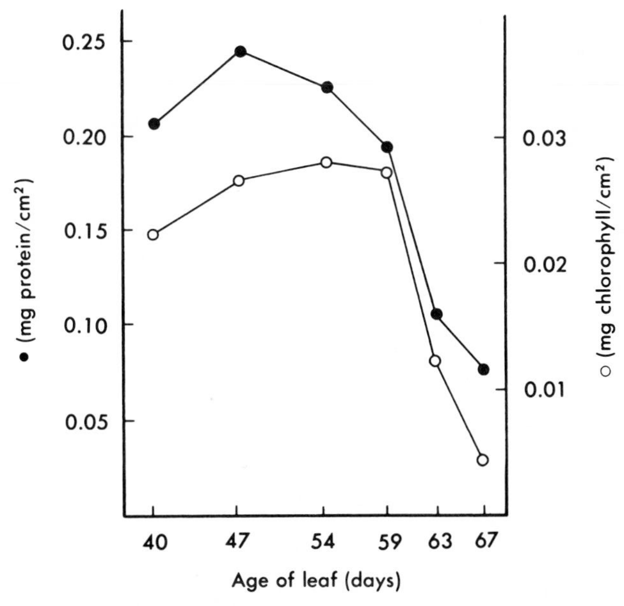

Fig. 20-3. Chlorophyll and protein content per unit area of the third pair of leaves of *Perilla frutescens* from the time of completion of expansion to the time of abscission. (From Woolhouse, H. W. 1967. Sympos. Soc. Exp. Biol. **21**:179-213.)

specific tRNA (Bick et al., 1970; Venkataraman and De Leo, 1972). Despite the general decrease in these substances, two histones increase in old tobacco leaves (Srivastava, 1971).

The changes in tree leaves are similar to the aging phenomena in leaves of herbaceous plants, but the protein decreases may commence before true senescence. In the case of *Ginkgo* leaves, development can be divided into three different periods (Specht-Jurgensen, 1967): a growing period until the end of June when nitrogen and chlorophyll contents increase; a second period until the end of September when nitrogen, protein, and chlorophyll slowly decrease; and a breakdown period from the end of September until leaf abscission, when 66% of the nitrogen is lost and 40% of the remaining nitrogen is soluble (mainly amino acids and amides). Only the last period can be considered true senescence. The decrease in protein observed in tobacco leaves with increasing physiological age is correlated with increase in specific activity of peptidase (Anderson and Rowan, 1965). In the case of pear fruit the ability of mitochondria to incorporate amino acids into their protein declines as the tissues age and reach their climacteric peak (Romani and Fisher, 1966). Senescence may therefore involve both an increase in protein breakdown and a decrease in protein synthesis.

The organelles show a loss of structure during senescence and may be seen to break down. In the case of fully senescent *Phaseolus* mesophyll cells, almost all the cytoplasmic contents are lost (Barton, 1966). Only the plasmalemma and some empty vesicles remain. Even the cell walls may be dissolved (e.g., in senescent fruit). Similarly, during senescence of barley leaves, polyribosomes and ribosomes are lost (Srivastava and Arglebe, 1967).

In leaves of deciduous trees in the fall, the chloroplasts are the first organelles to show signs of senescence and the last to remain after the other organelles have all broken down (Dodge, 1970; Stearns and Wagenaar, 1971). Their shape changes and their volume decreases to less than one fifth. Their system of membranes breaks down as lipid globules increase in size. A similar appearance of lipid bodies is the major structural change in aging cells of a green alga (McLean, 1968). Also in senescent tobacco leaves, osmiophilic globules become the predominant constituent of the chloroplast (Barr and Arntzen, 1969). All these observations point to a breakdown of membranes with the accumulation of their lipid components in the form of droplets. High free surface energy is in this way reduced to a minimum.

In detached, senescing wheat leaves, chloroplast ribosomes disappear before the cytoplasmic ribosomes, and chloroplast membranes become wavy (Mittelheuser and van Steveninck, 1971). Again, large lipid bodies appear in the cytoplasm and vacuole, and osmiophilic fibrils develop in the microbodies. Loss of ribosomes also occurs during the early stages of senescence of cucurbits (Eilam et al., 1971; Klyachko et al., 1971).

In old root cap cells, dense bodies that are believed to be plant lysosomes arise, releasing hydrolytic enzymes (Berjak, 1968). This conclusion is supported by electron micrographs of vacuoles showing autophagic activities in senescing corolla cells (Matile and Winkenbach, 1971). The last phase of senescence involves the breakdown of the tonoplast and complete digestion of the cytoplasmic constituents. In the case of aging embryos in grains of *Zea mays*, membrane damage apparently may be repaired at later stages of germination (Berjak and Villiers, 1972). All these changes seem to be caused by the appearance, either by synthesis or activation, of specific enzymes capable of solubilizing these structures. The thylakoids of the chloroplasts, for instance, are decomposed by an enzyme that increases in quantity and moves from the plastids to other organelles at a later stage (Barton, 1966).

Phosphatases, terminal oxidases, and hydrolases have been found to increase in activity (Table 20-3). The one most commonly reported to increase is ribonuclease. It has also been reported to decrease during senescence (Table 20-3). The discrepancy may, of course, be due to differences in plant material. It may also, however, be due to the method of calculating ribonuclease activity or to the time sam-

Table 20-3. Changes in enzyme activity associated with senescence

Enzyme	Plant	Organ	Reference
	Decrease in activity		
Alkaline inorganic pyrophosphatase	Four herbaceous species	Leaf	Rauser, 1971
Cytochrome oxidase Catalase Alkaline phosphatase	*Hedera helix*	Leaf	Ghosh et al., 1971
Glutamate dehydrogenase	*Daucus carota*	Root	Werner and Gogolin, 1970
Aldolase	*Acer pseudoplatanus*	Suspension culture	Simola and Sopanen, 1970
Leucyl tRNA synthetase	Soybean	Cotyledon	Bick and Strehler, 1971
Protein synthesizing Malic dehydrogenase Glutamic dehydrogenase Aspartate amino transaminase	Apple	Leaf	Spencer and Titus, 1972
Ribulose diphosphate carboxylase	Tobacco	Leaf	Kawashima and Mitake, 1969
Ribonuclease Chlorophyllase	*Phaseolus vulgaris*	Leaf	Phillips and Fletcher, 1969; Phillips et al., 1969
	High or increased activity		
Acid phosphatase	*Daucus carota*	Cultured cells	Halperin, 1969
Acid phosphatase	Banana	Fruit	De Leo and Sacher, 1970*a*
Acid phosphatase	*Hedera helix*	Leaf	Ghosh et al., 1971
Ribonuclease	Oat	Leaf	Lado et al., 1968; Wyen et al., 1971
Ribonuclease	Tomato	Leaf	McHale and Dove, 1968
Ribonuclease (and deoxyribonuclease)	Barley	Leaf	Srivastava, 1968
Phosphatases Ribonuclease Peroxidase	*Acer pseudoplatanus*	Suspension culture	Simola and Sopanen, 1970
Ribonuclease Polyphenol oxidase Proteolytic activity	Apple	Leaf	Spencer and Titus, 1972
Ribonuclease Deoxyribonuclease Peptidase α-Amylase	Barley	Leaf	Atkin and Srivastava, 1969
Maltase	Tobacco	Leaf	Abbott and Matheson, 1972
β-1,3-Glucan hydrolase	*Nicotiana glutinosa*	Leaf	Moore and Stone, 1972

pled. The ribonuclease may, for instance, decrease together with but more slowly than the general decrease in protein content, leading to an increase in specific activity (per unit total nitrogen) but to a decrease in absolute activity (Ecklund and Moore, 1969). The ribonuclease and deoxyribonuclease associated with the chromatin of barley leaves were found to decline at first during senescence, and later to increase (Srivastava, 1968). Even the RNA synthesizing activity increased, but less than the ribonuclease, so that there was a net breakdown. Many enzymes have, indeed, been shown to decrease in activity, particularly those involved in anabolic processes, but also some catabolic enzymes (Table 20-3).

It is now evident that many physiological changes accompany the aging process. However, senescence is not necessarily a single unique process for all plants and plant parts, in which all the aforementioned changes occur in exactly the same sequence. Even in the case of a single organ, senescence may not be a uniform process in all plants. Thus although the changes in aging leaves are general, there are actually two distinct kinds of leaf senescence. *Sequential aging* occurs when the leaves senesce in a definite order, usually the lowest leaf first and the others in an orderly succession. *Synchronous,* or *simultaneous, aging* occurs when they senesce all at once, for example, in the case of annual plants after fruiting or of deciduous plants in the fall. Translocation of substances from the leaf certainly plays a major role in sequential senescence since, as indicated earlier, it is initiated under normal conditions by competition between the mature leaves and the growing regions of the shoot. Translocation may also play a role in simultaneous senescence, for example, in the case of deciduous plants. Even in the case of excised leaves, translocation occurs from the blade to the base of the petiole (Simon, 1967).

The changes accompanying sequential aging of leaves all seem to be readily explained by translocation to the growing regions of the plant. The decrease in protein content would follow logically the export of amino acids, since this would shift the equilibrium to hydrolysis of further proteins in the mature leaf. Since the chloroplast proteins serving as photosynthetic enzymes (Fraction I) appear to be the first to suffer such breakdown, this would also explain the early decrease in rate of photosynthesis. The later breakdown of cytoplasmic proteins would similarly lead finally to a decrease in respiration. This would also eventually result from the export of carbohydrate—the substrate for respiration. Even the synchronous aging of leaves of deciduous plants could be explained in the same way, although the translocation would be to storage tissues instead of to growing regions.

Translocation must therefore play an important role in senescence. Yet it was shown earlier that senescence cannot be postponed indefinitely by preventing the translocation of substances from an organ. Furthermore, the export of amino acids and sugars from the mature leaves to the growing (or storage) regions would be expected to continue only until the concentration gradient is eliminated. The continuation of the changes characteristic of senescence requires something more than a simple decrease in reserves, and has led to a hormonal concept of senescence (Osborne, 1963). According to this point of view, a deficiency of one or more of the growth regulators leads to senescence. Thus the auxin content declines with the age of the leaf, and toward death it is too low to measure accurately. This decline in auxin content is accompanied by an export of other substances, perhaps because of a direct effect of auxin on phloem transport (Thrower, 1967). Therefore older leaves senesce more rapidly in the presence of young auxin-rich leaves above them. Paradoxically, auxin is produced during the later, irreversible senescent stages, because it is a normal product of autolyzing cells, whether they are from higher plants, yeast, or rat liver tissue (Sheldrake and Northcote, 1968). In the dwarf bean, this is due to a conversion of bound to free auxin (Wheeler, 1968). Dying cells may therefore be an important source of auxin for the living cells of a plant.

Leaf senescence can also be retarded by applications of GA in the case of *Rumex obtusifolius* (Whyte and Luckwill, 1966). GA causes a complete cessation of net chlorophyll and protein degradation for several days in leaf discs of *Rumex* species (Goldthwaite and Laetsch, 1968; Back and Richmond, 1969). It also delays the senescence of navel orange rind tissue (Coggins et al., 1969). In the case of soybean leaves, on the contrary, GA at all concentrations induced senescence and chlorophyll degradation (Pillay and Mehdi, 1968). It had no effect on the senescence of maize leaf discs (Knypl, 1970).

Ethylene induces senescence in carnation flowers (Nichols, 1968). This may be due to an induction of ABA activity, which in turn may control ethylene evolution by feedback induction (Mayak and Halevy, 1972). In general, since ethylene induces and speeds up the climacteric, it must also indirectly hasten the subsequent senescence.

ABA accelerates senescence in leaves of *Arabidopsis* (Sankhla and Sankhla, 1968*b*) and in leaf discs of the radish (Pearson et al., 1969). This effect is associated with a reduced capacity for protein and RNA synthesis. This is due partly to an increased RNA breakdown by the increase in ribonuclease activity and partly to a decreased synthesis as shown by reduced pyridine and amino acid incorporation (De Leo and Sacher, 1970*b*). On the other hand, coumarin and growth retardants that also inhibit RNA and protein synthesis prevent the senescence of maize leaf discs by arresting chlorophyll, protein, and RNA degradation (Knypl, 1970). It is suggested that they prevent the synthesis of a specific protein required for senescence.

The growth regulator par excellence for preventing senescence is CK. Innumerable investigations have reported this effect in leaves, fruit, and flowers. In the case of the intact plant, CK may retard the senescence of an organ by preventing the translocation of substances such as sugars and amino acids away from it (Fletcher et al., 1970) or by enhancing their translocation to it. In the case of excised organs, it must be explained by other effects of CK. Several such explanations have been proposed. (1) It suppresses the normal increase in ribonuclease activity that accompanies senescence (Dove, 1971). (2) It retards the rapid loss of polyribosomes and ribosomes in leaves floated on water (Srivastava and Arglebe, 1968). (3) It maintains a higher level of proteins, RNA, and chlorophyll (Fletcher, 1969).

A more fundamental explanation than any of the above is necessary to understand the basic mechanism. CK may conceivably accomplish any of the above three effects by (1) retarding the breakdown of RNA and proteins (Tavares and Kende, 1970) or (2) accelerating their synthesis (Atkin and Srivastava, 1970), or by both, for instance, by stimulating RNA synthesis and suppressing ribonuclease and peptidase activities (Srivastava and Arglebe, 1968). A more recent investigation proposes a simpler explanation than any of the above. According to this concept, senescence is initiated by the synthesis or release of a specific protein (Knypl, 1970), such as a protease (Martin and Thimann, 1972) that would hydrolyze the proteins and therefore set in motion the many other changes associated with senescence. For instance, if it hydrolyzed the membrane proteins, this would destroy the integrity of the layer, and membrane lipids would accumulate as droplets. The polysomes would break down to ribosomes, and even these would disappear due to loss of their proteins. The only enzymes remaining active would be the resistant ones such as ribonuclease, whose native structure is held tightly in place by four SS bonds. The specific activity (rather than the absolute activity) of the ribonuclease would therefore increase, and in the absence of the RNA synthesizing system, RNA would be quickly hydrolyzed. According to Martin and Thimann (1972), the CK prevents senescence by simply suppressing the formation of the protease enzyme. In support of this concept, CK maintains but does not increase the chloroplast and cytoplasmic ribosomes, whereas ABA accelerates the degenerative changes (Mittelheuser and Van Steveninck, 1971).

The initiation of senescence may therefore be triggered by a decline of some hormone in the leaves (e.g., CK) or the appearance of some hormone (e.g., ABA). Support for the first suggestion is the enhancement of senescence in detached cotyledons of radishes by the sucrose-induced suppression of root formation (Moore et al., 1972), since the roots are known to supply CK to the leaves.

Just as in the case of growth, however, the above concept is an oversimplification, since it does not take into account the interactions between growth regulators. The interactions may involve antagonism, synergism, or both. The antagonism by auxin of ethylene-induced senescence has been mentioned above. CK and GA may also antagonize ethylene (Wade and Brady, 1971). Similarly, CK almost completely eliminates the ABA-induced acceleration of aging (Sankhla and Sankhla, 1968*b*; Back and Richmond, 1971), but GA fails to do so. Conversely, ABA decreases the senescence-retarding effect of GA more than CK (Back and Richmond, 1971). On the other hand, a synergistic interaction between GA and CK may be a general phenomenon that is not detected in some systems because of the high endogenous levels of growth regulator (Back and Richmond, 1969). All three growth promoters (IAA, GA, and CK) in combination sometimes give the greatest effect (Kulaeva and Devyatko, 1969).

Hormonal control of aging also seems to operate in roots. When excised roots are grown in artificial culture, an externally applied auxin precursor (tryptophan) is important for the maintenance of meristematic activity (Street, 1967). Some other unidentified factor is also involved that critically limits the growth of most strains until eventually it comes to a stop. Street suggests that the shoot is the primary site of synthesis of such limiting factors (e.g., thiamine) and that this may explain the aging of excised roots. This explanation is supported by the antiaging effect of the illuminated shoot on the root of the intact plant.

Many environmental factors such as water, mineral nutrients, light, etc. may alter the rate of senescence. In the case of excised roots of groundsel *(Senecio vulgaris)* or roots of plants grown in the dark, each meristem functions for only a limited time and then dies (Wren and Hannay, 1963). Within one root the lateral meristems continue to grow after the main apex has ceased growth, but they in turn die while newer, secondary laterals continue to extend. When the plants are grown normally in the light, no such rapid aging and senescence occur.

Another hypothesis holds that senescence is genetical rather than a physiological wearing out (Curtis, 1963). According to this concept, aging or senescence is caused by somatic mutation. Thus organs having cells that frequently undergo meiotic cell division show little or no aging, whereas those in which cells seldom, if ever, divide meiotically have no opportunity to throw off harmful mutations. Radiation accelerates the aging process, presumably because of the increase in mutations. This theory of aging is supported in the case of seeds, since chromosome damage has been found to accumulate as they get older (Roberts et al., 1967). However, radiation damage seems to be of little importance in the senescence of growing plants (Woolhouse, 1967). Although it may conceivably explain the senescence of long-lived organisms, it certainly cannot explain that of short-lived organs, since this can often be reversed. The senescence of old leaves may play a definite role in the physiology of the plant and may have been selected during evolution as a survival factor. In many cases (e.g., cucumber, Hopkinson, 1966) these older leaves may be useful not as photosynthesizing organs but as reserve organs, able to redistribute both their mineral elements and organic substances to younger, growing organs. Shading hastens the senescence of such leaves.

Abscission

When individual organs senesce on an otherwise fully active plant, they must be discarded by the mother plant. This is accomplished by the process of *abscission* —the shedding of leaves, petals, seeds, fruit or other organs. Although there is still some controversy as to the details of the mechanism, a clearer understanding now exists of the many factors generally involved in abscission and the interrelationships between them (Addicott, 1970).

The separation usually occurs in an abscission zone at the base of the organ, consisting of small cells densely filled with cytoplasm. Abscission occurs at a layer of cells within this zone (called the *abscission layer*) as a result of partial or complete hydrolysis of their distal (outer) walls. In most species only the pectins of the middle lamella and a portion of the cellulose of the primary walls dissolve; in others the entire wall and cell contents disappear.

Hydrolysis of the cell wall must be due to the action of hydrolases. It is therefore not surprising that pectinase is synthesized, rising from an activity of zero before abscission to a maximum activity at a time of 50% abscission followed by a decrease in activity. Cellulase activity also increases during abscission, again apparently due to synthesis of a new protein (Lewis and Varner, 1970). On the other hand, pectin methyl esterase, which increases pectin rigidity, apparently declines in activity with the commencement of abscission. A climacteric rise in respiration accompanies abscission. A deficiency in oxygen, in fact, retards abscission. Presumably the increased respiration supplies the added free energy needed for the protein synthesis. Besides the cell wall hydrolyzing enzymes, others have also been found to increase in activity with abscission, perhaps as components of the respiratory increase.

Abscission is a correlation phenomenon that is controlled by growth regulators. A high auxin content is associated with inhibition and a low content with acceleration of abscission. The young leaf, for instance, has a high auxin content that decreases with leaf age to a minimum value at abscission. Mineral deficiencies lead to low auxin content and leaf abscission. Even artificial applications of auxin usually, but not always, retard abscission. A good deal of evidence indicates that the deciding factor is the auxin gradient across the abscission zone, so that a high *distal* (outward of the zone) auxin concentration retards, and a high *proximal* (inward of the zone) concentration accelerates abscission.

Ethylene is also involved in the control of abscission. The data indicate that this is due to an inhibition of auxin synthesis and transport, or enhancement of auxin destruction (Burg, 1968), for instance, by increasing IAA oxidase activity. Therefore ethylene promotes abscission, and unlike other growth regulators, it never inhibits abscission. It can stimulate RNA and protein synthesis leading to increased activities of pectinase and cellulase.

Abscisic acid, as its name implies, strongly accelerates abscission. Its concentration increases in organs as they mature, to a high value just before abscission (Fig. 20-4). Unlike auxin, it accelerates abscission whether applied distally or proximally to the abscission zone. Young tissues relatively high in IAA are relatively insensitive to ABA. ABA increases ethylene production and cellulase synthesis (Cracker and Abeles, 1969).

Gibberellic acid normally has no effect on abscission, although it may affect it in vitro through its effect on the level of endogenous auxin. Little work has been done with CK. It may affect abscission presumably by increasing the flow of nutrients from which IAA is synthesized.

The interrelations between the growth regulators are shown in Fig. 20-5. In summary, GA and CK appear to be of little importance in natural abscission and affect it in vitro only by changing the concentration of IAA. The main abscission controlling growth regulators are IAA, ethylene, and ABA. It seems possible that ABA functions by inducing ethylene formation, which in turn controls IAA concentration. It has even been suggested that cyclic AMP is the second messenger mediating the auxin-induced delay of abscission (Salomon and Mascarenhas, 1971). Abscission-promotion, however, still appears to be most closely related to a *decrease* in IAA. This does not explain the mechanism for synthesis of the hydrolases that dissolve the cell wall in the abscission layer.

Observations of the cells in the abscission zone of tobacco flower pedicels have pointed to a new link in the chain. There is an increase of approximately 30 times in rough endoplasmic reticulum within 5 hr of ethylene treatment (Valdovinos et al., 1971, 1972). This presumably reflects an increased protein synthesis or translocation, leading to the synthesis of the hydrolases. Just how the above ABA-ethylene-

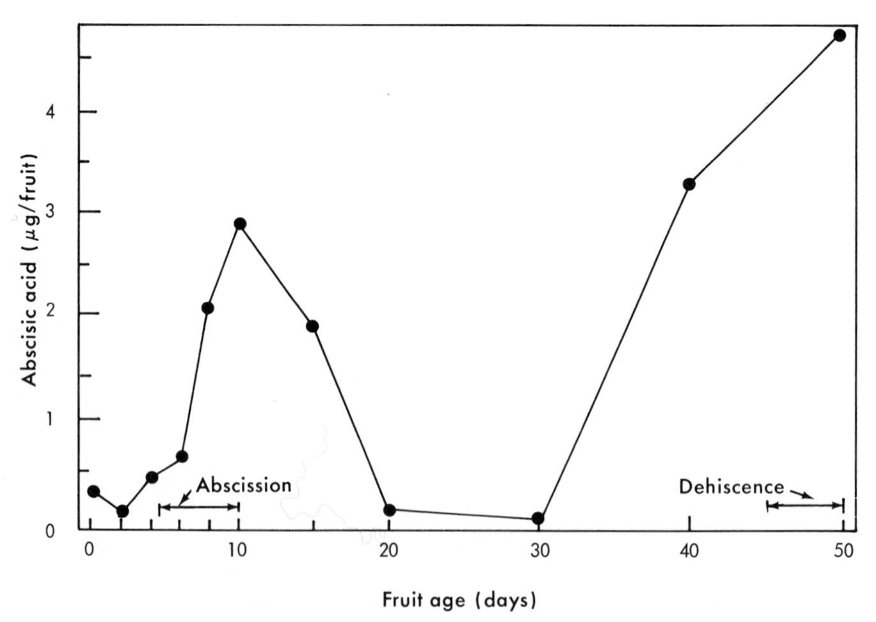

Fig. 20-4. Rise in abscisic acid content at time of young cotton fruit development and during senescence previous to abscission. (From Addicott, F. T., and J. L. Lyon. 1969. Ann. Rev. Plant Physiol. **20**:139-164.)

IAA interrelationship leads to this change is unknown.

The major relationships between the growth regulators and the stages of development are as follows. Those in parentheses are discussed under photoperiodism.

Developmental stage	*High concentration*	*Low concentration*
1. Germination	GA, CK	ABA
2. Juvenile phase	IAA (GA?)	GA
3. Reproductive stage	GA (IAA, CK, ETH)	(IAA, ABA)
4. Fruit formation (early)	CK, GA, IAA	ETH, ABA
Fruit maturation	ETH, ABA	CK, GA, IAA
5. Senescence	ETH, ABA	CK, IAA
6. Abscission	ETH, ABA	IAA

In general, then, the growth promoters favor the early stages of development involving vegetative or reproductive growth, and inhibit the later stages that bring all growth to a stop. The growth inhibitors inhibit the early growing stages and promote the later nongrowing stages. The role of the growth regulators in development is therefore similar to their role in growth itself. Since the roles are similar, the mechanism of action must be the same. As in the case of growth itself, some of the evidence indicates that it is the balance between the growth regulators, rather than the concentration of a single one, that controls the development of the plant.

Longevity

The longevity, or total life-span, of a plant or plant part varies between wide extremes. Plant organs such as fruit or leaves may have a life-span from a few weeks to 30 years (Woolhouse, 1967). Whole plants may also live for only a few weeks (ephemerals), for one growing season (annuals), for two growing seasons (biennials), or for as long as 3,000 years or more (some trees).

In these cases, however, regardless of

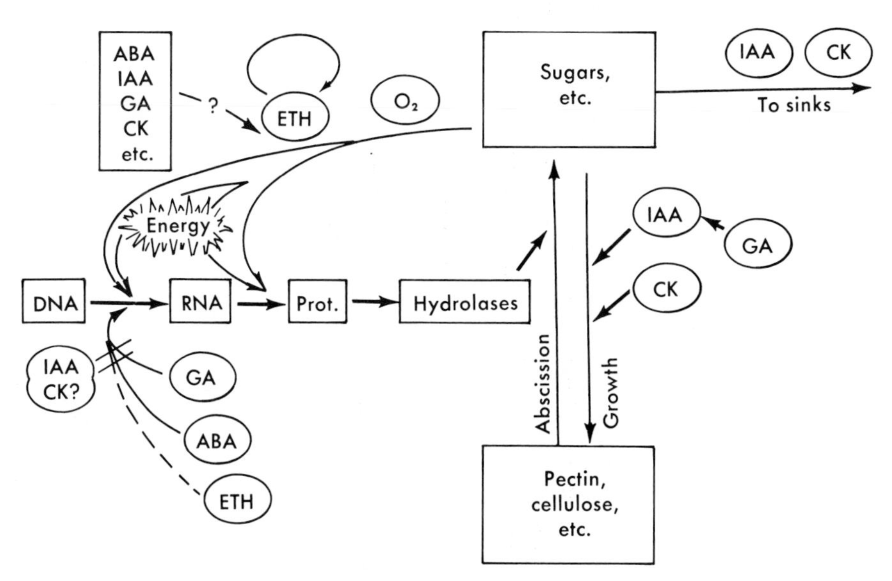

Fig. 20-5. Schematic representation of the controlling influences of plant hormones on the biochemical activity associated with abscission. Three major biochemical events are illustrated: (1) hydrolysis of pectin and cellulose to soluble sugars; (2) synthesis of the appropriate hydrolases by RNA; and (3) provision of energy for the synthetic reactions. (From Addicott, F. T. 1970. Biol. Rev. **45**:485-524.)

the longevity of the plant as a whole, the individual living cells may survive only one or a few years, although xylem parenchyma cells have been found to survive 100 years. The plant survives longer than its individual cells by possessing meristems with cells that continue to divide and produce new cells. Only in the case of seeds, which remain alive in the dormant state for long periods of time, do the measurements of longevity apply to the individual cells.

When measured on seed stored in the laboratory, the longevity of dormant seed is of much shorter duration than that of whole plants such as trees. Furthermore, usually only a few seed germinate in a large sample, for instance, in the case of 94-year-old wheat stored in hermetically sealed glass tubes (Ruckenbauer, 1971). In the case of strawberry seed stored at 40° F, however, 23-year-old seed germinated as well as 1-year-old seed (Scott and Draper, 1970). When seed lots contain both permeable and impermeable seed, the latter have much greater longevity. For instance, permeable seed of *Trifolium subterraneum* showed a decrease in germination from 99% after 1 year to 8% after 30 years, at which time 83% of the impermeable seed were still able to germinate (Carpenter, 1969).

In contrast to these relatively moderate longevities measured on laboratory-stored seed, measurements on seed found in nature lead to much higher longevity values. Seeds may be viable even after 1,700 years (Wesson and Wareing, 1967), although they may need to be stimulated by light in order to germinate. The longevity of seeds may be much greater than this under ideal conditions. Some seeds of *Lupinus arcticus* have been obtained from lemming burrows deeply buried in the permanently frozen silt of an arctic tundra at least 10,000 years old. They readily germinated in the laboratory and grew to normal, healthy plants (Porsild et al., 1967). Godwin (1968), however, doubts the validity of these measurements, although he accepts two observations indicating viability after 147 and 237 years, respectively.

The injurous effects of aging on wheat seed have been found both in embryo and endosperm by use of transplanting techniques (Floris, 1970). Even when they germinate, old seed generally show reduced viability and a slow growth rate accompanied by variable amounts of chromosome damage. The death of the embryo may be explained by the accumulation of nuclear damage, perhaps because of respiratory waste products (Kaul, 1969).

In all the foregoing cases, death eventually occurs either because of (1) externally produced stresses (Chapter 22) leading to "accidental" injury, or (2) internal changes in the normal development of the plant leading to senescence and death. If the stresses can be avoided, it is possible to produce a kind of immortality by vegetative reproduction. This can be achieved by means of stem cuttings, which can be rooted and produce new, vigorous plants, or even by excised roots (i.e., root tips), which can in some cases be maintained in the growing state essentially indefinitely by repeated excision of the tip and subculturing in artificial media. According to Molisch (1938), however, even such "rejuvenated" cuttings cannot be reproduced indefinitely. He believes that senescence continues, although at a slower rate, from one vegetative generation to another. In this way he explains the ultimate weakening and dying out of a variety after being reproduced vegetatively for many generations.

ENVIRONMENTAL CONTROL OF DEVELOPMENT

The plant physiologist attempts to answer the following two questions about these developmental stages:

1. What are the physiological changes within the plant that give rise to the transitions from one developmental stage to another? Or, what are the internal factors?

2. What are the environmental factors that control the plant's transitions from the vegetative to the reproductive stage? Or, what are the external factors?

The second question has led many of the Russian investigators to classify the developmental stages of the plant in terms of the environmental factor which is most important at this stage, that is, the thermostage or phase, the photostage or phase, etc. This kind of classification, however, is not based on specific morphological or physiological characteristics of the plant and is therefore not as fundamental as the preceding morphological and developmental classification. It is also difficult to apply. The physiology of development has been studied intensively for only a few decades. Until recently, nearly all the investigations were confined to the second of the two questions.

Photoperiodism

In 1920, Garner and Allard showed that the flowering of many plants could be induced or prevented simply by controlling the length of the daylight period (the "photoperiod"). Some plants flowered most rapidly when the day length was about 12 hr or less (short-day plants, Fig. 20-6, *A*), others when the day length was about 12 hr or more (long-day plants, Fig. 20-6, *B*); still others flowered at either day length (day-neutral plants). In some cases, the critical photoperiod may differ markedly from 12 hr. All types may occur in any one species, frequently producing so-called early (long-day) and late (short-day) varieties (Table 20-4). If the day length is unsuited to flowering, the plants may remain vegetative essentially indefinitely, or they may simply take much longer to flower (e.g., Peking and Biloxi soybean, Table 20-4).

The photoperiodic effect has been found in all kinds of plants (herbaceous and woody, annual, biennial, perennial) and even in animals. Beside reproduction, it affects many other phases of the physiology of the plant (Fig. 20-7). Some trees do not enter their rest period in the fall if long

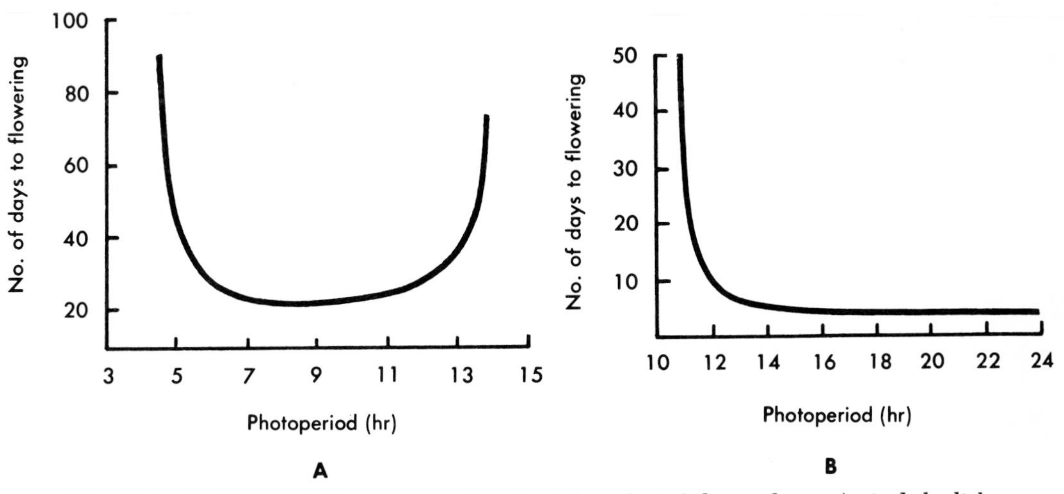

Fig. 20-6. A, Photoperiodic reaction of a short-day plant *(Chrysanthemum).* A daily light period of 7 to 11 hr is optimal for flowering. No flowering occurs above 14 hr and below 5 hr. **B,** Photoperiodic reaction of a long-day plant *(Hyoscyamus niger).* The critical day length is 11 hr. Flowering occurs more readily with lengthening of the photoperiod above this value. (From Bünning, E. 1948. *In* A. E. Murneek and R. O. Whyte. Vernalization and photoperiodism. The Ronald Press Co., New York.)

days are maintained. As a result, they fail to develop frost resistance and are readily killed by winter. Biennials remain in the rosette form if subjected to short days. Vegetative as well as sexual reproduction is affected. However, the optimum for flowering may be different from the optimum for vegetative reproduction. Maximum tuberization of potatoes occurs during shorter day length than that for maximum flowering.

Table 20-4. Flowering times for soybean varieties grown under different day lengths*

Day length (hr)	Time (days) from germination to blossoming		
	Mandarin (day neutral)	Peking (short day)	Biloxi (short day)
5	23	23	27
7	21	21	26
12	21	21	28
Full daylight (12½-15 hr)	26	62	110

*Modified from Garner, W. W., and H. Allard. 1920. J. Agric. Res. **18**:553-606.

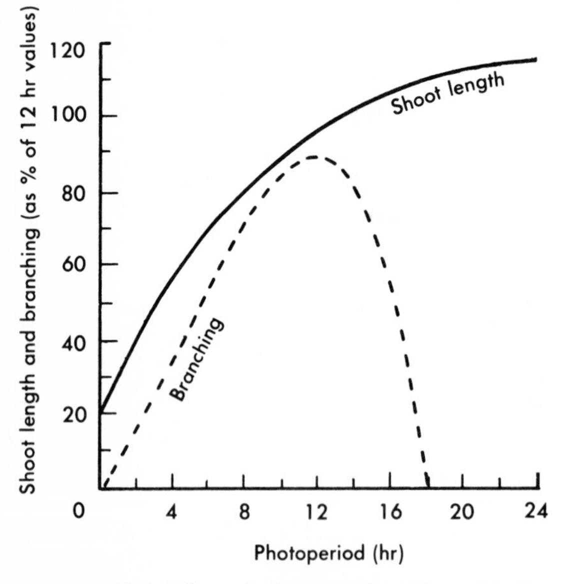

Fig. 20-7. Effect of photoperiod on the vegetative development (shoot growth and branching) in the fiber plant *Crotalaria juncea*. (Modified from Singh and Choudri. 1946. *In* R. O. Whyte. Crop production and environment. Faber & Faber, Ltd., London.)

The photoperiodic effect cannot substitute for photosynthesis. Consequently, the optimum day length for flowering will not be the optimum for yield in the case of short-day plants. To obtain maximum yield they must first be grown under a long day to enable the accumulation of reserves and enlargement of the plant. Similarly, the temperature factor cannot be ignored (Fig. 20-8).

In the case of a long-day plant, flowering can be induced in natural short days by lengthening the daylight period with weak light. This subsidiary light may be far too weak to induce a net photosynthesis. Similarly, long-day plants may be forced to flower and short-day plants may be prevented from flowering in short days by interrupting the dark period with a few minutes of light (Fig. 20-9). This and other evidence has proved that it is the dark period that inhibits flowering of long-day plants and stimulates flowering of short-day plants.

For this reason, Lang (1952) has suggested that all plants may be classified as short day or long day, provided that the two groups are defined as follows:

Short-day plants (SDP) are those in which flowering is induced or hastened by a daily dark period. Long-day plants (LDP) are those in which flowering is inhibited or retarded by a daily dark period. This would include day-neutral plants among the long-day plants. However, there are still some that are difficult to classify into one or the other group, for example, some that require both short-day and long-day treatments, the one following the other (SLDP).

Short-day plants will flower when exposed to long days, provided that they have already been exposed to a sufficient number of short days. Long-day plants will similarly flower during short days if previously exposed to a sufficient number of long days. This carryover is called the *photoperiodic aftereffect*, or *photoperiodic induction*. A plant is said to be photoinduced if it is exposed to a day length favorable for flowering for a sufficient time to result in subsequent flowering in an unfavorable photoperiod.

As shown earlier, there is a hormone control of development (Lang, 1952). This leads to formation of a so-called *floral stimulus*. The response occurs in the growing points, yet these do not have to be exposed to any particular day length. The region of perception is the leaf; and even a single one, if subjected to the necessary day length, may induce the growing points to flower. The floral stimulus may even be transferred from one plant to another by grafting. It is therefore generated in the

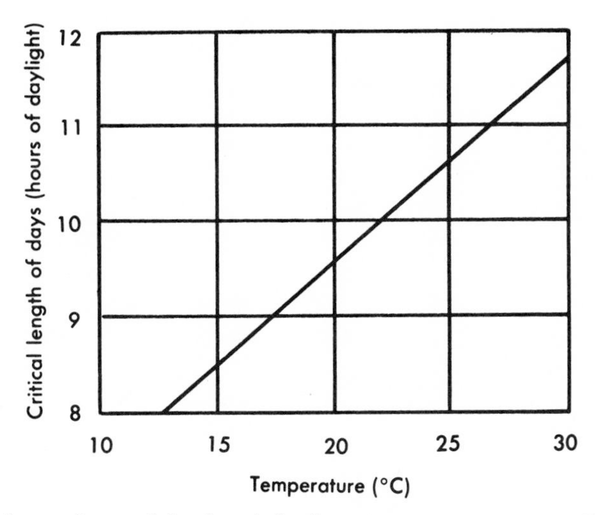

Fig. 20-8. Dependence of critical day length for flowering on temperature in *Hyoscyamus niger*. (Modified from Lang, A., and G. Melchers. 1943. Planta 33:653-702.)

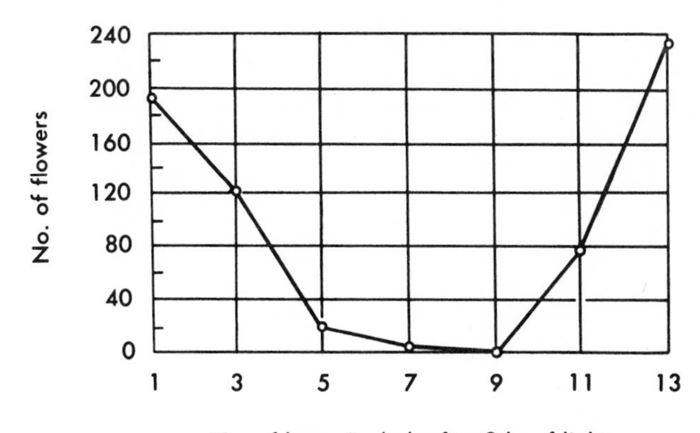

Fig. 20-9. Effect of a 1 min interruption of a long (15 hr) dark period by light on flowering of the short-day plant *Kalanchoe blossfeldiana*. When applied near the beginning or end of the dark period, there is no effect; when applied at the middle, it prevents flowering despite the favorable (9 hr) photoperiod. (From Bünning, E. 1948. *In* A. E. Murneek and R. O. Whyte. Vernalization and photoperiodism. The Ronald Press Co., New York.)

leaf and translocated to the growing points.

The mean velocity of transfer of the floral stimulus was found to be only 2 cm/hr in LDP, but 24 to 37 or even 51 cm/hr for SDP (Evans, 1971). Yet induced plants of the one type can act as donors for scions of the other type, indicating that a single floral stimulus is involved.

The floral stimulus or hormone is also able to multiply autocatalytically (Wellensiek, 1966). In this property it resembles certain nucleic acids that may be immortal and rapidly perpetuating. If the floral stimulus is a specific RNA, inhibition of flowering may result from the blocking of a specific chromosomal DNA particle, from which the specific RNA would originate. Induction of flowering may therefore simply be the removal of this inhibition by deblocking (derepressing) the DNA particle. The RNA produced may either be the floral hormone, or it may give rise to its synthesis. In favor of this concept, vegetative plants of *Sinapis alba* can be induced to flower by exposure to a single day of 20 hr. A sharp increase in total RNA occurred in the cells of the apical meristem as early as 18 hr after start of the long day (Bronchart et al., 1970). In further agreement, the SDP *Chenopodium rubrum* was inhibited from flowering by low concentrations (4×10^{-6} M and higher) of the inhibitor of NA synthesis 5-fluorodeoxyuridine (FDU) when applied during or immediately after photoinduction (Ullmann et al., 1971).

Evans (1971) suggests that although floral differentiation must at some stage involve calling new genes into play, the initial action of the photoperiodic stimulus at the shoot apex need not do so. It may simply activate all RNA species. In favor of this concept, the morning after exposure of a LDP to a single long day, there was a transient increase in RNA and protein synthesis in the shoot apices (Evans, 1971). No obvious changes occurred in the histones or in DNA.

RNA synthesis may be an essential component not only of floral evocation at the shoot apex, but also of induction in leaves of SDP (Evans, 1971). In LDP, on the contrary, inhibition of flowering on short days may be due to the synthesis of an inhibitor, requiring RNA synthesis.

Growth regulators may also affect flowering. It was long ago observed that the initiation of floral and fruit development is frequently associated with a decrease in growth. This led to the hypothesis that the same growth regulators that produce growth inhibition may also induce reproductive development. Many investigators have been able to show both an enhancement and a decrease in rate of flowering with increased auxin concentration, depending on the stage of the plant. It has even been possible in some cases (e.g., pineapple) to induce flowering commercially by application of an auxin (indolebutyric acid). But it is now well established that such effects are secondary and cannot explain the primary control. The evidence indicates that auxins, cytokinins, and gibberellins cannot transmit the photoperiodically induced stimulus from the leaves to the apex. A more specific flowering hormone has therefore been postulated, for which the name *florigen* has been proposed. Many attempts have been made to extract and isolate the florigen, and several reports of successful isolation are to be found in the literature (Lincoln et al., 1966).

Nevertheless, there definitely is an effect of photoperiod on growth-regulator content. Short-day conditions that lead to dormancy result in an increase in the level of growth inhibitors (Vegis, 1964, see Chapter 17), although most of these occur in the covering structures (e.g., bud scales). Conversely, long days may lead to the development of growth promoters.

Both the promotive effect of auxin on flowering of pineapple and its inhibitory effect on flowering of *Xanthium* are due to increased production of ethylene, which can induce flowering of SDP (Evans, 1971).

Gibberellic acid replaces the requirement for long days in many plants, but it fails in some and may even inhibit flowering in other LDP (Evans, 1971). It inhibits flowering in some SDP, but induces flowering in three SDP. Evans therefore concludes that GA is probably required, perhaps at different levels, for flower induction in all plants. Retardants promote flowering in some SDP but inhibit it in others.

Auxin levels, as well as GA, are higher on long days. CK levels may be higher on short days, and CK application can induce flowering in some SDP exposed to long days (Evans, 1971). CK has also been found to promote flowering in at least one LDP on short days. ABA may also be more abundant on short days and may be a component of the short-day stimulus in SDP, or of the short-day inhibitor in LDP.

Gibberellic acid has the greatest promoting effect, although only on LDP. But its promotion of flowering appears to be a secondary result of the stem elongation, which must precede flowering in SDP. Applied GA therefore replaces only the growth-promoting effect of long days (Cline and Agatep, 1970; Cleland and Zeevaart, 1970; Zeevaart, 1971). Endogenous GA appears to act in the same way, for the natural gibberellin content of plants increases in the morning, reaching a maximum in daylight, and falls to a minimum at night (Lozhnikova, 1966). This is appearently true of long-day, short-day, and day-neutral plants. It is not surprising, then, that in all cases the amount of gibberellin found in the leaves is greater when the plant is grown on a 16 hr day than on an 8 hr day (Chailakhyan and Lozhnikova, 1966). The normal increased growth of LDP during long days is therefore due to the increased biosynthesis of GA. This may also explain the existence of a LSDP, which normally requires a long day followed by a short day. It will flower in a short day without a preceding long day period if supplied with GA (Chailakhyan et al., 1968). In some cases, however, IAA may also be involved. The level of both IAA and GA was higher in long-day–grown *Begonia* plants than in short-day–grown plants (Heide, 1967).

A SDP has been reported to flower under long-day conditions in response to applied CK (Gupta and Maheshwari, 1970). In contrast, CK was found to decrease markedly in a SDP exposed to a photoinducing short day (Van Staden and Wareing, 1972). Abscisic acid inhibited flower induction in the long-day species *Lolium temulentum* and *Spinacea oleracea* when applied to the leaves during exposure to long days (El-Antably et al., 1967). Conversely, it promoted flowering in the short-day plants *Pharbitis nil, Betula nigrum,* and strawberry when applied under long-day conditions. However, it failed to induce flowering in certain other short-day plants. A natural inhibitor of flowering of the long-day plant *Silene armeria* is formed during short days at 20° C (Wellensiek, 1967). Induction by long days (20° C) or by short days (32° or 5° C) is a result of the removal of this inhibitor.

Phytochrome

As in any light-controlled process, photoperiodism is possible only if a substance is present that absorbs the light. This substance has now been isolated and is a pale blue protein that has been called *phytochrome.* The chromophore of phytochrome has been separated from the protein and appears to be a bilitriene (Siegelman et al., 1966) with a tetrapyrrole structure. Phytochrome differs from all other plant pigments by existing in two light-absorbing forms, each of which can be converted reversibly into the other as a result of light absorption:

$$\underset{\text{(blue-green pigment)}}{P_{660}} \underset{\text{Far red}}{\overset{\text{Near red}}{\rightleftharpoons}} \underset{\text{(light green pigment)}}{P_{730}}$$

The equation is also written

$$P_r \rightleftharpoons P_{fr} \text{ or } P_R \rightleftharpoons P_{FR}$$

where P_r or P_R = the near red absorbing form and P_{fr} or P_{FR} = the far red absorbing form.

The photochemical transformation of phytochrome may be a combination of an isomerization of the chromophore with a change in conformation of the protein (Kroes, 1970). The P_{660} (near red absorbing) form absorbs light maximally at a wavelength of 660 nm and, as a result, is converted to the P_{730} (far red absorbing) form, which absorbs maximally at 730 nm. The latter absorption process converts it back to the near red absorbing form. The process can be repeated over and over again apparently without limit. The conversion process is rapid; the half conversion time for P_r to P_{fr} is 2.3 sec (Hendricks and Borthwick, 1967). In contrast to these rapid conversions in the light, the P_{730} form changes back slowly in the dark to the P_{660} form, which is stable in the dark. Thus in dark-grown maize seedlings, phytochrome is present entirely as P_{660} (Butler et al., 1963). A single brief irradiation with red light converts it into P_{730}. On retransfer to the dark, the P_{730} reverts to P_{660}, and the total amount of photoreversible phytochrome decreases to about 30% of the original amount.

Recent results indicate that in monocots only decay occurs, whereas dicots exhibit both decay and reversion (Kendrick and Hillman, 1970). The P_{730} is the physiologically active form and is thought to be a highly active enzyme (Mohr, 1962).

Exactly how these reversals are related to the control of the plant's development is not known. During the photoperiod, P_{fr} presumably performs a flower promoting function that may be completed by the P_{fr} left in the plant at the beginning of the dark period (Cathey and Borthwick, 1971). Far red at this time removes P_{fr} before completion of its promotive action, and red light at a later hour of darkness reintroduces it at a time when its presence inhibits flowering.

Despite its obvious role in leaves, which act as the region of perception for the flowering response, phytochrome has been found in all parts of higher (vascular) plants (Siegelman, 1969) as well as in bryophytes and algae (Scheibe, 1972), although in these cases it does not affect flowering.

The effect of phytochrome has been interpreted in terms of modern concepts of molecular biology. Mohr has explained the phytochrome effect as the result of gene derepression. Scherf and Zenk (1967) suggest that as a consequence of illumination with high-energy blue light, a derepression of formerly repressed genes comes about; but only if there is phytochrome present in its far red absorbing form (P_{730}), can the derepressed gene induce protein synthesis. In agreement with Mohr's concept, several investigators have found a relation between the phytochrome effect and NA and protein synthesis (e.g., Okoloko et al., 1970). Phytochrome induces the formation of a number of enzymes: amylase (Drumm et al., 1971), inorganic pyrophosphatase and adenylate kinase (Butler and Bennett, 1969), and PAL (Schopfer, 1971; Bellini and Hillman, 1971).

Some results fail to support the gene derepression concept of the phytochrome control system and favor, instead, the older concept of permeability as the control mechanism because an early consequence of phytochrome action is a change in permeability (Hendricks and Borthwick, 1967). There is even evidence that phytochrome is localized at the plasmalemma (Haupt, 1970*a*). In the green alga *Mougeotia,* P_{660} is apparently parallel to the cell surface with P_{730} normal to it (Haupt, 1970*b*). In the case of nyctinastic movements in *Albizzia julibrissin,* the closure response is accompanied by an increased rate of electrolyte efflux from the cut pinna base (Jaffe and Galston, 1967). This result is not affected by actinomycin D and is therefore presumably independent of RNA metabolism and gene derepression, yet the nyctinastic movements are dependent on phytochrome. In support of this interpretation the light stimulation of lettuce seed by the action of P_{fr} is almost instantaneous and therefore more readily explained by a sudden permeability change than by the much slower gene action (Bewley et al., 1967). The same is true of barley root tips (Tanada, 1968).

Further experiments have led to the suggestion that acetylcholine acts as a local hormone that regulates the phytochrome-mediated phenomena (Jaffe, 1970). It acts, supposedly, as in the animal system, by mediating changes in ion flux across cell membranes by a rapid utilization of ATP pools (Yunghans and Jaffe, 1972). When applied to two different species of *Lemna,* acetylcholine suppressed flowering of the long-day species and promoted flowering of short-day species (Kandeler, 1972). As in the case of growth regulators, the gene repression theory and the permeability theory of phytochrome action are not necessarily mutually exclusive (Smith, 1970.)

The two theories of phytochrome action are identical to those proposed to explain the action of growth regulators. It is therefore logical to seek an interaction between phytochrome and the growth regulators. Several interactions have been found, especially with respect to photomorphogenesis (see following discussion). Phytochrome is involved in the light-induced opening of the hypocotyl hook in seedlings. CK inhibits the opening and GA promotes it, both in the light and the dark (Powell and Morgan, 1970). High concentrations of IAA caused closing in red light. In contrast to this result, red light induces a small stimulation of IAA transport that may be related to the light-induced opening (Rubinstein, 1971). CK may replace the red light induction of unrolling of etiolated wheat leaves, but GA is relatively ineffective, and ABA inhibits the unrolling response to both the hormones and red light (Beevers et al., 1970).

Even the growth process itself may be affected by phytochrome. When 100% of the phytochrome is in the P_{660} form, this apparently prevents growth of the oat coleoptile; when 100% is in the P_{730} form, maximal growth occurs (Hopkins and Hillman, 1966). This loss in growth capacity must be related to an observed decrease in diffusible auxin.

P_{730} may also affect growth qualitatively, for instance by preventing etiolation or by opening the hypocotyl hook. All these effects of phytochrome on growth, whether qualitative or quantitative, are called *photomorphogenesis.* Photosensitive seed, for instance, will not germinate unless the phytochrome is in the active, P_{730} form. If this is true of seed in general, then the surprise is not that some are photosensitive, but that any are not photosensitive. Germination in the dark can, however, be explained by the presence of a P_{730} pool that remains constant during imbibition while the total phytochrome increases (Boisard and Malcoste, 1970). Apparently, when kept in the dark, the pigment in these seeds shows the inverse reversion process: P_{660} reverts spontaneously to P_{730}.

The near red$\rightleftharpoons$far red phytochrome may not be the only one involved in plant development, since blue light (which, at least at low intensities, is not preferentially absorbed by phytochrome) may have a great effect on development in some cases. However, no other phytochromes have been isolated, and the preceding one has been shown to be important in so many processes (seed germination, sleep movements, coloring of fruit, etc.) that it is usually called *the* phytochrome of plant development. It occurs in all groups of plants, from algae to monocotyledons (Mohr, 1962). It is synthesized in the maturing leaves (Salisbury, 1965).

The many factors involved in photoperiodic control of flowering may be illustrated by the sequence of events in the flowering of the short-day plant *Pharbitis nil* (Zeevaart, 1962):

1. Exposure to a dark period of sufficient length
2. Change in the phytochrome system in the leaves
3. Translation by an unknown process into the synthesis of florigen (perhaps by the phytochrome acting as an effector)
4. Translocation of the florigen to the bud
5. Triggering of the events associated with the nucleic acid synthesis by florigen, which leads to flowering

Despite the striking effects of the photoperiod, it is not the only factor that controls flowering; temperature may also be a deciding factor.

Thermoperiodism

Tomatoes are day-neutral plants under ordinary conditions, but their flowering can be controlled by the day and night temperature. A night temperature of 15° C and a day temperature of 25° C are optimum for flowering. If these are changed too greatly from the optimum, flowering is reduced or even prevented. Other plants have shown similar responses. Went has given the name *thermoperiodism* to this control of development by alteration of day and night temperatures. Unlike photoperiodism, there is no evidence that correlation is involved in thermoperiodism. Thus there is no specific region of perception and no specific hormone known to control the plant's response. One of the major factors is apparently the quantitative relation between the photosynthate accumulated during the day and the respiratory breakdown at night. Therefore in the case of some plants, the effect of the thermoperiod is mainly on growth. The Sahara species *Hyoscyamus muticus* is normally exposed to a climate with a pronounced daily thermoperiod. Growth at a constant temperature of 27° C with either a 9 hr or 16 hr photoperiod for several generations led to a progressively smaller plant. This effect disappeared on cultivation under alternating temperatures (Saint-Firmin and Chouard, 1971). These results can be explained by the large respiratory loss of carbohydrate at the high (27° C) night temperature during the long nights of the Sahara. In the case of Engelmann spruce, however, increasing the night temperature within a range of 3° to 23° C increased growth even when the night temperature exceeded the day temperature (Hellmers et al., 1970). In this case, the logical explanation is that growth occurs mainly at night, and therefore the plant requires a relatively high night temperature. Other specific temperature effects (e.g., on starch $\rightleftharpoons$ sugar equilibrium) may also be involved.

Vernalization

Some winter annuals must be sown in the fall to flower the following summer. If they are sown in the spring, they either fail to flower the first year or flower much later than when fall sown. However, if the seeds are moistened (with enough water to increase their weight by about 60%) and kept at 0° to 5° C for about a month, they can then be spring sown and will flower at about the same time as the fall-sown seed. This method of inducing reproductive development by a long exposure to low temperature is known as *vernalization*. For some plants, vernalization alone is sufficient to induce flowering; other plants must still be subjected to a suitable photoperiod (usually a long day) to flower.

Some biennials show a similar behavior toward temperature, although at a later stage in their development. If the young seedlings (instead of the swollen seed) are exposed to low temperatures (about 5° to 10° C in the case of several species) for a week or two, they will subsequently flower the first year at a higher temperature, provided that the latter temperature is not too high. The actual temperature requirements vary considerably from species to species.

The mechanism of the temperature effect has not yet been discovered. However, the evidence again indicates a hormonal control, and the name *vernalin* has been coined for the unknown hormone. Lang (1952) suggested that vernalin is either a precursor of florigen or a catalyzer of its formation. According to Chailakhyan (1966), precursors of GA are produced and accumulated during vernalization. Under long-day conditions these become converted into gibberellins. These precursors of gibberellins are hormones and are therefore the vernalins, according to his interpretation. But unlike photoperiodism, the region of perception in the case of vernalization always contains dividing (mitotic) cells—the apical region of biennials or the slightly germinated embryo of seeds. Vernalin does not appear to be a true hormone, since it seems to move only with dividing cells. Recent results have shown that gibberellin may replace the vernalization treatment in the case of some plants but not in all. Uridylic acid has been found to replace vernalization in winter wheat (Suge and Yamada, 1963). Other pyrimidine bases have also been shown to induce flowering (Hirono and Rédei, 1966). Evans (1971) explains these results by a twofold action of vernalization: (1) activation of the meristems and (2) increased synthesis of GA. Neither alone is capable of inducing flowering. The activation of the meristem in the shoot apex may account for the occasional effects of applications of CK, NA bases, and extracts of vernalized plants. A third factor may be a reduction of inhibitor content.

QUESTIONS

1. What are the main stages of development of the plant?
2. Are all parts of the plant simultaneously at the same stage?
3. How are these stages related to growth regulators?
4. What specific regulators have been found to control senescence?
5. What major chemical change occurs during senescence?
6. Is there any observable change in the cells?
7. What changes lead to abscission?
8. How is abscission related to senescence?
9. What is believed to be the cause of senescence?
10. What is meant by photoperiodism?
11. What is a short-day plant?
12. What is a long-day plant?
13. What is a day-neutral plant?
14. Would an alternation of 14 hr light and 14 hr dark act as a short day, a long day, or neither?
15. Would 8 hr of light, ½ hr of dark, 7½ hr of light, and 8 hr of dark act as a short day, a long day, or neither?
16. What is a good definition of a short-day plant? Of a long-day plant?
17. Does photoperiodism involve anything besides flowering?
18. Give examples of the effects of photoperiodism.
19. What is meant by photoinduction?
20. What is the region of perception?
21. What evidence is there of hormonal control?
22. What is the relation between florigen and growth regulators?
23. What is phytochrome?
24. In what forms can phytochrome exist?
25. Which is the active form of phytochrome?
26. How is phytochrome believed to control plant responses to light?
27. What is thermoperiodism?
28. What is vernalization?
29. What kinds of plants respond to vernalization?
30. What relation is there between vernalization and photoperiodism?
31. What chemical substances have been found that can replace vernalization?

32. What evidence is there of a relation between nucleic acids and control of plant development?

SPECIFIC REFERENCES

Abbott, I. R., and N. K. Matheson. 1972. Starch depletion in germinating wheat, wrinkled-seeded peas and senescing tobacco leaves. Phytochemistry **11**:1261-1272.

Abeles, F. B., and J. Lonski. 1969. Stimulation of lettuce seed germination by ethylene. Plant Physiol. **44**:277-280.

Addicott, F. T. 1970. Plant hormones in the control of abscission. Biol. Rev. **45**:485-524.

Addicott, F. T., and J. L. Lyon. 1969. Physiology of abscisic acid and related substances. Ann. Rev. Plant Physiol. **20**:139-164.

Agrawal, P. K., and D. T. Canvin. 1971*a*. Respiration of developing castorbean seeds. Canad. J. Bot. **49**:263-266.

Agrawal, P. K., and D. T. Canvin. 1971*b*. The pentose phosphate pathway in relation to fat synthesis in the developing castor oil seed. Plant Physiol. **47**:672-675.

Amrhein, N., and M. H. Zenk. 1970. Activity of phenylalanine ammonia-lyase (PAL) and accumulation of phenylpropanoid compounds during the germination of buckwheat (*Fagopyrum esculentum* Moench) in the dark. Z. Pflanzenphysiol. **63**:384-388.

Anderson, J. W., and K. S. Rowan. 1965. Activity of peptidase in tobacco-leaf tissue in relation to senescence. Biochem. J. **97**:741-746.

App, A. A., M. C. Bulis, and W. J. McCarthy. 1971. Dissociation of ribosomes and seed germination. Plant Physiol. **47**:81-86.

Atkin, R. K., and B. I. S. Srivastava. 1969. The changes in soluble protein of excised barley leaves during senescence and kinetin treatment. Physiol. Plant. **22**:742-750.

Atkin, R. K., and B. I. S. Srivastava. 1970. Studies on protein synthesis by senescing and kinetin-treated barley leaves. Physiol. Plant. **23**:304-315.

Atsmon, D. 1968. The interaction of genetic, environmental, and hormonal factors in stem elongation and floral development of cucumber plants. Ann. Bot. **32**:877-882.

Back, A., and A. E. Richmond. 1969. An interaction between the effects of kinetin and gibberellin in retarding leaf senescence. Physiol. Plant. **22**:1207-1216.

Back, A., and A. E. Richmond. 1971. Interrelations between gibberellic acid, cytokinins and abscisic acid in retarding leaf senescence. Physiol. Plant. **24**:76-79.

Badr, S. A., H. T. Hartmann, and G. C. Martin. 1970. Endogenous gibberellins and inhibitors in relation to flower induction and inflorescence development in the olive. Plant Physiol. **46**: 674-679.

Baisted, D. J. 1971. Sterol and triterpene synthesis in the developing and germinating pea seed. Biochem. J. **124**:375-383.

Barendse, G. W. M. 1971. Formation of bound gibberellins in *Pharbitis nil*. Planta **99**:290-301.

Barker, G. R., and M. Rieber. 1967. The development of polysomes in the seed of *Pisum arvense*. Biochem. J. **105**:1195-1202.

Barr, R., and C. J. Arntzen. 1969. The occurrence of δ-tocopherylquinone in higher plants and its relation to senescence. Plant Physiol. **44**:591-598.

Barton, R. 1966. Fine structure of mesophyll cells in senescing leaves of *Phaseolus*. Planta **71**:314-325.

Bečka, J., and J. Sebanek. 1971. Effect of 2-chorethyl-trimethylammonium chloride on climacteric changes in apples. Biochem. Physiol. Pflanz. **162**:90-96.

Beevers, L. 1968. Protein degradation and proteolytic activity in the cotyledons of germinating pea seeds (*Pisum sativum*). Phytochemistry **7**: 1837-1844.

Beevers, L., B. Loveys, J. A. Pearson, and P. F. Wareing. 1970. Phytochrome and hormonal control of expansion and greening of etiolated wheat leaves. Planta **90**:286-294.

Beevers, L., and R. Poulson. 1972. Protein synthesis in cotyledons of *Pisum sativum* L. I. Changes in cell-free amino acid incorporation capacity during seed development and maturation. Plant Physiol. **49**:476-481.

Beevers, L., and W. E. Splittstoesser. 1968. Protein and nucleic acid metabolism in germinating peas. J. Exp. Bot. **19**:698-711.

Bellini, E., and W. S. Hillman. 1971. Red and far red effects on phenylalanine ammonia lyase in *Raphanus* and *Sinapis* seedlings do not correlate with phytochrome spectrophotometry. Plant Physiol. **47**:668-671.

Berjak, P. 1968. A lysosome-like organelle in the root cap of *Zea mays*. J. Ultrastruct. Res. **23**: 233-242.

Berjak, P., and T. A. Villiers. 1972. Ageing in plant embryos. II. Age-induced damage and its repair during early germination. New Phytol. **71**:135-144.

Bewley, J. D., and M. Black. 1972. Protein synthesis during gibberellin-induced germination of lettuce seed. Canad. J. Bot. **50**:53-59.

Bewley, J. D., M. Black, and M. Negbi. 1967. Immediate action of phytochrome in light-stimulated lettuce seeds. Nature (London) **215**:648-649.

Bhalla, P. R. 1971. Gibberellin-like substances in developing watermelon seeds. Physiol. Plant. **24**: 106-111.

Bick, M. D., H. Liebke, J. H. Cherry, and B. L. Strehler. 1970. Changes in leucyl-and tyrosyl-tRNA of soybean cotyledons during plant growth. Biochim. Biophys. Acta **204**:175-182.

Bick, M. D., and B. L. Strehler. 1971. Leucyl transfer RNA synthetase changes during soybean cotyledon senescence. Proc. Nat. Acad. Sci. USA **68**:224-228.

Black, M., and M. Richardson. 1968. Promotion of germination by inhibitors of protein synthesis: a clue to the germination mechanism? Bull. Soc. Fr. Physiol. Veg. **14**:73-81.

Blumenfeld, A., and S. Gazit. 1970. Cytokinin activity in avocado seeds during fruit development. Plant Physiol. **46**:331-333.

Bohmova, B., and Y. Guitton. 1971. Effects of nitrosomethylurea on RNA metabolism during germination of barley. Physiol. Veg. **9**:189-200.

Boisard, J., and R. Malcoste. 1970. Active form of phytochrome P_{730} and inverse reversion in seeds of some *Cucurbitaceae*. Physiol. Veg. **8**:565-572.

Brady, C. J., P. B. H. O'Connell, J. Smydzuk, and N. L. Wade. 1970*b*. Permeability, sugar accumulation, and respiration rate in ripening banana fruits. Aust. J. Biol. Sci. **23**:1143-1152.

Brady, C. J., J. K. Palmer, P. B. H. O'Connell, and R. M. Smillie. 1970*a*. An increase in protein synthesis during ripening of the banana fruit. Phytochemistry **9**:1037-1047.

Briggs, W. R., and H. V. Rice. 1972. Phytochrome: chemical and physical properties and mechanism of action. Ann. Rev. Plant Physiol. **23**:293-334.

Bronchart, R., G. Bernier, J. M. Kinet, and A. Havelange. 1970. RNA synthesis in the cells of the apical meristem of *Sinapis alba* during transition from the vegetative to the reproductive condition. Planta **91**:255-269.

Brunori, A. 1967. Relationship between DNA synthesis and water content during ripening of *Vicia faba* seed. Caryologia **20**:333-338.

Burg, S. P. 1968. Ethylene, plant senescence and abscission. Plant Physiol. **43**:1503-1511.

Burg, S. P. 1973. Ethylene in plant growth. Proc. Nat. Acad. Sci. USA **70**:591-597.

Burrows, W. J., and D. J. Carr. 1970. Cytokinin content of pea seeds during their growth and development. Physiol. Plant. **23**:1064-1070.

Butler, L. G., and V. Bennett. 1969. Phytochrome control of maize leaf inorganic pyrophosphatase and adenylate kinase. Plant Physiol. **44**:1285-1290.

Butler, W. L., H. C. Lane, and H. W. Siegelman. 1963. Nonphotochemical transformations of phytochrome in vivo. Plant Physiol. **38**:514-519.

Cameron, R. J. 1970. Light intensity and the growth of *Eucalyptus* seedlings. I. Ontogenetic variation in *E. fastigata*. Aust. J. Bot. **18**:29-43.

Carpenter, J. A. 1969. The interrelation of age, permeability and viability of subterranean clover seeds. Aust. J. Exp. Agr. Anim. Husb. **9**:513-516.

Cathey, H. M., and H. A. Borthwick. 1971. Phytochrome control of flowering of *Chrysanthemum morifolium* on very short photoperiods. J. Amer. Soc. Hort. Sci. **96**:544-546.

Chailakhyan, M. K. 1966. Role of gibberellins in the processes of photoperiodism and vernalization of plants. [Transl. title.] Biol. Zh. Armenii **19**:3-14.

Chailakhyan, M. K., and V. N. Lozhnikova. 1966. Effect of interruption of darkness by light and plant gibberellins. Fiziol. Rast. **13**:833-841 [734-741, Transl. ed.].

Chailakhyan, M. K., L. I. Yanina, and I. A. Frolova. 1968. Effect of day length and gibberellin on flowering of *Bryophyllum* plants of different age. Dokl. Akad. Nauk. SSSR **183**:230-233.

Chailakhyan, M. K., L. I. Yanina, and I. A. Frolova. 1970. Flowering of *Bryophyllum* plants lacking roots. Fiziol. Rast. **17**:709-711.

Chalmers, D. J., and K. S. Rowan. 1971. The climacteric in ripening tomato fruit. Plant Physiol. **48**:235-240.

Chen, D., and D. J. Osborne. 1970. Hormones in the translational control of early germination in wheat embryos. Nature **226**:1157-1160.

Chen, D., G. Schultz, and E. Katchalski. 1971. Early ribosomal RNA transcription and appearance of cytoplasmic ribosomes during germination of the wheat embryo. Nature (New Biol.) **231**:69-72.

Chen, Y.-M., and C.-Y. Lin. 1971. Certain aspects of control mechanism of stored foods degradation during seed germination. Taiwania **16**:143-155.

Chen, S. S. C., and J. E. Varner. 1969. Metabolism of ^{14}C-maltose in *Avena fatua* seeds during germination. Plant Physiol. **44**:770-774.

Chou, K. H., and W. E. Splittstoesser. 1972. Changes in amino acid content and the metabolism of γ-aminobutyrate in *Cucurbita moschata* seedlings. Physiol. Plant. **26**:110-114.

Cleland, C. F., and J. A. D. Zeevaart. 1970. Gibberellins in relation to flowering and stem elongation in the long day plant *Silene armeria*. Plant Physiol. **46**:392-400.

Cline, M. G., and A. O. Agatep. 1970. Control of stem elongation and flowering in *Scrophularia marilandica*. Physiol. Plant. **23**:993-1003.

Coggins, C. W., Jr., R. W. Scora, L. N. Lewis,

and J. C. F. Knapp. 1969. Gibberellin-delayed senescence and essential oil changes in the naval orange rind. J. Agr. Food. Chem. **17**:807-809.

Come, D. 1971*a*. Gases released from seed coats during imbibition. I. General case. Physiol. Veg. **9**:439-446.

Come, D. 1971*b*. Degazage des enveloppes seminales lors de leur imbibition. II. Cas des graines de Pommier. Physiol. Veg. **9**:447-452.

Cooper, T. G., and H. Beevers. 1969. Mitochondria and glyoxysomes from castor bean endosperm. J. Biol. Chem. **244**:3507-3513.

Cracker, L. E., and F. B. Abeles. 1969. Role of abscisic acid. Plant Physiol. **44**:1144-1149.

Curtis, H. J. 1963. Biological mechanisms underlying the ageing process. Science **141**:686-694.

Dale, J. E. 1969. Gibberellins and early growth in seedlings of *Phaseolus vulgaris*. Planta **89**:155-164.

De Leo, P., and J. A. Sacher. 1970*a*. Association of synthesis of acid phosphatase with banana ripening. Plant Physiol. **46**:208-211.

De Leo, P., and J. A. Sacher. 1970*b*. Control of ribonuclease and acid phosphatase by auxin and abscisic acid during senescence of *Rhoeo* leaf sections. Plant Physiol. **46**:806-811.

Deltour, R. 1967. Action de l'azote minéral sur la croissance et la mise à fleurs de plantes issues d'apex de *Sinapis alba* L. cultivés in vitro. Compt. Rend. Acad. Sci. (Paris) **265**:1932-1935.

Deltour, R., and R. Bronchart. 1971. Changes in the ultrastructure of the root cells of the *Zea mays* embryo at the beginning of germination. Planta **97**:197-207.

Dilley, D. R., and I. Klein. 1969. Proteins synthesis in relation to fruit ripening. Qual. Plant. Mat. Veg. **19**:55-65.

Dodge, J. D. 1970. Changes in chloroplast fine structure during the autumnal senescence of *Betula* leaves. Ann. Bot. **34**:817-824.

Dove, L. D. 1971. Short term responses and chemical control of ribonuclease activity in tomato leaflets. New Phyt. **70**:397-401.

Drumm, H., I. Elchinger, J. Moeller, K. Peter, and H. Mohr. 1971. Induction of amylase in mustard seedlings by phytochrome. Planta **99**:265-274.

Dull, G. G., R. E. Young, and J. B. Biale. 1967. Respiratory patterns in fruit of pineapple, *Ananas comosus,* detached at different stages of development. Physiol. Plant. **20**:1059-1065.

Durzan, D. J., A. J. Mia, and P. K. Ramaiah. 1971. The metabolism and subcellular organization of the jack pine embryo *(Pinus banksiana)* during germination. Canad. J. Bot. **49**:927-938.

Eastwood, D., and D. L. Laidman. 1968. Mineral translocation in germinating wheat grain. Biochem. J. **109**:9p-10p.

Eastwood, D., and D. L. Laidman. 1971. The hormonal control of inorganic ion release from wheat aleurone tissue. Phytochemistry **10**:1459-1467.

Eastwood, D., and D. L. Laidman. 1971. The mobilization of macronutrient elements in the germinating wheat grain. Phytochemistry **10**:1275-1284.

Ecklund, P. R., and T. C. Moore. 1968. Quantitative changes in gibberellin and RNA correlated with senescence of the shoot apex in the Alaska pea. Amer. J. Bot. **55**:494-503.

Ecklund, P. R., and T. C. Moore. 1969. RNA and protein metabolism in senescent shoot apices of Alaska peas. Amer. J. Bot. **56**:327-334.

Eilam, Y., R. D. Butter, and E. W. Simon. 1971. Ribosomes and polysomes in cucumber leaves during growth and senescence. Plant Physiol. **47**:317-323.

El-Antably, H. M. M., P. F. Wareing, and J. Hillman. 1967. Some physiological responses to D,L abscisin (dormin). Planta **73**:74-90.

Evans, L. T. 1969. The induction of flowering: some case histories. Cornell University Press, Ithaca, N. Y.

Evans, L. T. 1969. Inflorescence initiation in *Lolium temulentum* L. XIII. The role of gibberellins. Aust. J. Biol. Sci. **22**:773-786.

Evans, L. T. 1971. Flower induction and the florigen concept. Ann. Rev. Plant. Phys. **22**:365-394.

Evenari, M. 1961. A survey of the work done in seed physiology by the department of botany, Hebrew University, Jerusalem (Israel). Proc. Int. Seed Test. Asso. **26**:597-658.

Feldman, L. J., and E. G. Cutter. 1970*a*. Regulation of leaf form in *Centaurea solstitialis* L. I. Leaf development on whole plants in sterile culture. Bot. Gaz. **131**:31-39.

Feldman, L. J., and E. G. Cutter. 1970*b*. Regulation of leaf form in *Centaurea solstitialis* L. II. The development potentialities of excised leaf primordia in sterile culture. Bot. Gaz. **131**:39-49.

Fletcher, R. A. 1969. Retardation of leaf senescence by benzyladenine in intact bean plants. Planta **89**:1-8.

Fletcher, R. A., G. Hofstra, and N. O. Adepipe. 1970. Effects of benzyladenine on bean leaf senescence and the translocation of ^{14}C-assimilates. Physiol. Plant. **23**:1144-1148.

Floris, C. 1970. Ageing in *Triticum durum* seeds: behaviour of embryos and endosperms from aged seeds as revealed by the embryo-transplantation technique. J. Exp. Bot. **21**:462-468.

Frenkel, C. 1972. Involvement of peroxidase and

indole-3-acetic acid oxidase isozymes from pear, tomato, and blueberry fruit in ripening. Plant Physiol. **49**:757-763.

Garner, W. W., and H. Allard. 1920. Effect of the relative length of day and night and other factors of the environment on growth and reproduction in plants. J. Agric. Res. **18**:553-606.

Gaspar, T., R. Verbeek, and H. van Onckelen. 1969. Variations in some enzymatic activities (peroxidase, catalase, IAA oxidase) and polyphenol content during barley germination: effect of kinetin. Physiol. Plant. **22**:1200-1206.

Gepstain, S., and I. Ilan. 1970. A promotive action of kinetin on amylase activity in cotyledons of *Phaseolus vulgaris*. Plant Cell Physiol. **11**:819-822.

Gerhardt, B. P., and H. Beevers. 1970. Development studies on glyoxysomes in *Ricinus* endosperm. J. Cell Biol. **44**:94-102.

Ghosh, B. N., A. Glowinkowska, and D. F. Millikan. 1971. Enzymatic activities associated with maturation and senescence in *Hedera helix*. Bull. Acad. Pol. Sci. Ser. Sci. Biol. **19**:605-609.

Gifford, E. M., Jr. 1969. Initiation and early development of the inflorescence in pineapple (*Ananas comosus* "Smooth Cayenne") treated with acetylene. Amer. J. Bot. **56**:892-897.

Gilad, T., I. Ilan, and L. Reinhold. 1970. The effect of kinetin and of the embryo axis on the level of reducing sugars in sunflower cotyledons. Isr. J. Bot. **19**:447-450.

Godwin, H. 1968. Evidence for longevity of seeds. Nature **220**:708-709.

Goldthwaite, J. J., and W. M. Laetsch. 1968. Control of senescence in *Rumex* leaf discs by gibberellic acid. Plant Physiol. **43**:1855-1858.

Goren, R. 1971. Nucleic acids in developing fruits and other tissues of Shamouti orange (*Citrus sinensis* L.) Osbeck. J. Amer. Soc. Hort. Sci. **96**:10-14.

Goren, R., and E. E. Goldschmidt. 1970. Regulative systems in the developing citrus fruit. I. The hormonal balance in orange fruit tissues. Physiol. Plant. **23**:937-947.

Grellett, F., R. Julien, and H. Milhomme. 1968. Ribonuclease activity and RNA content of *Vicia sativa* L., during germination. Physiol. Veg. **6**:11-17.

Groat, J. I., and D. E. Briggs. 1969. Gibberellins and α-amylase formation in germinating barley. Phytochemistry **8**:1615-1627.

Gumilevskaya, N. A., E. B. Kuvaeva, L. V., Chumikina, and V. L. Kretovich. 1971. Ribosomes of dry pea seeds. Biokhimiya. **36**:277-288.

Gupta, S., and S. C. Maheshwari. 1970. Growth and flowering of *Lemna paucicostata*. II. Role of growth regulators. Plant. Cell. Physiol. **11**:97-106.

Haber, A. H., D. E. Foard, and S. W. Perdue. 1969. Action of gibberellic and abscisic acids on lettuce seed germination without actions on DNA synthesis. Plant Physiol. **44**:463-467.

Hale, C. R., B. G. Coombe, and J. S. Hawker. 1970. Effects of ethylene and 2-chloroethylphosphonic acid on the ripening of grapes. Plant Physiol. **45**:620-623.

Halevy, A. H., D. R. Dilley, and S. H. Wittwer. 1966. Senescence inhibition and respiration induced by growth retardants and ^{6}N-benzyl adenine. Plant Physiol. **41**:1085-1089.

Hallam, N. D., B. E. Roberts, and D. J. Osborne. 1972. Embryogenesis and germination in rye. II. Biochemical and fine structural changes during germination. Planta **105**:293-309.

Halperin, W. 1969. Ultrastructural localization of acid phosphatase in cultured cells of *Daucus carota*. Planta **88**:91-102.

Harris, G. P., B. Jeffcoat, and J. F. Garrod. 1969. Control of flower growth and development by gibberellic acid. Nature **223**:1071.

Hartmann, C. 1969. Observations on pentose phosphate cycle participation in the carbohydrate catabolism of certain fruits. Qual. Plant Mat. Veg. **19**:67-77.

Harwood, J. L., and P. K. Stumpf. 1970. Fat metabolism in higher plants. XL. Synthesis of fatty acids in the initial stage of seed germination. Plant Physiol. **46**:500-508.

Hattori, H., and S. Marumo. 1972. Monomethyl-4-chloroindolyl-3-acetyl-L-aspartate and absence of indolyl-3-acetic acid in immature seeds of *Pisum sativum*. Planta **102**:85-90.

Haupt, W. 1970*b*. The dichroism of phytochrome P_{660} and P_{730} in *Mougeotia*. Z. Pflanzenphysiol. **62**:287-298.

Haupt, W. 1970*a*. Localization of phytochrome in the cell. Physiol. Veg. **8**:551-563.

Heide, O. M. 1967. The auxin level of *Begonia* leaves in relation to their regeneration ability. Physiol. Plant. **20**:886-902.

Hellmers, H., M. K. Genthe, and F. Ronco. 1970. Temperature affects growth and development of Engelmann spruce. Forest Sci. **16**:447-452.

Hendricks, S. B., and H. A. Borthwick. 1967. The function of phytochrome in regulation of plant growth. Proc. Nat. Acad. Sci. USA **58**:2125-2130.

Heslop-Harrison, J. 1964. The control of flower differentiation and sex expression. Colloq. Int. Cent. Nat. Rech. Sci. **123**:649-664.

Hewish, D. R., J. F. Wheldrake, and J. R. E. Wells. 1971. Incorporation of ^{32}P into ribosomal RNA, transfer RNA and inositol hexaphosphate in germinating pea cotyledons. Biochim. Biophys. Acta **228**:509-516.

Hirono, Y., and G. P. Rédei. 1966. Early flowering in *Arabidopsis* induced by DNA base analogs. Planta **71**:107-112.

Hodson, H. K., and K. C. Hamner. 1970. Floral inducing extract from *Xanthium*. Science **167**: 384-385.

Hopkins, W. G., and W. S. Hillman. 1966. Relationship between phytochrome state and photosensitive growth of *Avena* coleoptile segments. Plant Physiol. **41**:593-598.

Hopkinson, J. M. 1966. Studies on the expansion of the leaf surface. VI. Senescence and the usefulness of old leaves. J. Exp. Bot. **17**:762-770.

Horton, R. F., and D. J. Osborne. 1967. Senescence, abscission and cellulase activity in *Phaseolus vulgaris*. Nature (London) **214**:1086-1088.

Hulme, A. C., M. J. C. Rhodes, L. S. C. Wooltorton, and T. Galliard. 1969. Biochemical changes associated with ripening of apples. Qual. Plant. Mat. Veg. **19**:1-18.

Humphries, E. C., and A. W. Wheeler. 1963. The physiology of leaf growth. Ann. Rev. Plant Physiol. **14**:385-410.

Ingle, J., D. Beitz, and R. H. Hageman. 1965. Changes in composition during development and maturation of maize seeds. Plant Physiol. **40**:835-839.

Jacques, M. 1969. Different morphological aspects of the flowering of *Blitum capitatum* and *virgatum* in relation to the methods of floral induction. Compt. Rend. Acad. Sci. (Paris) **268**:1045-1047.

Jaffe, M. J. 1970. Evidence for the regulation of phytochrome-mediated processes in bean roots by the neurohumor, acetylcholine. Plant Physiol. **46**:768-777.

Jaffe, M. J., and A. W. Galston. 1967. Phytochrome control of rapid nyctinastic movements and membrane permeability in *Albizzia julibrissin*. Planta **77**:135-141.

Jones, R. L., and J. E. Armstrong. 1971. Evidence for osmotic regulation of hydrolytic enzyme production in germinating barley seeds. Plant Physiol. **48**:137-142.

Jones, T. W. A., and J. L. Stoddart. 1970. Gibberellin-induced changes in protein synthesis and enzyme activity in shoot apices of *Trifolium pratense*. J. Exp. Bot. **21**:452-461.

Julien, R., F. Grellet, and Y. Guitton. 1970. RNA synthesis during radish germination. Physiol. Plant. **23**:323-334.

Kandeler, R. 1972. Die Wirkung von Acetylcholin auf die photoperiodische Steuerung der Blütenbildung bei Lemnaceen. Z. Pflanzenphysiol. **67**: 86-92.

Kasai, T., and Y. Obata. 1967. Changes in amino acid composition during germination of soybean. III. Changes in γ-glutamyltranspeptidase activity. Agric. Biol. Chem. **31**:127-129.

Kaufmann, M. R. 1969. Effects of water potential on germination of lettuce, sunflower, and citrus seeds. Canad. J. Bot. **47**:1761-1764.

Kaul, B. L. 1969. Aging in relation to seed viability, nuclear damage and sensitivity mutagens. Caryologia **22**:25-34.

Kawashima, N., and T. Mitake. 1969. Studies on protein metabolism in higher plants. VI. Changes in ribulose diphosphate carboxylase activity and fraction 1 protein content in tobacco leaves with age. Agr. Biol. Chem. **33**:539-543.

Kendrick, R. E., and W. S. Hillman. 1970. Dark reversion of phytochrome in *Sinapis alba* L. Plant Physiol. **46**:596-598.

Kessler, B., S. Spiegel, and Z. Zolotov. 1967. Control of leaf senescence by growth retardants. Nature (London) **213**:311-312.

Khan, A. A. 1967. Antagonism between cytokinins and germination inhibitors (*Lactuca sativa* cv. 'Grand Rapids'). Nature **216**:166-167.

Khan, A. A. 1971. Cytokinins: permissive role in seed germination. Science **171**:853-859.

Klyachko, N. L., L. A. Yakovleva, and O. N. Kulaeva. 1971. Effect of age on protein synthesis in gourd cotyledons. Fiziol. Rast. **18**:1225-1231.

Knypl, J. S. 1970. Arrest of yellowing in senescing leaf discs of maize by growth retardants, coumarin and inhibitors of RNA and protein synthesis. Biol. Plant. **12**:199-207.

Kolloffel, C. 1970. Oxidative and phosphorylative activity of mitochondria from pea cotyledons during maturation of the seed. Planta **91**:321-328.

Kramer, P. J. 1969. Plant and soil water relationships: a modern synthesis. McGraw-Hill Book Co., New York.

Kroes, H. H. 1970. The structure of the pigment phytochrome. Physiol. Veg. **8**:533-549.

Ku, L. L., and R. J. Romani. 1970. The ribosomes of pear fruit: their synthesis during the climacteric and the age-related compensatory response to ionizing radiation. Plant Physiol. **45**: 401-407.

Kulaeva, O. N., and O. I. Devyatko. 1969. Delay of yellowing of barley leaves due to phytohormone treatment. Fiziol. Rast. **16**:288-292.

Kurasawa, H., T. Hayakawa, and S. Watanabe. 1969. Change of inositol phosphate in rice seed during germination in dark. Nippon Nogei. Kagaku Kaishi. **43**:55-59.

Lado, P., F. R. Caldogno, and A. Pennacchioni. 1968. On the development of RNA activity in oat leaves and in fragments thereof. Ist. Lombardo Accad. Sci. Lett. Rend. Sci. Biol. Med. B. **102**:277-284.

Laiche, A. J., Jr., and C. O. Box. 1970. Response of Easter lily to bulb treatment of precooling, packing media, moisture, and gibberellin. Hortscience **5**:396-397.

Laidman, D. L., and R. J. A. Tavener. 1971. Triglyceride mobilization in the germinating wheat grain. Biochem. J. **124**:4p-5p.

Lang, A., and G. Melchers. 1943. Die photoperiodische Reaktion von *Hyoscyamus niger*. Planta **33**:653-702.

Leopold, A. C. 1967. The mechanism of foliar abscission. Sympos. Soc. Exp. Biol. **21**:507-516.

Lewis, L. N., and J. E. Varner. 1970. Synthesis of cellulase during abscission of *Phaseolus vulgaris* leaf explants. Plant Physiol. **46**:194-199.

Lezica, R. F. P. 1969. Effect of seven different gibberellins on stem elongation and flower formation in *Rudbeckia bicolor* Nutt. grown under noninductive conditions. Phyton. Rev. Int. Bot. Exp. **26**:185-190.

Lezica, R. F. P. 1970. Formation of gibberellin-like substances in potato plants during tuberization in relation to day length and temperature. Potato Res. **13**:323-331.

Lincoln, R. G., A. Cunningham, B. G. Carpenter, J. Alexander, and D. L. Mayfield. 1966. Florigenic acid from fungal culture. Plant Physiol. **41**:1079-1080.

Linskens, H. F. 1967. Isolation of ribosomes from pollen. Planta **73**:194-200.

Longo, C. P., and G. P. Longo. 1970. The development of glyoxysomes in peanut cotyledons and maize scutella. Plant Physiol. **45**:249-254.

Looney, N. E. 1969. Control of apple ripening by succinic acid 2,2-dimethyl hydrazide, 2-chloroethyltrimethylammonium chloride and ethylene. Plant. Physiol. **44**:1127-1131.

Lozhnikova, V. N. 1966. Dynamics of natural gibberellins under various photoperiodic cycle conditions. Dokl. Akad. Nauk. S. S. S. R. **168**: 223-226.

Luckwill, L. C., P. Weaver, and J. MacMillan. 1969. Gibberellins and other growth hormones in apple seeds. J. Hort. Sci. **44**:413-424.

Maasch, H. J., and U. Ruge. 1970. Phenolic substances in air-dried, swelled and germinating organs of *Helianthus* achenes and their germinative-physiological effect. Z. Pflanzenphysiol. **63**: 337-343.

MacGregor, A. W., D. E. La Berge, and W. O. S. Meredith. 1971. Changes in barley kernels during growth and maturation. Cereal Chem. **48**:255-269.

MacLeod, A. M. 1969. The utilization of cereal seed reserves. Sci. Progr. **57**:99-112.

Makower, R. V. 1969. Changes in phytic acid and acid-soluble phosphorus in maturing pinto beans. J. Sci. Food. Agric. **20**:82-84.

Mandal, N. C., and B. B. Biswas. 1970*a*. Metabolism of inositol phosphates. I. Phytase synthesis during germination in cotyledons of mung beans, *Phaseolus aureus*. Plant Physiol. **45**:4-7.

Mandal, N. C., and B. B. Biswas. 1970*b*. Metabolism of inositol phosphates. II. Biosynthesis of inositol polyphosphates in germinating seeds of *Phaseolus aureus*. Indian J. Biochem. **7**:63-67.

Martin, C., and K. V. Thimann. 1972. The role of protein synthesis in the senescence of leaves. I. The formation of protease. Plant Physiol. **49**: 64-71.

Martin, G. C., M. Iona, R. Mason, and H. I. Forde. 1969. Changes in the endogenous growth substances in the embryos of *Juglans regia* during stratification. J. Amer. Soc. Hort. Sci. **94**:13-17.

Matile, P., and K. Winkenbach. 1971. Function of lysosomes and lysosomal enzymes in the senescing corolla of the morning glory. J. Exp. Bot. **22**:759-771.

Mayak, S., and A. H. Halevy. 1972. Interrelationships of ethylene and abscisic acid in the control of rose petal senescence. Plant Physiol. **50**:341-346.

Mayer, A. M., and A. Poljakoff-Mayber. 1963. The germination of seeds. The Macmillan Co., New York.

McHale, J. S., and L. D. Dove. 1968. Ribonuclease activity in tomato leaves as related to development and senescence. New Phytol. **67**:505-515.

McLean, R. J. 1968. Ultrastructure of *Spongiochloris typica* during senescence. J. Phycol. **4**: 277-282.

Melera, P. W. 1971. Nucleic acid metabolism in germinating onion. I. Changes in root tip nucleic acid during germination. Plant Physiol. **48**:73-81.

Milletti, G., and L. De Capite. 1968. Photosynthesis and respiration in *Hydrangea macrophylla* Ser. treated with CCC and B-995. Ann. Fac. Agr. Univ. Perugia **23**:183-193.

Millikan, D. F., and B. N. Ghosh. 1971. Changes in nucleic acids associated with maturation and senescence in *Hedera helix*. Physiol. Plant. **24**: 10-13.

Mittelheuser, C., and R. F. M. van Steveninck. 1971. The ultrastructure of wheat leaves. I. Changes due to natural senescence and the effects of kinetin and ABA on detached leaves incubated in the dark. Protoplasma **73**:239-252.

Mohr, H. 1962. Primary effects of light on growth. Ann. Rev. Plant Physiol. **13**:465-488.

Moore, A. E., and B. A. Stone. 1972. Effect of senescence and hormone treatment on the activity of a β-1,3-glucan hydrolase in *Nicotiana glutinosa* leaves. Planta **104**:93-109.

Moore, K. G., A. Cobb, and P. H. Lovell. 1972. Effects of sucrose on rooting and senescence in detached *Raphanus sativus*. L. cotyledons. J. Exp. Bot. **23**:65-74.

Mori, K., M. Matsui, I. Yoshimura, and K. Saeki. 1970. Synthesis of compounds with juvenile hormone activity. VI. A mixture of (+)– dehydrojuvabione and its stereoisomer. Agr. Biol. Chem. **34**:1204-1209.

Morohashi, Y., and M. Shimokoriyama. 1972. Physiological studies on germination of *Phaseolus mungo* beans. J. Exp. Bot. **23**:45-53.

Mukherji, S., B. Dey, A. K. Paul, and S. M. Sircar. 1971. Changes in phosphorus fractions and phytase activity of rice seeds during germination. Physiol. Plant. **25**:94-97.

Mukherji, S., B. Dey, and S. M. Sircar. 1968. Changes in nicotinic acid content and its nucleotide derivatives of rice and wheat seeds during germination. Physiol. Plant. **21**:360-368.

Narayanan, A., and A. Pradet. 1968. Evolution of adenosine-5-mono-, di- and triphosphates during the germination of lettuce seeds. Indian J. Plant. Physiol. **11**:201-206.

Nichols, R. 1968. The response of carnations *(Dianthus caryophyllus)* to ethylene. J. Hort. Sci. **43**:335-349.

Nomura, T., Y. Kono, and T. Akazawa. 1969. Enzymatic mechanism of starch breakdown in germinating rice seeds. II. Scutellum as the site of sucrose synthesis. Plant Physiol. **44**:765-769.

Okoloko, G. E., L. N. Lewis, and B. R. Reid. 1970. Changes in nucleic acids in phytochrome-dependent elongation of the Alaska pea epicotyl. Plant Physiol. **46**:660-665.

Opik, H. 1968. Development of cotyledon cell structure in ripening *Phaseolus vulgaris* seeds. J. Exp. Bot. **19**:64-76.

Osborne, D. J. 1963. Hormonal control of plant death. Discovery (London) **24**:31-35.

Owens, J. N., and R. P. Pharis. 1971. Initiation and developement of western red cedar cones in response to gibberellin induction and under natural conditions. Can. J. Bot. **49**:1165-1175.

Palmiano, E. P., and B. O. Juliano. 1972. Biochemical changes in the rice grain during germination. Plant Physiol. **49**:751-756.

Paul, A. K., S. Mukherji, and S. M. Sircar. 1971. Metabolic changes in developing rice seeds. Physiol. Plant. **24**:342-346.

Payne, E. S., A. Brownrigg, A. Yarwood, and D. Boulter. 1971. Changing protein synthetic machinery during development of seeds of *Vicia faba*. Phytochem. **10**:2299-2303.

Pearson, J. A., L. Beevers, and P. F. Wareing. 1969. Effects of abscisic acid on RNA metabolism. Proc. Soc. Exp. Biol. **23**: London Conference.

Pharis, R. P., and W. Morf. 1969. Precocious flowering of coastal and giant redwood with gibberellin A_3, $A_{4/7}$, and A_{13}. BioScience **19**:719-720.

Phillips, D. R., and R. A. Fletcher. 1969. Ribonuclease in leaves of *Phaseolus vulgaris* during maturation and senescence. Physiol. Plant. **22**: 764-768.

Phillips, D. R., R. F. Horton, and R. A. Fletcher. 1969. Ribonuclease and chlorophyllase activities in senescing leaves. Physiol. Plant. **22**:1050-1054.

Pillay, D. T. N., and R. Mehdi. 1968. Chemical regulation of soybean leaf senescence. Phyton. Rev. Int. Bot. Exp. **25**:75-79.

Podol'nyi, V. Z., and V. M. Loboda. 1971. The content of auxins and auxin oxidase activity in leaves and cotyledons of cocklebur in the juvenile phase. Fiziol. Rast. **18**:86-91.

Porsild, A. E., C. R. Harington, and G. A. Mulligan. 1967. *Lupinus arcticus* Wats. grown from seeds of Pleistocene age. Science **158**:113-114.

Powell, R. D., and P. W. Morgan. 1970. Factors involved in the opening of the hypocotyl hook of cotton and beans. Plant Physiol. **45**:548-552.

Pradet, A., A. Narayanan, and J. Vermeersch. 1968. Study of adenosine-5′-mono-, di- and triphosphate in plant tissues. III. Energy metabolism during the 1st stages of lettuce seed germination. Bull. Soc. Fr. Physiol. Veg. **14**:107-114.

Price, C. E., and A. W. Murray. 1969. Purine metabolism in germinating wheat embryos. Biochem. J. **115**:129-133.

Rabechault, H., G. Guenin, and J. Ahee. 1970. Water absorption by African oil palm nuts (*Elaeis guineensis* Jacq. var. *dura* Becc.): IV. Water migration during storage. Cah. Off. Rech. Sci. Tech. Outre-Mer Ser. Biol. **13**:25-39.

Ram, H. Y. M., and M. Batra. 1970. Stimulation of flower formation by cytokinins in the excised immature inflorescences of *Cyperus rotundus*. Phytomorphology **20**:22-29.

Rauser, W. E. 1971. Inorganic pyrophosphatase in leaves during plant development and senescence. Canad. J. Bot. **49**:311-316.

Reid, J. S. G. 1971. Reserve carbohydrate metabolism in germinating seeds of *Trigonella foenum-*

graecum L. (Leguminosae). Planta **100**:131-142.

Rijven, A. H. G. C., and L. T. Evans. 1967*a*. Inflorescence initiation in *Lolium temulentum* L. IX. Some chemical changes in the shoot apex at induction. Aust. J. Biol. Sci. **20**:1-12.

Rijven, A. H. G. C., and L. T. Evans. 1967*b*. Inflorescence initiation in *Lolium temulentum* L. X. Changes in ^{32}P incorporation into nucleic acids of the shoot apex at induction. Aust. J. Biol. Sci. **20**:13-24.

Robbins, W. J. 1960. Further observations on juvenile and adult *Hedera*. Amer. J. Bot. **47**: 485-491.

Roberts, E. H., F. H. Abdalla,and R. J. Owen. 1967. Nuclear damage and the ageing of seeds. Sympos. Soc. Exp. Biol. **21**:65-99.

Romani, R. J., and L. K. Fisher. 1966. Decreased synthesis of mitochondrial protein in the senescent cells of pear fruit. Life Sci. **5**:1187-1190.

Rowan, K. S., W. B. McGlasson, and H. K. Pratt. 1969. Changes in adenosine pyrophosphate in cantaloupe fruit ripening normally and after treatment with ethylene. J. Exp. Bot. **20**:145-155.

Rubinstein, B. 1971. Auxin and red light in the control of hypocotyl hook opening in beans. Plant Physiol. **48**:187-192.

Ruckenbauer, P. 1971. Winter wheat capable of germination from 1877: observations and experiments. Bodenkultur **22**:372-386.

Sacher, J. A. 1967. Control of synthesis of RNA and protein in subcellular fractions of *Rhoeo discolor* leaf sections by auxin and kinetin during senescence. Exp. Gerontol. **2**:261-278.

Sachs, R. M., D. M. Kotranek, and S. Y. Shyr. 1967. Gibberellin-induced inhibition of floral initiation in Fuchsia. Amer. J. Bot. **54**:921-929.

Saint-Firmin, A., and P. Chouard. 1971. *L'hyoscyamus muticus* L. et quelques facteurs de l'environment, température. Compt. Rend. Acad. Sci. (Paris) **272**:1507-1510.

Salisbury, F. B. 1965. The initiation of flowering. Endeavour **24**:74-80.

Salomon, D., and J. P. Mascarenhas. 1971. The effect of cyclic 3′,5′-adenosine monophosphate on abscission of *Coleus* petioles. Z. Pflanzenphysiol. **65**:385-388.

Sandstedt, R. 1971. Cytokinin activity during development of cotton fruit. Physiol. Plant. **24**: 408-410.

Sankhla, N., and D. Sankhla. 1968*a*. Interaction between growth regulators and (+)–abscisin II in seed germination. Z. Pflanzenphysiol. **58**:402-409.

Sankhla, N., and D. Sankhla. 1968*b*. Abscisin II-kinetin interaction in leaf senescence. Experientia **24**:294-295.

Saski, S., and G. N. Brown. 1971. Polysome formation in *Pinus resinosa* at initiation of seed germination. Plant Cell Physiol. **12**:749-758.

Scheibe, J. 1972. Photoreversible pigment: occurrence in a blue-green alga. Science **176**: 1037-1039.

Scherf, H., and M. H. Zenk. 1967. Der Einfluss des Lichtes auf die Flavonoidsynthese und die Enzyminduktion bei *Fagopyrum esculentum* Moench. Z. Pflanzenphysiol. **57**:401-418.

Schnarrenberger, C., A. Oeser, and N. E. Tolbert. 1971. Development of microbodies in sunflower cotyledons and castor bean endosperm during germination. Plant Physiol. **48**:566-574.

Schopfer, P. 1971. Phenylalanine ammonia-lyase (PAL EC 4.3.1.5) of mustard seedling (*Sinapis alba* L.), an electrophoretically homogeneous enzyme. Planta **99**:339-346.

Schwabe, W. W. 1971. Physiology of vegetative reproduction and flowering, p. 233-411. *In* F. C. Steward (Ed.). Plant Physiology VI A. Physiology of development: plants and their reproduction. Academic Press, Inc., New York.

Scott, D. H., and A. D. Draper. 1970. A further note on longevity of strawberry seed in cold storage. Hortscience **5**:439.

Sheldrake, A. R., and D. H. Northcote. 1968. The production of auxin by autolysing tissues. Planta **80**:227-236.

Siegelman, H. W. 1969. Phytochrome, p. 489-506. *In* M. B. Wilkins (Ed.). The physiology of plant growth and development. McGraw-Hill Book Co., New York.

Siegelman, H. W., E. C. Turner, and S. B. Hendrix. 1966. The chromophore of phytochrome. Plant Physiol. **41**:1289-1292.

Simola, L. K., and T. Sopanen. 1970. Changes in the activity of certain enzymes of *Acer pseudoplatanus* L. cells at four stages of growth in suspension culture. Physiol. Plant. **23**:1212-1222.

Simmonds, J. A., and G. M. Simpson. 1971. Increased participation of pentose phosphate pathway in response to after-ripening and gibberellic acid treatment in caryopses of *Avena fatua*. Canad. J. Bot. **49**:1833-1840.

Simon, E. W. 1967. Types of leaf senescence. Sympos. Soc. Exp. Biol. **21**:215-230.

Sircar, S. M. 1967. Biochemical changes of rice seed germination and its control mechanism. Trans. Bose. Res. Inst. **30**:189-198.

Skene, K. G. M., and A. J. Antcliff. 1972. A comparative study of cytokinin levels in bleeding sap of *Vitis vinifera* (L.) and the two grapevine rootstocks, 'Salt Creek' and '1613.' J. Exp. Bot. **23**:283-293.

Smith, H. 1970. Phytochrome and photomorphogenesis in plants. Nature **227**:665-671.

Smith, O. E., and C. E. Palmer. 1970. Cytokinin-induced tuber formation on stolons of *Solanum tuberosum*. Physiol. Plant. **23**:599-606.

Snyder, M. J., N. C. Welch, and V. E. Rubatzky. 1971. Influence of gibberellin on time of bud development in globe artichoke. Hortscience **6**:484-485.

Somme, R. 1971. The correlation between the mono-, oligo- and polysaccharides and the glycosidases present in clover seeds. Acta Chem. Scand. **25**:759-761.

Spaulding, D. W., and G. L. Steffens. 1969. Elimination of light requirements for tobacco seed germination with gibberellic acid, indole-3-acetic acid and N^6 benzyladenine. Tob. Sci. **13**:156-159.

Specht-Jurgensen, I. 1967. Studies on nitrogen compounds and chlorophyll during the senescence of the leaves of *Ginkgo biloba*. I. Leaves ageing on the tree. [Transl. title.] Flora **157**: 426-453.

Spencer, P. W., and J. S. Titus. 1972. Biochemical and enzymatic changes in apple leaf tissue during autumnal senescence. Plant Physiol. **49**:746-750.

Srivastava, B. I. S. 1968. Studies on the chromatin of barley leaves during senescence. Biochem. J. **110**:683-686.

Srivastava, B. I. S. 1971. Relationships of histones to RNA synthesis in plant tissues. Physiol. Plant. **24**:27-33.

Srivastava, B. I. S., and C. Arglebe. 1967. Studies on ribosomes from barley leaves; changes during senescence. Plant Physiol. **42**:1497-1503.

Srivastava, B. I. S., and C. Arglebe. 1968. Effect of kinetin on ribosomes of excised barley leaves. Physiol. Plant. **21**:851-857.

Stearns, M. E., and E. B. Wagenaar. 1971. Ultrastructural changes in chloroplasts of autumn leaves. Canad. J. Genet. Cytol. **13**:550-560.

Stein, D. B. 1971. Gibberellin-induced fertility in the fern *Ceratopteris thalictroides* (L). Brongn. Plant Physiol. **48**:416-418.

Street, H. E. 1967. The ageing of root meristems. Sympos. Soc. Exp. Biol. **21**:517-542.

Suge, H., and N. Yamada. 1963. Chemical control of plant growth and development. 4. Promotion of flowering induced by uracil, uridylic acid, and several growth regulators in winter wheat. Proc. Crop Sci. Soc. Japan **32**:77-80.

Tanada, T. 1968. A rapid photoreversible response of barley root tips in the presence of 3-indoleacetic acid. Proc. Nat. Acad. Sci. USA **59**: 376-380.

Tanaka, K., K. Watanabe, K. Asada, and Z. Kasai. 1971. Occurrence of myo-inositol monophosphate and its role in ripening rice grains. Agr. Biol. Chem. **35**:314-320.

Taillandier, J. 1971. The effect of 5-fluorodeoxyuridine and actinomycin D on the flowering of *Anagallis arvensis* L. var. *phoenicea*. Compt. Rend. Acad. Sci. (Paris) **272**:557-560.

Tavares, J., and H. Kende. 1970. The effect of 6-benzylaminopurine on protein metabolism in senescing corn leaves. Phytochemistry **9**:1763-1770.

Thorpe, T. A., and H. Z. Hield. 1970. Gibberellin and rejuvenation in *Citrus*. Phyton. Rev. Int. Bot. Exp. **27**:63-68.

Thrower, S. L. 1967. The pattern of translocation during leaf ageing. Sympos. Soc. Exp. Biol. **21**: 483-506.

Tsai, C. Y., F. Salamini, and O. E. Nelson. 1970. Enzymes of carbohydrate metabolism in the developing endosperm of maize. Plant Physiol. **46**:299-306.

Turner, J. F. 1969. Starch synthesis and changes in uridine diphosphate glucose pyrophosphorylase and adenosine diphosphate glucose pyrophosphorylase in the developing wheat grain. Aust. J. Biol. Sci. **22**:1321-1327.

Ueda, M., and R. S. Bandurski. 1969. A quantitative estimation of alkali-labile indole-3-acetic acid compounds in dormant and germinating maize kernels. Plant Physiol. **44**:1175-1181.

Ullmann, J., F. Seidlova, and J. Krekule. 1971. 5-Fluorodeoxyuridine inhibition of photoperiodically induced flowering in *Chenopodium rubrum* L. Biol. Plant. **13**:305-312.

Valdovinos, J. G., and L. C. Ernest. 1967. Effect of protein synthesis inhibitors, auxin, and gibberellic acid on abscission. Physiol. Plant. **20**: 1027-1038.

Valdovinos, J. G., T. E. Jensen, and L. M. Sicko. 1971. Ethylene-induced rough endoplasmic reticula in abscission cells. Plant Physiol. **47**: 162-163.

Valdovinos, J. G., T. E. Jensen, and L. M. Sicko. 1972. Fine structure of abscission zones. IV. Effect of ethylene on the ultrastructure of abscission cells of tobacco flower pedicels. Planta **102**:324-333.

van Onckelen, H. A., and R. Verbeek. 1969. Formation of α-amylase during germination of barley. Planta **88**:255-260.

van Staden, J., and P. F. Wareing. 1972. The effect of photoperiod on levels of endogenous cytokinins in *Xanthium strumarium*. Physiol. Plant. **27**:331-337.

Vendrell, M. 1969. Reversion of senescence: effects of 2,4-dichlorphenoxyacetic acid and indoleacetic acid on respiration ethylene production, and ripening of banana fruit slices. Aust. J. Biol. Sci. **22**:601-610.

Venkataraman, R., and P. De Leo. 1972. Changes in leucyl tRNA species during ageing of detached soybean cotyledons. Phytochem. **11**:923-927.

Verbeek, R., H. A. van Onckelen, and T. Gaspar. 1969. Effects of gibberellic acid and kinetin on α-amylase production during the germination of barley. Physiol. Plant. **22**:1192-1199.

Vold, B. S., and P. S. Sypherd. 1968. Changes in soluble RNA and ribonuclease activity during germination of wheat. Plant Physiol. **43**:1221-1226.

Wade, N. L., and C. J. Brady. 1971. Effects of kinetin on respiration, ethylene production, and ripening of banana fruit slices. Aust. J. Biol. Sci. **24**:165-167.

Wade, N. L., P. B. H. O'Connell, and C. J. Brady. 1972. Content of RNA and protein of the ripening banana. Phytochemistry **11**:975-979.

Wahab, A. H., and J. S. Burris. 1971. Carbohydrate metabolism and respiratory activity of soybean seedlings. Iowa State J. Sci. **45**:595-605.

Walbot, V. 1971. RNA metabolism during embryo development and germination of *Phaseolus vulgaris*. Dev. Biol. **26**:369-379.

Walton, D. C., and G. S. Soofi. 1969. Germination of *Phaseolus vulgaris*. III. The role of nucleic acid and protein synthesis in the initiation of axis elongation. Plant Cell Physiol. **10**:307-315.

Wang, C. Y., and E. Hansen. 1970. Differential response to ethylene in respiration and ripening of immature 'Anjou' pears. J. Amer. Soc. Hort. Sci. **95**:314-316.

Wang, C. Y., S. Y. Wang, and W. M. Mellenthin. 1972. Identification of abscisic acid in Bartlett pears and its relationship to premature ripening. J. Agric. Food. Chem. **20**:451-453.

Wang, D. 1968. Metabolism of amino acids and amides in germinating seeds. Contrib. Boyce Thompson Inst. **24**:109-115.

Wareing, P. F., and A. K. Seth. 1967. Ageing and senescence in the whole plant. Sympos. Soc. Exp. Biol. **21**:543-558.

Weeks, D. P., and A. Marcus. 1971. Preformed messenger of quiescent wheat embryos. Biochim. Biophys. Acta **232**:671-684.

Wejnar, R. 1969. Plant physiology: research on the acid metabolism of ripening berries of *Vitis vinifera* L. Flora Abt. A. Physiol. Biochem. (Jena) **160**:211-216.

Wellensiek, S. J. 1966. The mechanism of flower formation in *Silene armeria* L. Naturwissenschaften **53**:411.

Wellensiek, S. J. 1967. The relations between the flower inducing factors in *Silene armeria* L. Z. Pflanzenphysiol. **56**:33-39.

Wellensiek, S. J. 1970. The floral hormones in *Silene armeria* L. and *Xanthium strumarium* L. Z. Pflanzenphysiol. **63**:25-30.

Werner, D., and D. Gogolin. 1970. Characterization of root initiation and senescence in callus and organ cultures of *Daucus carota* by means of glutamate dehydrogenase activity. Planta **91**:155-164.

Wesson, G., and P. F. Wareing. 1967. Light requirement of buried seed. Nature (London) **213**:600-601.

Wheeler, A. W. 1968. Changes in auxins in expanding and senescent primary leaves of dwarf French bean *(Phaseolus vulgaris)*. J. Exp. Bot. **19**:102-107.

Whyte, P., and L. C. Luckwill. 1966. A sensitive bioassay for gibberellins based on retardation of leaf senescence in *Rumex obtusifolius* (L.) Nature (London) **210**:1360.

Williams, S. G. 1970. The role of phytic acid in the wheat grain. Plant Physiol. **45**:376-381.

Wilson, S. B., and W. D. Bonner, Jr. 1971. Studies of electron transport in dry and imbibed peanut embryos. Plant Physiol. **48**:340-344.

Wiltbank, W. J., and A. H. Krezdorn. 1969. Determination of gibberellins in ovaries and young fruits of navel oranges and their correlation with fruit growth. J. Amer. Soc. Hort. Sci. **94**: 195-201.

Woolhouse, H. W. 1967. The nature of senescence in plants. Sympos. Soc. Exp. Biol. **21**:179-213.

Wren, M. J., and J. W. Hannay. 1963. Ageing in roots of groundsel (*Senecio vulgaris* L.). New Phytol. **62**:249-256.

Wyen, N. V., S. Erdei, and G. L. Farkas. 1971. Isolation from *Avena* leaf tissues of a nuclease with the same type of specificity towards RNA and DNA accumulation of the enzyme during leaf senescence. Biochim. Biophys. Acta **223**: 472-483.

Yoneda, Y., and T. Stonier. 1966. Elongation of stem internodes in the Japanese morning glory *Pharbitis nil* in relation to auxin destruction. Physiol. Plant. **19**:977-981.

Yoo, B. Y. 1970. Ultrastructural changes in cells of pea embryo radicles during germination. J. Cell Biol. **45**:158-171.

Yoshida, K., K. Umemura, K. Yoshinaga, and Y. Oota. 1967. Specific RNA from photoperiodically induced cotyledons of *Pharbitis nil.* Plant Cell Physiol. **8**:97-108.

Yunghans, H., and M. J. Jaffe. 1972. Rapid respiratory changes due to red light or acetylcholine during the early events of phytochrome-mediated photomorphogenesis. Plant Physiol. **49**:1-7.

Zeevaart, J. A. D. 1962. Physiology of flowering. Science **137**:723-731.

Zeevaart, J. A. D. 1971. Effects of photoperiod on growth rate and endogenous gibberellins in the long-day rosette plant spinach. Plant Physiol. **47**:821-827.

Zimmerman, R. H. 1971. Flowering in crabapple seedlings: methods of shortening the juvenile phase. J. Amer. Soc. Hort. Sci. **96**:404-411.

GENERAL REFERENCES

Barton, L. V. 1961. Seed preservation and longevity. Leonard Hill Books, Ltd., London.

Carns, H. R. 1966. Abscission and its control. Ann. Rev. Plant Physiol. **17**:295-314.

Hillman, W. S. 1962. The physiology of flowering. Holt, Rinehart & Winston, Inc., New York.

Hillman, W. S. 1967. The physiology of phytochrome. Ann. Rev. Plant Physiol. **18**:301-324.

Hillman, W. S. 1969. Photoperiodism and vernalization, p. 559-601. *In* M. B. Wilkins (Ed.). The physiology of plant growth and development. McGraw-Hill Book Co., New York.

Laetsch, W. M., and R. E. Cleland. 1967. Plant growth and development. Little, Brown & Co., Boston.

Lang, A. 1952. Physiology of flowering. Ann. Rev. Plant Physiol. **3**:265-306.

Molisch, H. 1938. The longevity of plants. Science Press, Lancaster, Pa.

Murneek, A. E., and R. O. Whyte. 1948. Vernalization and photoperiodism. The Ronald Press Co., New York.

Salisbury, F. B. 1963. The flowering process. Pergamon Press, Inc., New York.

Sax, K. 1962. Aspects of aging in plants. Ann. Rev. Plant Physiol. **13**:489-506.

Torrey, J. G. 1967. Development in flowering plants. The Macmillan Co., New York.

Whyte, R. O. 1946. Crop production and environment. Faber & Faber, Ltd., London.

21 Plant rhythms

RELATION TO VIBRATORY, OR PERIODIC, MOTION

The rates and directions of many plant processes show regular patterns of change. The most obvious are the nyctinastic (sleep) movements of many plants—the closing of leaves or flowers at night and their opening at daybreak. Similar, although less obvious, patterns are to be found in many of both the physical and chemical processes in plants. In succulents, for instance, organic acids regularly accumulate at night and just as regularly disappear during daylight (Chapter 3). The changes may occur daily, weekly, monthly, annually, etc., and they are therefore called rhythms. Since they are repeated at regular intervals of time known as periods, they are also sometimes called periodisms (or periodicities). The latter terminology has the advantage of indicating the similarity between plant rhythms and the periodic motions of physical systems. Because of the similiarity, the terminology for plant rhythms has been borrowed from the accepted terminology for physical systems. Therefore it is first necessary to be familiar with the fundamentals of periodic motion before discussing plant rhythms.

A periodic or vibratory motion may occur in any body that is able to oscillate. When, for instance, a weight is suspended from a spring and a downward force is exerted on the weight, the weight will oscillate in a vertical direction. This oscillation is called vibratory, or periodic, motion, and the quantitative relations are shown in Fig. 21-1.

There are two distinct kinds of oscillation.

1. The capacity to oscillate may be passive—the rhythm being driven by an external forcing agent (the synchronizer). Thus if an external force repeatedly pushes the weight up and down, a *forced vibration* occurs, resulting in a *forced period.* The periodic motion is then said to be entrained by the external force.
2. On the other hand, the capacity to oscillate may be active, resulting in spontaneously oscillating systems with natural periods. If such a system is allowed to oscillate freely, a *natural vibration* occurs, and the time for one complete cycle will be the *natural period.*

When the impressed force is in unison with the natural vibration, this is called *resonance. Sympathetic vibrations* are produced when there is resonance. The *phase* is the position of the vibrating body at any

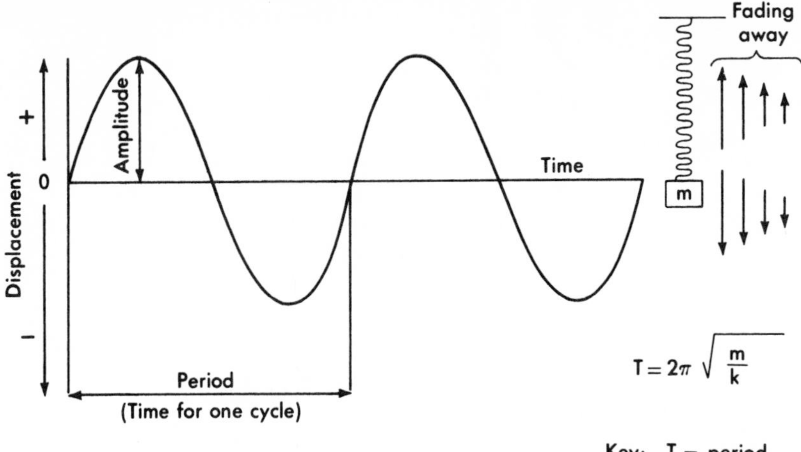

Fig. 21-1. Periodic or vibratory motion. When a weight is displaced on the end of a spring (top right), a periodic motion is initiated, but it fades away due to frictional resistance.

instant plus the direction in which it is moving. It is measured as an angle with reference to a circle. Two objects are in phase when they are in the same position relative to the circle and moving in the same direction.

For a natural vibration to occur, there must be:

1. An impressed force, or an initial displacement
2. A return force
3. Inertia (otherwise the body will come to a stop when the force is zero)
4. Not too large a frictional resistance to movement

Since friction always exists, the amplitude of the motion in the absence of the impressed force is decreased by the friction in each successive period until the motion comes to a stop. This is called *fading away* (Fig. 21-1). All the preceding terms and concepts are used in describing plant rhythms.

Some plant rhythms (such as the sleep movements of leaves, the opening and closing of flowers) are literally periodic motions. Other rhythmic phenomena, such as the luminescence of dinoflagellates and metabolic rhythms are not motions; nevertheless, the terminology of periodic motion may be applied to them. The plant's environment is also characterized by rhythms, and the plant's rhythms are normally found to follow the environmental rhythms. They are therefore classified according to the environmental rhythm that they parallel.

ANNUAL RHYTHMS

There are many annual plant rhythms, some of which are obviously controlled by the annual environmental temperature rhythm. Others, however, are not so obviously related.

Growth rhythm

In temperate climates, at least, growth must come to a stop during winter when the plant freezes. However, growth actually ceases well before the environmental conditions become unfavorable. This is caused by dormancy.

Dormancy or rest period rhythm

Many plants enter a rest period every fall and emerge from it in late winter or early spring. This rhythm is not so obviously related to the environment since the plant enters its rest period (and therefore stops growth) while the environment is still capable of supporting growth. Similarly, the rest period is broken before the environmental temperatures are capable of supporting growth. In opposition to this rhythm in bud dormancy, with maximum dormancy in winter, trunk fragments show maximum proliferation of callus tissue when sampled during the October to December period (Scaramuzzi et al., 1971). Tissues taken during spring and summer showed practically no proliferation. Even the ability of the plant to open its stomata under favorable environmental conditions shows this annual rhythm (Fig. 21-2).

Hardiness rhythm (see Chapter 22)

Again, the hardiness rhythm is not a simple, forced rhythm, since a well-adapted plant shows an increase in frost resistance long before any frosts occur. This is accompanied by other annual rhythms, for example, a winter maximum in sugar content and therefore in osmotic pressure, and a minimum in starch content. In xylem tissues of some trees there may, however, be a starch maximum in winter. Other substances may also show a maximum in winter, for example, soluble proteins. There are similar annual rhythms in resistance to drought injury, heat injury, and according to recent results (Biebl and Hofer, 1966), even radiation injury.

Water content rhythm

In temperate climates some tree trunks show maximum water contents in winter and minimum water contents in summer (Gibbs, 1939).

Reproductive rhythms

Many plants reproduce only at one season in the year, for example, in spring.

Annual rhythms are only beginning to be investigated under artificial conditions (Fig. 21-2) because of the difficulties in growing plants under controlled conditions for a year and because of the time required to complete an experiment. It is therefore not

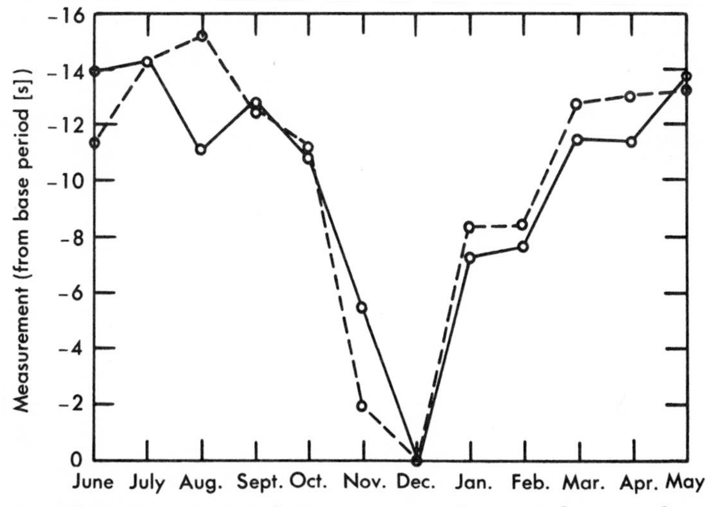

Fig. 21-2. Seasonal pattern in stomatal opening in pinto beans. Ordinate = decrease (sec) in time for the air to flow through the leaf, compared to a minimum opening in December; —, plants grown in a growth chamber and a photoperiod of 16 hr; - - -, plants grown in a greenhouse and the normal photoperiod for the time of the year. (From Seidman, G., and W. B. Riggan. 1968. Nature [London] **217**:684-685.)

known to what extent they are forced rhythms or natural rhythms.

LUNAR RHYTHMS

Lunar rhythms have not been studied extensively and apparently are not as common as annual rhythms. In brown algae (e.g., *Dictyota*) there is a lunar periodicity in the release of sexual cells. This periodicity can be repeated under artificial conditions (Muller, 1962). If the daily light-dark periods are interrupted every 28 days by giving artificial moonlight during one dark period, the production of eggs reaches a maximum on the tenth and twenty-sixth days after the stimulus. A "moonlight" intensity of 3 lux is fully effective. After this periodicity is induced, the plants will show their next maximum release of sexual cells at 16- to 17-day intervals without any further exposure to the moonlight stimulus. This is evidence of the existence of a natural rhythm.

CIRCADIAN RHYTHMS

Circadian rhythms are the commonest and have been studied the most intensively. They have also been called diurnal rhythms. The term *circadian* is preferred because, as will be seen later, the periods are only approximately 24 hr in length (circa means approximately, dies means day).

Leaf (sleep) movements were the earliest studied circadian rhythms (Fig. 21-3). That they are responses to the daily light rhythm can be easily demonstrated by darkening the plant during daylight. This induces the sleep movement that otherwise occurs only at night. On the other hand, the rhythmic movement will continue even in continuous, weak light (Fig. 21-3). When this occurs, the period is seen to be a little longer than 24 hr. The opening of flowers is also a typical circadian rhythm in many plants. Other well-known circadian rhythms are the luminescence of dinoflagellates (*Gonyaulax* sp.) and even metabolic processes such as carbon dioxide evolution.

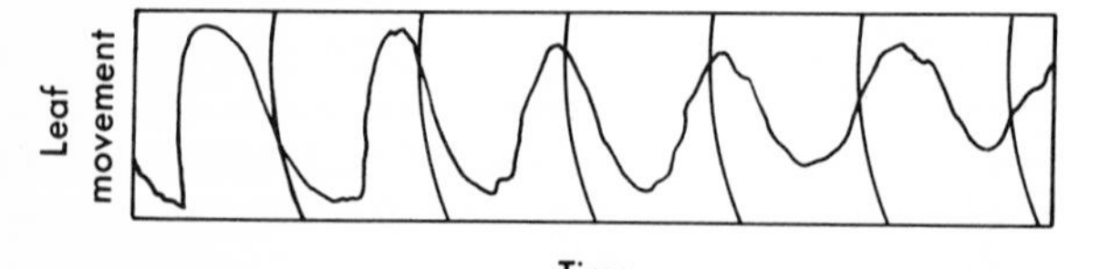

Fig. 21-3. Sleep movements of *Phaseolus* leaves in continuous, weak light—a circadian movement. Guide lines are 24 hr apart, and the length of periods is 27 hr. (From Bünning, E. 1964. The physiological clock. Springer-Verlag, Berlin.)

There are some differences between the rhythmic leaf movements of plants from different geographical regions (Mayer, 1966). In general, under constant conditions the oscillations in arctic plants persist for a shorter time than in plants of other regions. The length of the natural ("free running") period, however, is the same in arctic plants (24 hr) as in plants from Central Europe. On the other hand, the natural periods of some tropical plants are much longer than 24 hr. In several of these tropical species the period decreases greatly with increasing temperature. Consequently, if the tropical plants are tested at 27° C and the Central European and arctic plants at 17° C, the periods always approach 24 hr.

In recent years an artificially induced diurnal rhythm has become common. When organisms or tissues are grown in artificial culture, it is frequently desired to have all, or nearly all, the cells pass through the same stage (e.g., cell division) simultaneously. These are called *synchronous cultures,* and they can be produced by a rhythmic daily alternation of some environmental factor (e.g., light, temperature, etc.).

TIDAL RHYTHMS

Tidal rhythms have a period of 12½ hr and have been investigated in marine plants (Sweeney, 1963). There are many rhythms shorter than the tidal rhythms, such as (1) circular rhythms in twining shoots, tendrils, and some pulvini, with periods of 1 to 2 hr; (2) the reversal of the direction of cytoplasmic streaming, which may have a period of 60 sec. These rhythms have been called ultradian (Bailloud, see Aschoff, 1965).

EXOGENOUS AND ENDOGENOUS RHYTHMS

The classification by periods indicates that all plant rhythms (with the possible exception of the ultradian) are basically controlled by external rhythms in the plant environment. The period of plant rhythms seems to be that of an external oscillator arising because of the motions of the earth (annual and circadian rhythms) or of the moon (lunar and tidal rhythms). These external oscillators have their own external rhythms (Sollberger, 1965) and are classified by the strength of their effects on plant rhythms as strong or dominant (light, temperature) and weak synchronizers (sound, vibration, atmospheric humidity, atmospheric pressure). A single oscillator may produce more than one plant rhythm. Thus a single light perturbation produces similar phase shifts in the flowering rhythm and the leaf movement rhythm of Biloxi soybeans (Brest et al., 1971).

Plant rhythms that are purely passive or forced by an external oscillator are called *exogenous*. Spontaneous or active plant rhythms with a natural period but that are entrained by the external oscillator are called *endogenous*. To distinguish between exogenous and endogenous rhythms, it is necessary to remove the plant from the action of the external oscillator. Thus in the case of an external oscillator caused by the daily alternation of light and dark, the plant must be placed in continuous light or continuous dark. An exogenous rhythm either stops immediately or fades away after a few oscillations, with the same period as that of the external oscillator. An endogenous rhythm (1) continues to oscillate for a considerable time, but (2) with a period different from that of the external oscillator. This shows that the external oscillator entrains the plant rhythm, giving it an artificial period; and as soon as the environmental rhythm is removed, the plant reverts to its own natural period. An endogenous rhythm can therefore be defined as one capable of self-sustained oscillations after the external entraining rhythm is removed and capable of having its own natural period.

Besides the two criteria just mentioned, three more have been proposed for establishing the endogenous nature of a rhythm (Pittendrigh, see Wilkins, 1969). (3) It should be possible to shift the phase to a new one, which is then retained under uniform environmental conditions. (4) The rhythm must be initiated by a single stimulus. (5) The phase should be delayed under hypoxia.

Some plant rhythms seem to belong clearly to one or other of these two groups. An example of an exogenous rhythm is the daily alternation of carbon dioxide absorption and evolution. If the alternation of light and darkness ceases and the plant is kept in continuous darkness, photosynthetic carbon dioxide absorption also ceases. There are many other examples of such exogenous rhythms. On the other hand, the circadian sleep movements of leaves are as pronounced in continuous weak light as when exposed to normal daily alternations of light and darkness (Fig. 21-3) and have periods close to, although not exactly equal to, 24 hr. They are therefore endogenous. The rhythmic movements continue, in fact, in bean plants grown from seed in an environment kept free of all major fluctuations by controlling the levels of carbon dioxide, relative humidity, temperature, light, nutrient concentration, and water tension (Alford and Tibbits, 1970). They may continue for up to 1 to 2 wk in uniform environments, compared to animal rhythms that may continue for several months (Wilkins, 1969). However, even endogenous rhythms must be initiated by an external stimulus, just as natural vibrations must be given a single displacement by an impressed force.

Seedlings raised in complete darkness fail to show their normal rhythms (e.g., sleep movement of leaves). Yet one brief light signal may be sufficient to initiate the rhythms. Even some of the apparently exogenous rhythms may have an endogenous component. The daily rhythm of photosynthetic carbon dioxide assimilation is exogenous, since it immediately and completely comes to a stop in continuous darkness and cannot recommence in the absence of light; nevertheless, the *capacity* to photosynthesize follows an endogenous rhythm. The plant will therefore photosynthesize at a greater rate at one time of the day than at another, even under constant environmental conditions. Similarly, the opening and closing of stomata may sometimes be strictly controlled by alternations of light and dark, but the "opening ability" in response to light may be altered in a diurnal periodicity, depending on the length of the preceding dark period. Even in continuous darkness, a diurnal rhythmic opening and closing of the stomata occurs (Stålfelt, 1963). A sharp line between exogenous and endogenous rhythms may therefore not always be possible.

Although most endogenous rhythms identified so far are circadian or ultradian, recent evidence indicates that some annual rhythms may also be endogenous. Species of eight genera of trees, when grown under continuous light and a constant temperature of 25° C, were still able to enter a rest period and to spontaneously resume their growth (Lavarenne et al., 1971).

There are two possible kinds of endogenous rhythms:

1. The rhythms may be similar to sympathetic vibrations. This would involve a kind of resonance, and the plant would have its own natural rhythm, which happened to have the same period as that of the environmental rhythm or synchronizer. Removal of the environmental rhythms would immediately reduce the amplitude of the plant rhythm but would not alter its period.
2. The observed period might be entrained by the environment, and as soon as the environmental rhythm is removed, the plant would revert to its own natural period.

In both cases the amplitude of the rhythm would decrease steadily with time; that is, the rhythm would fade away. Circadian rhythms are commonly of the second kind, and as their name implies, their natural period is only approximately 24 hr in length (Fig. 21-3). The natural periods of the different kinds of plant rhythms may vary from 0.001 sec to a day, a month, a year, or more (Sollberger, 1965); but they most commonly are close to the period of some environmental factor. This common similarity between the natural periods and those of the entraining external oscillator is not likely to result from mere coincidence. The explanation is that it arose by natural selection assuming that the major role of plant rhythms is for optimum adaptation to the environmental rhythms. This is most evident in the annual growth rhythm of temperate plants. The role of rhythmic leaf movements is not so clear. That it does play a role in plant development has been indicated by experimental hindering of leaf movement with wires (Bünning and Moser, 1969). This strongly inhibited flowering in two SDP and one LDP.

Further evidence of the endogenous nature of the plant rhythms can sometimes be obtained after the rhythm has come to a stop because of removal of the environmental rhythm. A single brief exposure to the environmental factor (e.g., a flash of light) may supply the "displacement" to initiate the plant rhythm (e.g., in the case of sleep movements). The unalterability of the plant's natural period can be demonstrated by entraining it to another period (e.g., a 14 hr day). As soon as the environmental rhythm is removed, the plant returns to its own natural period, even after an entrainment for as long as a year.

According to Bünning (1948, see Chapter 20), photoperiodism involves an endogenous rhythm explained in Fig. 21-4. When in the photophile phase, the plant is highly synthetic and when in the scotophile phase, highly hydrolytic. The difference between SDP and LDP is that in the former the endogenous rhythm is changed to the photophile phase soon after the beginning of illumination. In the LDP the rhythm is not changed until 12 hr after the beginning of illumination. Thus in the LDP a long light period is unnecessary, provided that the plant is illuminated after the photophile phase is initiated—preferably in the middle of the photophile period. Some evidence seems to agree with his concept. Tomatoes, for instance, are indeterminate, yet they will not survive continuous illumination (Highkin and Hanson, 1954). The leaves that develop are small, stiff, and yellow with dark necrotic spots. The same effect is obtained when a 24 hr light period alternates with a 24 hr dark period. This is explained by assuming that the endogenous rhythm of alternating photophile and scotophile phases has a period of 24 hr. Only when the photophile phase coincides with the light period and the scotophile phase with the dark period can normal growth and development occur. Other experiments oppose this concept, since 4 hr of light during the postulated scotophile phase hastens flowering (Hussey, 1954). More recent results (Brest et al., 1970; King and Cumming, 1972) are in agreement with Bünning's concept. However, no explanation has been proposed for the assumed photophile and scotophile phases.

BIOLOGICAL CLOCKS

Since so many rhythms appear to be endogenous, the plant is said to possess a "biological clock," which controls the period of the rhythms. According to most investigators, this clock has an intrinsic mechanism (i.e., within the plant) with a precise and dependable method of time measure-

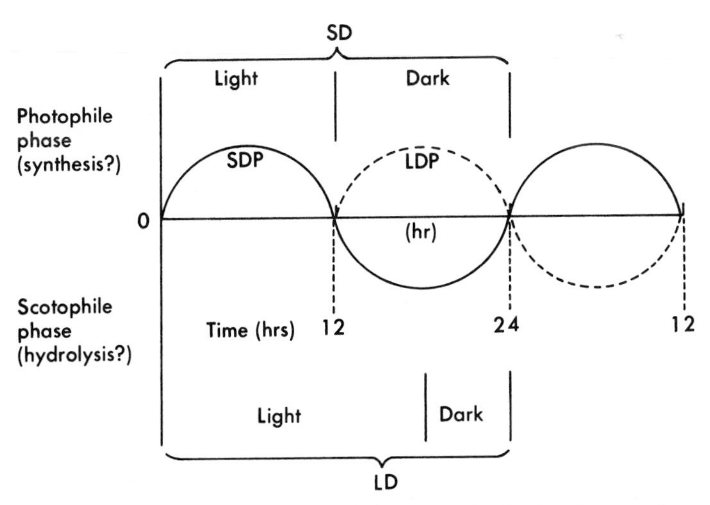

Fig. 21-4. Bünning's explanation of photoperiodism on the basis of an endogenous rhythm of alternating photophile (light-loving) and scotophile (dark-loving) phases. In short-day plants (SDP) the photophile phase begins immediately on illumination; in long-day plants (LDP) it begins only about 12 hr after the beginning of illumination. **SD,** Short day; **LD,** long day. (From Murneek, A. E., and R. O. Whyte. 1948. Vernalization and photoperiodism. The Ronald Press Co., New York.)

ment that does not require continuous information from the environment (Sweeney, 1963). From this point of view the environment supplies only the secondary "impressed force," which may change the periods or amplitude of the natural rhythm. This would lead to the conclusion that the rhythm of plant development is controlled primarily by the internal clock and only secondarily by external rhythms. The latter interpretation is in agreement with the foregoing evidence of natural rhythms in the plant.

The location of the timer, as well as the photoreceptor, is the pulvinus, in the case of clover *(Trifolium repens).* The same was true for responses to rhythmic temperature variations under constant illumination (Scott and Gulline, 1972). On the other hand, small cubes of leaf mesophyll, from which the epidermis was removed, still exhibit circadian rhythms of carbon dioxide metabolism, regardless of the region of the leaf from which they are taken (Wilkins, 1969). It is not clear whether all plant rhythms must be due to a single clock, or whether more than one is required. In the green alga *Gonyaulax polyedra,* rhythms in photosynthetic capacity, luminescence, and cell division all have fixed phase relations to each other, indicating a single clock. Nevertheless, the basic clock may not be the same for all organisms or even for all rhythms of a single plant (Wilkins, 1969).

Many investigators are now devoting their efforts to an elucidation of the nature of the biological clock. Some say it is physical, others that it is chemical. A distinction between the two is sometimes attempted by an investigation of the temperature effect. When this is negligible, the clock is thought to be physical; when pronounced, it is thought to be chemical. Both kinds of results have been obtained. The periodicity of nutation of pea epicotyls, for instance, bore a linear relation to temperature from 15° to 27° C (Heathcote, 1969), the cycle length decreasing with temperature increase. Hastings (Aschoff, 1965), believes that the rhythmicity is in some way dependent on the cell's normal ability to synthesize a specific RNA. Evidence for this concept is the effect of inhibitors of nucleic acid or protein synthesis on the rhythm. In *Gonyaulax* (Sweeney in Aschoff, 1965) the rhythmic activity of luciferase was abolished by actinomycin D and puromycin—inhibitors of RNA and protein synthesis. A similar result has been obtained with the unicellular green alga *Acetabularia.* The rhythms in photosynthetic capacity and in chloroplast shape were dramatically inhibited by actinomycin D (Van den Driessche, 1966); but this was true only of intact algae. Anucleate algae were able to maintain their rhythms, and in this case actinomycin D did not affect them. These results have been corroborated by Sweeney and co-workers (1967). In the case of anucleate *Acetabularia,* inhibitors of protein synthesis (actinomycin D, chloramphenicol, puromycin) had no effect on either the period or the phase of the photosynthetic rhythm despite their large effect on incorporation of ^{14}C-leucine. According to van den Driessche's hypothesis, the mechanism is dependent on nuclear DNA, but the oscillations are possible only in the presence of a light-dependent and nucleus-dependent substance that triggers the biological clock. The nucleus controls the synthesis of messenger RNAs (mRNA) carrying the genetic information for the oscillator and for the light-dependent substance. The nucleus is not immediately required for maintenance of the rhythm because the nucleus-dependent mRNA is particularly stable and therefore has a long life. More recently, rifampicin was shown to have no effect on the circadian rhythm of both whole and enucleate *Acetabularia,* although it dramatically inhibits RNA synthesis (van den Driessche et al., 1970). Contradictory results have also been obtained with duckweed (*Lemna gibba* G3). Nuclear DNA, newly synthesized in the light, was apparently intimately related to the reappearance of the "light interruption rhythm" (Nakashima and Mori, 1970). On the other hand, an inhibitor of DNA synthesis (FUDR) did not eliminate the oxygen uptake rhythm (Miyata, 1971).

The endogenous rhythm in the rate of carbon dioxide output by *Bryophyllum* has a different explanation (Wilkins, 1967). Nucleic acids are apparently not involved, nor is there a variation in enzyme content (Wilkins, 1969). It is caused by a rhythm in the dark fixation of carbon dioxide, which continues in prolonged darkness, fading away in 5 days. This is explained by a periodic change in activity of the enzyme phosphoenol-pyruvate carboxylase (which fixes carbon dioxide in the dark). In this case the change in enzyme activity appears to be controlled by an inhibitor that may actually be a product of carbon dioxide fixation.

A relation to growth promoters has also been suggested. The daily rhythm in nitrate reductase activity in tobacco leaves may be controlled by the endogenous levels of CK and GA (Roth-Bejerano and Lips, 1970). Similarly, the timing of the cell-division rhythm in *Adiantum* gametophytes seemed to be controlled by phytochrome and a blue-light-absorbing pigment (Wada and Furuya, 1972).

During synchronous (i.e., uniformly rhythmic) growth of *Chlorella,* there is a periodism in the rate of increase in the cellular level of DNA (Johnson and Schmidt, 1966). A similar periodism in the level of the enzyme deoxythymidine monophosphate kinase precedes the DNA rise by 1 hr, suggesting that this enzyme may be limiting the rate of DNA synthesis. Similar results with another enzyme indicate that perhaps the synthesis of all the enzymes on the deoxythymidine triphosphate (dTTP) pathway may be coordinated and in control of the synchronous growth. Several such enzyme rhythms have been reported (Table 21-1).

Despite all the foregoing evidence, we still do not know how the biological clock is timed. We do not even know for certain whether the primary timing oscillators are in the organism or in its environment (Brown, 1960). As indicated earlier, most investigators assume that the organism is an independently oscillating system, with one

Table 21-1. Enzyme or metabolic rhythms

Rhythm	Plant	Enzyme or metabolic relation	Reference
Growth of mycelium	*Podospora anserina*	Enhancement of oxidative sugar degradation	Lysek, 1971
Respiratory metabolism	*Lemna gibba*	Glyc-3-P DH and acid phosphatase both varied diurnally	Miyata and Yamamoto, 1969
Light-on rhythm in respiration	*Lemna gibba*	Two components of respiration	Miyata, 1971
Diurnal enzyme activity	*Cactus phylloclades*	Some highest at noon, others at midnight	Khan et al., 1970
Luminescence	*Gonyaulax*	Luciferase varies diurnally	Wilkins, 1969
Photosynthesis	*Gonyaulax*	RUDPC same amplitude and phase as photosynthesis	Wilkins, 1969
CO_2 metabolism	*Bryophyllum*	Not due to rhythmic variation in enzyme content or to NA involvement	Wilkins, 1969
NO_3 reductase activity	Tobacco leaves	Summer: maximum at midday Winter: minimum at midday	Roth-Bejerano and Lips, 1970

of its own natural periods closely approximating a day. An alternative hypothesis (Brown in Aschoff, 1965) is that the organism under natural conditions possesses no intrinsic daily rhythmicity but is merely responding to extrinsic (i.e., environmental) rhythms. The organism is exposed not only to the gross rhythms of light and temperature but also to the subtler geophysical rhythms of cosmic rays, magnetism, barometric pressure, etc. Unlike the diurnal rhythms, the latter are always present and uncontrolled by the experimenter. Brown therefore suggests that the plant is like a sundial or electric clock: it draws its timing capacity from outside itself. He calls this an extrinsic living clock mechanism. He admits, however, that both kinds of timing mechanism may conceivably occur.

Perhaps the best evidence for the possibility of endogenous rhythms in living organisms is their existence in nonliving systems. An example of the latter are the periodic precipitation rings known as Liesegang rings, which may be seen in ordinary gels. They occur when one electrolyte diffuses into the solution of another with which it interacts to form a precipitate (for instance, silver nitrate diffusing into potassium bichromate). The deposition of this precipitate is rhythmical, appearing as bands that gradually widen and separate outward as the gradient of diffusion falls (Scarth and Lloyd, 1930). Since the rhythmic precipitation is independent of any environmental rhythms outside the gel, it is a true endogenous rhythm. Circadian, Liesegang-like rhythms of zonation in growth have been found in cultures of some fungi (Jerebzoff in Aschoff, 1965). In this case, all the factors that promote the rhythms of zonation are amino acids connected with protein synthesis. Many other zonation rhythms have been found in fungi, not necessarily in those with circadian periods. The sporulation rhythm in *Nectria cinnabarina* is expressed by either concentric rings or spirals (Bourret et al., 1971). The period is inversely proportional to the temperature and does not approach 24 hr. The period of zonation rhythm is lengthened by culturing the fungus on a dialysis membrane over the agar surface. These results suggest diffusion-dependent Liesegang rings. Similarly, a rhythm of coremia production by *Penicillium claviforme* was induced by substances capable of changing their cell permeability (Faraj Salman, 1971). Rhythmic changes in the pool of organic acids in starving aerobic yeast were explained by changes in the permeability of the inner mitochondrial membrane (Gottschalk, 1971). A zonation rhythm with a period of 1.3 to 0.7 hr was released in cultures of *Neurospora crassa* by addition of tryptophan to the minimal medium (Nysterakis et al., 1969). It is therefore perhaps conceivable that all the rhythms are caused by a rhythmic change in availability of substances required to produce the particular phenomenon. In the case of Liesegang rings, diffusion of one of the reacting substances is the limiting factor; in the case of the plant, perhaps synthesis of RNA, amino acids, etc., or in some cases a physical factor such as diffusion is the limiting factor. When the limiting material is used up, the process stops or is reversed until the deficiency is overcome by diffusion or regeneration of the limiting factor.

Regardless of the origin of plant rhythms, they do exist. Because of their existence, optimal plant development *requires* an alternation in environmental conditions, for example, light and temperatures, at least in the case of some plants. Tomato plants, for instance, suffered interveinal chlorosis when subjected to photoperiods of 20 hr or longer, whereas 18 hr was optimal (Ketellapper, 1969). Peas, peanuts, and soybeans, however, did not require a daily periodicity. *Hyoscyamus muticus,* which grows normally in a climate with a daily thermoperiodicity, becomes progressively more dwarf when grown in a constant temperature (Saint-Firmin and Chouard, 1971).

Rhythms are therefore of practical as well as theoretical importance.

QUESTIONS

1. What are rhythms or periodisms?
2. What is the difference between a forced period and a natural period, a passive rhythm and an active rhythm?
3. When is a rhythm entrained?
4. What is fading away?
5. What annual rhythms occur in plants?
6. Are any lunar rhythms known in plants?
7. What is a circadian rhythm?
8. Give an example.
9. Is the period for a specific circadian rhythm constant or variable? Explain.
10. Are there any rhythms in plants with periods shorter than one day? Explain.
11. What is the difference between exogenous and endogenous rhythms? Give examples of each.
12. What is meant by weak and strong synchronizers of plant rhythms?
13. How can we determine whether the plant rhythm is forced or natural?
14. How can a plant rhythm be initiated?
15. When does it fade away?
16. Can the plant's natural rhythm be changed? Explain.
17. What is meant by a biological clock?
18. Is it physical or chemical in nature?
19. What are the current concepts of its mechanism?
20. How can it be prevented from functioning?
21. Is the timing mechanism internal or external? Explain.
22. For optimum growth and development, should plants be grown under constant or alternating environmental conditions? Why?

SPECIFIC REFERENCES

Alford, D. K., and T. W. Tibbitts. 1970. Circadian rhythms of leaves of *Phaseolus angularis* plants grown in controlled carbon dioxide and humidity environment. Plant Physiol. **46**:99-102.

Biebl, R., and K. Hofer. 1966. Tages- und Jahresperiodik der Strahlenresistenz pflanzlicher Zellen. Radiat. Bot. **6**:225-250.

Bourret, J. A., R. G. Lincoln, and B. H. Carpenter. 1971. Modification of the period of a non-circadian rhythm in *Nectria cinnabarina*. Plant Physiol. **47**:682-684.

Brest, D. E., T. Hoshizaki, and K. C. Hamner. 1970. Circadian leaf movements in Biloxi soybeans. Plant Physiol. **45**:647-648.

Brest, D. E., T. Hoshizaki, and K. C. Hamner. 1971. Rhythmic leaf movements in Biloxi soybean and their relation to flowering. Plant Physiol. **47**:676-681.

Brown, F. A. 1960. Response to pervasive geophysical factors and the biological clock problem. Sympos. Quant. Biol. **25**:57-71.

Bünning, E., and I. Moser. 1969. Einfluss der Blattlage auf die Blütenbildung. Naturwiss. **56**:519.

Faraj Salman, A.-G. 1971. On the induction of an endogenous rhythm in mutants of *Penicillium claviforme* Bainier. II. Action of detergents. Biochem. Physiol. Pflanz. **162**:470-473.

Gibbs, R. D. 1939. Studies in tree physiology. I. General introduction; water contents of certain Canadian trees. Canad. J. Res. [C] **17**: 460-482.

Gottschalk, H. J. 1971. Metabolic rhythms of aerobic yeasts. Arch. Mikrobiol. **79**:249-262.

Heathcote, D. G. 1969. Some effects of temperature on the nutation of young *Phaseolus* epicotyls. J. Exp. Bot. **20**:849-855.

Highkin, H. R., and J. B. Hanson. 1954. Possible interaction between light-dark cycles and endogenous daily rhythms. Plant Physiol. **29**:301-302.

Hussey, G. 1954. Experiments with two long day plants designed to test Bünning's theory of photoperiodism. Physiol. Plant. **7**:253-260.

Johnson, R. A., and R. R. Schmidt. 1966. Enzymatic control of nucleic acid synthesis during synchronous growth of *Chlorella pyrenoidosa*. I. Deoxythymidine monophosphate kinase. Biochim. Biophys. Acta **129**:140-144.

Ketellaper, H. J. 1969. Diurnal periodicity and plant growth. Physiol. Plant. **22**:899-907.

Khan, A. A., C. P. Tewari, P. S. Krishman, and G. G. Sanwal. 1970. Diurnal variations in enzymic activities in subcellular fractions of *Cactus phylloclades*. Phytochemistry **9**:2097-2104.

King, R. W., and B. G. Cumming. 1972. Rhythms as photoperiodic timers in the control of flowering in *Chenopodium rubrum* L. Planta **103**:281-301.

Lavarenne, S., P. Champagnat, and P. Barnola. 1971. Rhythmic growth of some woody plants of temperate regions cultivated in air-conditioned rooms at a high and constant temperature and under various photoperiods. Bull. Soc. Bot. Fr. **118**:131-162.

Lysek, G. 1971. Rhythmic mycelial growth in *Pondospora anserina.* III. Effect of metabolic inhibitors. Arch. Mikrobiol. **78**:330-340.

Mayer, W. 1966. Besonderheiten der circadianen Rhythmik bei Pflanzen verschiedener geographischer Breiten. Planta **70**:237-256.

Miyata, H. 1971. Endogenous light-on rhythm in respiration of a long-day duckweed, *Lemna gibba* G3. II. On basic and rhythmic components of the rhythm. Plant Cell Physiol. **12**: 517-524.

Miyata, H., and Y. Yamamoto. 1969. Rhythms in respiratory metabolism of *Lemna gibba* under continuous illumination. Plant Cell Physiol. **10**:875-889.

Moorby, J., and P. F. Wareing. 1963. Ageing in woody plants. Ann. Bot. **27**:291-308.

Muller, D. 1962. One year lunar periodicity phenomena in several brown algae. Bot. Marina **4**:140-155.

Nakashima, H., and H. Mori. 1970. DNA synthesis as related to the reappearance of "light interruption rhythm" in a long-day duckweed, *Lemna gibba* G3. Plant Cell Physiol. **11**:805-816.

Nysterakis, F., S. Jerebzoff, and S. Jerebzoff-Quintin. 1969. Release by tryptophan of an endogenous zonation rhythm of a very short period in *Neurospora crassa* 74-A-8 Srb. Compt. Rend. Acad. Sci. (Paris) **268**:1498-1501.

Roth-Bejerano, N., and S. H. Lips. 1970. Seasonal characteristics of the daily fluctuation in nitrate reductase activity. Physiol. Plant. **23**:530-535.

Saint-Firmin, A. and P. Chouard. 1971. *Hyoscyamus muticus* L. and some environmental factors: temperature. Compt. Rend. Acad. Sci. (Paris) **272**:1507-1510.

Scaramuzzi, F., V. Procelli-Armenise, and A. De Gaetano. 1971. Research on the behavior and annual rhythm of callogenesis in trunk fragments from certain trees. Compt. Rend. Acad. Sci. (Paris) **272**:2544-2547.

Scarth, G. W., and F. E. Lloyd. 1930. Elementary course in general physiology. John Wiley & Sons, Inc., New York.

Scott, B. I. H., and H. F. Gulline. 1972. Natural and forced circadian oscillations in the leaf of *Trifolium repens.* Aust. J. Biol. Sci. **25**:61-76.

Seidman, G., and W. B. Riggan. 1968. Stomatal movements; a yearly rhythm. Nature (London) **217**:684-685.

Stålfelt, M. G. 1963. Diurnal dark reactions in the stomatal movements. Physiol. Plant. **16**:756-766.

Sweeney, B. M. 1963. Biological clocks in plants. Ann. Rev. Plant Physiol. **14**:411-440.

Sweeney, B. M., C. F. Tuffli, Jr., and R. H. Rubin. 1967. The circadian rhythm in photosynthesis in *Acetabularia* in the presence of actinomycin D, puromycin, and chloramphenicol. J. Gen. Physiol. **50**:647-659.

van den Driessche, T. 1966. The role of the nucleus in the circadian rhythms of *Acetabularia mediterranea.* Biochim. Biophys. Acta **126**:456-470.

van den Driessche, T., S. Bonotto, and H. Brachet. 1970. Inability of rifampicin to inhibit circadian rhythmicity in *Acetabularia* despite inhibition of RNA synthesis. Biochim. Biophys. Acta **224**:631-634.

Wada, M., and M. Furuya. 1972. Phytochrome action on the timing of cell division in *Adiantum* gametophytes. Plant Physiol. **49**:110-113.

Wilkins, M. B. 1967. An endogenous rhythm in the rate of carbon dioxide output of Bryophyllum. V. The dependence of rhythmicity upon aerobic metabolism. Planta **72**:66-77.

Wilkins, M. B. 1969. Circadian rhythms in plants, p. 647-671. *In* M. B. Wilkins (Ed.). Physiology of plant growth and development. McGraw-Hill Book Co., New York.

GENERAL REFERENCES

Aschoff, J. 1965. Circadian clocks. North-Holland Publishing Co., Amsterdam.

Bünning, E. 1964. The physiological clock. Springer-Verlag, Berlin.

Sollberger, A. 1965. Biological rhythm research. Elsevier Publishing Co., Amsterdam.

22

Stress resistance

ENVIRONMENTAL STRESSES

Although the growth and development of the plant are internal processes, they are under the control of the environment. Temperature, moisture, radiation, nutrients, and gases can all either enhance or retard growth and development. However, their effects may transcend the quantitative—they may also act as stresses, leading to injury and death of the plant.

In the physical sense a force accompanied by its counterforce constitutes a stress. Quantitatively, the stress is expressed as the force per unit area and therefore has the dimensions of pressure. As a result of the stress, a body undergoes a strain, for example, an increase in length, a change in volume or shape, etc. The strain may be elastic and reversible, or it may be plastic, giving rise to an irreversible or permanent set. In the physiological sense a stress is a potentially injurious force or pressure acting on the plant that may lead to a reversible strain or to an irreversible strain (injury or death). The plant is constantly being subjected to the stresses of its environment, and it survives only because of its adaptation to them. There are two main kinds of adaptation possible:

1. *Elastic resistance* enables the plant to grow and develop under environmental conditions that do not permit normal growth and development in the case of unadapted plants. This may mean simply the development of lower temperature minima or higher temperature maxima for its physiological processes.
2. *Plastic resistance* enables the plant to survive environmental stresses that kill unadapted plants.

In the second case the stress acts more directly and quickly. In the first case death may also occur, but only after an extended exposure to the environmental stress, sufficient to produce a disturbance in the metabolism of the plant. Elastic resistance has been studied more extensively in animals than in plants, and plastic resistance, more in plants than in animals. We will therefore discuss the latter only. Strains must occur as a result of the environmental stress, but in the resistant plants these are always reversible. There are many kinds of environmental stresses, and each may produce its own kind of irreversible strain or stress injury.

Temperature stress

1. Most plants in the growing state are killed by freezing; and even dormant overwintering plants may be injured. This is called *freezing*, or *frost, injury*.

2. Some plants (mainly from tropical regions) are also injured or killed by exposure to low temperatures above the freezing point. This is called *chilling injury*.
3. Temperatures of 60° C or higher will kill most living organisms, and even more moderate high temperatures may injure some plants. This is called *heat injury*.

Water stress

Water may injure or kill plants if present in excess, but this is actually due to lack of oxygen and is called a *flooding injury*. A more common injury, however, is caused by a water deficit, and this is called *drought injury*.

Radiation stress

Visible radiations (light) are seldom directly responsible for death, although they are sometimes capable of causing *radiation injury*. Radiations below the wavelength of light may, however, be highly injurious or fatal. This does not commonly occur in the natural habitat of the plant, since only the ultraviolet radiations are of shorter wavelength than light in the radiation received on earth from the sun, and they are usually of too low an intensity to injure most plants. However, because of the modern interest in nuclear energy and in outer space, injury by radiations of lower wavelength is now receiving greater attention. Injury caused by infrared radiation is actually heat injury.

Salt stress

Salt stress can be produced by an excess of any one of a large number of salts. It is important in nature, since there are many saline (sodium salt) soils. Plants showing salt (sodium) resistance are called *halophytes*.

Gas stress

Because of modern industrialization, pollution injury is daily becoming more important. This is commonly caused by toxic gases and vapors.

There are also other stresses that are beginning to receive attention—pressure, electric, magnetic, sound, etc.

KINDS OF STRESS RESISTANCE

To survive, the plant must develop resistance toward potentially injurious environmental stresses. Although there are many kinds of stresses, there are only two basic resistance mechanisms.

1. The plant may exclude the stress from its tissues and therefore possess *stress avoidance*.
2. The plant may survive the penetration of its tissues by the stress and therefore possess *stress tolerance* (or "hardiness").

Thus a cold-avoiding organism is one that remains warm at low temperatures, and a cold-tolerant one becomes as cold as its environment but because of its tolerance or hardiness is not injured by the cold. The same terminology may be used for the other factors (Table 22-1).

The two kinds of resistance may not always be developed to the same degree. Tolerance seems to be the more primitive of the two adaptations because it is the only kind of resistance developed by lower plants in the case of the three factors most intensively studied (Table 22-2). Similarly, drought resistance seems to be the most advanced of the three resistances, since higher plants show the highest development of avoidance in the case of the drought stress.

Freezing (or frost) injury and resistance

Plant protoplasm can survive the lowest temperatures attainable (0° K approximately) if no ice forms in the tissues. Even if frozen, it may survive the lowest temperatures, provided that the freezing process is of a specific kind, for example, if the ice crystals are so small as to require x-ray analysis for detection. Under normal conditions, however, only air-dry plant parts

Table 22-1. Stress resistance

Environmental stress	Condition of resistant plant possessing	
	Avoidance	**Tolerance**
1. Chilling temperatures (cold but above freezing)	Warm	Cold
2. Freezing temperatures	Unfrozen	Frozen
3. High temperatures	Cool	Hot
4. Drought	High vapor pressure	Low vapor pressure
5. Radiation	Low absorption	High absorption
6. Salt	Low salt concentration	High salt concentration
7. Flooding (oxygen deficiency)	High oxygen content	Low oxygen content

(e.g., seeds) can show the first kind of survival, and the second kind occurs only under artificial conditions (see later discussion). Frost resistance of all except dehydrated plant parts (with a few exceptions) is therefore tolerance, since the vast majority of plants cannot avoid freezing on exposure to extreme subfreezing temperatures. But this tolerance exists only toward extracellular (i.e, intercellular) ice formation. No plant can survive the formation of microscopically visible crystals within its living cells (intracellular freezing), at least not if the ice is formed within the protoplasm.

Since the extracellular freezing removes water from the cell, the cell becomes more and more dehydrated as the temperature falls further and further below the freezing point (Fig. 22-1). In many plants it is not until the temperature drops to –20° to –30° C that 95% of the cell's water may be removed in this way. If the cell cannot survive such a profound dehydration, it will be killed. It is therefore logical to conclude that when plants are frozen extracellularly, any injury that occurs is likely to be a dehydration injury. If dehydration injury and resistance can be explained, this can therefore also be expected to explain frost injury and resistance. Once the cell has nearly all of its water frozen (at about –30° C) without injury, it can also survive subsequent immersion in liquid air or liquid nitrogen and may logically be expected to survive absolute zero.

Table 22-2. Relative importance of tolerance and avoidance in lower and higher plants in the vegetative state

Stress	Resistance in lower plants	Resistance in higher plants
Frost	Solely tolerance	Nearly always tolerance
Heat	Solely tolerance	Mainly tolerance
Drought	Solely tolerance	Mainly avoidance

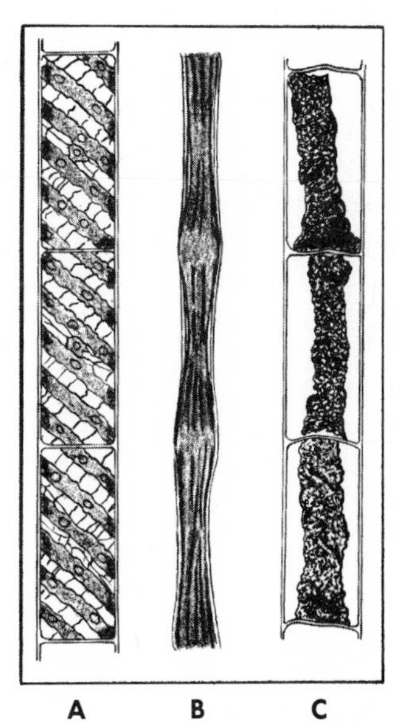

Fig. 22-1. Contraction of *Spirogyra* cell caused by extracellular freezing (and therefore dehydration). Thawed cell shows coagulated (killed) protoplasm. **A,** Normal; **B,** frozen; **C,** thawed. (From Molisch, H. 1897. Untersuchungen über das Erfrieren der Pflanzen. Gustav Fischer Verlag, Jena.)

Frost resistance varies seasonally, so that even the most resistant plants of midwinter, which survive temperatures of –50° C (or even –190° C under artificial conditions), are killed by about –5° C in early spring (Fig. 22-2). During the fall the plant "hardens"; that is, it slowly develops a greater and greater tolerance or hardiness, until the maximum is reached in midwinter, and then it dehardens slowly until the minimum is reached in spring. Hardening can be induced artificially by exposure to low temperatures (e.g., 0° to 5° C), dehardening by exposure to high temperatures (e.g., above 10° C).

The problem of freezing effects has become so important that in recent years it has led to the development of a whole new science—*cryobiology* (literally, the biology of freezing temperatures). Many biologists from different biological disciplines are cryobiologists: the surgeon who freezes tissues in order to remove them, the microbiologist who freezes cultures in order to keep them alive for long periods without subculturing them, the biophysicist who is interested in the nature of the freezing process, and the biochemist who investigates the effect of freezing on the rates and kinds of chemical reactions. There are therefore branches of cryobiology called cryosurgery, cryopreservation, cryobiophysics, cryobiochemistry, etc.

Drought injury and resistance

Unlike frost resistance, drought resistance may be caused by either avoidance or tolerance. Plants such as succulents are "water savers" and may survive long periods of drought by cutting down water loss to a minimum. This is accomplished by a thick and highly impermeable cuticle and the closure of their stomata during a large part of the day. This forces them to develop a special kind of metabolism (Chapter 15) to survive. Other plants are "water spenders" and maintain their water contents at a high level by means of a large and active water-absorbing system relative to the water-losing portion (Fig. 22-3). As is true in other societies, the spenders get along better than the savers when liquid water is to be found somewhere (e.g., at considerable depths), but only the savers can survive if there is no water to be absorbed for long periods of time.

Drought avoidance seems to be the major mechanism of survival in vegetative higher plants. Exceptions that owe their resistance to tolerance have recently been found in South Africa (Gaff, 1971). However, in vegetative lower plants and even in dor-

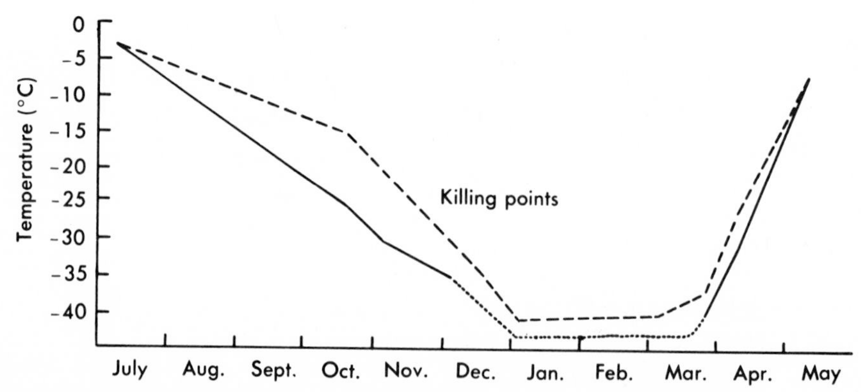

Fig. 22-2. Annual change in freezing resistance of the twigs of two apple varieties. (From Hildreth, see Levitt, J. 1966. *In* H. T. Meryman. Cryobiology. Academic Press, Inc., New York.)

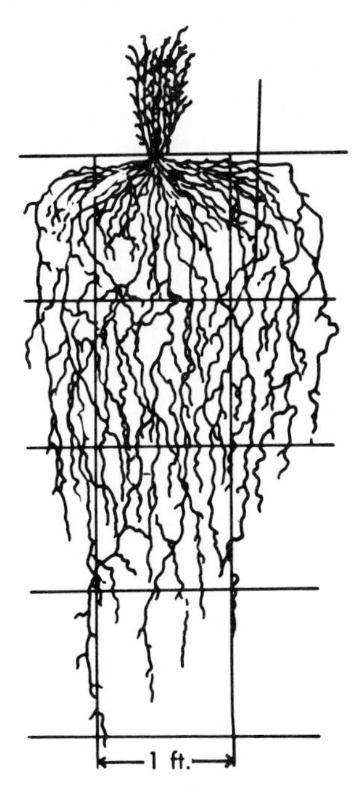

Fig. 22-3. A water spender. Root system is considerably larger than the shoot system. (From Weaver, J. E. 1919. Carnegie Institution of Washington, Pub. No. 286, Washington, D. C.)

mant parts of higher plants (e.g., seeds, pollen grains, etc.), the main method of survival is drought tolerance. Even the vegetative parts of higher plants must possess some tolerance, slight though it may be. Thus there are (1) intolerant avoiders, (2) tolerant avoiders, and (3) tolerant nonavoiders. As mentioned earlier, drought tolerance and frost tolerance may depend on similar mechanisms, and, in fact, the two have often proved to be correlated. A plant in midwinter, for instance, has its highest degree of frost tolerance and in some cases, at least, also of drought tolerance.

Heat injury and resistance

The range of high temperatures survived by plants is small compared to the range of low temperatures. The highest temperature that air-dry cells have been able to survive is about 140° C, and normally moist vegetative cells are usually killed by 40° to 50° C. As in the case of frost resistance, heat resistance is nearly always tolerance. Exceptions have been found among plants of the Sahara Desert. Some of these normally survive air temperatures of 50° C by absorbing and transpiring such tremendous quantities of water that their leaves are cooled as much as 10° C below that of the air (Lange, 1959, see Chapter 11). Although heat injury is not a dehydration process, heat tolerance is frequently (but not always) correlated with frost and drought tolerance.

MECHANISM OF STRESS INJURY AND RESISTANCE

Although it is relatively simple to find out whether a plant possesses stress resistance, and if it does, whether this results from avoidance or tolerance, it is more difficult to discover the mechanisms by which these two kinds of resistance operate. Even the mechanism of injury is not usually known.

In the case of temperature stresses, injury may depend on the membrane lipids. Chilling-sensitive plants have membrane lipids with high melting points that are presumably due to a low degree of unsaturation. As a result, the membrane lipids apparently solidify at chilling temperatures, leading to membrane damage and death. Chilling-resistant plants have more unsaturated lipids with lower melting points. Therefore no solidification occurs at low temperatures.

The correlations between frost, drought, and heat tolerance encourage attempts to explain the injury and tolerance in all three cases by a single theory. Both freezing and drought injury in plants are the result of the dehydration process. But it is not known how the dehydration produces the injury. Many suggestions have been made. The removal of water may conceivably concentrate the solutes to the point of toxicity to protoplasmic constituents. The collapse of the cell may produce a physical tension on the protoplasm that would injure some sensitive component. Many other possibilities have been proposed. A recent concept suggests that the proteins are denatured reversibly at the low temperature and that the dehydration brings the proteins close enough together to form intermolecular bonds. This results in irreversible aggregation of the proteins and therefore death. Even heat injury may be thought of as a kind of dehydration, since the increased kinetic energy of the water molecules may convert previously adsorbed molecules to the free form and may permit the "naked" protein molecules to approach each other close enough for aggregation to occur. All three injuries would then be caused by bringing the protein molecules so close to each other by removing the protective water coats that new chemical bonds would form between them.

Most of the soluble plant proteins are apparently unaffected by many of the injurious stresses. This and many other lines of evidence, points to the membrane proteins as the probable source of injury. In fact, both the lipid and the protein results point to the membranes as the locus of stress injury.

Whatever the cause of the dehydration injury, resistance could be caused by either avoidance of dehydration or tolerance of it. Partial avoidance would result if any property of the plant decreased the degree of dehydration at any one freezing temperature. It is not surprising, then, that the cell sap concentration of plants commonly increases during hardening (Table 6-5). In winter it may be 2 to 4 times as high as in spring and therefore only 50% or even 25% as much ice would form in the plant when frozen in winter as when frozen at the same temperature in spring. This increase in cell sap concentration is primarily caused by a conversion of starch to sugars at low temperatures during the hardening period in the fall.

However, the partial avoidance of dehydration is only one component of the stress resistance. There are plants with high cell sap concentrations that possess no stress resistance (e.g., sugarcane) and plants with low sap concentrations that possess high resistance (some evergreens and some succulents). Tolerance of dehydration is therefore the primary component of such stress resistance. In other words, the protoplasm of the resistant cells possesses a specific ability to become highly dehydrated without suffering injury—the dehydration strain is fully reversible. The exact mechanism of this dehydration tolerance is still unknown.

However, the physical properties of the protoplasm change during hardening, presumably because of changes in the proteins, since these account for the largest fraction of the protoplasm's dry matter. This interpretation has been supported by biochemical investigations of the proteins. Soluble and even total proteins have been found to increase during the hardening period. Changes in other protoplasmic components (e.g., RNA, lipids) also occur. It appears likely, therefore, that the dehydration tolerance of the hardy plant depends on the properties of its proteins. For instance, if the injury is caused by denaturation followed by irreversible aggregation of the proteins, then tolerance that has developed during the hardening process would be caused by changes in the nature of the proteins, which would prevent this irreversible aggregation.

Despite the great deal of information accumulated by investigators of freezing, drought, and heat stresses, neither the mechanism of the stress injury nor that of the stress resistance is fully understood. In the case of other stresses even less is known.

QUESTIONS

1. What is the physiological concept of a stress?
2. What is the physiological concept of a strain?
3. Compare elastic resistance with plastic resistance.

4. To what kinds of temperature stress may a plant be exposed?
5. To what other kinds of stress may it be exposed?
6. What is meant by stress resistance?
7. What two kinds are there?
8. Explain the difference between them.
9. Which is the more primitive?
10. What is the maximum possible freezing resistance of plants?
11. How is this related to the moisture content of the plant?
12. Compare the plant's tolerance of extracellular and intracellular freezing.
13. What effect does the normal freezing of plants have on the water content of its cells?
14. What is meant by hardening?
15. How can it be induced artificially?
16. What is cryobiology?
17. Is freezing resistance of plants the result of tolerance or avoidance? Explain.
18. Is drought resistance of plants the result of tolerance or avoidance? Explain.
19. How high a temperature can plant cells survive?
20. Is heat resistance the result of tolerance or avoidance? Explain.
21. Are any kinds of stress resistance correlated with each other? Explain.
22. What is the probable mechanism of injury in the case of freezing stresses?
23. What is probably involved in the mechanism of resistance?

SPECIFIC REFERENCES

Gaff, D. F. 1971. Desiccation-tolerant flowering plants in Southern Africa. Science **174**:1033-1034.

Weaver, J. E. 1919. The ecological relation of roots. Carnegie Institution of Washington, Pub. No. 286. Washington, D. C.

GENERAL REFERENCES

Asahina, E. (Ed.). 1967. Cellular injury and resistance in freezing organisms. Institute of Low Temperature Science, Hokkaido University, Sapporo, Japan.

Biebl, R. 1962. Protoplasmatische Ökologie der Pflanzen; Wasser und Temperatur. Protoplasmatologia **12**(**1**). Springer-Verlag, Vienna.

Levitt, J. 1956. The hardiness of plants. Academic Press, Inc., New York.

Levitt, J. 1958. Frost, drought, and heat resistance. Protoplasma-monographien **8**(**6**). Springer-Verlag, Vienna.

Levitt, J. 1972. Responses of plants to environmental stresses. Academic Press, Inc., New York.

Meryman, H. T. (Ed.). 1966. Cryobiology. Academic Press, Inc., New York.

Prosser, C. L. (Ed.). 1967. Molecular mechanism of temperature adaptation. Amer. Assoc. Adv. Sci. Pub. No. 84, Washington, D. C.

Troshin, A. S. (Ed.). 1967. The cell and environmental temperature. Pergamon Press, Inc., New York.

23

Laws of plant physiology

The aim of plant physiology is to discover and describe the natural laws controlling all life processes in the plant. We are now in a position to ask: What laws have so far emerged? The logical answer is the laws of physics and chemistry, since the results have so far upheld the basic assumption (Chapter 1) that all physiological phenomena are explainable by these laws. But does it follow that there are no additional, specifically physiological laws? If we are seeking laws that can describe life processes quantitatively, in the form of a mathematical equation, we must admit, unfortunately, that none has yet been established. Nevertheless, it must be evident by now that the explanations of physiological phenomena are not always directly deducible from the laws of physics and chemistry. Physicochemical theory must be supported by experimental evidence. In many cases such evidence has revealed which physicochemical laws must be selected and how they must be combined to explain a specific physiological phenomenon. As a result, a number of fundamental relations have emerged that are now beyond the stage of theory and may tentatively be stated in the form of natural laws.

LAWS OF TRANSFER OF MATERIALS

Law of cell permeability. The living protoplasm of a cell is surrounded by a differentially permeable lipid membrane that controls the diffusion of substances into and out of the cell. It permits some substances to diffuse rapidly into or out of the cell and slows down or prevents the diffusion of others. This law also applies to individual organelles within the protoplasm.

Law of water transfer. Water movement in the plant is a passive or purely physical process. The water moves into and out of the plant by diffusion and within the plant by diffusion and bulk flow, commonly aided by cohesive and adhesive forces. Active processes (directly dependent on metabolic energy) can affect the movement only indirectly by altering the factors that affect diffusion or bulk flow.

Law of solute transfer. Solute movement

in the plant is both active and passive. Apolar solutes move into the plant purely passively; polar substances move into the living part of the plant actively, although aided by passive processes. Movement within the plant is passive in the xylem, passive and active in the phloem.

LAWS OF NUTRITION AND METABOLISM

Law of autotrophism. The normal green plant is capable of synthesizing all the organic substances needed for completion of its life cycle. The abnormal green plant, or a part of it, may be deficient in ability to synthesize one or more of these substances.

Law of nutrient essentiality. All the elements that participate in the life processes of a plant must be supplied by its environment. For the plant to complete its normal life cycle, the quantity must be sufficient to supply all the metabolic and other needs.

Law of enzyme control. All the metabolic processes in the plant are controlled by enzymes, which are organic catalysts consisting of a protein and commonly a nonprotein cofactor.

Law of electron transfer. In all living protoplasm, chemical energy is constantly being released by electron transfer between substances undergoing oxidation-reduction reactions.

Law of energy transfer. All the energy requirements of the plant are supplied by the conversion of light energy into chemical energy in the form of high energy phosphates. The light energy may be stored in more stable forms of chemical energy and later again converted into high energy phosphates in the process of respiration.

Law of cardinal points. All factors that affect a plant process may be present in quantities related to the three cardinal points. The minimum quantity of the factor is the smallest amount capable of supporting a measurable rate of the process. The optimum quantity of the factor is the amount capable of supporting the most rapid rate of the process. The maximum quantity of the factor is the amount above which the process cannot occur at a measurable rate.

Corollary (law of the minimum). The rate of a plant process is controlled by the physical or chemical factor that is present in minimum quantity relative to its optimum quantity.

Law of template synthesis. All the proteins and nucleic acids for each kind of plant are synthesized on a template nucleic acid molecule that can reproduce the same pattern of molecule repeatedly. The patterns of the templates differ for each protein or nucleic acid produced.

LAWS OF GROWTH AND DEVELOPMENT

Law of enlargement. All plant growth is a result of cell enlargement in response to the force of turgor pressure.

Law of plant movement. Some movements of plant organs are controlled by growth and plant turgor, others by changes in permeability or active absorption.

Law of regulation of growth and development. The growth and development of a plant is regulated by the presence in low concentrations of hormones known as growth regulators. These substances may enhance (or inhibit) growth at low concentrations and inhibit it at high concentrations. They may affect cell division, enlargement, or differentiation, depending on the balance between the kinds of substances.

Law of plant development. The plant passes successively through stages or phases of development that differ both morphologically and physiologically. The passage from one stage to another is subject to external control by the environment (e.g., the light and temperature regimen) to which it is exposed and to internal control by the balance of both growth and flowering hormones.

Law of endogenous rhythm. Plant processes exhibit periodic changes in rate, controlled by either an intrinsic or extrinsic clock capable of measuring time. Most of these have approximately diurnal periods, but some have tidal, lunar, or annual periods.

Law of stress resistance. Plants may survive stresses by two kinds of resistance: avoidance or tolerance or a combination of the two. Avoiders prevent the stress from penetrating their tissues; tolerant plants develop an internal resistance that enables them to survive the penetration of their tissues by the stress.

• • •

Evidence is accumulating for other physiological laws, but it is too soon to attempt formulating them. Even the suggested laws are not as firmly established as the scientist would desire. Further research is necessary, and a more quantitative, mathematical presentation of these laws is to be hoped for.

Index

Q

T